Organic Chemistry
A Guided Inquiry

A Moog/Spencer Guided Inquiry Course

Andrei Straumanis

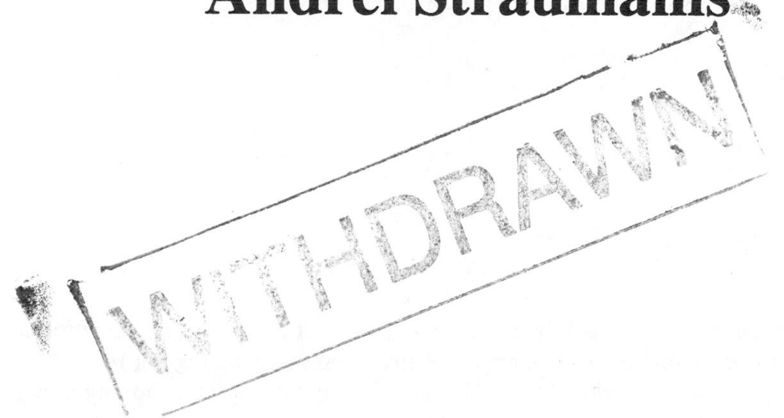

Houghton Mifflin Company Boston New York

Publisher: Charles Hartford
Executive Editor: Richard Stratton
Assistant Editor: Danielle Richardson
Editorial Assistant: Rosemary Mack
Senior Project Editor: Fred Burns
Manufacturing Manager: Florence Cadran
Marketing Manager: Katherine Greig

Printed in the U.S.A.

ISBN: 0-618-30852-0
2 3 4 5 6 7 8 9-VH-07 06 05 04 03

For my new son, Milo.
Now he has it in writing that *teaching by telling* doesn't work.
And we may be in big trouble.

Foreword on Guided Inquiry Learning

by Rick Moog, Jim Spencer and John Farrell, authors of *Chemistry: A Guided Inquiry*[1] (for General Chemistry) and two volumes of *Physical Chemistry: A Guided Inquiry*[2].

These guided activities were written because much research has shown that more learning takes place when the student is actively engaged, and ideas and concepts are developed by the student, rather than being presented by an *authority* — a textbook or an instructor.[3] The *ChemActivities* presented here are structured so that information is presented to the reader in some form (an equation, a table, figures, written prose, etc.) followed by a series of Critical Thinking Questions which lead the student to the development of a particular concept or idea. Whenever possible, data are presented *before* a theoretical explanation, and the Critical Thinking Questions lead the student through the thought processes which results in the construction of a particular theoretical model. This is what makes these guided activities a "guided inquiry." We have tried to mimic the scientific process as much as possible throughout. Students are often asked to make predictions based on the model that has been developed up to that point, and then further data or information is provided which can be compared to the prediction. It is important that these predictions be made BEFORE proceeding to get the full appreciation and benefit of this way of thinking. In this way, models can be confirmed, refined, or refuted, using the paradigm of the scientific method.

There are several articles in the literature by Moog, Spencer, and Farrell about guided inquiry.[4,5,6] *Of particular relevance is a recent article by Jim Spencer describing the philosophical and pedagogical underpinnings of guided inquiry.*[7]

---

[1] Moog, R.S, Farrell, J.J., *Chemistry: A Guided Inquiry, 2nd Edition,* John Wiley & Sons, New York, 2003.
[2] Moog, R.S, Spencer, J.N., Farrell, J.J., *Physical Chemistry: A Guided Inquiry*: *Atoms, Molecules, and Spectroscopy*, and *Physical Chemistry: A Guided Inquiry: Thermodynamics,* Houghton Mifflin Co., Boston, 2003.
[3] Johnson, D. W. ; Johnson, R. T. *Cooperative Learning and Achievement.* In Sharon, S. (Ed.), *Cooperative Learning: Theory and Research*, pp 23-37, New York: Praeger.
[4] Lloyd, B. N.; Spencer, J. N. *J. Chem. Educ.* 1994, *71*, 206.
[5] Gillespie, R. J.; Spencer, J. N.; Moog, R. S. "Demystifying Introductory Chemistry" *J. Chem. Educ.* 1996, *73*, 617. Earlier versions appeared in *New Directions for General Chemistry*, Lloyd, B. W., ed., Washington, DC: Division of Chemical Education of the American Chemical Society, 1994.
[6] Farrell, J.J., Moog, R.S., Spencer, J.N., "A Guided Inquiry General Chemistry Course." *J. Chem. Educ.*, 1999. 76: 570-574.
[7] Spencer, J. N. "New Directions in Teaching Chemistry." *J. Chem. Educ.* 1999, *76*, 566.

To the Instructor

The worksheets contained in *Organic Chemistry: A Guided Inquiry* are designed to guide students through discovery of the material found in a standard organic chemistry course. These materials are intended to be used in conjunction with any standard textbook.

Available Course Materials

- *Organic Chemistry: A Guided Inquiry*, containing the ChemActivity worksheets
- *Solutions Manual*, containing solutions to the homework "Exercises"
- *Instructor's Edition*[†], containing answers to the in-class "Skill Development Exercises"
- *Instructor's Guide*[†], containing suggestions on how to use the materials, reading assignments for several popular texts, information about each ChemActivity, and sample quizzes

Suggested Uses

The guided inquiry worksheets are a backbone that can support many different teaching styles. Listen to your instincts and to your students. Build on the suggestions in the *Instructor's Guide* to create your own personal version of guided inquiry.

- The most popular and effective strategy for using these materials involves devoting most of each class meeting to student group work on a ChemActivity worksheet (groups of three or four students). The instructor serves as facilitator: observing student group work, answering questions, and intervening when necessary. Common interventions include: presenting an interesting answer, leading a whole-class discussion, or delivering a 3-minute summary mini-lecture. The larger your class, and the faster you wish to push your students through the activity, the more often you will need to address the class (more on this in the *Instructor's Guide*).
- A good way to get your feet wet is to replace a lecture with group work on a ChemActivity. (Cautionary note: if you portray this as an "experiment" it could undermine its effectiveness.) From here, many faculty never look back; others continue to cover some topics using a ChemActivity and some topics using interactive lectures. Several faculty have tried mixing lectures and group work by having a summary lecture at the end of several class periods of group work. At first glance, this is an appealing option. The danger in doing this is that students may get the message that they can hold out for "the answers" which are eventually coming from "the authority," and this may degrade their willingness to struggle toward answers of their own.
- If giving up a whole lecture is too drastic, you may choose to excerpt key questions from a ChemActivity and use them during your lecture as short group-work problems.

For further reading, you may want to consult the following excellent resources on the subjects of cooperative learning and group problem solving.[8,9,10]

[†] Available as an Acrobat download at www.hmco.com (for instructors only).

[8] Bruffee, Kenneth A. *Collaborative Learning: Higher Education, Interdependence, and the Authority of Knowledge.* Baltimore: Johns Hopkins Press, 1993.

[9] Johnson, D. W.; Johnson, R. T; Smith, K. A. *Active Learning: Cooperation in the College Classroom*; Interaction: Edina, MN, 1991.

[10] A variety of useful information on cooperative learning is available through Pacific Crest Software, Corvallis, OR. They offer *A Handbook on Cooperative Learning* along with other resources including workshops and teaching institutes. http://www.pcrest.com

Comments from Faculty Who Have Used These Materials

- *I appreciate the way the ChemActivities develop a concept and then apply it to other situations. I like the focus on conceptual understanding and developing students' ability to reason. This year was a test year for me, and I intend to use the materials again based on my experience.*
  -Dr. Laura Parmentier, **Beloit College**
- *I really enjoyed teaching the course using the ChemActivities. It only took about a week or two for me, and the students, to get used to the group work approach. My overall impression is that the students are orders of magnitude better at "thinking like organic chemists." Their ability to defend an answer with specific data...is much improved compared to my other classes. I'm very much planning to adopt this approach in my general chemistry and introductory chemistry sequences next year.*
  -Chuck Leland, **Black Hawk College**
- *I feel that this method helps students "own" the material themselves. My gut reaction so far is that this method will produce students better able to learn for themselves. Thanks for providing me with this opportunity to use GI. I can't begin to explain how refreshing it is for me to teach with this method. I feel the enthusiasm I remember having when I first started! Muchas gracias.*
  -Dr. Karen Glover, **Clarke College**
- *Never before have I felt so fulfilled as a teacher. In addition, never before have I seen so many students so thoroughly engaged in the classroom! I strongly urge all professors with relatively small class sizes to try this approach.*
  -Dr. Glen Keldsen, **Purdue U. North Central**
- *I am convinced that incorporating the activities into the teaching of organic chemistry helps students gain a far deeper understanding of the structures and reactions that make up organic chemistry, I will certainly be using the materials with future classes!*
  -Dr. Clare Gutteridge, **University of New Hampshire** (now at Navy)

Comments from Guided Inquiry Organic Chemistry Students

- *Class time was actually learning time, not just directly-from-ear-to-paper-and-bypass-brain-writing-down time. Learning the material over the whole term is far easier than not "really" learning it until studying for the tests.*
- *Overall, I was far less stressed than many of my friends who took the lecture class. They basically struggled through everything on their own.*
- *It was hugely beneficial to be able to discuss through ideas as we were learning them; this way it was easy to immediately identify problem areas and work them out before going on.*
- *Shifting the groups was a good idea, that way we were continually exposed to different ways of thinking.*
- *I didn't get tired during class because I was constantly thinking and working instead of in a lecture class where I just listen and get easily tired.*
- *The method of having us work through the material for ourselves – as opposed to being told the information and trying to absorb it - makes it seem natural or intuitive. This makes it very nice for learning new material because then we can reason it out from what we already know.*
- *I felt like I was actually learning the information as I received it, not just filing it for later use. The format helped me retain much more material than I have ever been able to in a lecture class, and the small, group atmosphere allowed me to feel much more comfortable asking questions of both other students and the professor.*
- *Through working in groups it was nice to see where everyone else was in understanding the material (i.e. to know where other people were having trouble too).*
- *We were able to discover how things happened and why for ourselves...instead of being told "this is what will happen if this mixes with this."*

To the Student

Doing Organic Chemistry

Imagine a coach who spends every practice demonstrating good technique, but never lets her students hold the racket or figure out how to hit the ball. Most people learn by doing. In fact, research on teaching and learning has shown that *teaching by telling* does not work for most students. This class is about *figuring out* the concepts of organic chemistry not memorizing what your teacher tells you. The activities in this class are designed to mimic what a real scientist does in the laboratory, so you will be asked to draw conclusions based on observations and to test your conclusions via conversations with peers. You will learn organic chemistry, but you will also gain valuable experience thinking logically and interacting successfully as part of a small group. In the research laboratory and in the real world there is no answer key. To be successful after school you must be able to analyze a situation that is new to you and draw your *own* valid conclusions.

In-Class Group Work

Every instructor has a different way of using these materials. One of the most popular involves replacing each lecture with group work on a ChemActivity from this book. Students typically work in groups of three or four and the instructor acts as facilitator, walking around the room observing and answering questions. Each day, the instructor will likely interrupt group work for one or more brief whole-class discussions. Please pause group work during these discussions. It is your collective responsibility to make sure your group is productive. The most common impediment to effective group-work is a member who either over or under contributes. If you feel you are not getting the most out of your group-work experience, talk to your instructor outside of class about strategies for remedying the situation.

Asking and Answering Questions

A huge part of this course will be learning to ask good questions along the way. It is imperative that you do not put off your questions until right before the exam. Organic chemistry is a cumulative course. Understanding today's activity will give you a leg up on understanding the activity for next class. While you work on an activity, talk out the difficulties with your group mates. At any point you can raise your hand and ask for help from your instructor, but you will find that, in many cases, a fellow student is better at explaining the concept to you than an expert who is almost too familiar with the concept.

Research has shown that answering questions is just as important to the learning process as asking them. Many people say you never really understand a subject until you have to explain it to someone else. You will be amazed at how much you learn while trying to explain a concept to a classmate who is still in the dark.

When you are working on a ChemActivity, your job goes beyond simply answering the Critical Thinking Questions. Most every CTQ is designed to teach you something. Ask yourself: "Why is this question here? What is its purpose? What concept or distinction is it designed to convey?" If you can answer these "meta" questions you are well on your way to understanding the big picture, and figuring out where the ChemActivity is going.

Texts and Models

This book is very different from most texts and study guides. It is workbook written to guide student groups in their exploration of organic chemistry. You may find it useful to use a text or study guide in conjunction with this book. The course has been taught using several different texts. Your instructor will choose a text that you can use for supplemental reading and problems. A model set is also highly recommended. Most students find it best to work through the assigned sections in the textbook as a review after they have completed the appropriate ChemActivities and homework problems.

Group Work Outside of Class

You will be successful and enjoy this course only if you put in significant hours on a regular basis outside of class. You are strongly encouraged to work outside of class as part of a group. Many students find studying in a group less boring and less frustrating than studying alone, but make sure your out-of-class group work is productive. Try to avoid excessive social conversation. A good focus for your study group is to complete the most recent ChemActivity worksheet, including the homework Exercises at the end of each chapter. (A *Solutions Manual* for the end-of-chapter Exercises is available for purchase.) During a study group session you might spend the majority of your time quietly working on your own, but for those times when you are stuck or frustrated, your group mates are there to help you. Organic chemistry problems are like puzzles, and often you will find that your study partner has the piece you are missing, and vice versa. **Find a study partner or start a study group. This is the single most important thing you can do to make yourself successful in organic chemistry.**

Guided Inquiry Learning vs. Lecture

There is a growing body of evidence that guided inquiry is more effective than traditional teaching methods such as lecture, but the reasons I teach using guided inquiry are that I enjoy it *and* my students enjoy it. Anonymous surveys have found that the vast majority of guided inquiry students say they prefer guided inquiry to lecture. I hope you will also find this course refreshingly fun and rewarding.

If you have suggestions for improving this book, please contact me at:
a.r.s@Stanfordalumni.org

Dr. Andrei Straumanis, November 2002

Acknowledgements

Thanks to the faculty who volunteered their students to test these materials. These include: Christine Gaudinski, Arapahoe Community College; Laura Parmentier, Beloit College; Chuck Leland, Black Hawk College; Karen Glover, Clarke College; James Stevenson, Concordia University; Christina Mewhinney, Dallas County Community College; Susan Phillips, Holy Family College; Dave Bugay, Kilgore College; Lew Fikes, Ohio Wesleyan University; Glenn Keldsen, Purdue North-Central University; Gary Hollis, Roanoke College; Cristi Hunnes, Rocky Mountain College; Bruce Heyen, Tabor College; Clare Gutteridge, University of New Hampshire; Bob Ludt, Virginia Military Institute.

This work was greatly improved by my conversations with the following faculty regarding both pedagogy and organic chemistry: R. Daniel Libby, Moravian College; Leroy C. Butler, Norwich University; John J. Farrell, Franklin and Marshall College; Tricia A. Ferrett, Carleton College; David G. Alberg, Carleton College; and Paul E. Papadopoulos, University of New Mexico.

Thanks to Richard Stratton, Danielle Richardson and the staff at Houghton Mifflin Company for making this book available to the growing number of guided inquiry enthusiasts.

Thanks to Laura Parmentier of Beloit College for agreeing to error-check the final draft and for encouraging me to replace "attack language" with more pedagogically accurate terms for describing nucleophile-electrophile reactions.

This work would not have been possible without the support of each of the following individuals: Frank J. Creegan of Washington College, James N. Spencer of Franklin and Marshall College, and Jerry R. Mohrig of Carleton College. At every step in the process they offered guidance, opened doors, and were instrumental in procuring the support of administrators and funding agencies. I have learned that nothing substitutes for having a few heavyweights on your team.

Special thanks to Richard S. Moog of Franklin and Marshall College. Rick's contagious enthusiasm for guided inquiry inspired me to embark on this project. More than anyone else I know, he has succeeded in changing the way chemists think about teaching chemistry, and that is no small feat.

Funding for this work was provided by the National Science Foundation New Traditions Project at the University of Wisconsin, the National Science Foundation Post-Doctoral Fellowship in SMET Education, and the Houghton Mifflin Company, Boston.

Finally, many thanks to the organic chemistry students at Washington College, Carleton College and the University of New Mexico where I have taught the course.

Contents

ChemActivity 1

Part A: Bond Angles

(What is the H–C–H bond angle of CH_4?)

Model 1: Electron Pairs

Figure 1: Coulombic Force

Two particles of like charge repel one another.

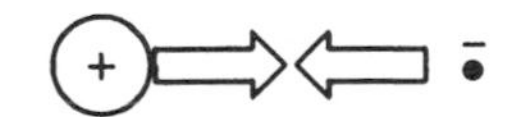

Two particles of opposite charge attract one another.

(+)	= proton (+)
•	= electron (-)
•• or ↿⇂	= an electron pair

Electrons tend to form pairs even though two particles of like charge repel one another. This unexpected phenomenon is attributed to **spin pairing**, which we will not discuss in this course. For now, we will use the model that for an electron "two is company but three is a crowd." That is, two electrons can form a pair, but this pair of electrons will repel other electrons and electron pairs.

Model 2: Planetary Model of an Atom

One way of drawing an atom is called the planetary model. In this model electrons are arranged in **shells** that look like orbits.

valence shell = outermost occupied shell
valence electrons = electrons in valence shell (involved in bonding)
core electrons = electrons in interior shells (don't participate in bonding)

Figure 2: Valence Shell Representations of Various Atoms

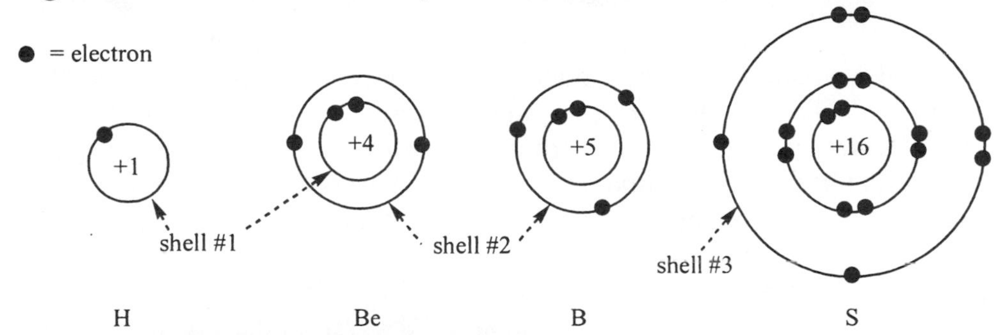

Critical Thinking Questions

1. According to Model 2 how many
   a) core electrons does an oxygen atom have?
   b) valence electrons does a carbon atom have?

2. According to Model 2, what is the maximum number of electrons that can fit in
   a) shell #1?
   b) shell #2?

3. Explain how the periodic table contains the answer to…
   a) Critical Thinking Question **1b** (**CTQ 1b**).
   b) CTQ 2.

Model 3: Bonds

One definition of a bond: a **shared** pair of valence electrons that is **localized** between two core atoms. The space that these bonding electrons occupy is sometimes called a **bonding electron domain**.

Figure 3: Four Different Representations of a Bond

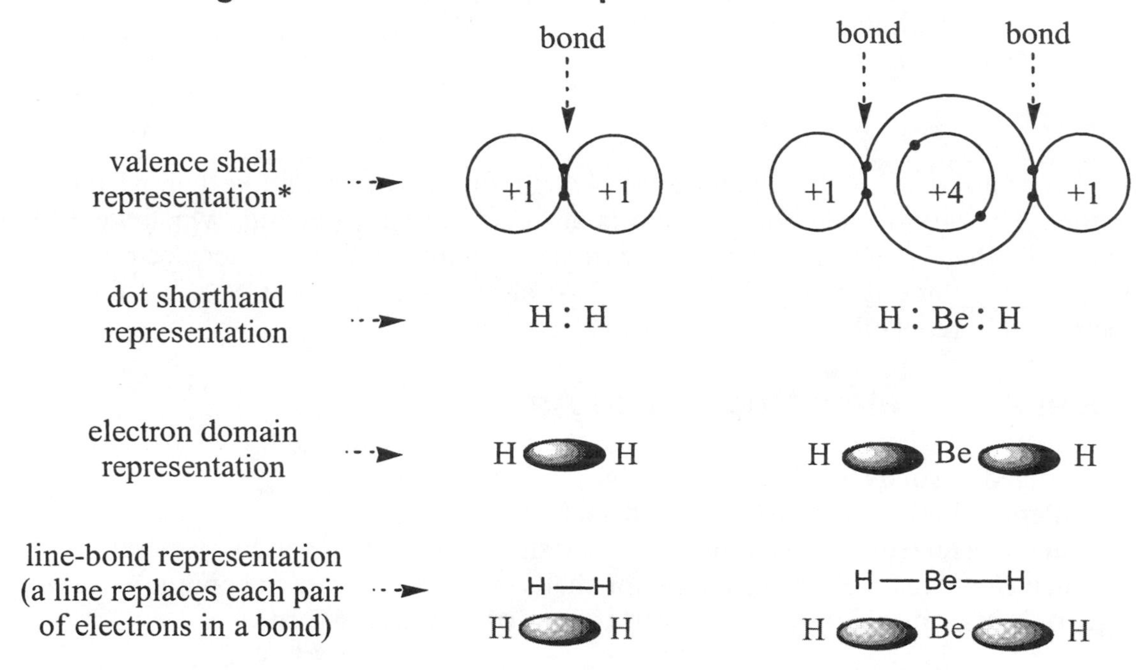

*The planetary model is misleading since modern experiments show that electrons **DO NOT** "orbit" the nucleus like the earth orbits around the sun. The electron domain representation is more accurate, in which electrons are confined to a 3D region of space between two core atoms. It is useful to imagine that the + charged core atoms are mutually attracted to the electron cloud between them. **Electron glue**, if you like.

Critical Thinking Questions

4. Draw a dot shorthand and line-bond representation of a BH_3 molecule.

Model 4: Bond Angles

The **bond angle**, ∠XYZ, is the angle defined by the three atoms X, Y, and Z.

Table 4: Bond Angles of Selected Atoms.

Molecular Formula	Line-bond Structure	Experimentally Determined Bond Angle
BeH_2	H—Be—H	∠ HBeH = 180°
BH_3	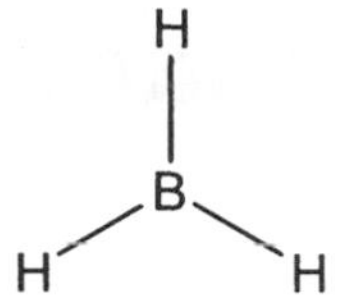 H above B; B bonded to H (lower left) and H (lower right)	∠ HBH = 120°

Critical Thinking Questions

5. Note that the H's of both structures in Table 4 are spread out as much as possible around the central atom (Be or B). Construct an explanation, based on the information in this ChemActivity, for why the H's of BeH_2 and BH_3 spread themselves out as much as possible around the central atom.

6. Consider the following drawing of methane (CH_4).

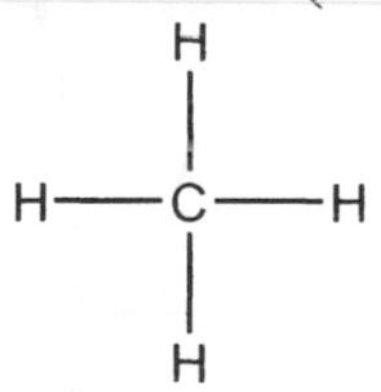

```
      H
      |
  H — C — H
      |
      H
```

   a) What is the ∠HCH bond angle implied by this drawing?

   b) In this drawing are the H's of methane spread out as much as possible around the central carbon?

7. Use the model materials provided to **make a model of methane**. The holes in the black carbon piece are pre-drilled to give the correct ∠HCH bond angles for CH_4. This angle is 109.5°.

   a) In which representation, **the drawing in CTQ 6** or **the model in your hand** [circle one] are the H's of CH_4 more spread out around the central carbon?

   b) You will often see methane drawn as it is in CTQ 6. Why is this drawing misleading, and what is left to the viewer's imagination when looking at this drawing?

c) Confirm that your model of methane looks like the following drawing. **Wedges** represent bonds coming out of the page and **dashes** represent bonds going into the page.

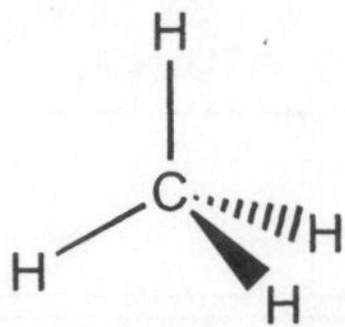

d) Explain why this drawing of methane is better than the drawing in CTQ 6.

Part B: Molecular Shape

(How do you determine the shape of a molecule?)

Model 5: Non-bonding Electrons

Recall from Part A that a pair of valence electrons localized between two core atoms is called a **bonding pair**, and that the region of space they occupy is called a **bonding domain**.

In some cases, a pair of valence electrons is localized in a region of space near a core atom, but is not involved in a bond. This pair of electrons is said to occupy a **non-bonding domain**. This non-bonding pair of electrons is usually called a **lone pair**.

Table 5A: Representations of Selected Atoms with Lone Pair/s.

dot representation	H •• H : N : H ••	H •• H : O : ••	•• H : F : ••	•• : Ne : ••
line-bond representation	H | H—N—H ••	H | H—O : ••	•• H—F : ••	•• : Ne : ••
electron domain representation	H H N H	H H O	H F	Ne
common name	ammonia	water	hydrofluoric acid	neon

Critical Thinking Questions

8. Circle each lone pair in Table 5. How many lone pairs does ammonia have?

9. Based on the hypothesis that the H's of NH_3 will spread out as much as possible, a student predicts that the ∠HNH bond angles of ammonia should be 120°.

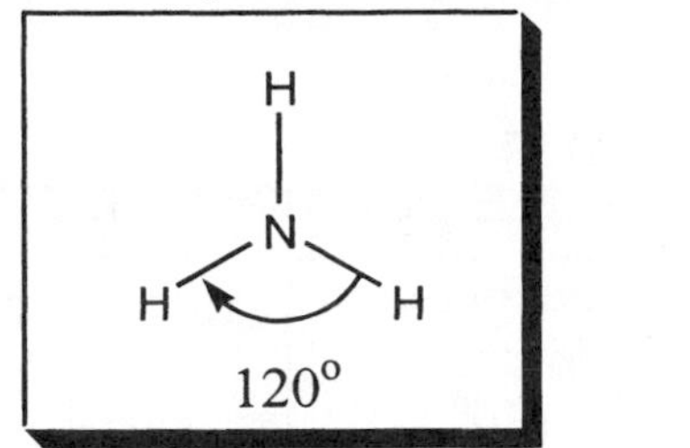

Student's Incorrect Hypothesis

lone pair in non-bonding domain

N H H H

close to 109.5°

In fact, NH_3's bond angles are much smaller, close to 109.5°. What repulsive force, which squeezes the H's of ammonia together, did this student neglect to consider.

10. Which of the following hypotheses better explains the bond angles of ammonia?
    A. Electron pairs repel one another.
    B. H atoms repel one another.

Table 5B: Bond Angles of Selected Atoms.

Common Name	Molecular Formula	Lewis Structure	Experimentally Determined Bond Angle
methane	CH_4	H H—C—H H	∠HCH = 109.5°
ammonia	NH_3	H H—N—H ··	∠HNH = 108.9°
water	H_2O	H H—O : ··	∠HOH = 103.5°

Critical Thinking Questions

11. Is your answer to CTQ 10 consistent with the fact that methane, ammonia and water have about the same ∠HXH bond angle (close to 109.5°)? Explain.

12. You **do not** need to memorize the exact bond angles of ammonia and water. You are expected to memorize the number 109.5, and to know that the bond angles of ammonia and water are close to this number. According to the data in Table 5b, which takes up more space, a bonding pair of electrons or a lone pair?

Model 6: Geometric Shape

* **"Total number of valence electron domains around central atom"** includes both bonding and non-bonding domains.

Table 6: Factors Effecting the Geometry of Selected Molecules.

Column #1	*Column #2*	*Column #3*	*Column #4*	*Column #5*
molecular formula (common name)	total number of atoms around central atom	total number of valence electron domains around central atom*	bond angle is closest to: 180°, 120°, or 109.5°	shape
BeH_2				linear
BH_3				trigonal planar
CH_4 (methane)				tetrahedral
NH_3 (ammonia)				pyramidal
H_2O (water)				bent

Critical Thinking Questions

13. Complete Table 6.
14. Memorize the five shape names listed in Table 6. Each shape is derived by considering the atoms "dots" and playing "**connect the dots**," as shown below. (Note the difference between tetrahedral and pyramidal).

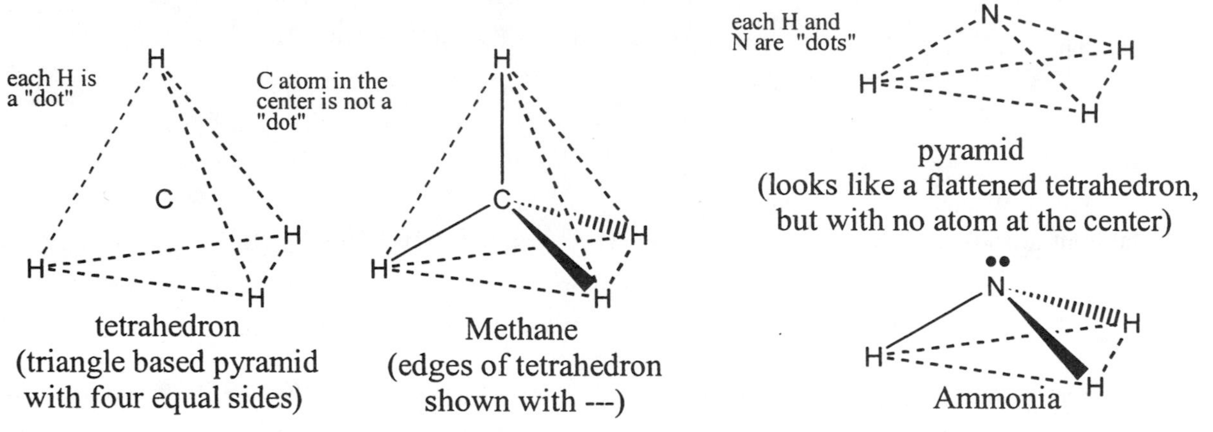

15. Is the bond angle (column 4) dependent on column 2, column 3 or both?

16. Is the shape of a molecule (column 5) dependent on column 2, column 3 or both?

17. A student who missed this class needs to know how to predict the bond angles and shape of a molecule from looking at its line-bond representation. Write a concise but complete explanation for this student.

Exercises for Part A

1. Draw a valence shell representation of a
   a) carbon atom.
   b) fluorine atom.

2. Consider the fluorine atom you drew in Exercise 1.
   a) How many electrons must be added to fill the fluorine atom's valence shell?
   b) What would be the total charge on such a fluorine atom? Explain your answer.

3. Consider the incomplete valence shell representation below.

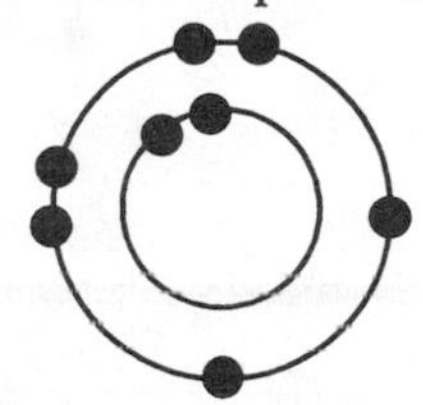

   a) Assume the atom is neutral and write the correct nuclear charge at the center of the atom.
   b) What is the identity of this neutral atom?

4. How many valence electrons does a neutral
   a) K atom have?
   b) C atom? N atom? O atom?
5. Consider the compounds $AlCl_3$ (aluminum chloride) and CF_4 (carbon tetrafluoride).
   a) Draw the valence shell representation of each.
   b) Predict the value of the XYX bond angle. Explain your reasoning.

6. Read the assigned pages in the text, and do the assigned problems (*see* reading list provided by your instructor).

Exercises for Part B

7. Wedge and dash line-bond drawings are a good way to convey the 3D shape of a non-planar models such as a model of methane (see CTQ 7c). Draw wedge and dash bond representations of the following molecules. Be sure to include the following in each drawing: **bond angles, lone pairs, the shape.**

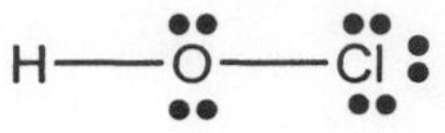

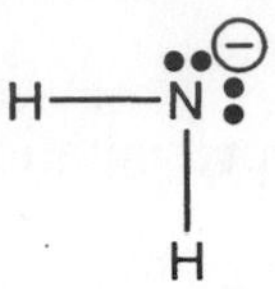

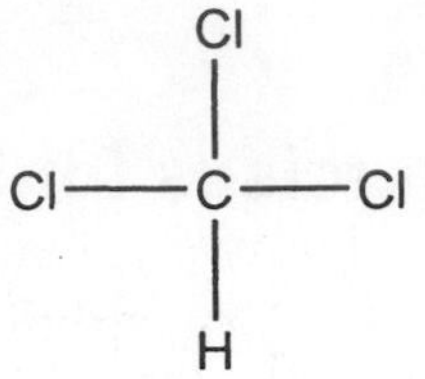

8. For each of the following molecules indicate the geometry at each central atom as in the example below. For molecules with more than one central atom, each central atoms is considered to have its own shape.

Example

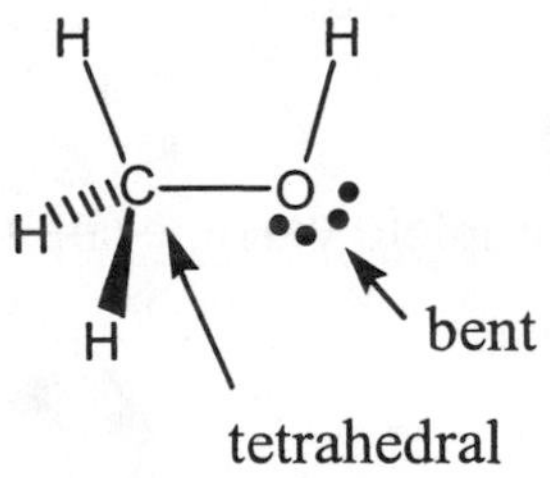

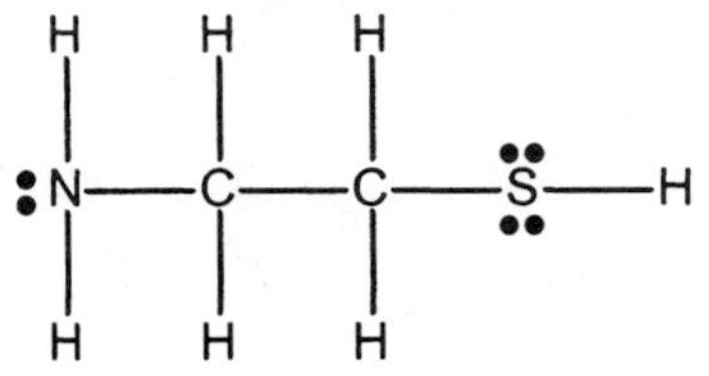

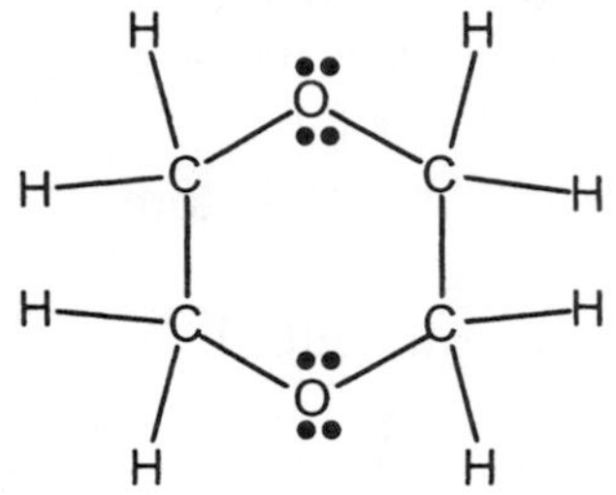

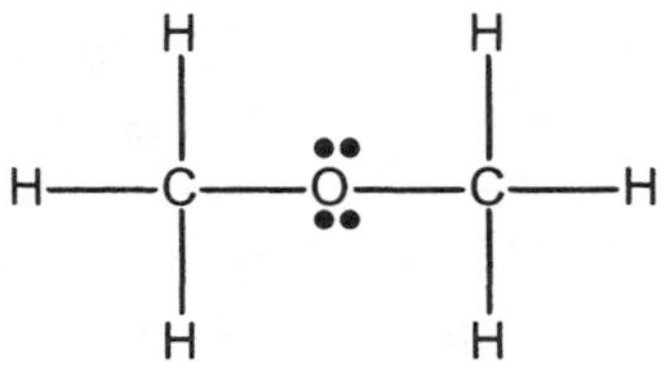

tetrahedral

bent

9. Read the assigned pages in the text, and do the assigned problems (*see* reading list provided by your instructor).

ChemActivity 2

Part A: Lewis Structures

(How do I draw a legitimate Lewis structure?)

Dot and line-bond representations which follow certain rules will be called "legitimate" Lewis structures.

Model 1: G. N. Lewis' Method for Predicting Molecular Properties

In the early part of this century, a chemist named Gilbert N. Lewis devised a system for diagramming atoms and molecules. The system is still used today because predictions made from these simple diagrams often match experimentally determined ion charges, bond angles, molecular shapes, bond orders and other chemical properties.

Lewis proposed the following representations for the first ten elements with their valence electrons.

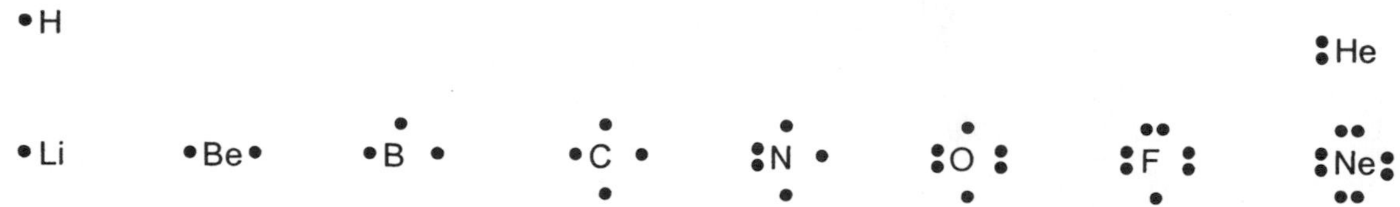

Of these 12 elements, only He and Ne are found in nature as single neutral atoms (as shown). The other elements are found as ions or as parts of molecules.

Lewis proposed that any atom, ion or molecule that can be represented by a "legitimate" Lewis structure should exist in nature or be possible to make in the laboratory.

A "legitimate" Lewis structure is a dot or line bond representation in which:

1. *The sum of the electrons around a hydrogen is two (a bonding pair).*
2. *The sum of the electrons (bonding pairs + lone pairs) around a **<u>carbon, nitrogen, oxygen or fluorine atom</u>** is eight–an **octet**.* **(The "octet rule.")**

For example:

O 2 H = O H H

"lone pair" "bonding pair"

O 2 H = O—H H

we will usually use this line-bond notation

Figure 1: Shell and Lewis Representations of Selected Compounds

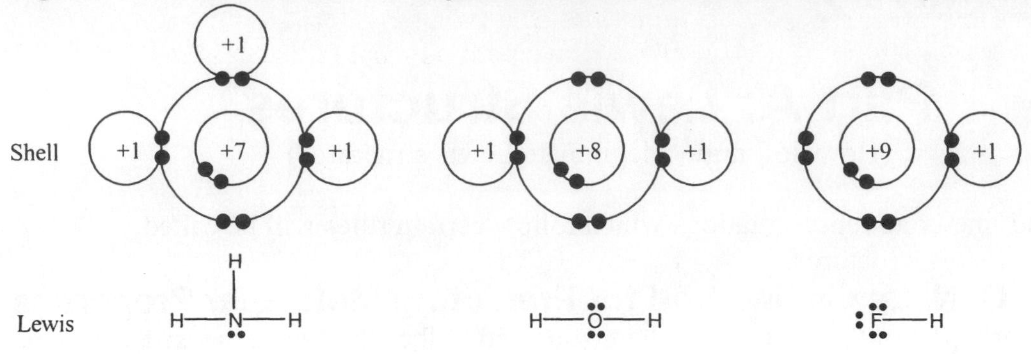

Critical Thinking Questions

1. The valence shell of an atom in a legitimate Lewis structure (e.g. N, O, F or H, *above*) has what in common with the valence shells of He, Ne and all elements in the last column of the periodic table?

2. Draw a shell representation and Lewis structure for the ion of fluorine that you predict is most likely to be stable. Explain your reasoning.

3. Draw a Lewis structure of a neutral molecule that is a naturally occurring combination of fluorine atoms and one carbon atom.

4. What is the shape of the molecule in CTQ 3?

5. Make a checklist that can be used to determine if a Lewis structure for a molecule is legitimate.

6. a) How many valence electrons does one nitrogen atom have?
   b) How many valence electrons does one hydrogen atom have?
   c) Hypothetically, how many valence electrons would a (neutral) NH_4 molecule have if it could exist?
   d) How many valence electrons does one NH_4^+ ion have?
   e) Draw the Lewis structure for NH_4^+

7. Describe how to calculate the total number of valence electrons in an ion (+ or -)?

Model 2: Multiple Bonds

Table 2a: Bond Information for Second Row Diatomic Molecules

Formula	Lewis Structure	Bond Order	Type of Bond	Number of Bonding Domains
F_2	:F̤̈—F̤̈:	1	single bond	1
O_2	Ö=Ö (with lone pairs)	2	double bond	1
N_2	:N≡N:	3	triple bond	1

Critical Thinking Questions

8. Does the Lewis structure of O_2 satisfy the checklist you made in CTQ 5?

9. How many electrons are involved in a triple bond?

10. Consider the molecule H_2CO.
    a) Draw the Lewis structure.

    b) How many bonding domains are there around carbon, the central atom?
    c) What is the shape of this molecule?

    d) Are your answers above consistent with a HCO bond angle very close to 120°?

Part B: Formal Charge

(How can you choose between two possible Lewis structures?)

Model 3: Two Lewis Structures for CO_2.

$\ddot{\underset{..}{O}}=C=\ddot{\underset{..}{O}}$ $:\ddot{\underset{..}{O}}-C\equiv O:$

I   II

Experimentally, we find that both carbon-oxygen bonds in CO_2 are identical.

Critical Thinking Questions

11. Are both structures in Model 3 *legitimate* Lewis structures?

12. Which Lewis structure in Model 3 best fits the experimental data?

Model 4: Formal Charge.

For many molecules we can draw more than one legitimate Lewis structure. In these situations it can be helpful to calculate each atom's **formal charge.** This will help you determine "hot spots" of + or – charge on a structure.

For each atom…

formal charge = core charge* – number of assigned electrons

Rules for assigning electrons:

1. *Non-bonding electrons are assigned to the attached atom.*
2. *Shared electrons are evenly divided between the bonded atoms.*

(*core charge = nuclear charge – number of core electrons)

Example:

Electron accounting for a Lewis structure of ammonia, H_3N

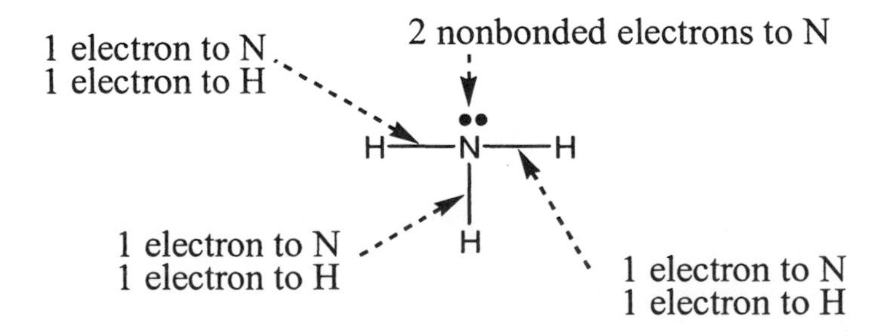

Formal charge on the nitrogen = +5 (core charge) – 5 (assigned electrons) = 0

Formal charge on each hydrogen = +1 (core charge) – 1 (assigned electron) = 0

Zero formal charges are not written on Lewis structures so we do not need to add any formal charges to the Lewis structure of ammonia shown above.

- From now on, a Lewis structure is not complete unless the non-zero formal charges are indicated. (Place a circled + or – next to the atom holding that formal charge, as in CTQ 18, structure II, on the next page.)
- Fewer non-zero formal charges = a better Lewis structure for making predictions.
- Formal charges greater than ±1 are never found in good Lewis structures.

Critical Thinking Questions

13. Calculate the formal charge on each atom of the following Lewis structures, and write in these formal charges where appropriate.

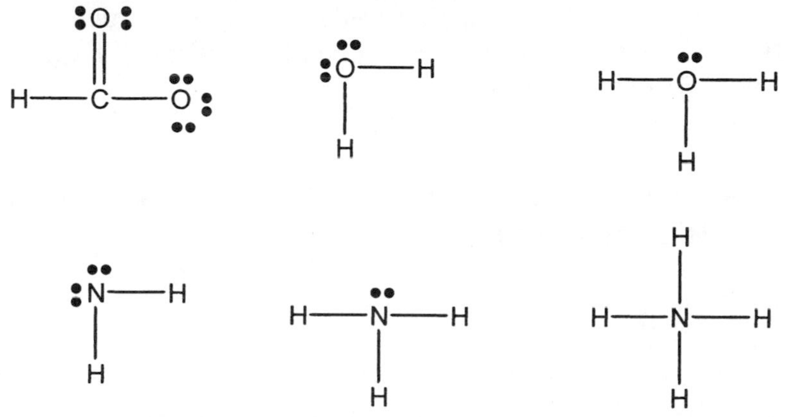

14. Add formal charges to each Lewis structure of CO_2 in Model 3.
    a) Based on the concept of formal charge, which is the better Lewis structure for CO_2 (in Model 3)— I or II? Explain your reasoning.

    b) Is your choice consistent with the experimental data?

15. Draw a Lewis structure (including any non-zero formal charges) of carbon monoxide (CO). Hint: there is no Lewis structure for CO with all zero formal charges.

16. T or F: The sum of the formal charges on a Lewis structure is equal to the total charge on the molecule or ion.

17. T or F: If the net charge on a molecule is zero, the formal charge on every atom in the molecule must equal zero.

18. Two Lewis structures for glycine (an amino acid) are given below.

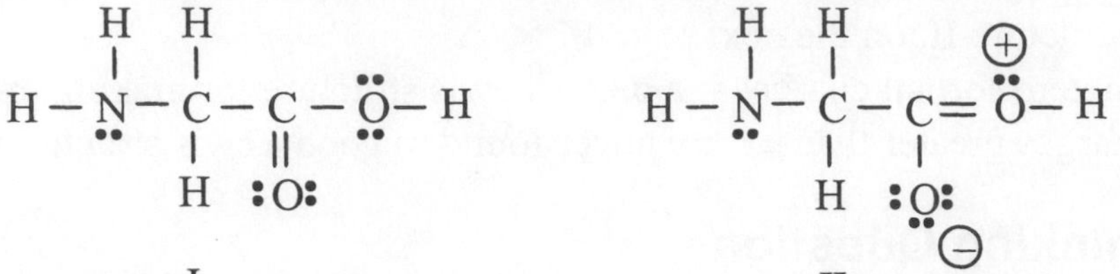

```
     H   H                              H   H
     |   |                                  |   |        ⊕
                  ••                                     ••
H – N – C – C – O – H              H – N – C – C = O – H
    ••   |   ||    ••                   ••   |   |
         H  :O:                              H  :O:
                                                 ••
                                                  ⊖
         I                                   II
```

a) Predict the ∠COH bond angle based on the Lewis structure on the left.
b) Predict the ∠COH bond angle based on the Lewis structure on the right.
c) Which prediction do you expect to be more accurate? Explain why.

19. Make a checklist that can be used to determine if a Lewis structure is correct and that it is the best Lewis structure.

Exercises for Part A

1. Draw the Lewis structure of a neutral molecule that is a naturally occurring combination of hydrogen atoms and one sulfur atom. What is the shape of this molecule?

2. Draw legitimate Lewis structures of the following species and predict the geometry about the central atom (shape).

a)	NH_3	d)	NO_2^+
b)	CCl_4	e)	CO_3^{2-}
c)	N_2	f)	N_2O (try with N or O at center)

3. For each element, predict (and draw a Lewis structure of) the most commonly occurring ion:

a)	sulfur	c)	magnesium
b)	iodine	d)	oxygen

4. Predict which of the following species is NOT commonly found in nature.
   CH_2 NO^+ HO^-

5. The molecules BH_3 and SF_6, and the ion SO_4^{2-} are found in nature. Is this consistent with Lewis' predictions? Explain why or why not.

6. The following are NOT legitimate Lewis structures. Show (as in the example) how a single pair of electrons can be moved to make the Lewis structure legitimate.

(curved arrow shows where the electron pair was moved from and to)

legitimate Lewis structure

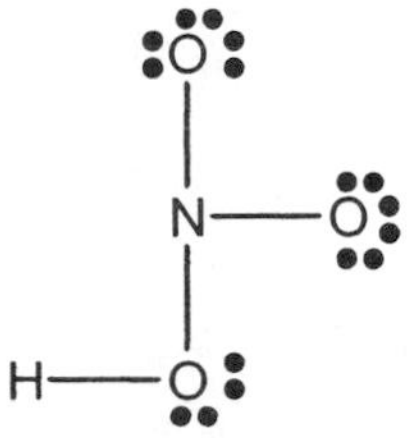

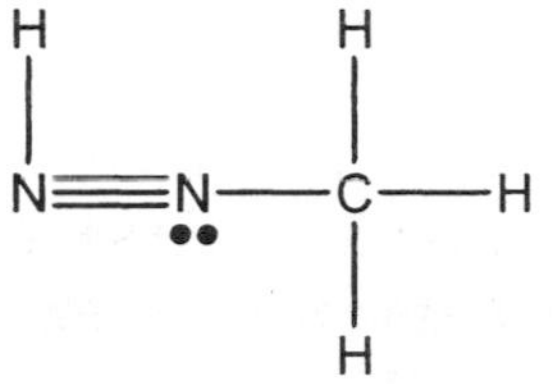

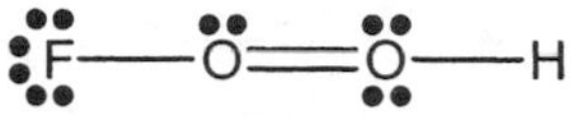

7. Read the assigned pages in your text, and do the assigned problems.

Exercises for Part B

8. Some of the following Lewis structures are missing formal charges. Fill in the formal charges (other than zero) where needed. Then use your list of factors (from CTQ 19) to verify that each completed Lewis structure is correct.

a) 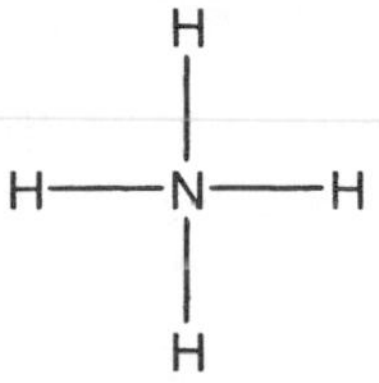 NH_4^+

b) 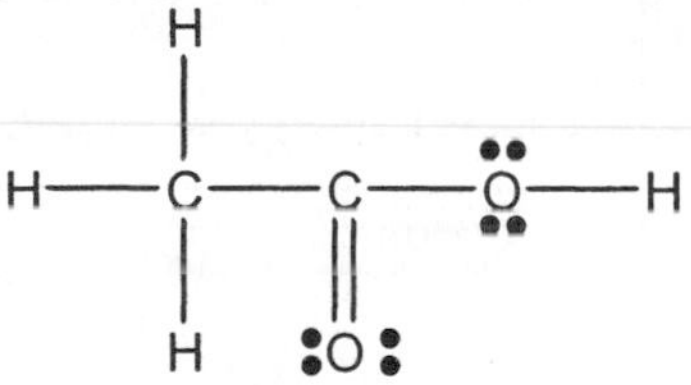 CH_3COOH

c) 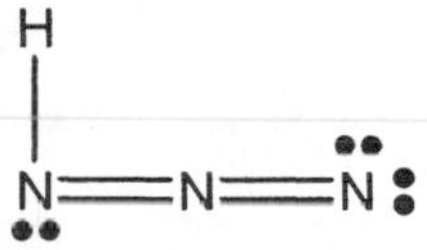 HN_3

d) 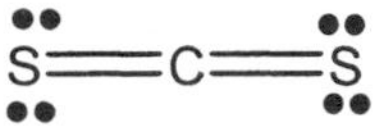 CS_2

e) 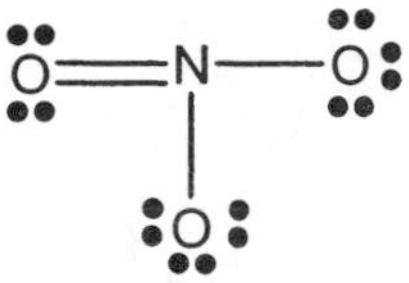 NO_3^-

f) 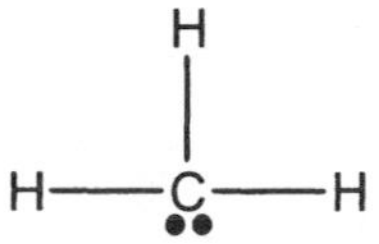 CH_3^-

9. For a-f in Exercise 8, predict the shape of the molecule or the geometry about each central atom, as appropriate.

10. The following Lewis structure for CO has no formal charges. Explain why this is not a valid Lewis structure.

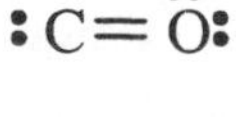

:C=Ö:

11. Complete each row on the following table by drawing an example of a molecule containing the atom listed with +1, 0 and –1 formal charge and the correct number of hydrogens and lone pairs. (**Note: entry for H_nC^+ is an exception to the octet rule.**)

	+1	0	-1
C	Exception to octet rule		
N			
O			
F	(uncommon)		

12. Each H on the table above can be replaced with any atom and the formal charges will stay the same. This means, for example, that any oxygen with three bonds will have a +1 formal charge.
    a) What is the formal charge on any N with 4 bonds to it?
    b) What is the formal charge on any O with one bond to it?
    c) What is the formal charge on any C with two single bonds and one double bond to it (as in H_2CO)?
    d) What is the formal charge on any O with one single bond and one double bond to it?

e) How many total bonds (a double bond counts as two) must a neutral N atom have going to it?

f) Draw an example of a molecule containing a neutral N with a double bond to that N.

13. Carbon is a little strange in that it does not always follow the octet rule. We will learn about this later in the course. For now know that:
    - C with three bonds and a lone pair must have a –1 formal charge.
    - C with three bonds and <u>no lone pair</u> must have a +1 formal charge.

a) A molecule containing a C with a –1 charge must have how many bonds and lone pairs to that C?

b) A molecule containing a C with a +1 charge must have how many bonds and lone pairs to that C?

14. Read the assigned pages in your text, and do the assigned problems.

ChemActivity 3

Part A: Resonance Structures

(What do we do when a single Lewis structure fails to yield accurate predictions?)

Model 1: Two Lewis Structures for Ozone (O_3)

Table 1: Bond Data for Typical Oxygen-Oxygen Bonds

	bond length (10^{-10} m)	bond order	# of electrons per bond
typical O–O single bond	1.48	1	2
ozone*	1.28		
typical O=O double bond	1.21	2	4

*Experimentally, we find that both oxygen-oxygen bonds of ozone are identical.

Critical Thinking Questions

1. Is each proposed structure in Model 1 a legitimate Lewis structure? Explain why or why not.

2. Explain why the structure below is not a good description of ozone.

3. Does structure I taken by itself, (or structure II taken by itself) explain the experimental fact written at the bottom of Table 1? Explain why or why not.

4. According to the bond length data in Table 1, the **bond order** of each oxygen-oxygen bond in ozone is best described as…
   A. bond order = 1
   B. bond order = 2
   C. bond order = between 1 and 2

Model 2: Hybrid Structure

I II

Taken separately, neither structure I nor structure II provides an accurate picture of bonding in ozone. As the bond length data in Table 1 indicates, the oxygen-oxygen bond of ozone is somewhere between a single bond and a double bond.

A different picture of bonding in ozone is the **hybrid structure** shown below. It is the average of structures I and II.

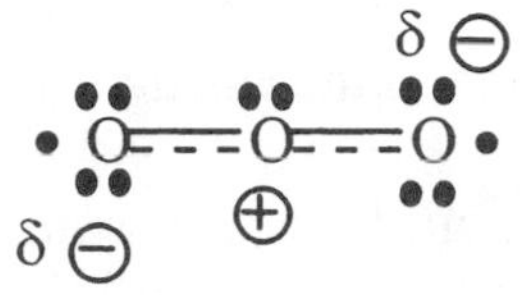

- - - - = half a bond (one electron)

$\delta\ominus$ = partial negative charge

hybrid structure

Critical Thinking Questions

5. Delta (δ) is used to indicate any amount less than one. Use the rules for calculating formal charge to determine the exact formal charge on each oxygen of ozone.

6. According to the hybrid structure, what is the bond order of each bond in ozone?

7. According to the hybrid structure, how many electrons participate in each bond of ozone?

8. In what ways is the hybrid structure of ozone better than either structure I or II, alone?

Model 3: Resonance Structures

A set of resonance structures is an alternate way of representing a hybrid structure.

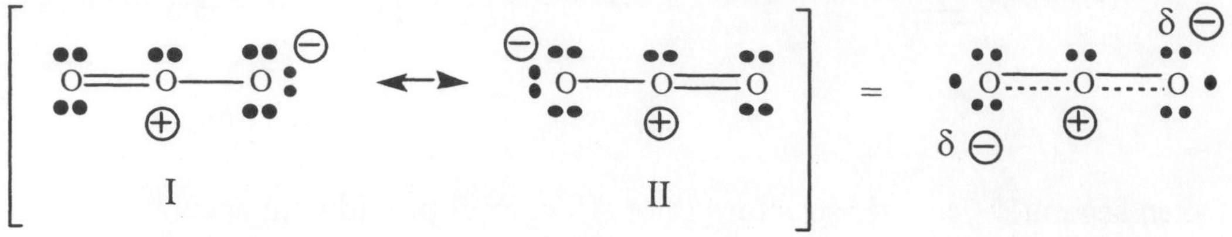

I II

set of resonance structures **hybrid structure**

Square brackets are used to indicate that the reader must consider the average of all the **contributing resonance structures** within the brackets in order to gain an accurate picture of bonding.

At no time are the electrons of ozone arranged according to the pattern shown in resonance structure I or II. In this sense, a single contributing resonance structure is a misrepresentation!

The special double-headed arrow does not indicate a reaction; however, some students find it useful to think of the double headed arrow as indicating an infinitely fast conversion back and forth between contributing resonance structures. Therefore, even over a very short time period we would "see" the average of the contributing resonance structures, not one structure or the other.

Critical Thinking Questions

9. The better the arrangement of electrons in a resonance structure, the more **important** that resonance structure is for making predictions. For example, resonance structures with formal charges $> \pm 1$ are **not important**. The following is the complete set of **important** resonance structures for nitrate ion:

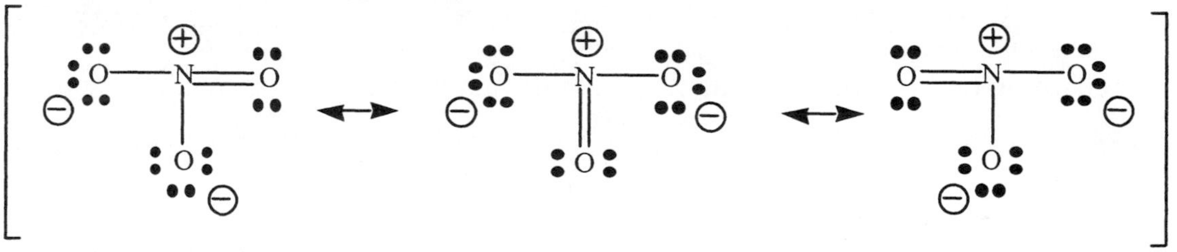

   a) What is the nitrogen-oxygen bond order of nitrate ion?

   b) What is the exact formal charge on each oxygen of nitrate ion?

   c) What is the total charge on the nitrate ion?

   d) Explain why the following is not included in the set of important resonance structures for nitrate ion?

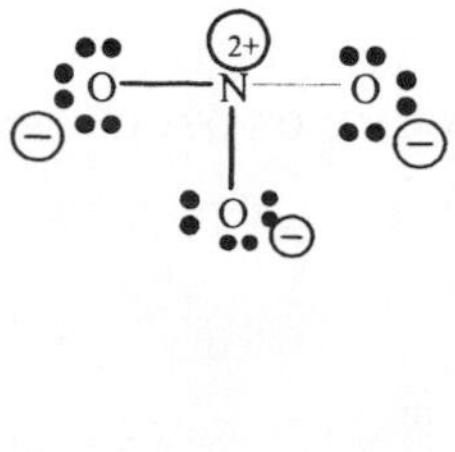

Important note: If you draw the hybrid structure of nitrate ion (above) you will find each oxygen is assigned a 1/3 electron. It is hard to make sense of such oddities and they have little physical significance. In this course, we will make many predictions using resonance structures rather than a hybrid structure.

Part B: Curved Arrow Notation

(How do organic chemists keep track of electrons?)

The organic chemist's most valued tool is neither beaker nor Bunsen burner, but a **curved arrow**. We will use curved arrows extensively in this course to keep track of actual and imaginary electron movement.

Model 4: Drawing Resonance Structures (Imaginary Reactions)

The following are the three "legal" ways of moving electrons to generate a set of resonance structures. Please study these arrows carefully. They are the key to drawing resonance structures.

I. You may move a lone pair of electrons on an atom so as to form a new double bond to that atom (*see* **Arrow 1**).

II. You may break part of a double bond and make those two electrons a lone pair on one of the atoms originally involved in the double bond (*see* **Arrow 2**).

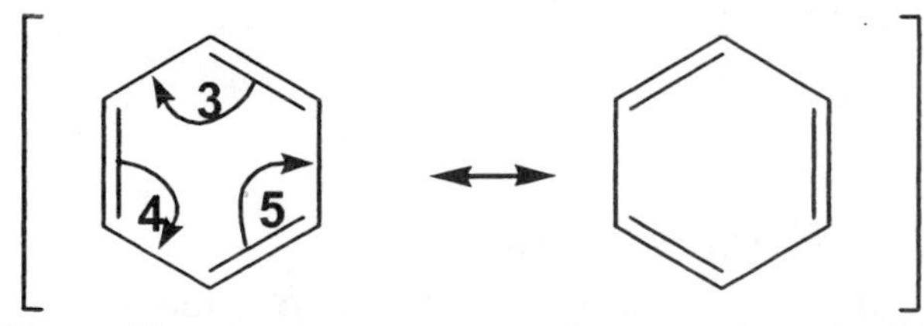

III. You may move two electrons in a double bond so as to form a new double bond next door (**Arrows 3, 4 & 5**).

In addition:

- You may not break any single bonds.
- You may not move any nuclei.
- You may not change the total number of electrons.
- The resulting resonance structure **must be a legitimate Lewis structure**.
- Arrows always BEGIN AT THE SOURCE OF ELECTRONS!!

For example we can generate the second resonance structure of nitrate ion from the first.

Figure 4A: Resonance Structures of Nitrate Ion

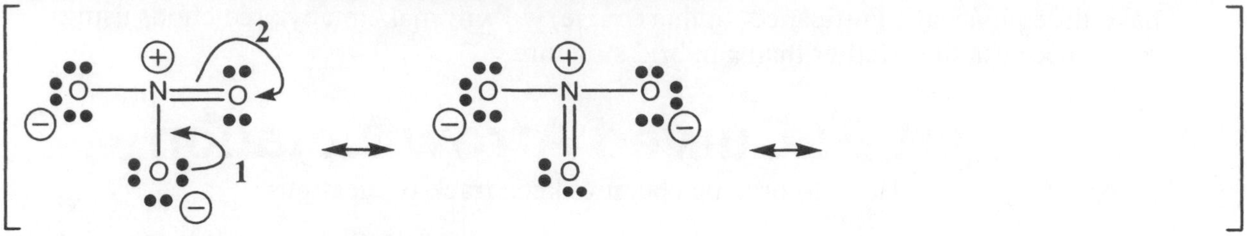

Critical Thinking Questions

10. Add curved arrows to the second resonance structure of nitrate ion, showing how it can be changed to generate the third resonance structure (draw this in the space above.)
11. Draw the structure that would result if you performed the electron movement described by **Arrow 1** in Figure 4A, but <u>not</u> **Arrow 2**, and explain why the resulting structure is NOT a legitimate Lewis structure.

12. For the following ion, use curved arrows to generate two other important resonance structures.

Model 5: Reaction Mechanisms (Real Reactions)

Recall that transformations between resonance structures are totally imaginary, although it can be useful to think of such transformation as infinitely fast reactions.

Curved arrows can also be used to depict **real reactions**. For example:

Figure 5: Mechanism of an Acid-Base Reaction

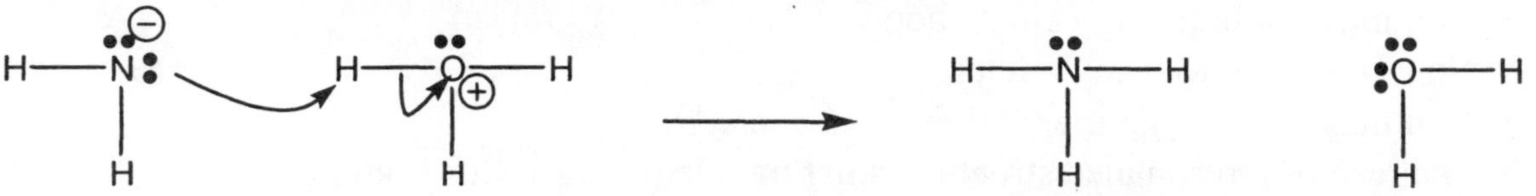

This accounting of bonds formed and broken is called **a mechanism of the reaction**. There can be more than one reasonable mechanism for a particular reaction.

Critical Thinking Questions

13. Each arrow in a mechanism can be read as a sentence. For example the first arrow in Figure 5 says "The electrons of a lone pair on N are used to make a new single bond to an H on H_3O^+." Write a sentence describing what the second arrow in Figure 5 tells you.

14. Deuterium (D) is an isotope of hydrogen (H) with nearly identical reactivity. Use curved arrows to show a reasonable mechanism for the following acid-base reaction.

Exercises for Part A

1. For each of the following, complete the Lewis structure. Include...
   i) all important resonance structures
   ii) all non-zero formal charges

(Skeleton structures and molecular formulas are given.)

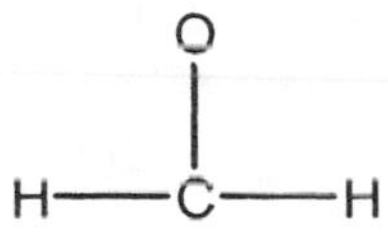

H_2CO

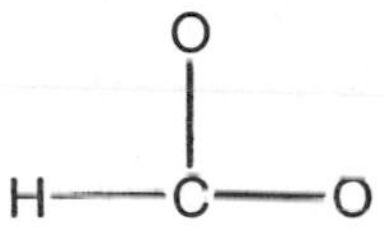

HCO_2^-

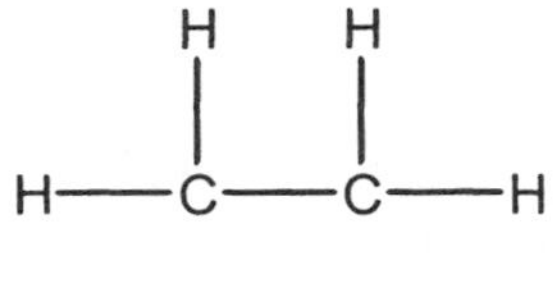

CH_2CH_2

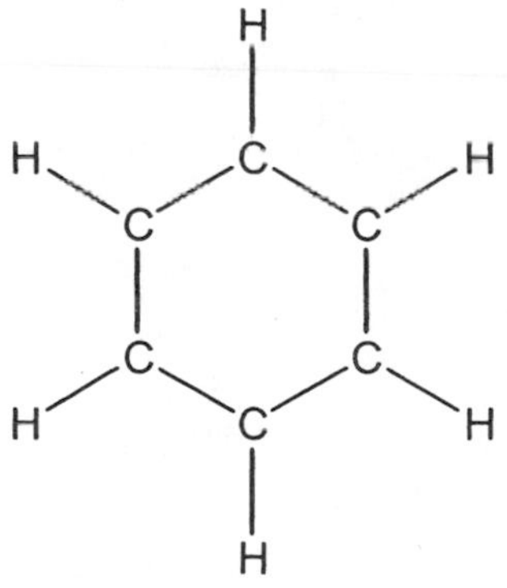

$(CH)_6$

2. For each of the following skeleton structures: complete the Lewis structure including any important resonance structures and predict a value for the bond angles requested.

   a) Nitrous acid (HNO_2)

   H—O—N—O ∠HON = ∠ONO =

b) Enolate ion ($C_2H_3O^-$)

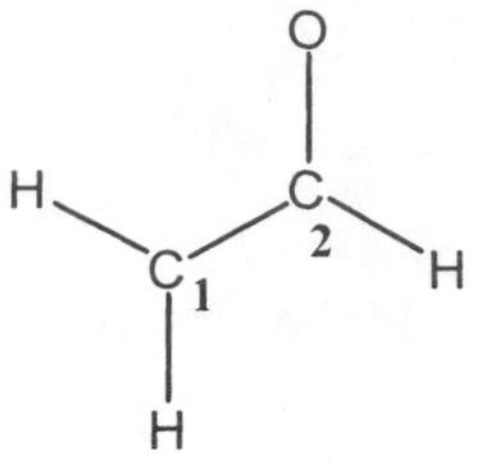

$\angle C_1C_2H =$ $\angle HC_1C_2 =$

Explain your reasoning.

3. For each of the following sets of structures:
   i) Add missing formal charges.
   ii) Cross out any unimportant resonance structures.

a)

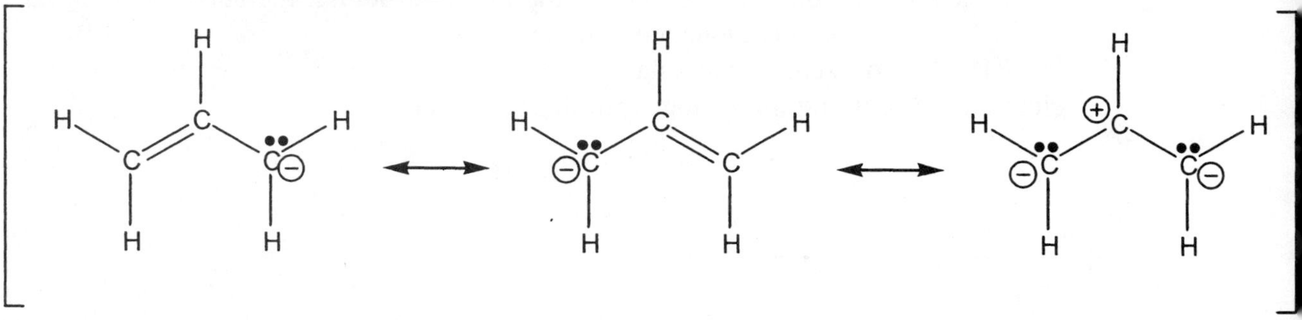

b)

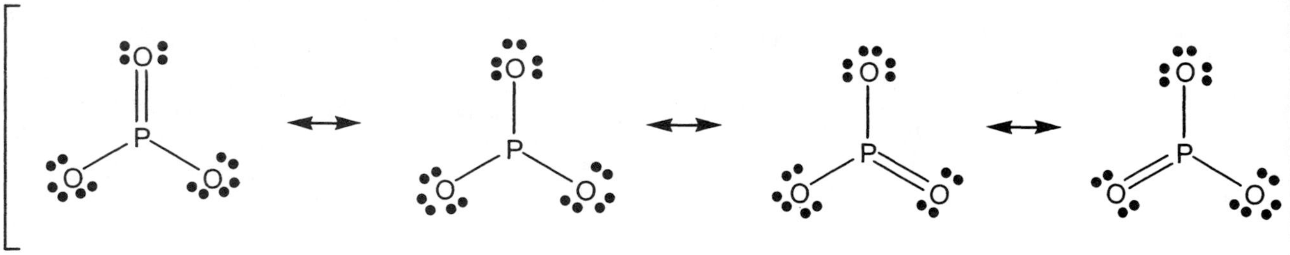

4. Read the assigned sections in your text and do the assigned problems.

Exercises for Part B

5. For each set of resonance structures in Exercise 3 (above) use curved arrows to show the movement of electrons that would accompany imaginary (infinitely fast) conversion of one resonance structure to the next (exclude crossed out structures).

6. Each of the following is one of a set of resonance structures. Use curved arrows to help keep track of electrons and complete each set of resonance structures.

a)

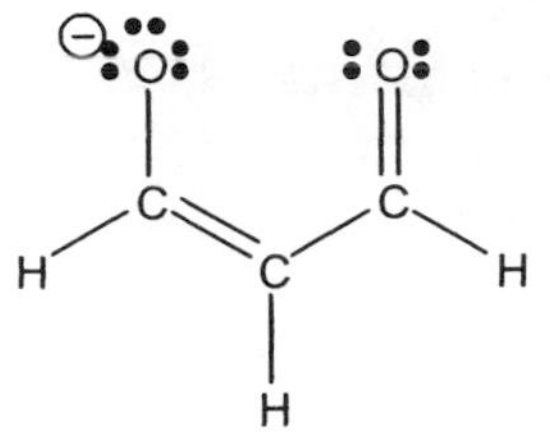

b)

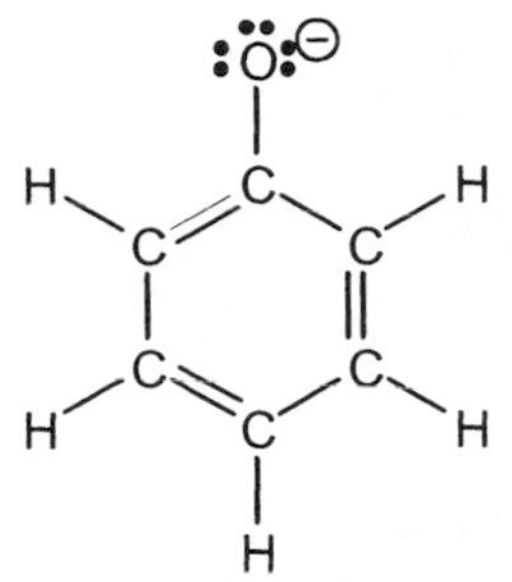

7. If the ion in CTQ 12 is treated with D_3O^+ two distinct products are observed (Products A & B).

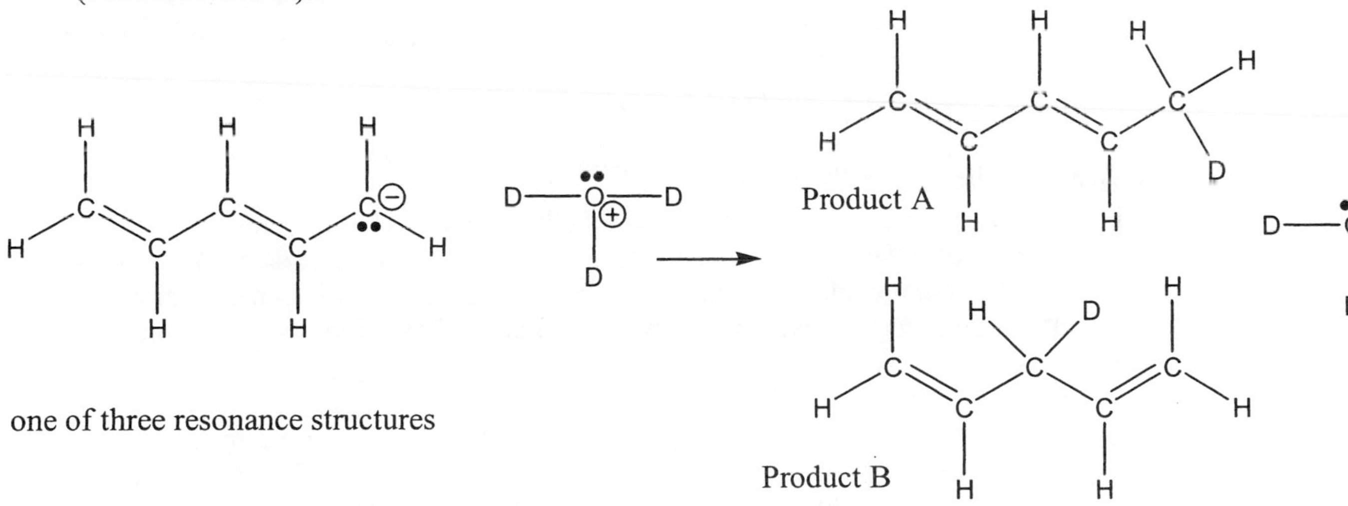

one of three resonance structures

Product A

Product B

a) Use curved arrows to show a reasonable mechanism for formation of Product B.

b) A student predicts the ratio of A:B in the product mixture will be 2:1. Construct an explanation for this prediction.

8. Read the assigned sections in your text and do the assigned problems.

ChemActivity 4

Conformers of Carbon Structures

(What is the lowest energy conformation of butane?)

Model 1: Alkane Conformations

Figure 1: Representations of Ethane in Its Most Favored Conformation

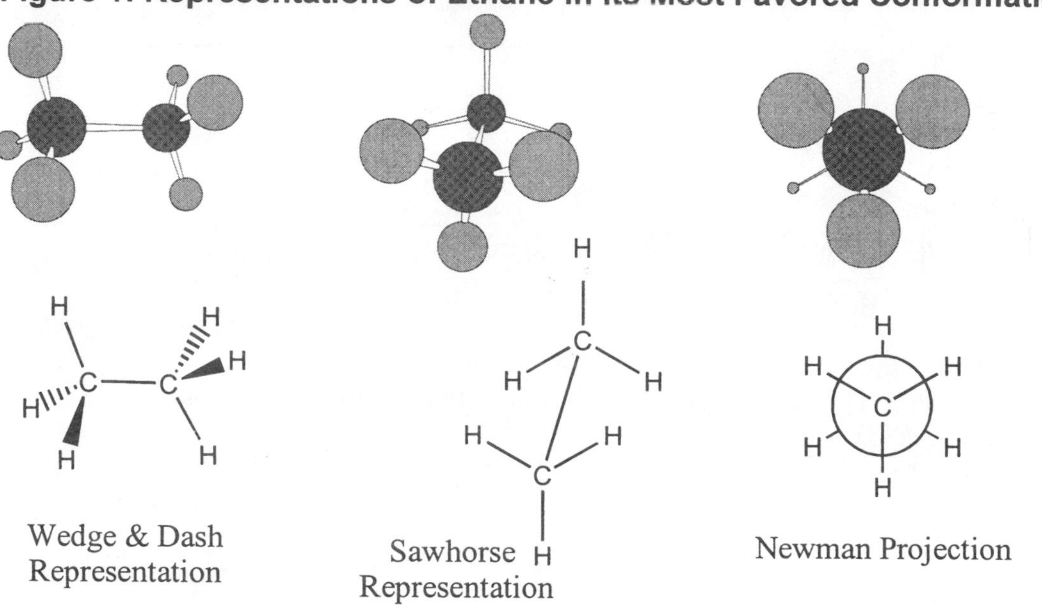

Critical Thinking Questions

1. Each representation in Figure 1 shows ethane (CH_3CH_3) in its lowest potential energy (most favorable) conformation. If you have a model set available, make a model of ethane and rotate the single bonds until it is in this conformation.
   a) Construct an explanation for why this conformation is most favorable.

   b) Consider the Newman Projection in Figure 1. Newman projections are confusing partly because an imaginary disc is inserted in the drawing to create the illusion of near and far. What single atom of ethane is hidden from view behind this disc?

2. Consider the following Newman Projection of ethane in its least favorable conformation and construct an explanation for why it is so un-favorable.

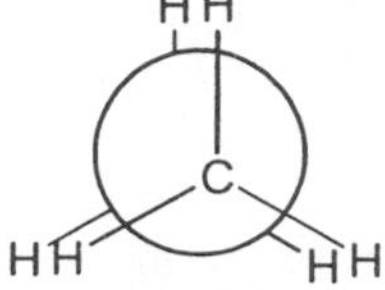

3. Construct an explanation for why the conformation in Figure 1 is called **staggered** and the conformation in CTQ 2 is called **eclipsed**.

Model 2: Potential Energy

A marble at the top of a hill is high in **potential energy** due to gravitational attraction between itself and the Earth. Unless it is perfectly balanced, gravity will cause the marble to roll to the bottom of the hill where it is low in **potential energy**. Most processes in Nature tend toward lower potential energy.

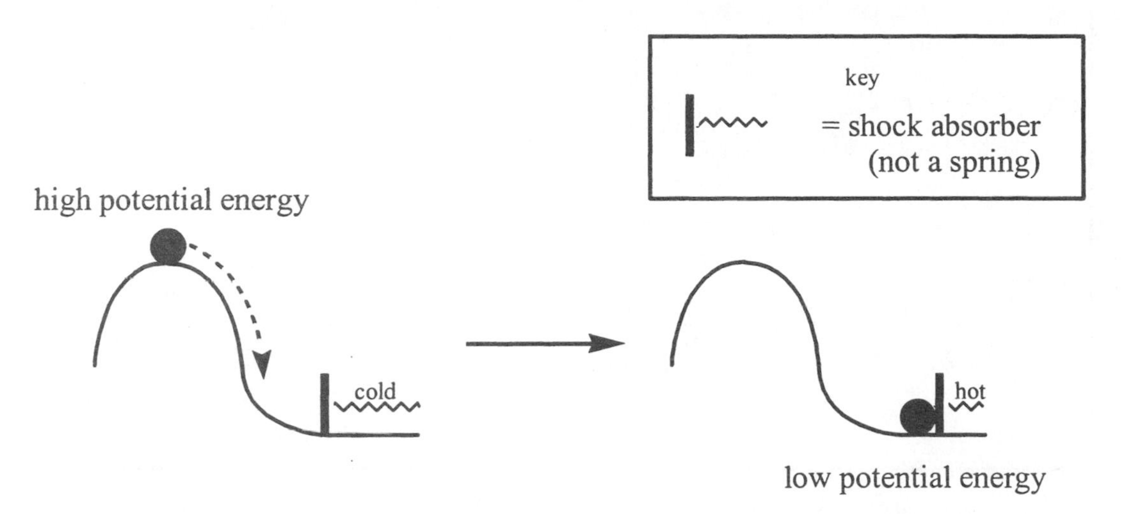

Critical Thinking Questions

4. In which location is the ball more stable (unlikely to change position)?

5. In general, **high potential energy** systems are **unstable** and **likely to change**. For example, ethane in the eclipsed conformation (see Newman Projection in CTQ 2).

   a) What conformation of ethane is "downhill" in terms of potential energy from the eclipsed conformation?

   b) What repulsive forces will cause ethane in the eclipsed conformation to quickly adopt this new conformation?

6. Complete the following graph of **potential energy** vs. **rotation of the C_1–C_2 bond** of ethane. Do not worry about assigning numerical values to the y axis. The point drawn for you indicates the potential energy at 0° (eclipsed).

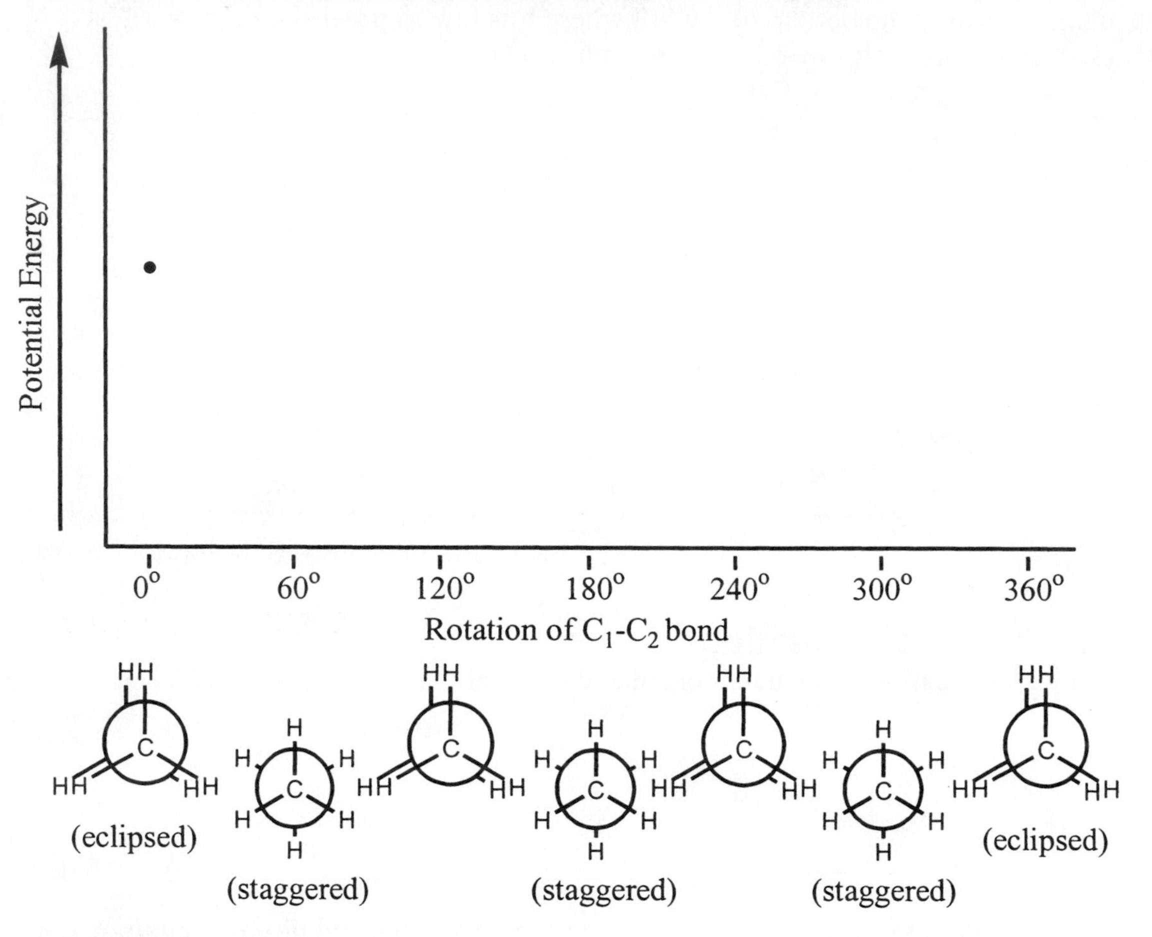

Model 3: Ethane in Motion

At room temperature, the C_1-C_2 bond of ethane is spinning rapidly. If you could observe this spinning you would see that the molecule zips past the high energy eclipsed conformations and slows down to take advantage of the low energy staggered conformations. Chemists explain this phenomenon by saying that the molecule gets stuck in the potential energy wells of the staggered conformations. (These wells are the valleys on your diagram above.)

Critical Thinking Question

7. If you could devise a way to take an instantaneous snapshot of ethane, in what conformation would you be most likely to “catch it?” Explain your reasoning.

Model 4: Newman Projections of Butane

A Newman Projection of a complex molecule focuses on one bond. In this case it is the C_2-C_3 bond, so C_1 and C_4 are drawn as CH_3 groups (**methyl groups**). Since they are spinning rapidly, these two methyl groups can be viewed as spheres about 5 times the size of a hydrogen (H).

Staggered conformation with methyl groups **gauche** to one another

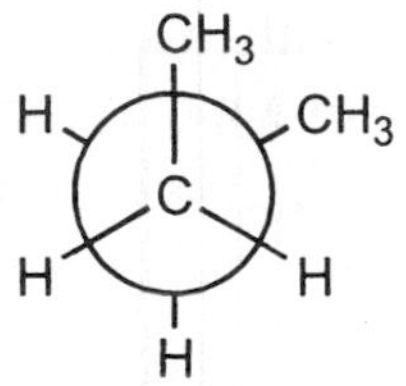

Critical Thinking Questions

8. What single atom of butane is hidden from view by the imaginary disc of the Newman Projection in Model 4?

9. In the staggered conformation shown above, the two methyl groups are **gauche** to one another. The **staggered-gauche** conformation is NOT the most favorable conformation for butane.

    a) Draw a Newman Projection of butane in its most favorable conformation.

    b) This most favorable (lowest potential energy) conformation which you drew in part a) is called **staggered-anti**. Construct an explanation for both parts of this name.

    c) The staggered-gauche conformation shown in Model 4 is not the most favorable, or the least favorable conformation. Draw a Newman Projection of butane in its least favorable (highest potential energy) conformation.

Exercises

1. Make a model of butane ($CH_3CH_2CH_2CH_3$).
   a) Rotate the single bonds until you find the most favorable **conformation**.
   b) Draw a wedge and dash-bond representation of butane in this conformation.

2. Consider the molecule 1-bromo-2-methylbutane. C_3 and C_4 should be drawn as CH_2CH_3 as in the example. This group is called an **ethyl group** and can be considered a sphere about twice the size of a methyl group. Draw the following Newman projections **sighting down the C_1—C_2 bond...**

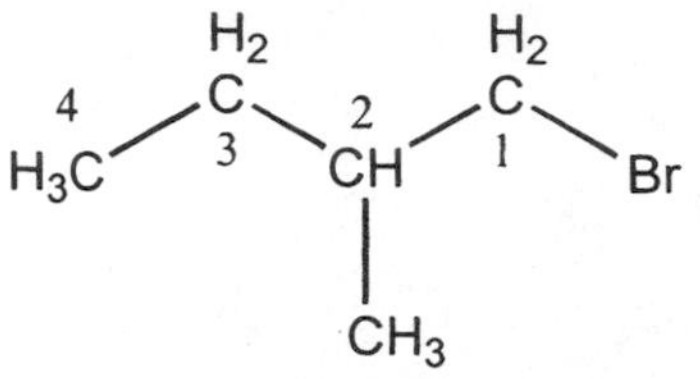

H2 H2
4 C 2 C
H3C 3 CH 1 Br
CH3

1-bromo-2-methylbutane

example Newman Projection

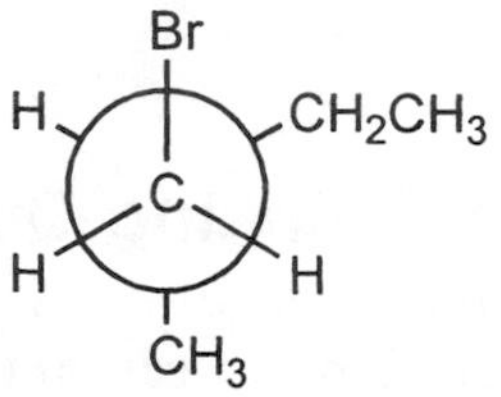

Br
H CH2CH3
C
H H
CH3

   a) The lowest potential energy conformation.
   b) The highest potential energy <u>staggered</u> conformation.

3. Biologists often represent sugars and other complex molecules using **Fisher Projections**. In a Fisher projection all horizontal bonds should be assumed to come out of the page toward you (wedge bonds) and all vertical bonds into the page, away from you (dash bonds). Draw wedge and dash representations of the following Fisher projections.

O H
C
HO—C—H
CH2OH

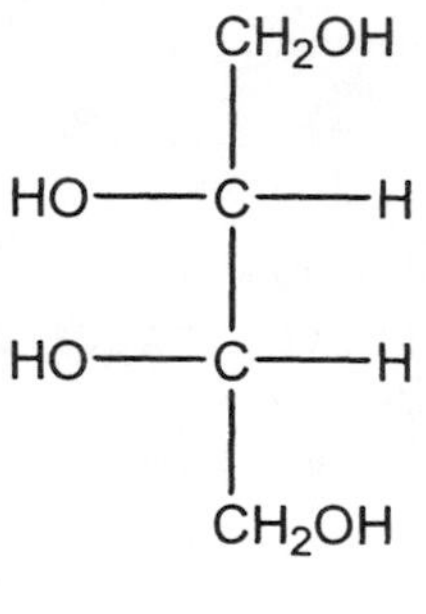

CH2OH
HO—C—H
HO—C—H
CH2OH

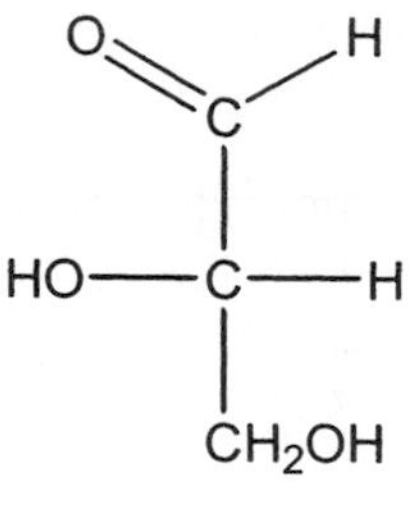

4. Read the assigned sections in your text and do the assigned problems.

ChemActivity 5

Part A: Constitutional Isomers

(Are they the same, or are they isomers?)

Model 1: Representations of Carbon Structures

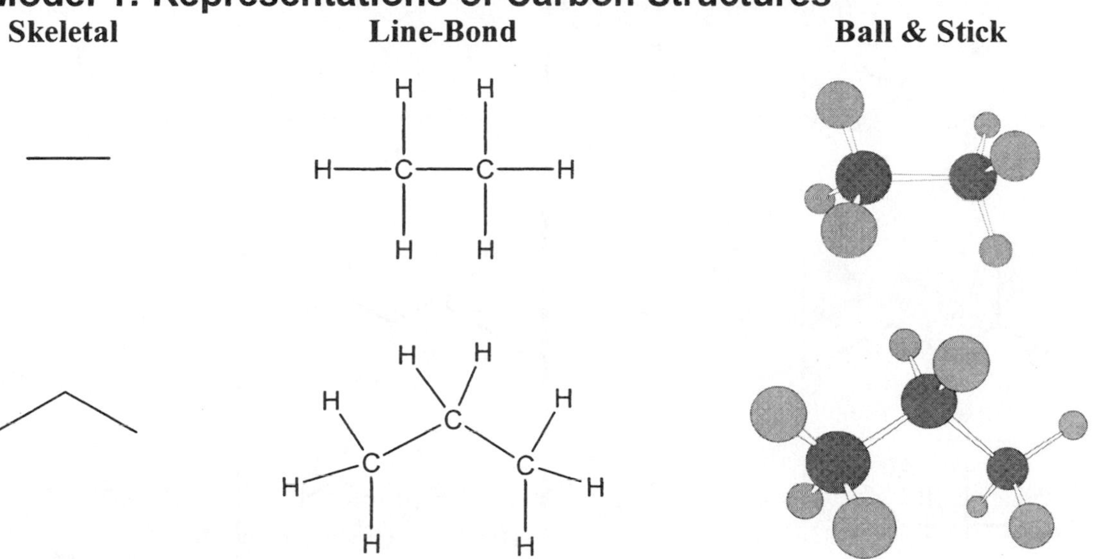

Critical Thinking Questions

1. Draw a line-bond representation of the molecule below shown as a skeletal representation.

2. In skeletal representations the hydrogens are not shown. Is it still possible to tell how many hydrogens there are on a particular carbon? Explain how?

Model 2: Constitutional Isomers

Column 1		Column 2	
structure	molecular formula	structure	molecular formula
	C_6H_{14}	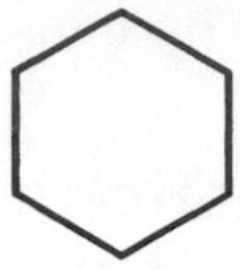	C_6H_{12}
	C_6H_{14}		
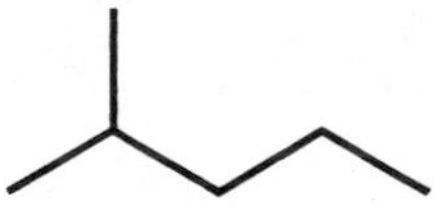			
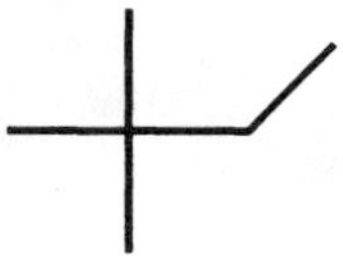			

Critical Thinking Questions

3. Complete the table in Model 2 by writing in the missing molecular formulas.

4. What do the molecules in a given column (1 or 2, above) have in common with the other molecules in that column?

5. What do the molecules in a given column **NOT** have in common with the other molecules in that column?

6. All the structures in a given column are **constitutional isomers** of one another, but the structures in column 1 are not constitutional isomers of structures in column 2. Based on this information, write a definition for the term **constitutional isomers**.

7. Below each of these structures, draw the structure from Model 2 to which it is identical. Note: For our purposes, two structures are considered identical if they are conformers of one another. Recall from ChemActivity 4 that **conformers** are structures that can be inter-converted via rotation of single bonds.

8. Draw the constitutional isomer that is missing from column 1 of Model 2. (Exactly one is missing.)

Part B: Stereoisomers

(What is the difference between *cis* and *trans* stereoisomers?)

Model 3: Isomers of Cyclic Hydrocarbons

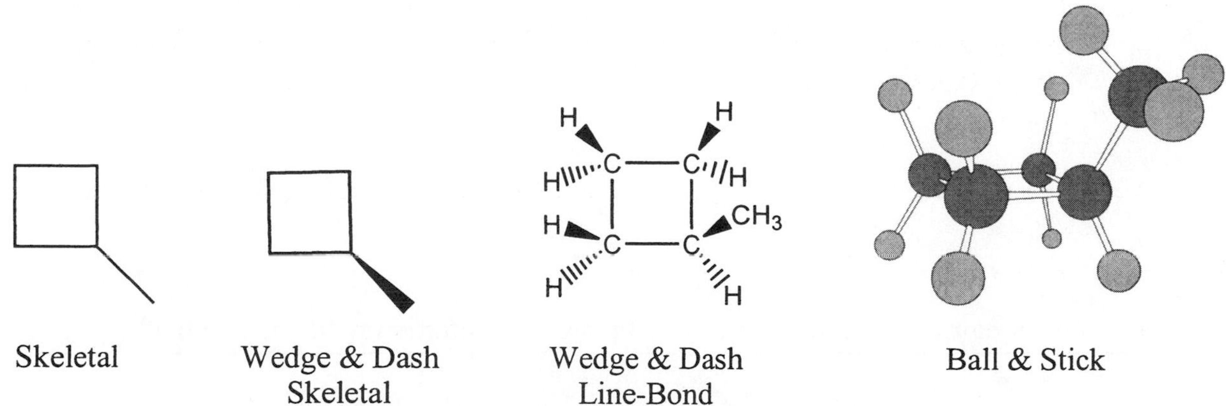

Figure 3a: Representations of Methylcyclobutane

The above are equivalent representations of methylcyclobutane.

Figure 3b: 1,2-dimethylcyclobutane

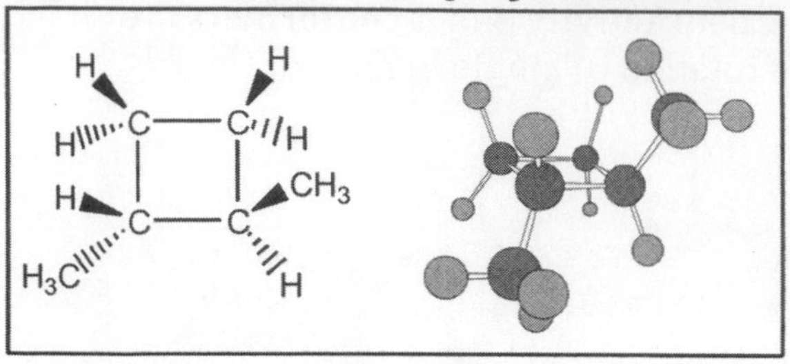

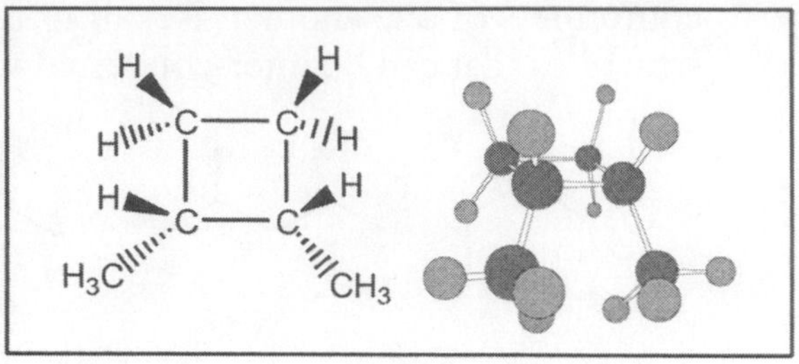

Critical Thinking Questions

9. Is the molecule in the left box identical to the molecule in the right box? (Our criterion for identicalness is: "Two molecules are the same if models of the molecules can be superimposed without breaking any bonds."

Information

Imagine that the four carbons of the ring define a plane. If two groups (such as the two methyl groups in 1,2-dimethylcyclobutane) are on the same side of this plane they are considered ***cis*** to one another. The groups are ***trans*** to one another if they are on opposite sides of the ring-defined plane.

Critical Thinking Questions

10. Label one box in Figure 3b with "*cis*-1,2-dimethycyclobutane" and the other with "*trans*-1,2-dimethylcyclobutane," as appropriate.

11. Draw wedge and dash skeletal representations of *cis* and *trans* 1,3-dimethylcyclobutane.

12. Which pair of molecules, below, have more in common with one another?

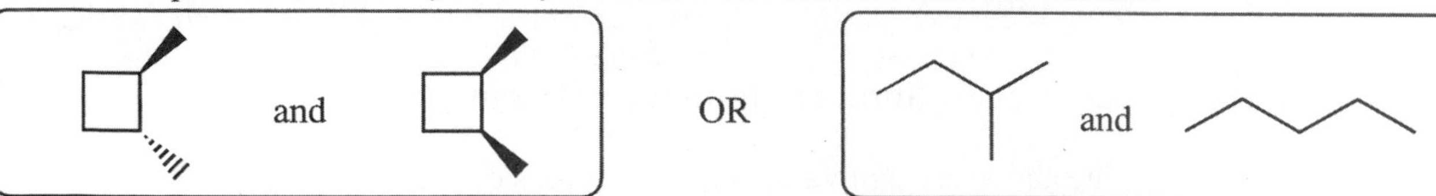

13. What do the pair on the left have in common that the pair on the right do not?

Model 4: Stereoisomers

A pair of structures are…

- **constitutional isomers** if they have the same molecular formula but <u>different</u> *atom connectivity* (see example in right box at bottom of previous page).
- **conformers** if they can be inter-converted via single bond rotation. (Conformers are considered the same compound.)
- **stereoisomers** if they have the same molecular formula AND the <u>same</u> *atom connectivity*, but are not the same compound. (Stereoisomers cannot be inter-converted via single bond rotation. See example in left box at bottom of previous page.)

Figure 4: Various Isomers of Dimethylcyclopentane

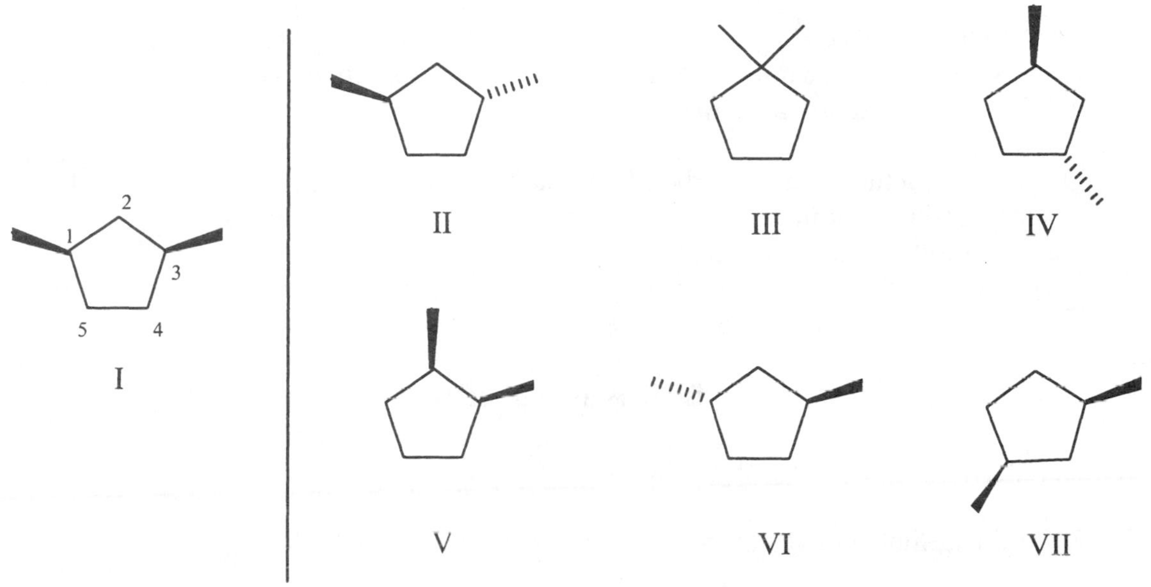

14. Circle each stereoisomer of Molecule I on the right of the line in Figure 4.
15. What is the relationship between Molecule I and each of the un-circled molecules on the right of the line in Figure 4? Label each.

16. Draw a stereoisomer of Molecule V.

17. There is <u>no free rotation of double bonds</u> at room temperature. This means the two molecules below are not the same (cannot be interconverted via rotation of single bonds).

   a) Label one of these molecules *cis*-2-butene and the other *trans*-2-butene by analogy to the ring systems shown in Model 3.

b) According to the definitions in Model 4, what is the relationship between *cis*-2-butene and *trans*-2-butene?

c) Draw skeletal representations of *cis* and *trans*-3-hexene.

Exercises For Part A

1. Draw skeletal representations of six constitutional isomers missing from column 2 in Model 2. Are there more than six?

2. Draw the structure of a six carbon hydrocarbon with one ring and one double bond. (hydrocarbon = a molecule containing only carbon and hydrogen.)
   a) Draw a constitutional isomer of the structure you drew above with no rings.
   b) Explain the following statement found in many text books: "In terms of molecular formula, a ring is equivalent to a double bond."

3. Draw as many constitutional isomers as you can with the formula $C_5H_{11}F$.

4. Complete Nomenclature Worksheet I.

5. Read the assigned pages in the text and do the assigned problems.

Exercises for Part B

6. The words *cis* and *trans* are usually used to describe the relationship between two identical groups (as in CTQ 16). However, chemists also use these words to describe the relationship between two like groups. For example: *trans*-2-hexene (below).

   Draw a representation of *cis*-2-hexene.

7. Draw 1-butene. Why does it not make sense to specify either *cis* or *trans* 1-butene, while you must specify *cis* or *trans* 2-butene to draw the correct molecule.

8. Make up an example (not appearing in this ChemActivity) of a pair of molecules that are a) constitutional isomers, b) conformers, c) stereoisomers.

9. Label the following compounds with the words *cis* and/or *trans* as appropriate

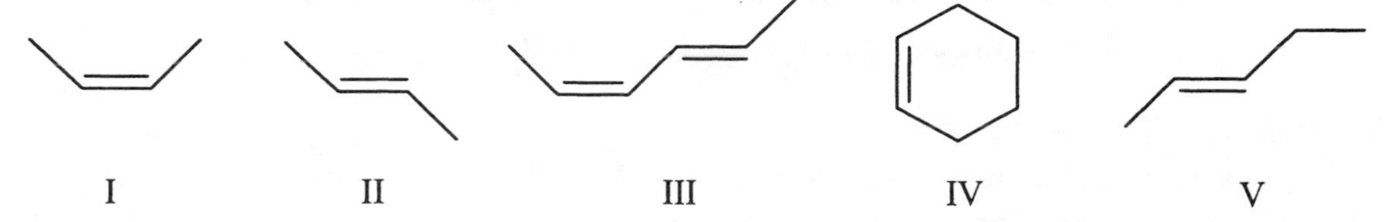

10. What is the relationship between Compounds III and IV above?

11. A double bond can be rotated by 1) adding enough potential energy to break the double bond (E_{act}), 2) rotating, then 3) reforming the double bond. The energy diagram below depicts this reaction. Draw *trans*-2-butene in one box and *cis*-2-butene in the other box and explain your reasoning.

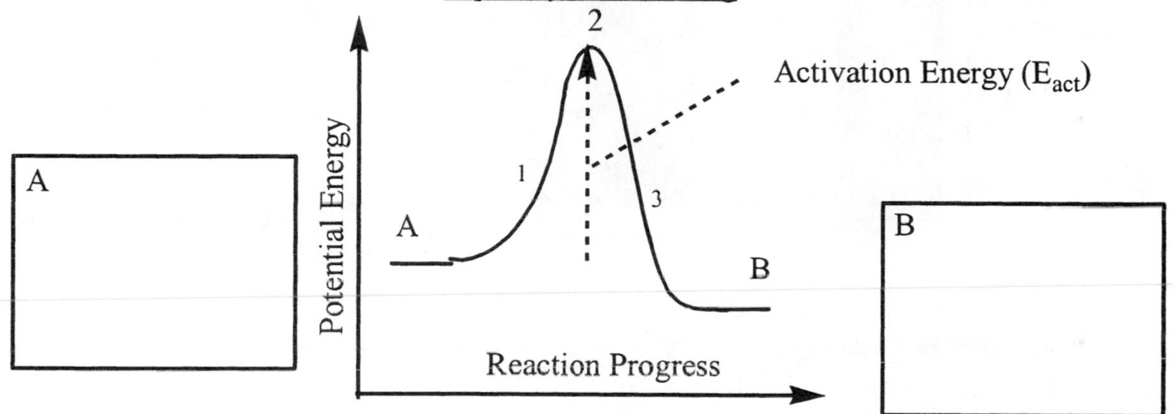

12. Read the assigned pages in the text and do the assigned problems.

ChemActivity 6

Part A: Energy Diagrams

(What is the sign of $\Delta H_{reaction}$ for an exothermic reaction?)

Model 1A: The Slot Machine

Imagine a gambler sitting at a slot machine. Over the course of an hour the gambler puts $100 into the machine and the machine pays out $90. We know a net change (Δ) of ten dollars occurred, but what is the sign of this change in money (Δ$). To determine if Δ$ = +10 or –10 we need to decide if we calculate Δ$ from the perspective of the gambler or from the perspective of the slot machine.

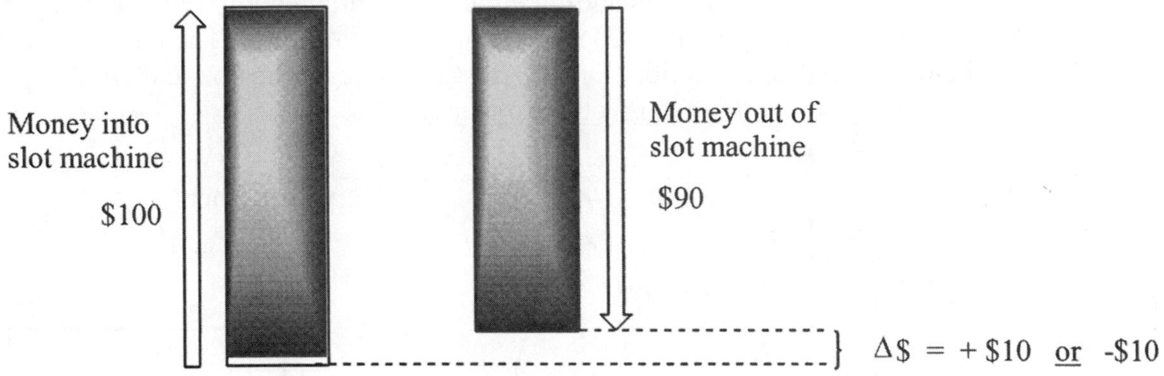

To translate this analogy into chemistry remember that:

- **Money = energy**.
- The **slot machine** = the **reaction**.
- The **gambler** = the **experimenter** (you).
- **All changes (Δ) will be calculated from the perspective of the reaction** (a.k.a. the slot machine).

Critical Thinking Questions

1. Using the convention listed above for calculating changes (Δ), what is the sign of Δ$ in the example in Model 1A?

2. Based on this same convention, what sign must be attached to the $100 that the gambler put into the slot machine in Model 1A? The sign of the $90 that comes out is the opposite. **Add the appropriate sign to each value in Model 1A.**

3. On a different day the gambler puts in a total of $20 and the slot machine pays out a total of $100. What is the sign and value of Δ$ for this day?

Model 1B: Burning of Natural Gas (Methane)

- Breaking bonds requires you to put energy into the molecule (think "slot machine").
- Making bonds releases energy from the molecule, which is felt by the experimenter (think "gambler") as heat.

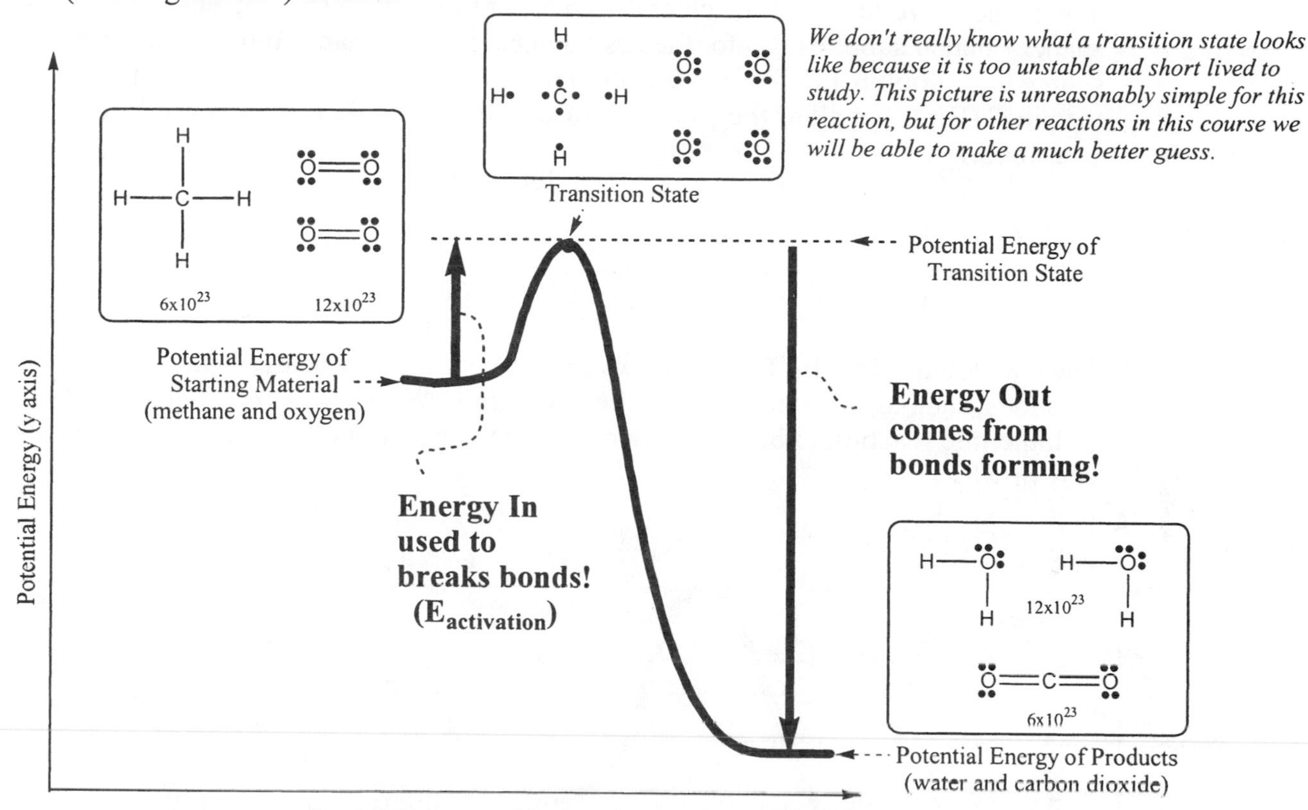

Critical Thinking Questions

4. According to the diagram, which has more stored energy (higher in potential energy): **methane and oxygen** or **water and carbon dioxide** [circle one]? Is your answer consistent with the fact that one of these pairs is commonly used as a fuel?

5. According to our convention, what is the sign of $E_{activation}$ (the energy put into the reaction to break bonds).

6. What is the sign of ΔE, the total net change in energy, for this reaction?

Information

The **net** energy change (ΔE) in an organic reaction is often called the **heat of reaction** (ΔH)

7. Draw an arrow on the diagram above indicating the **heat of reaction** for the burning of methane. **Be sure to assign a direction to your arrow indication the sign of this change in heat/energy.**

Recall that energy in = up arrow = + energy; and energy out = down arrow = -energy.

8. On a gas stove, what supplies the **activation energy ($E_{activation}$)** to start the reaction?

9. You do not have to hold a match under your pot to keep a gas flame burning. A spark or match supplies E_{act} for the first molecules of methane. What is the energy source that supplies E_{act} for subsequent molecules of methane coming up through the pipe? (Hint: not all of the energy released by burning methane goes to heating your food.)

10. The reaction in Model 1B is called **exothermic** because it releases heat. Some reactions, such as the one below, are **endothermic**. Would you expect an endothermic reaction to be self-sustaining like the burning of methane? Explain why or why not.

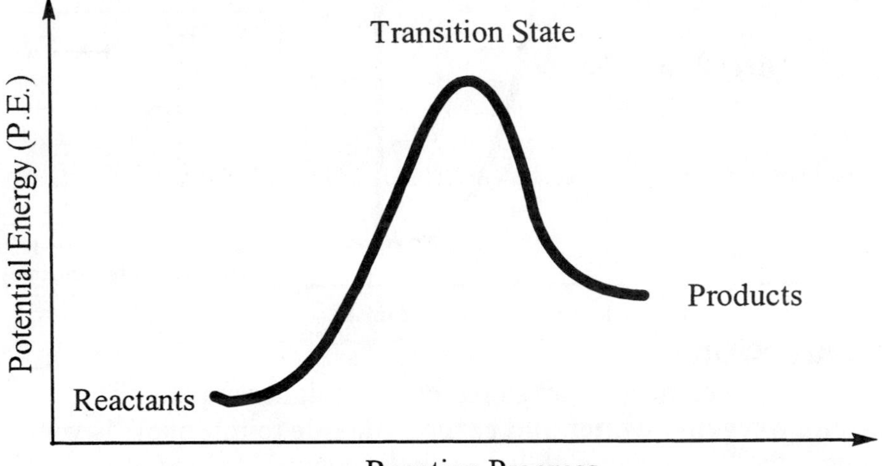

a) Add an arrow to the energy diagram above indicating the size and direction of the **activation energy** and label this arrow "E_{act}."

b) Add an arrow to the energy diagram above indicating the size and direction of the **heat of reaction** and label this arrow "$\Delta H_{reaction}$." (Note: Δ means "the change in" or "net" and H stands for heat.)

Model 2: Ring Strain

In previous ChemActivities you observed that your molecular model set can help predict the best conformation for a molecule. This is because the holes in the atom pieces are consistent with experimentally determined bond angles for carbon (109.5°).

If you tried, you would find it very difficult to build a model of cyclopropane. This is because the 60° bond angles of cyclopropane are very far from carbon's preferred 109.5° bond angles. As you might expect, cyclopropane is unstable and high in potential energy. Organic chemists call the extra energy associated with cyclopropane's strained bonds "**ring strain**."

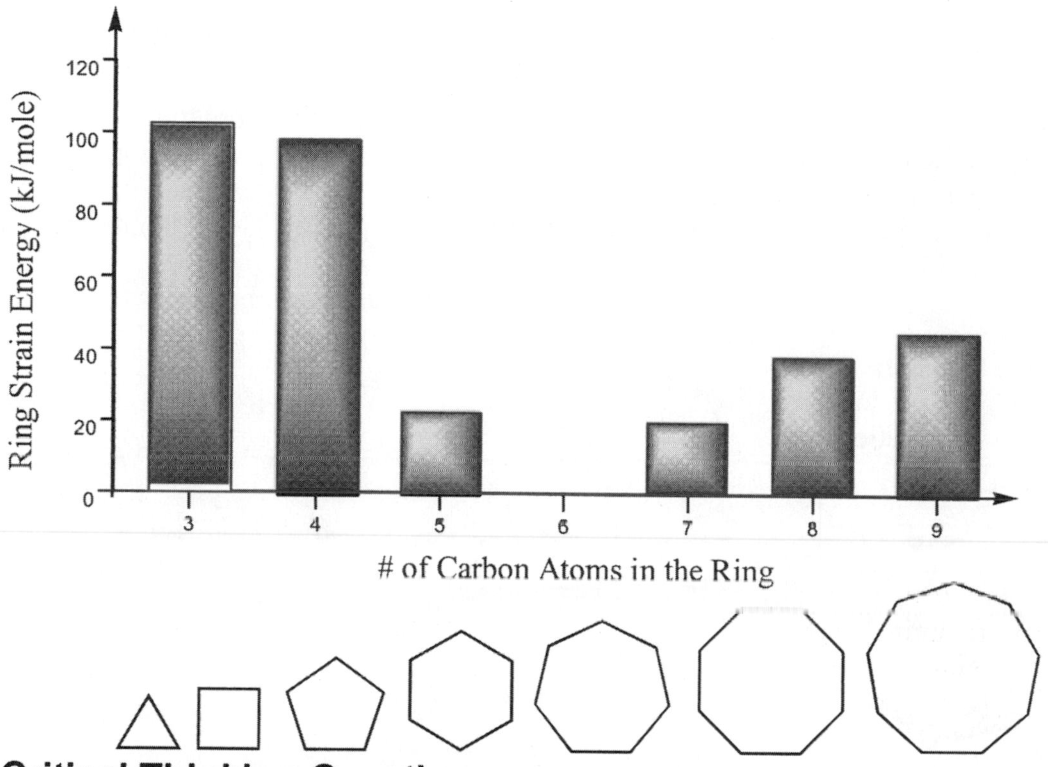

Critical Thinking Questions

11. Circle the cycloalkane in Model 2 which you expect is most commonly found in nature, and explain your reasoning.

12. Cyclopropane has a molecular formula of C_3H_6 and a molecular weight of 42 g/mole, making it exactly half the size of cyclohexane, whose molecular formula is C_6H_{12} and molecular weight is 84g/mole.

- Burning 1 mole (84g) of cyclohexane releases 3900 kJ of heat energy.
- Burning 2 moles (84g) of cyclopropane releases 4100 kJ of heat energy.

Both reactions produce 6 moles of H_2O and 6 moles of CO_2. On the same set of axes, draw an energy diagram for each of these two reactions and explain why the cyclopropane reaction releases more heat.

Part B: Cyclohexane Conformations

(How do I represent cyclohexane in its lowest potential energy conformation?)

Model 3: The Chair Conformation of Cyclohexane

With zero ring strain, the six member ring is the most stable alkane ring. Not surprisingly, six member rings are very common in Nature and the laboratory. Previously, we have drawn cyclohexane as though the carbons all lie in the same plane, but a flat hexagon has internal angles of 120°, not 109.5°.

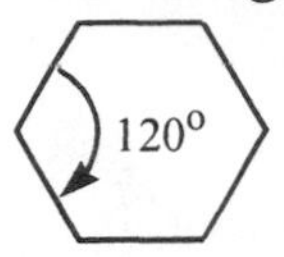

The carbons of cyclohexane do NOT form a flat hexagon!

As we will see in the following section, cyclohexane adopts a low potential energy conformation called the **chair conformation**. In this conformation a cyclohexane molecule has perfect 109.5° bond angles and zero ring strain.

The chair conformation is characterized by three sets of parallel C—C bonds and a "tripod" of C—H bonds on each face (*see below*).

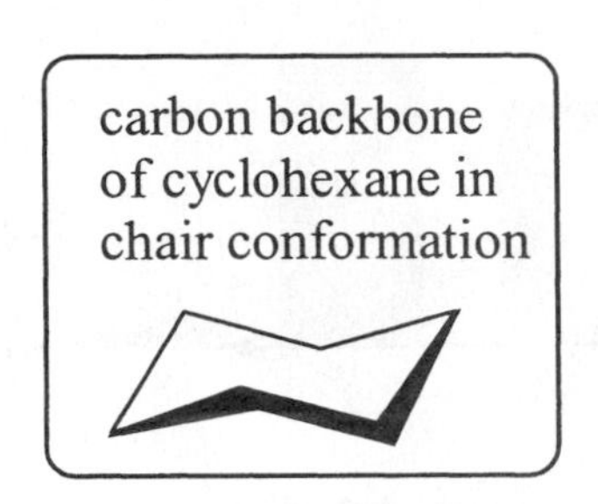

Looks like a reclining chair!

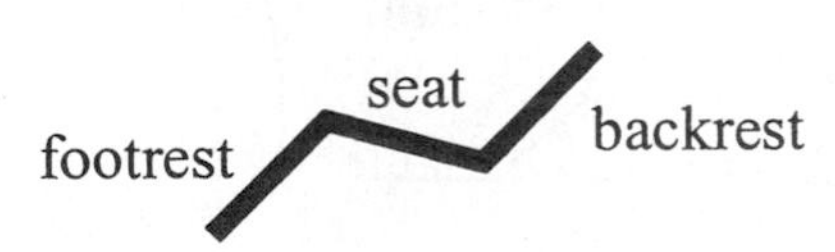

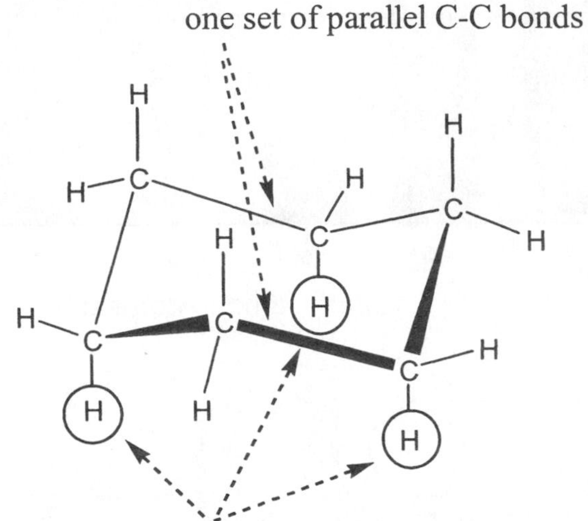

Critical Thinking Questions

13. Build a model of cyclohexane in the chair conformation. Check with your instructor or another group to confirm that your model is correct.

14. Six of the H's of cyclohexane are called "**axial,**" the other six are called **"equatorial."** Based on these names, find the axial (think axis of the globe) and equatorial (think equator) H's. Label each H of chair cyclohexane in the diagram above: "**A**" or "**E**."

15. With an input of 45 kJ/mole cyclohexane can undergo a change in conformation such that each equatorial H becomes axial and vice versa.

    a) Without breaking any bonds, perform this conformation change on your model of cyclohexane. This is often called a "cyclohexane chair flip."

b) Complete the following energy diagram for the chair flip process.

16. Cyclohexanes are often drawn as partial skeletal structures, showing the H's but not the C's. On the left, below, is a drawing methylcyclohexane in a chair conformation with the methyl group in an axial position. Complete the drawing on the right, showing this same structure after a chair "flip" has occurred. Carbon 1 is marked in each drawing.

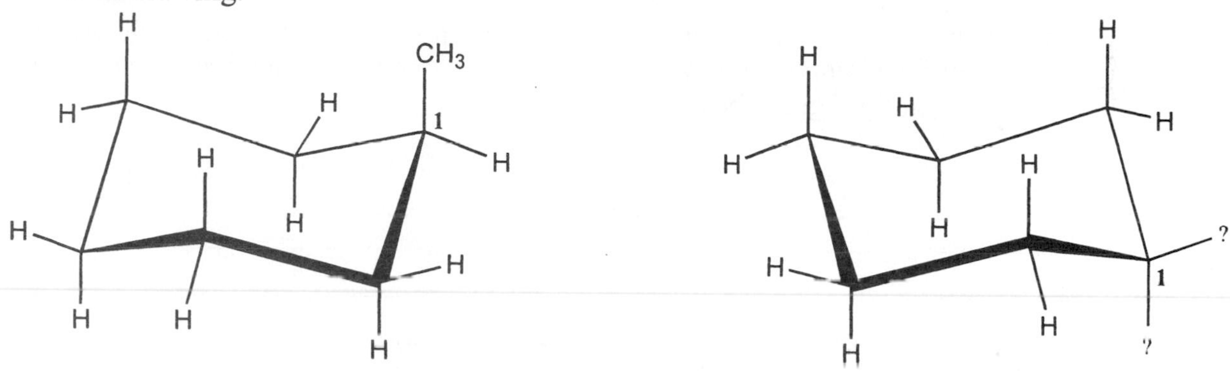

17. Make a model of methylcyclohexane with the methyl group in an equatorial position. Construct an explanation for why this is a more favorable conformation than the conformation with CH_3 in an axial position.

18. It is important to know that, in general, it is favorable to have the largest group on a cyclohexane ring in an equatorial position. Draw a **chair representation** of... (Represent the methyl group as CH_3 and the *t*-butyl group as $C(CH_3)_3$.)
   a) *cis*-1-methyl-4-*tert*-butylcyclohexane in its most favorable conformation.
   b) *trans*-1-methyl-4-*tert*-butylcyclohexane in its most favorable conformation.
   c) Which is lower in potential energy: ***cis*** or ***trans*** [circle one] and explain your reasoning.

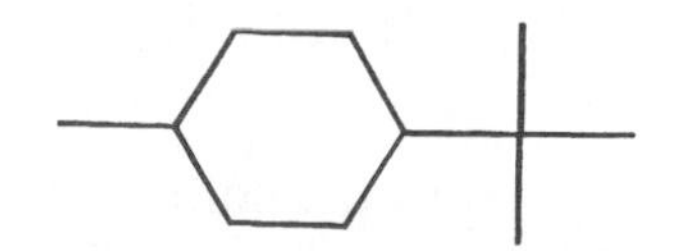

1-methyl-4-*tert*-butylcyclohexane

19. Label each of the following dimethylcyclohexanes as *cis*, *trans* or neither.

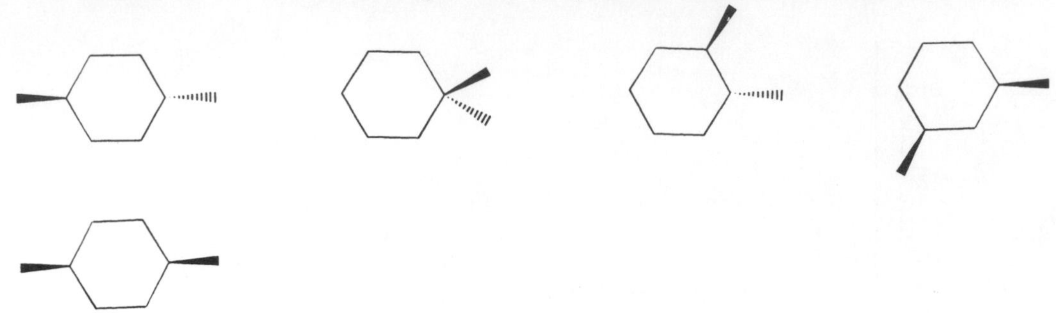

a) Below each structure that is *cis* draw the *trans* stereoisomer and vice versa. The first one is done for you.

b) **True of False**: When you perform a "chair flip" on a *cis* stereoisomer, the resulting structure is the *trans* stereoisomer, and vice versa.

c) Draw <u>chair representations</u> of the lowest potential energy (most favorable) conformation of the two 1,4-dimethylcyclohexane structures shown on the left in the figure above.

Exercises For Part A

1. In the reactants box in Model 1B, why must there be twice as much oxygen as methane?

2. According to Model 1B, in going from the Reactants to the Transition State, were any electrons or atoms gained or lost?

3. By convention, on an energy diagram up arrows are positive (heat is added to the reaction) and down arrows are negative (heat is coming out of the reaction).
   a) According to the above statement, the sign of all activation energies (E_{act}) must be **positive** <u>or</u> **negative** [circle one].
   b) The sign of $\Delta H_{reaction}$ for an <u>exothermic</u> reaction is **positive** <u>or</u> **negative** [circle one].
   c) The sign of $\Delta H_{reaction}$ for an <u>endothermic</u> reaction is **positive** <u>or</u> **negative** [circle one].

4. Butane ( ) is the fuel used in cigarette lighters. To start the combustion reaction, a small amount of energy must be put into the system to get over the initial hump. What is the source of this **activation energy** in a standard cigarette lighter?

5. Add arrows indicating heats of reaction ($\Delta H_{reaction}$) to the energy diagram below.

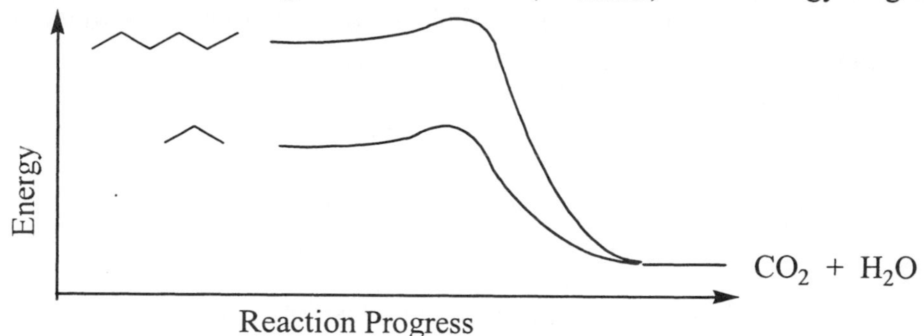

6. Propose a reason for why burning a mole of hexane ( ) gives off about twice as much heat as burning a mole of propane ( ).

7. Use the data below to estimate the heat of combustion of a —CH_2— group.

Structure	Heat of Combustion (kJ/mole)
CH_3CH_3	-1560
$CH_3CH_2CH_3$	-2220
$CH_3CH_2CH_2CH_3$	-2870
$CH_3CH_2CH_2CH_2CH_3$	-3530

8. For now, neglect ring strain and use your estimate to predict the heat of combustion in kJ/mole of cyclopropane, cyclobutane cyclopentane and cyclohexane. That is, assume cyclopropane is three CH_2 groups. Then fill in the table below.

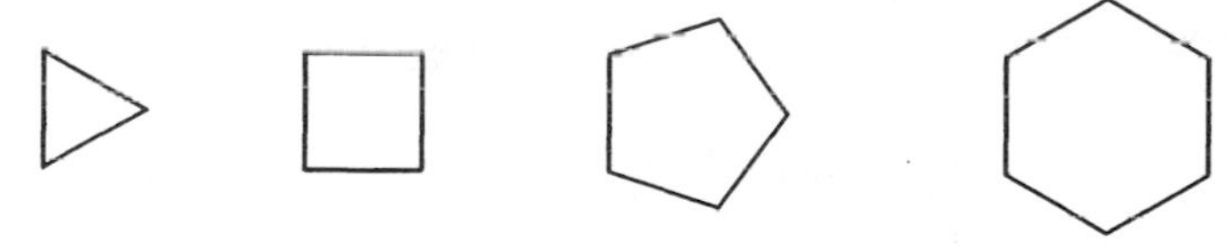

Table A. Heat of Combustion of Various Cycloalkanes

Compound	Experimental ΔH_{rxn} (kJ/mole)	Calculated ΔH_{rxn} using your estimate for ΔH_{rxn} for CH_2 (kJ/mole)	Difference between Calc. and Exp. ΔH_{rxn} (kJ/mole)
cyclopropane	-2090		
cyclobutane	-2740		
cyclopentane	-3320		
cyclohexane	-3940		
cycloheptane	-4620	-4600	off by 20 kJ
cyclooctane	-5320	-5260	off by 60 kJ

a) Which is higher in potential energy: cyclopropane or three –CH_2– groups?
b) Construct an explanation for this difference in potential energy.
c) Explain why Column 4 in the table above could be labeled "ring strain."
d) For which molecule is the calculated value closest to the experimental value?
e) Are your answers to part a-d consistent with the graph in Model 2? Explain.

9. For each pair, explain why the first molecule listed releases more heat when burned (has a larger heat of reaction arrow and a more negative ΔH_{rxn}).
   a) ethanol more than methanol
   b) cis-2-butene more than trans-2-butene

10. Read the assigned pages in your text and do the assigned problems.

Exercises For Part B

11. Draw a chair representation of each of the dimethylcyclohexane structures shown in CTQ 19 in its lowest potential energy conformation. (You already did the first two in CTQ 19c.)

12. Draw chair representations of the two different chair conformations of *trans*-1,4-dimethylcyclohexane. Note: one has both methyl groups in axial positions, the other has both methyl groups in equatorial positions.) Circle the conformation that this molecule is more likely to adopt and explain your reasoning.

13. Draw *trans*-1-methyl-3-*tert*-butylcyclohexane in its most favorable chair conformation. Explain your reasoning.

14. Which has a larger heat of reaction arrow (more negative ΔH_{rxn}): *cis*-1,4-dimethylcyclohexane or *trans*-1,4-dimethylcyclohexane? Explain your reasoning.

15. Build a model of methylcyclohexane and use the model to complete the following Newman projections:

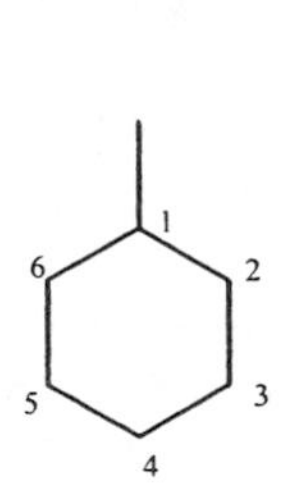

methylcyclohexane

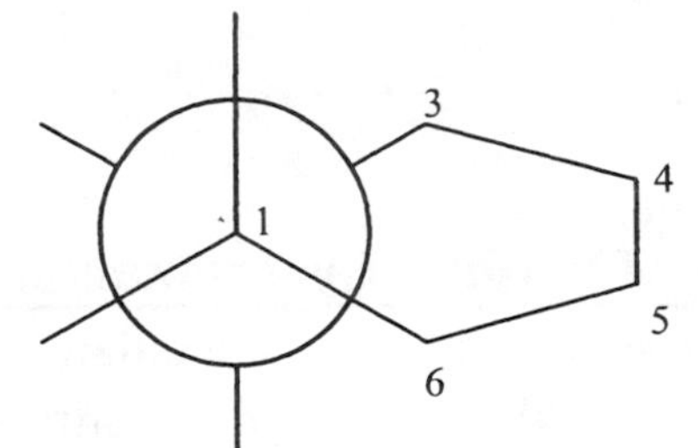

methylcyclohexane
(show CH_3 in axial position)

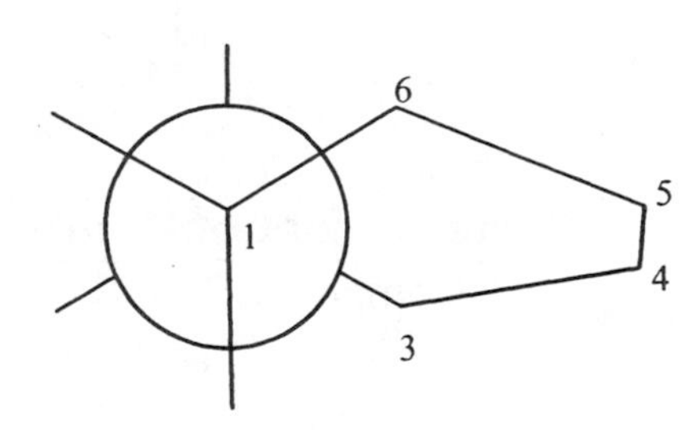

methylcyclohexane
(show CH_3 in equatorial position)

   a) Circle the conformation that is lower in potential energy.
   b) Use the terms **anti** and **gauche** to describe the relationship between the methyl group and the rest of the ring in each Newman projection.
   c) Are your answer to parts a) and b) consistent with one another? Explain.

16. Read the assigned pages in your text and do the assigned problems.

ChemActivity 7

Part A: Chirality

(How can you tell if an object is chiral?)

Model 1: Criterion for Identical-ness

Two molecules are **identical** if models of the two molecules can be superimposed without breaking any bonds.

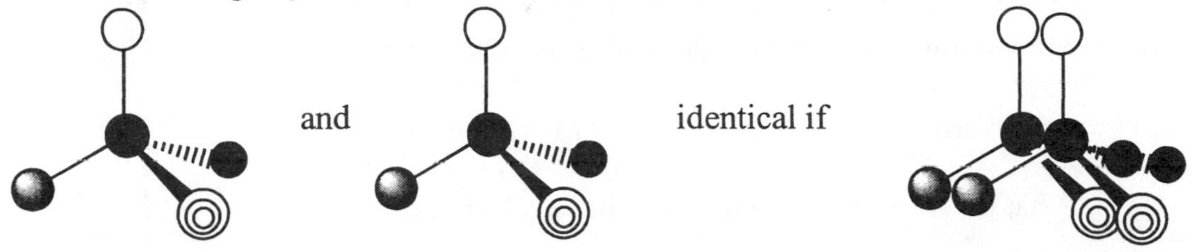

super-imposable

Critical Thinking Questions

1. Make two identical models of the following molecule and confirm that they can be superimposed on one another. Use different colored balls for each of the four substituents. If your set has green, orange and purple, use the following color code.

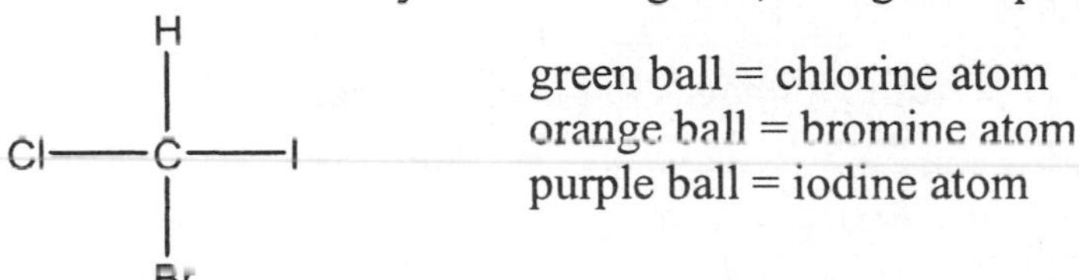

green ball = chlorine atom
orange ball = bromine atom
purple ball = iodine atom

2. Switch any two balls on one of your models (leave the other model unchanged). Is this new model identical to your original model?

3. Which of the following words describe the relationship between these two models? Circle more than one choice, if appropriate.

   A. Identical
   B. Stereoisomers (same atom connectivity, but not identical)
   C. Conformers (can be made identical via single bond rotation)
   D. Constitutional isomers (same formula, different atom connectivity)
   E. Mirror images (look like reflections of one another in the mirror)

4. Consider the following set of molecules. Note: assume each molecule is in its simplest conformation, even if it is not the lowest energy conformation. In this case, assume each cyclohexane ring is in a planar conformation (rather than a chair).

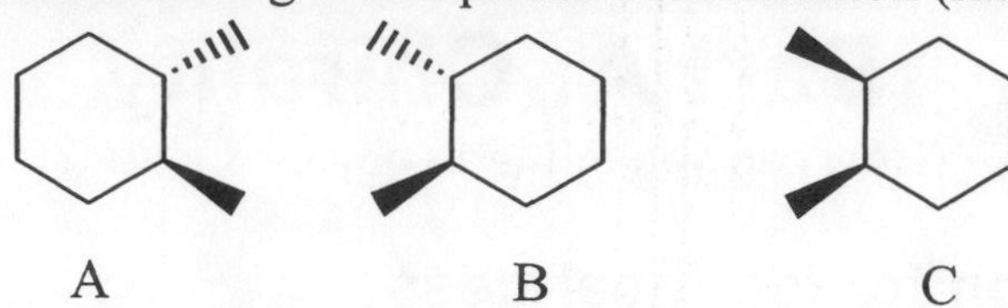

A B C

a) Label each of A-C with the word *cis* or *trans*, as appropriate.

b) Are any two of the set an identical pair? If so which ones.

c) Which are more similar **A&B** or **B&C** [circle one]?

d) What do A&B have in common that B&C do not?

Definitions:

chiral = not identical to its mirror image.
(a property of an object) (Chiral is Latin for "handed.")

enantiomers = a pair of objects that are mirror images but not identical.
(a relationship between two objects) (Enantiomers is Latin for "opposites.")

diastereomers= a pair of objects that are stereoisomers but not enantiomers
(a relationship between two objects)

racemic mixture = a 1:1 mixture of a pair of enantiomers.
(a special sample of objects)

Critical Thinking Questions

5. Give an example of an everyday object that is chiral and one that is not chiral.

6. Give an example of a pair of everyday objects that are enantiomers.

7. Circle each structure in CTQ 4 that is chiral (it may help to draw the mirror image of each).

   a) What is the relationship between A&B (be as specific as possible)?

   b) What is the relationship between B&C (be as specific as possible)?

   c) What is the relationship between A&C (be as specific as possible)?

8. T or F: All chiral objects have exactly one enantiomer. If false, give an example from CTQ 4.

9. T or F: All diastereomers are chiral. If false, give an example from CTQ 4.

Model 2: Internal Plane of Symmetry (Mirror Plane)

It is always true that an object (or molecule) with an **internal plane of symmetry** is not chiral (not chiral = **achiral** = identical to its mirror image).

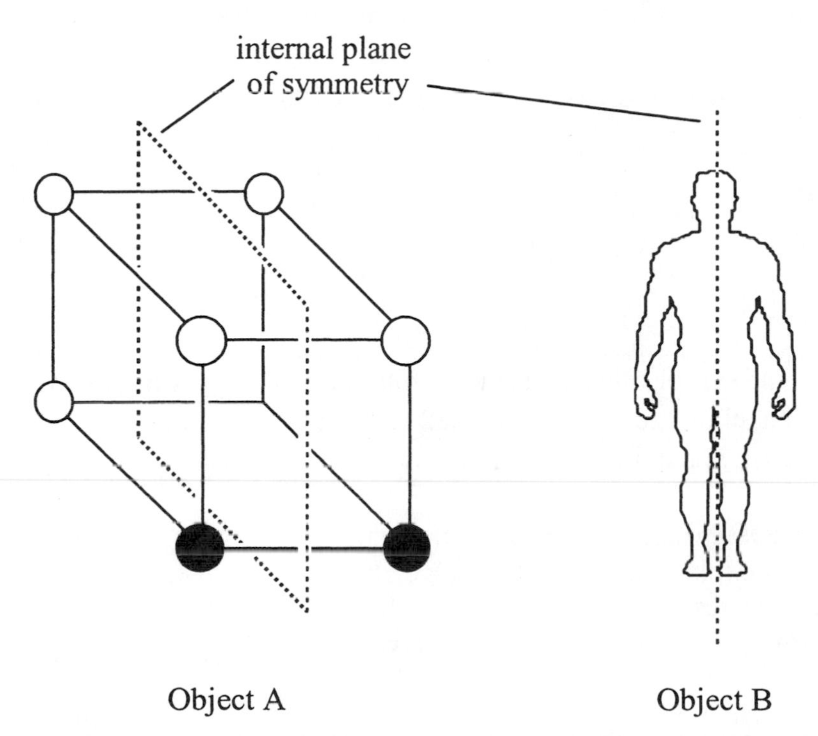

Critical Thinking Questions

10. Object C has a very obvious plane of symmetry (marked with a dotted line), and a not so obvious one. Where is this second plane of symmetry? (Assume the balls are uniform spheres.)

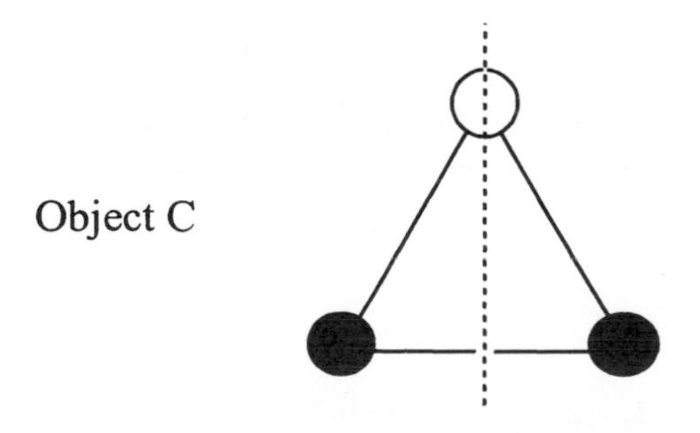

11. Objects D and E are NOT chiral. Indicate the internal mirror plane in each.

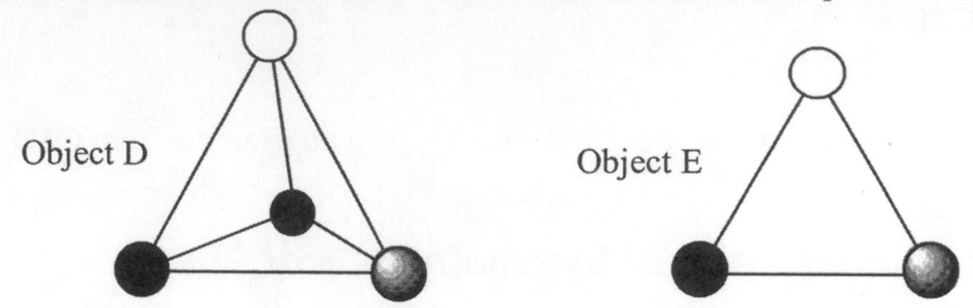

Part B: Absolute Configuration

(How can you tell if a chiral center is right handed "**R**" or left handed "**S**"?)

Model 3: A Trick for Determining If a Molecule Is Chiral

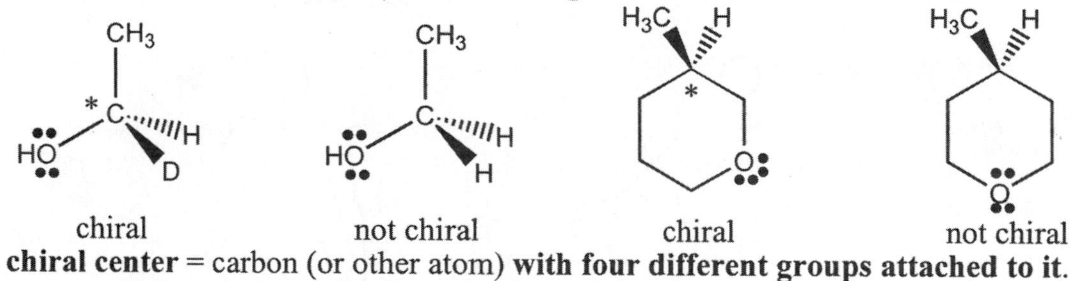

chiral not chiral chiral not chiral

chiral center = carbon (or other atom) **with four different groups attached to it.**

- By convention, chiral centers are marked with an *.
- A molecule with one chiral center is always chiral.

Critical Thinking Questions

12. Construct an explanation for why each carbon indicated with a "1" is considered to have **two identical groups**, while the carbons with an * are chiral, with four different groups. (Hint: It has to do with weather the ring is symmetrical or not.)

Not Chiral

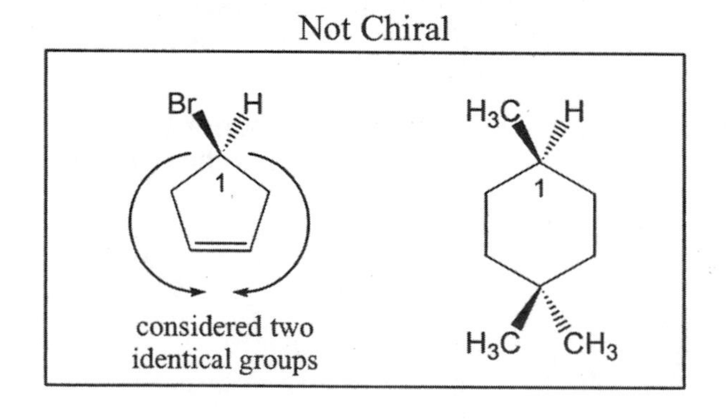

Chiral

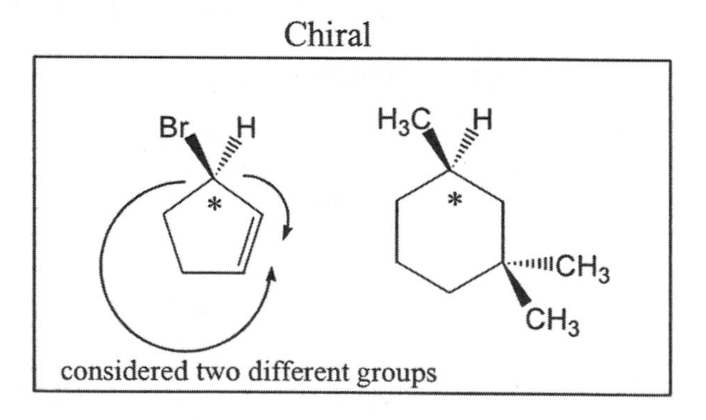

13. Consider the following structures, some of which are chiral.

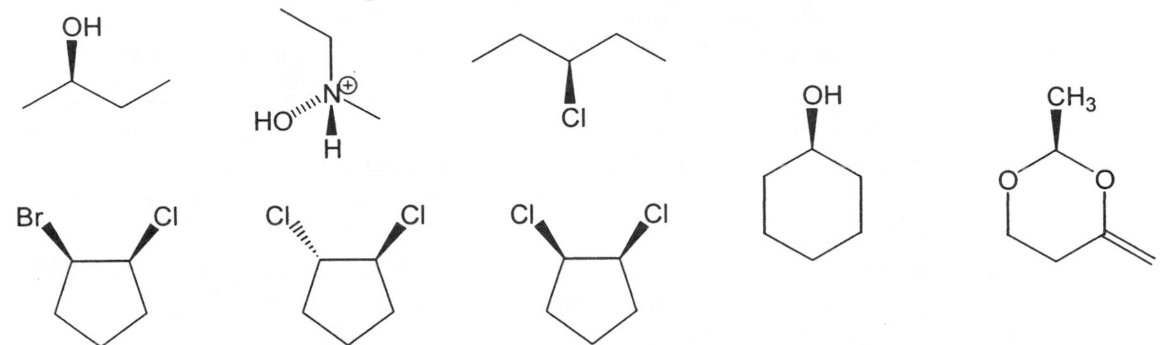

a) For some of the chiral centers above (for example, the first one), the H is not shown. How can you tell if an H is going into the page or coming out of the page?

b) Mark each chiral center on the structures above with an * (there are 9).

c) Circle the **three** structures at the top of the page that are NOT chiral, and for each circled structure, indicate the plane of symmetry (mirror plane).

d) One of the structures you circled is an example of the rare case where a structure **contains chiral centers but is not chiral**. These molecules are called **meso compounds**. Label the meso compound above.

Information

Like gloves each chiral center is either right or left handed. The convention for determining handedness requires we rank four groups attached to a chiral center.

Rules for Ranking the Four Groups from highest (3) to lowest (0):

- The higher the atomic number of an element, the higher its rank. For example: N (atomic # = 7) beats C (atomic # = 6), and everything beats H (atomic # = 1)
- In the first round, look at the 1st "shell." This is comprised of the four atoms directly attached to the chiral center of interest. (*see* "1st shell", below)
- If there is a tie in the first round, a second round is necessary in which we look to break the tie by examining atoms attached to the tied groups that are in the 2nd "shell". (*see* example).
- If there is a tie in the 2nd round, atoms in the third "shell" are compared, etc.

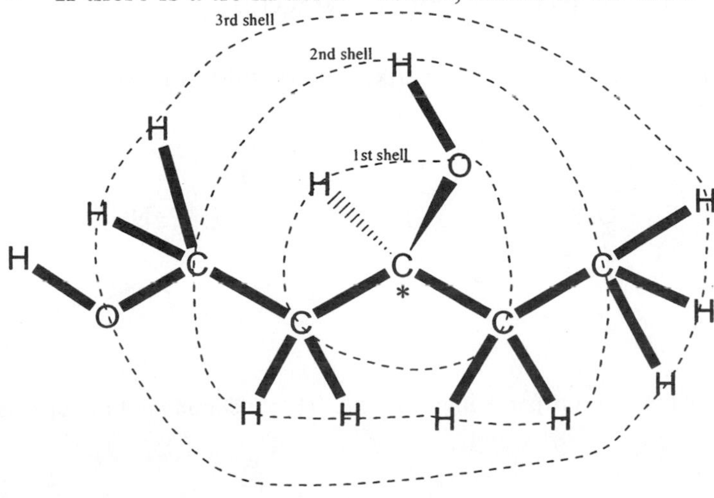

Critical Thinking Questions

14. For the example on the previous page,
    a) Based on examination of the first shell, an overall winner (3) and an overall loser (0) can be determined. Mark these atoms with **3** and **0**, respectively.
    b) In the first round there is a tie for ranks 2 and 1. Identify the two atoms that are tied based only on examination of the 1st shell.
    c) In this case, examination of the 2nd shell does NOT break the tie and help determine ranks 2 and 1. Explain.
    d) Based on the 3rd shell. Mark the two carbons in the 1st shell as "**2**" and "**1**" to complete your ranking of the four atoms attached to the chiral center.

Model 4: Nomenclature for Chiral Centers

- Each chiral center is designated as either **right handed** (R) or **left handed** (S).
  Right handed chiral centers are called **R** from the Latin word for right: ***rectus***.
  Left handed chiral centers are called **S** from the Latin word for left: ***sinister***.
- The R or S assignment is called the **absolute configuration** of that chiral center.
- To determine the absolute configuration of a chiral center such as the one below:

I. Rank the four groups attached to the chiral center 3, 2, 1 and 0, respectively.
II. With your RIGHT hand, grab the bond to the lowest rank group (use a model at first). Be sure that the 0 group (usually H) is at the pinky end, and the three larger groups are sticking out at the thumb end like a bouquet of flowers.
III. If the natural curl of your fingers matches the progression from rank = 1, to rank = 2, to rank = 3 (highest) than the center is right handed or R.
IV. If the curl of the fingers on your right hand does not match this progression (1-2-3) **then the chiral center is left handed (S).** Confirm the S assignment by grabbing this same bond with your left hand and checking that the progression from 1-2-3 (highest rank) matches the curl of your fingers.

Try your technique on the following examples. Rankings are shown using the numbers 0 = lowest, 3 = highest.

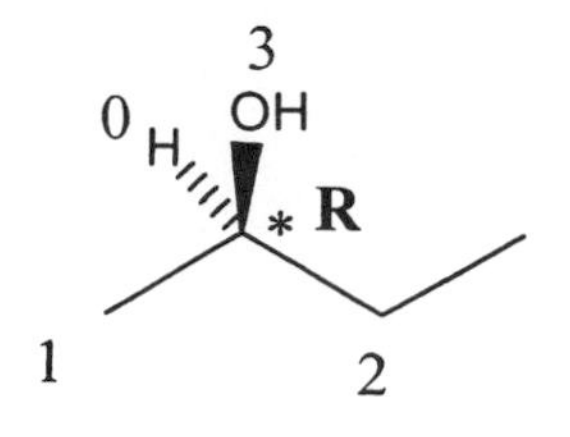

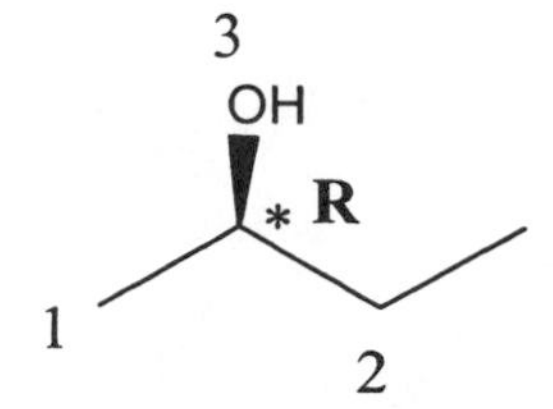

H (#4) going into paper is implied

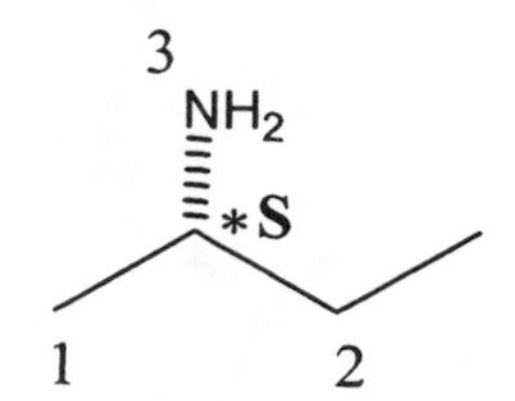

H (#4) coming out of paper is implied

2 F
0 H C* R CH3 1
Cl 3

H3C 2 CH2
0 H C* R CH3 1
Cl 3

H3C 2 CH2
3 Cl C* S CH3 1
H 0

Critical Thinking Questions

15. Determine the absolute configuration of each chiral center in the following structures.

F
Cl C* CH3
H

H3C CH2
Cl C* CH3
H

H3C CH2
H C* CH3
Cl

16. Note that the structures in the previous question are the enantiomers of the row of examples at the top of the page. What happens to the absolute configuration of a molecule if you switch exactly two groups (e.g. Cl and H) attached to a chiral center?

17. Determine the absolute configuration (R or S) of each chiral center in CTQ 14.

Exercises for Part A:

1. Draw an example of a pair of stereoisomers that are NOT enantiomers. What is the name for the relationship between such structures?

2. Shade or mark the balls on the object below to generate a chiral object.

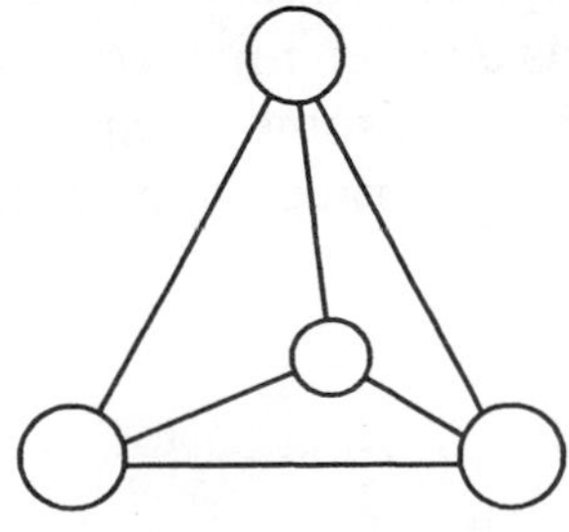

3. Consider the structures below.

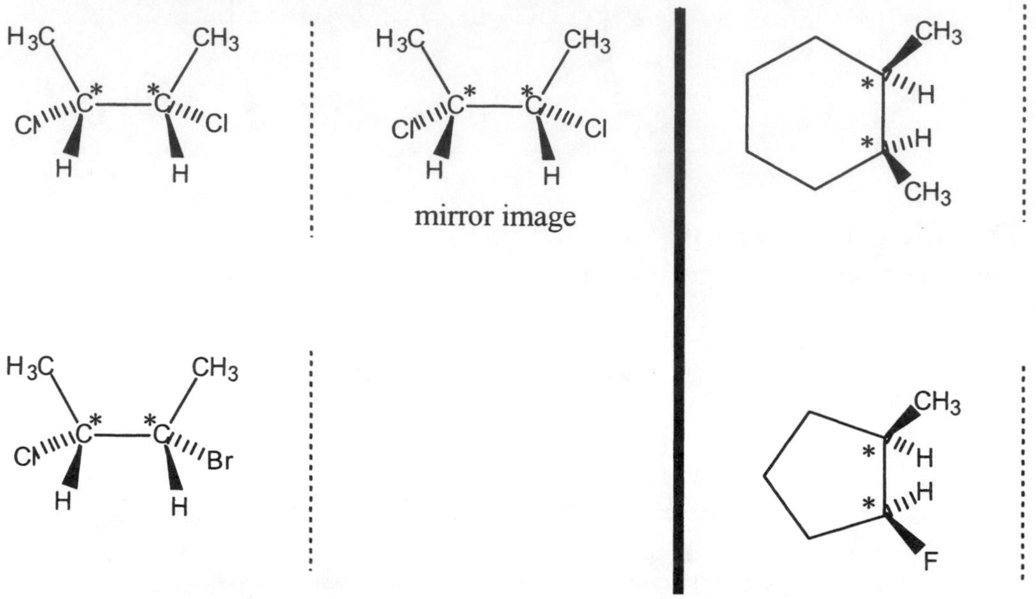

mirror image

a) Draw the mirror image of each molecule (the first one is done for you.)
b) Circle the structures that are not chiral and use a dotted line to show any <u>internal</u> symmetry planes. Label these structures **meso compound**.

4. Read the assigned sections in your text and do the assigned problems.

Exercises for Part B

5. Sometimes a molecule with one or more chiral centers is drawn without giving enough information to determine if the chiral center/s are R or S. Mark each chiral center below with an *.

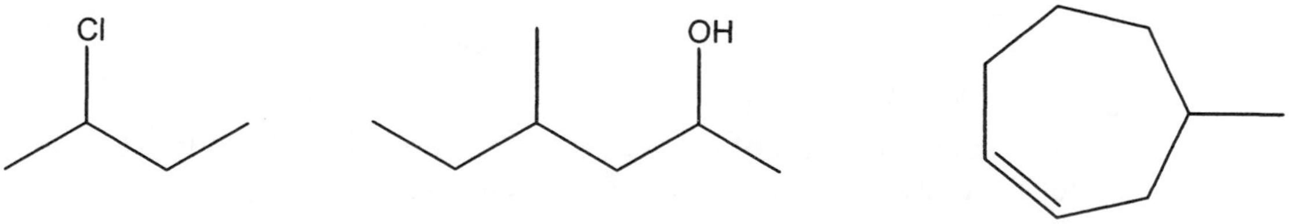

By convention, if there is not enough information given to assign R or S, we are to assume that the structure represents an even mixture of all stereochemical possibilities. For each structure above, draw all stereoisomers implied by the drawing. Circle those structures that imply a racemic mixture (**racemic mixture** = 1:1 mixture of exactly two enantiomers).

6. Determine the absolute configuration of each chiral center in Exercise 3. Explain why a meso compound must always have a 1:1 ratio of R chiral centers to S chiral centers. (Hint: consider what happens to an R chiral center when it is reflected in an internal mirror plane.)

7. Label every chiral center on the following molecules with an * and an R or S; and state whether the molecule as a whole is chiral.

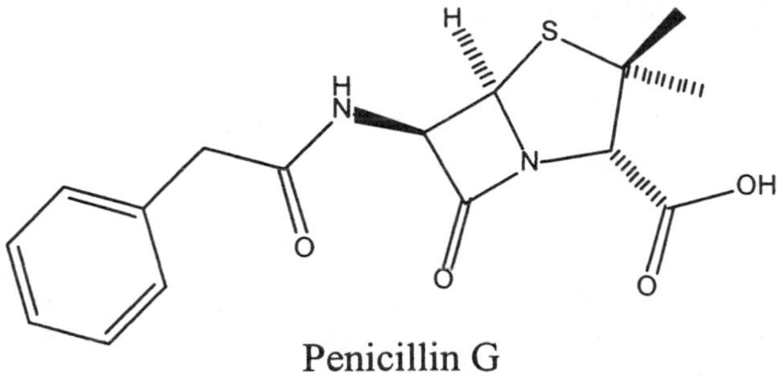

Penicillin G

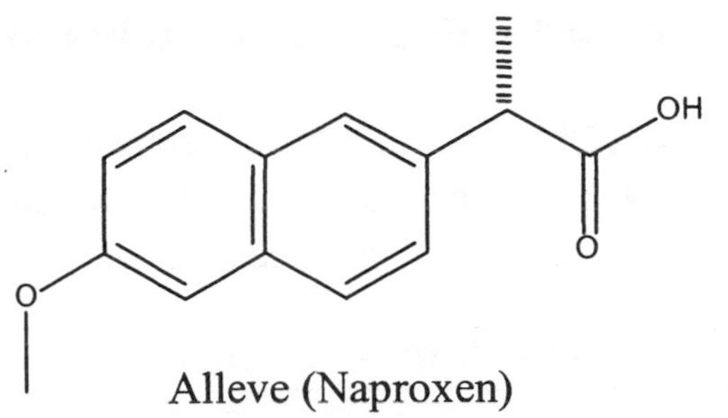

Alleve (Naproxen)

8. How many different stereoisomers are there of each molecule in previous question?

9. Most books teach a different and equally good way of figuring out R and S. Find this section in your textbook by looking in the index under chiral center or absolute configuration.

10. Read the assigned sections in your text and do the assigned problems.

ChemActivity 8

Acid-Base Reactions

(How can you predict if an acid-base reaction will be up-hill or down-hill?)

Model 1: Brönsted-Lowry Acids and Bases

Figure 1a: Examples of Acid/Base Conjugate Pairs

- **ACID:** has an H^+ to give **(proton donor)** (H^+ = a proton)
- **BASE:** can make a bond to H^+ **(proton acceptor)**
- The stronger the conjugate acid, the weaker its conjugate base.

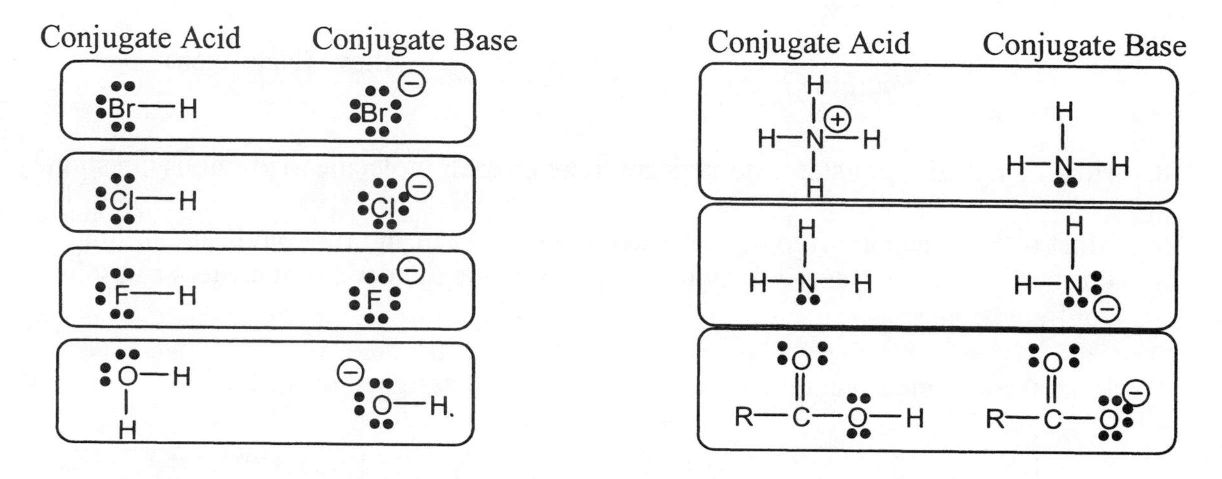

Figure 1b: Bronsted-Lowry Acid-Base Reaction ("Proton Exchange")

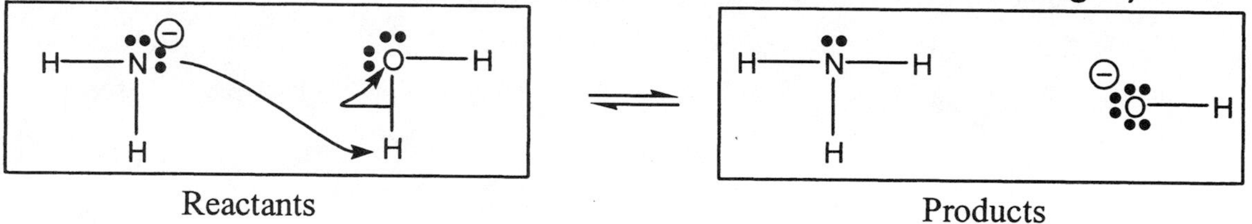

Critical Thinking Questions

1. Write the reverse of the reaction in Figure 1b. **Show the forming and breaking of bonds using curved arrows.**

Model 2: pK_a as a Measure of H–A Bond Strength

Imagine an acid (H–A) in a reaction with water to form its conjugate base (A^-).

The pKa of the acid (H–A) tells you two things. **You are responsible for both!!**

- pK_a = the pH of a water solution where $[A^-] = [H—A]$ (half reacted with water)
- $pK_a \approx$ energy (in pK_a units) required to break the H—A bond

The derivation of the first statement can be found in most textbooks. Look in the index under Henderson-Hasselbach. The derivation for the second statement is provided in the Exercises of this activity. You are responsible for the two results above, but not their derivations.

Example: Hydrofluoric acid (H—F) $pK_a = 3.2$

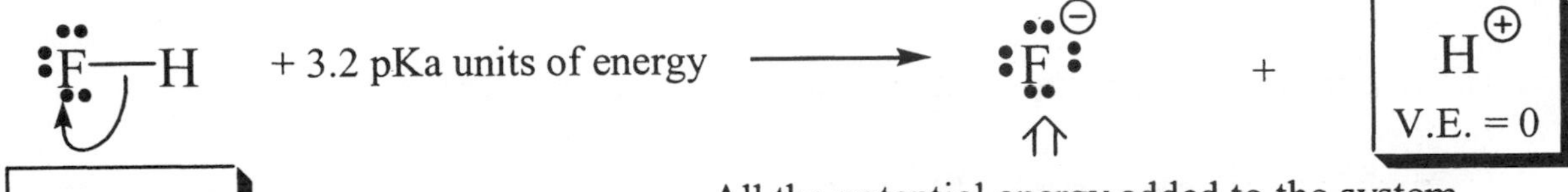

All the potential energy added to the system goes into the conjugate base! This means F^- is 3.2 pKa units higher in P.E. than HF.

Critical Thinking Questions

2. A good acid is often defined as a good H^+ donor.
   a) Which do you expect to give up an H^+ more easily; an acid with a **high pKa** or a **low pKa** [circle one]?

   b) Is your answer above consistent with the fact that the six strong acids we will study in this course have pKa values close to zero? Explain.

   c) Explain why it is very difficult to remove an H^+ from a compound with a very high pKa. (Is such a compound a strong acid or a weak acid?)

Information

- There are six common strong acids: H_3O^+, HCl, HBr, HI, H_2SO_4 and HNO_3.
- You must memorize the name of each acid along with its structure and the structure of its conjugate base. (Names, structures and conjugate bases are shown on the last page of this activity.)
- For simplicity we will assume that all strong acids have a pKa = 0.

Table 2a: pK_a Values for Some Acid/Base Conjugate Pairs

Acid		pK_a	Conjugate Base
hydrofluoric acid	$H{-}F$	3.2	F^-
water	$H{-}O{-}H$	15.7	$^-O{-}H$
ammonia	NH_3	35	$^-NH_2$
methane	CH_4	~50	$^-CH_3$

Figure 2b: Graphical Representation of pK_a's from Table 2a.

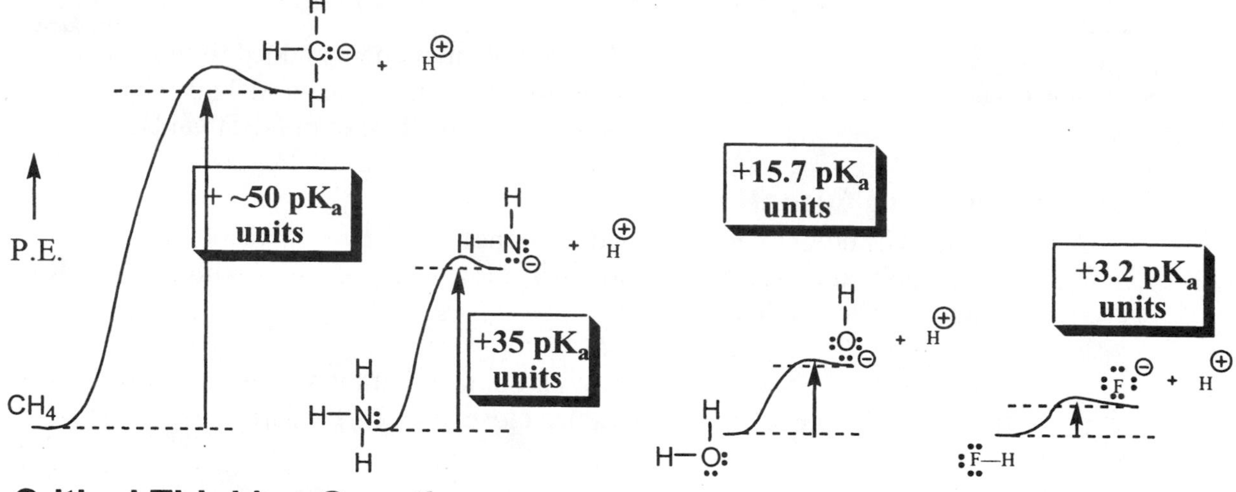

Critical Thinking Questions

3. According to Model 2...
    a) How much energy (in pK_a units) must be added to OH_2 in order to break the first O—H bond? Specify the sign of energy that is added to a molecule.

    b) How much energy (in pK_a units) is released when a lone pair on NH_2^- forms a bond to H^+? Specify the sign of energy that is added to a molecule.

    c) Which species in Table 2a is the strongest acid? (**strongest acid** = species which requires the least energy to remove an H^+ from it.)

    d) Which species in Table 2a is the strongest base? (**strongest base** = species that releases the most energy when it is combined with an H^+.)

Model 3: Potential Energies of Electrons on Negative, Neutral and Positive Atoms

The chart below shows potential energy changes for many conjugate acid-base pairs we will encounter in this course. **Commit these pKa's to memory.**

Note: any species with a -2 charge is extremely high in V.E.(off the chart)

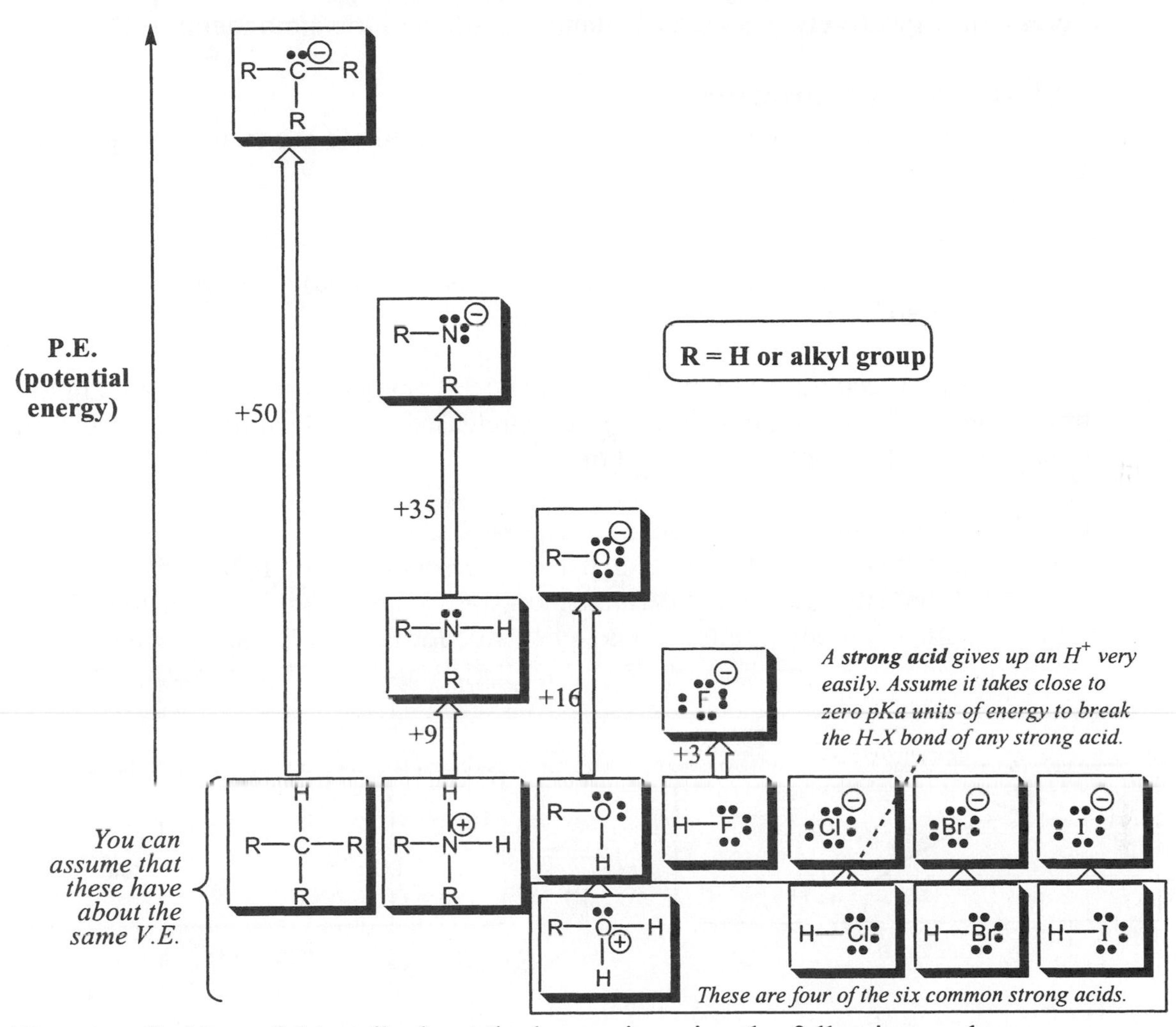

You may find it useful to talk about the lone pairs using the following analogy:

- ***happy* electron pair** = unreactive, low potential energy, unlikely to combine with H^+
- ***unhappy* electron pair** = reactive, high potential energy, wants desperately to combine with H^+. (A strong acid or base is *angry*. If you get it on your hand, it hurts!)

Critical Thinking Questions

4. In general, which is "happier" (read "lower in P.E./weaker base") a species with a **negatively charged atom** or a species with only **neutral atoms** [circle one]? Cite an example of a pair of species from Model 3 to demonstrate this generalization.

Model 4: The Formal charge effect

The following generalizations are often referred to as the **formal charge effect**.

- In general, a species with a **negatively** (-) charged atom is **higher in potential energy than** a species with only **neutral** atoms (regardless of the atom identities).
- In general, a species with only **neutral** atoms is **higher in potential energy than** a species with a **positively** (+) charged atom (regardless of the atom identities).

Critical Thinking Questions

5. Cite a pair of species from Model 3 that are an exception to the formal charge effect generalization.

6. By convention, energy added to a molecule (breaking bonds) is +energy, and energy released by a molecule (making bonds) is -energy.

   (It may help to recall the slot machine analogy: (+) energy experimenter puts in to break bonds = $ into slot machine; (-)energy released by molecules as heat from the making of bonds = $ paid out by slot machine.)

   Use the pKa information in Model 3 to…
   a) Estimate the change in potential energy associated with half-reaction #1.
   b) Estimate the change in potential energy associated with half-reaction #2.
   c) Estimate the total potential energy change for this acid-base reaction.

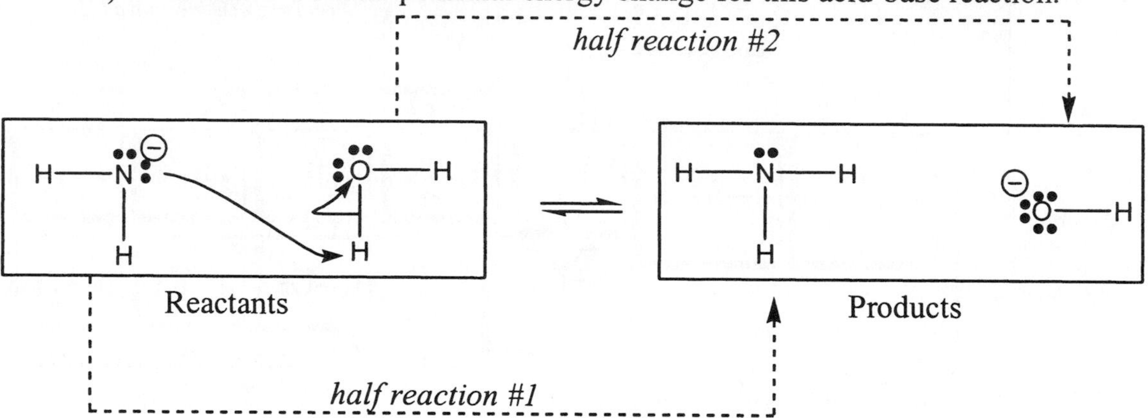

7. Gravitational potential energy is the energy given off when an object (like a roller-coaster car) falls from a height. The higher your altitude, the higher your gravitational potential energy.
   a) Which is higher in gravitational potential energy a **100kg person in Denver** (alt: 1200m) <u>or</u> a **100kg person in Albuquerque** (alt: 1100m) [circle one]?

   b) Is a trip from Albuquerque to Denver **uphill** <u>or</u> **downhill** [circle one] in terms of gravitational potential energy?

8. It can be useful to picture gravitational potential energy when you are thinking about chemical potential energy (which we call "potential energy" in this course).
   a) Is the reaction in CTQ 6 **up-hill** or **down-hill** [circle one] from left to right, in terms of potential energy?

   b) Draw an energy diagram for the reaction in CTQ 6. Include an arrow showing the change in potential energy for the reaction ($\Delta P.E._{rxn}$).

   c) For most organic reactions, released potential energy is released as heat (H). This means that $\Delta P.E._{rxn} \approx \Delta H_{rxn}$. Label the arrow you drew in part b) ΔH_{rxn} (the change in heat for the reaction).

Note: total change in P.E. is sometimes called Gibbs free energy (ΔG). $\Delta G = \Delta H - T(\Delta S)$, where ΔS is the change in entropy. Since $\Delta S \approx 0$ for most organic reactions, we will assume in this course that $\Delta G \approx \Delta H$.

   d) Specify the **value** and **sign of ΔH_{rxn}** (in pKa units) on your energy diagram.

9. The equilibrium constant for the reaction from CTQ 6 is given below.

$$K_{eq} = \frac{[NH_3] \times [HO^-]}{[NH_2^-] \times [H_2O]} = 10^{19}$$

   a) For this reaction at equilibrium, will there be more **reactants** or **products**?

   b) Chemical reactions, like water in a riverbed, tend to go **downhill**. Is your answer in a) consistent with your answer to CTQ 8a? Explain.

   c) Which of the following best describes the exact mathematical relationship between ΔH (in pK_a units) and K_{eq}?

$K_{eq} = \Delta H$ $\quad$ $K_{eq} = \Delta H^{10}$ $\quad$ $K_{eq} = 10^{\Delta H}$ $\quad$ $K_{eq} = 10^{-\Delta H}$

Exercises

1. Summarize the relationship between pK_a and acid strength by completing the following sentences:
   i) The <u>higher</u> the pKa of an acid, the **stronger** <u>or</u> **weaker** the acid.
   ii) The <u>lower</u> the pKa of an acid, the **stronger** <u>or</u> **weaker** the acid.

2. Summarize the relationship between pK_a and base strength by completing the following sentences:
   i) For a given base, the <u>higher</u> the pKa of its <u>conjugate acid</u>, the **stronger** <u>or</u> **weaker** the base.
   ii) For a given base, the <u>lower</u> the pKa of its <u>conjugate acid</u>, the **stronger** <u>or</u> **weaker** the base.

3. Based on the information in Model 3, for each pair circle the stronger base. If you think they are about the same strength, write "about the same P.E.."

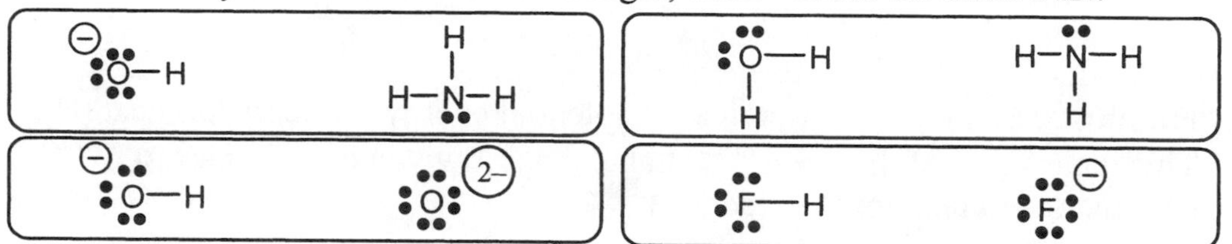

4. For each of the following pairs of species:
   a) Use curved arrows to show the most likely acid-base reaction.
   b) Draw the products that would result from this reaction.
   c) Calculate ΔH (in pK_a units) for the reaction (use the attached table of pKa's).
   d) Draw an energy diagram for the reaction.
   e) State whether the reaction is endothermic (uphill), exothermic (downhill) or thermoneutral (thermoneutral = little heat is consumed or released).
   f) Calculate the equilibrium constant, K_{eq} and state whether there will be more reactants or products at equilibrium.

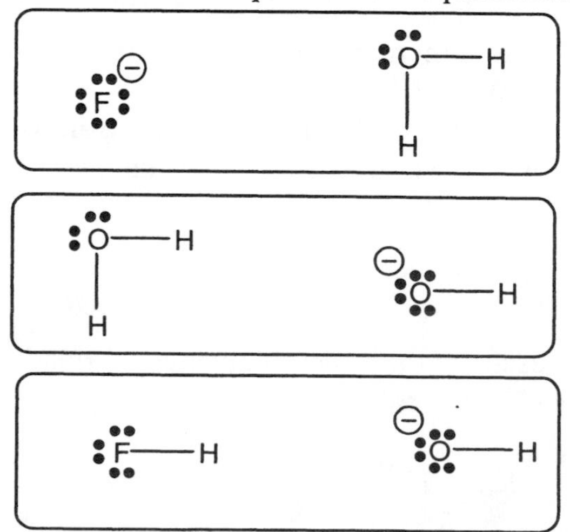

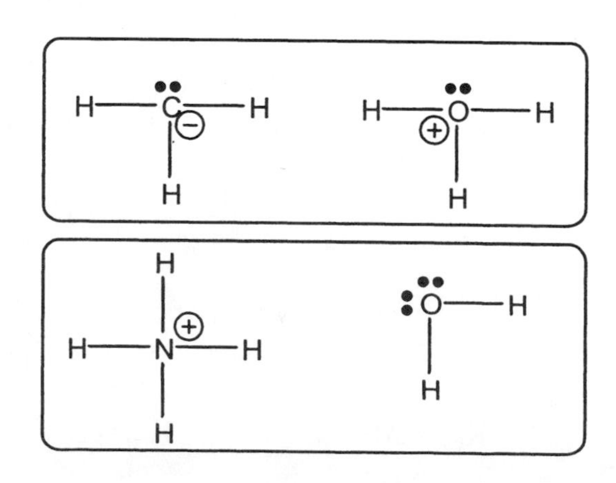

5. A base that is high in potential energy is very reactive and is called a **strong base**. But **strong acids** are low in potential energy. Explain how the strong acids listed in Model 3 can be low in potential energy but still be very reactive (give up an H^+).

6. Read Sections the assigned sections in your text and do the assigned problems.

7. The derivation of the statement that pKa ≈ the energy required to break the H–A bond is as follows:

$\mathbf{pK_a = -log\ K_a}$

($\propto$ = "proportional to")

$K_a \propto K_{eq}$ (K_{eq} = equilibrium constant for the reaction H—A + $H_2O \rightarrow A^- + H_3O^+$)

$\mathbf{\Delta G = -RT\ \mathit{ln}K_{eq}}$

since pK_a = -log K_a and (log x) $\propto$ ($ln\ x$) it follows that $pK_a \propto \Delta G$

ΔG is a measure of energy therefore...

pK_a is a measure of energy!

8. On the next page is a Table of pK_a's.
   - Please memorize the pK_a's of the **weak acids** listed (along with names and structures).
   - Also know the names and structures of the six most common strong acids (acids which dissociate almost completely in water).
   - You don't need to memorize the pK_a's of the strong acids. For simplicity we will assume that all of their pK_a's = zero. This makes nice conceptual sense since it means that it takes zero energy units to remove an H^+ from a strong acid. In other words, it is very easy to remove an H^+ from a strong acid.
   - Note that, if a pK_a of zero means that breaking the H—A bond (in water) requires no input of energy, it follows that a pK_a of less than zero means breaking the H—A bond in water gives off energy (exothermic). Indeed, strong acids dissociate very readily in water—and the process gives off a lot of heat. However, for now assume pKa = 0 for all strong acids.

Strong Acids

(dissociate >99.9% in water)

Know the 6 most common strong acids and that their pKa's are negative!

Name	Conjugate Acid	pKa	Conjugate Base
hydroiodic acid	$I{-}H$	-5.2	I^{-}
hydrobromic acid	$Br{-}H$	-4.7	Br^{-}
hydrochloric acid	$Cl{-}H$	-2.2	Cl^{-}
sulfuric acid	$(HO)_2SO_2$	-5.0	$HOSO_2O^{-}$
nitric acid	$HO{-}N^{+}(=O){-}O^{-}$	-1.4	$^{-}O{-}N^{+}(=O){-}O^{-}$
hydronium ion	H_3O^{+}	-1.7	H_2O

Weak Acids

(dissociate partially in water)

Know the approximate pKa's of these weak acids

R = H or an alkyl group such as $-CH_3$

Name	Conjugate Acid	pKa	Conjugate Base
hydrofluoric acid	$F{-}H$	~3	F^{-}
carboxylic acid	$R{-}C(=O){-}O{-}H$	~5	$R{-}C(=O){-}O^{-}$
ammonium ion	$R{-}NH_3^{+}$	~9	$R{-}NH_2$
phenol	$C_6H_5{-}OH$	~10	$C_6H_5{-}O^{-}$
alcohol (or water)	$R{-}O{-}H$	~16	$R{-}O^{-}$
amine	$R{-}NH_2$	~35	$R{-}NH^{-}$
alkane	$R{-}CH_3$	~50	$R{-}CH_2^{-}$

ChemActivity 9

Part A: The Core Charge Effect

(Which atoms have lower potential energy electron pairs?)

Model 1: Core Charge Revisited

● = electron

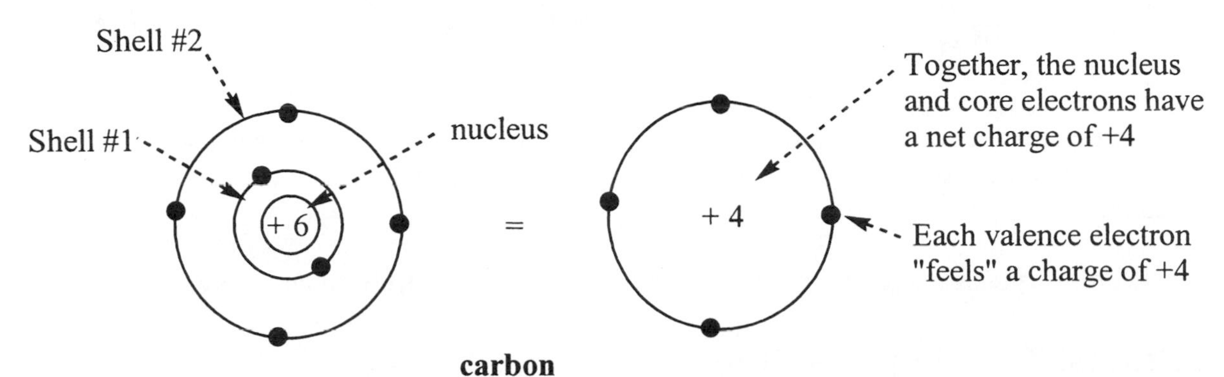

Critical Thinking Questions

1. Draw a core charge representation (like the one above right) for nitrogen, oxygen and fluorine.

nitrogen   oxygen   fluorine

2. Based on the core charge model, explain why fluorine holds its valence electrons more tightly than oxygen (and O holds electrons more tightly than N, and N holds electrons more tightly than C).

Model 2: Core Charge Representations of Ions

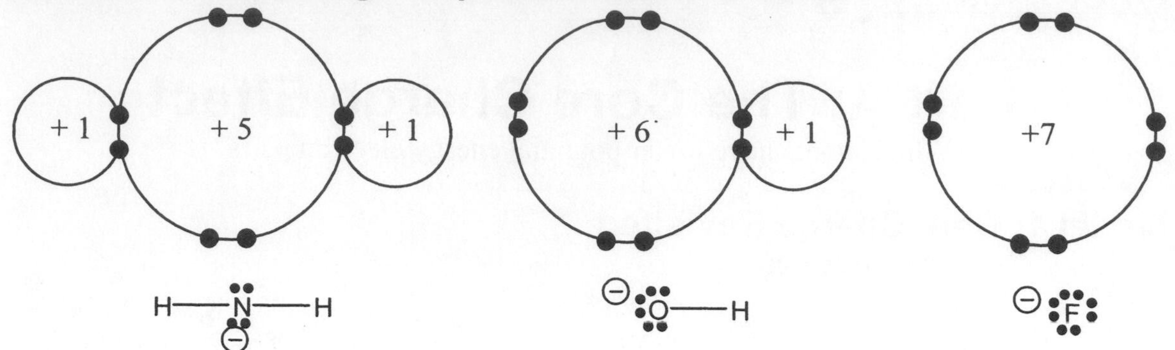

The valence electrons of a fluoride ion (F^-)see a core charge of +7. This is used to explain the fact that electrons on this ion are "happier" than electrons on the other ions shown in Model 2.

(**"happier"** = less reactive, lower in potential energy, **less likely to combine with H^+.**)

Critical Thinking Questions

3. Based on the information in Model 2, which do you expect to be "happier" an electron on hydroxide (HO^-) or an electron on H_2N^-? Explain your reasoning.

4. Rank the four bases below from strongest to weakest.

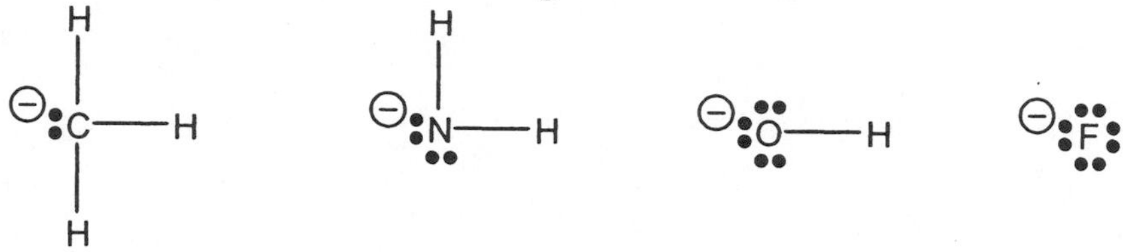

5. Are the pK_a data below consistent with your answer to CTQ 4?

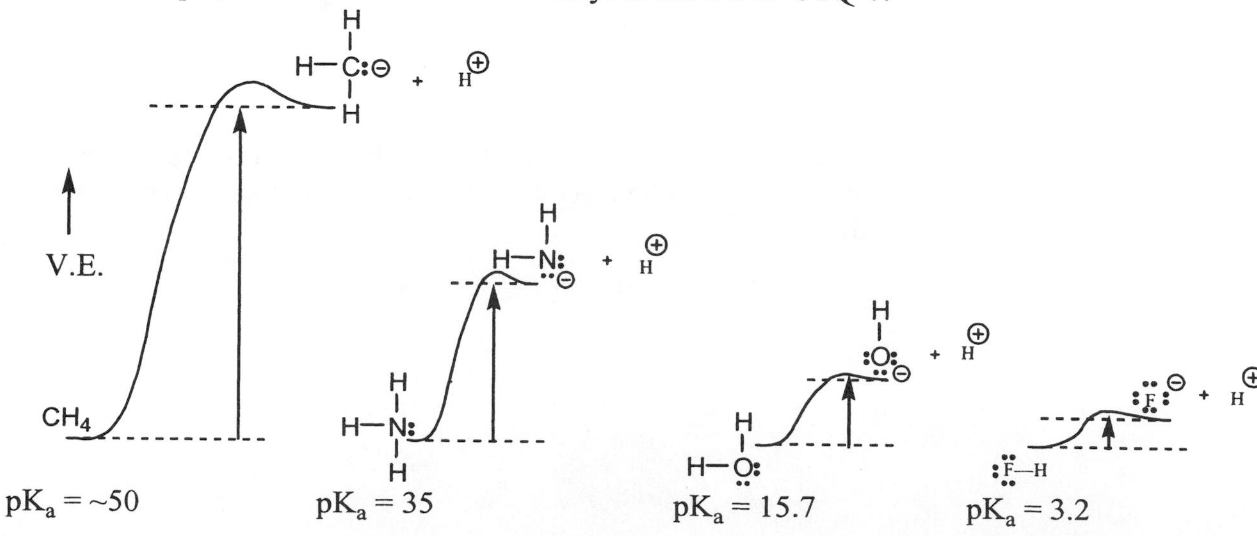

Model 3A: Electronegativity

Previously, we discussed why a fluorine atom holds electrons more tightly than a carbon atom–and that the tighter an electron pair is held, the lower its potential energy. This capacity for certain atoms to lower the potential energy of electrons is so important that chemists give it a name: **electronegativity**.

Table 3: Electronegativities of First Row Elements

C	N	O	F
2.5	3.1	3.5	4.1

- The higher the electronegativity number, the tighter that atom holds its electrons.

Model 3B: Electronegativity Effect vs. Formal Charge Effect

- **Formal Charge Effect** = lone pair electrons on an atom with formal charge = zero are much happier than lone pair electrons on an atom with formal charge = –1.
- **Electronegativity Effect** (a.k.a the core charge effect) = an electron pair is happier on an atom with higher core charge.
  (most happy) F > O > N > C (least happy).
- In most cases the formal charge effect outweighs the electronegativity effect.

For example: a lone pair on NH_3 is happier than a lone pair on HO^-

The important exceptions to this rule are that NH_3 is a stronger base (higher P.E. lone pair) than any of the halide ions (F^-, Cl^-, Br^- or I^-) and H_2O is a stronger base than than three of the halide ions (Cl^-, Br^- or I^-). See Model 3 in ChemActivity 8

Critical Thinking Questions

6. Which of the following TRUE statements illustrates that the formal charge effect is stronger than the electronegativity effect?

F^- higher in P.E. than H_2O; H_2O higher in P.E. than Cl^-; H_2N^- higher in P.E. than NH_3

Part B: The Inductive Effect

(Which is stronger the electronegativity effect or the inductive effect?)

Model 4: Bond Polarization

- An electronegative atom such as F attracts electrons toward its high core charge. This means it "hogs" more than its share of nearby electrons.
- This creates **polarized bonds** with more negative charge at one end (δ–) and less negative charge at the other end (δ+).
- δ+ means **"a partial charge between 0 and +1."** (The Greek letter δ means "small amount.")
- A polarized bond is often represented by a dipole arrow with a + sign at the positive end of the bond (*see examples below*).

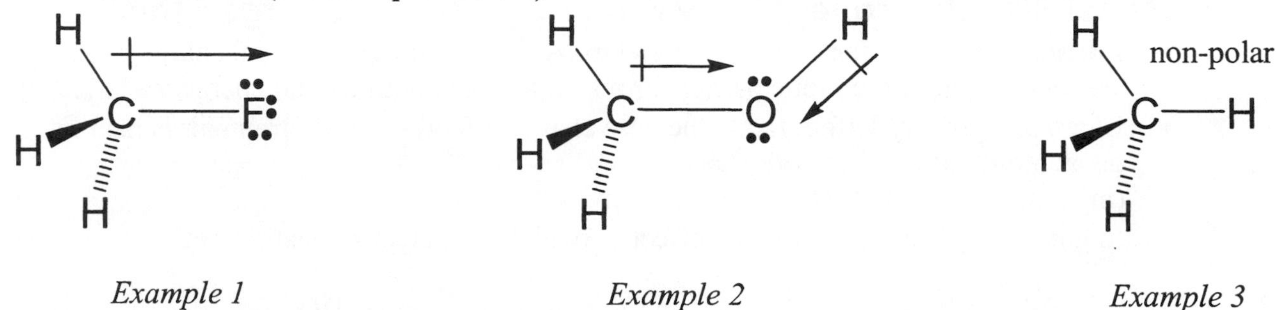

Example 1 *Example 2* *Example 3*

Critical Thinking Questions

7. Place a δ+ next to any atom in Model 4 that you expect will have a partial positive charge, and a δ– next to any atom in Model 4 that you expect will have a partial negative charge.

8. Why is the arrow in Example 2 smaller than the one in Example 1?

Model 5: Inductive Effect (= Long Range Electronegativity Effect)

The effects of an electronegative atom are felt on neighboring atoms, AND can be felt more than one atom away. This is called an **inductive effect**. It is like the ripple effect that is caused when you throw a brick into a still pond. The waves are intense near the impact site, but get smaller farther away.

Figure 5a: Graphical Illustration of the Inductive Effect

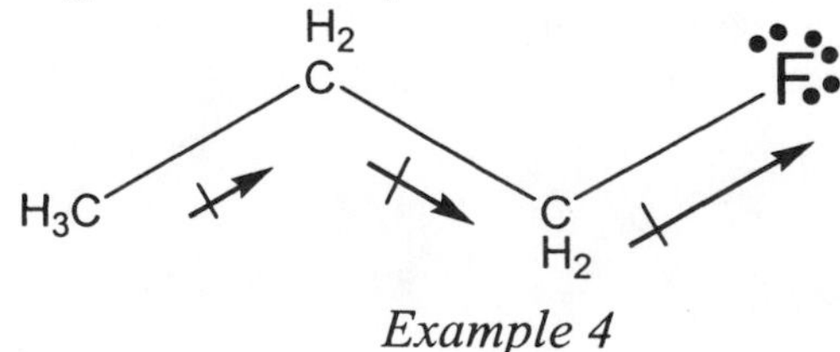

Example 4

9. Draw the three possible conjugate bases of 1-fluoropropane (shown above). Hint: they result from removal of an H^+ from C_1, C_2, and C_3, respectively.

Figure 5b: Energy Required for H^+ Removal from 1-Fluoropropane

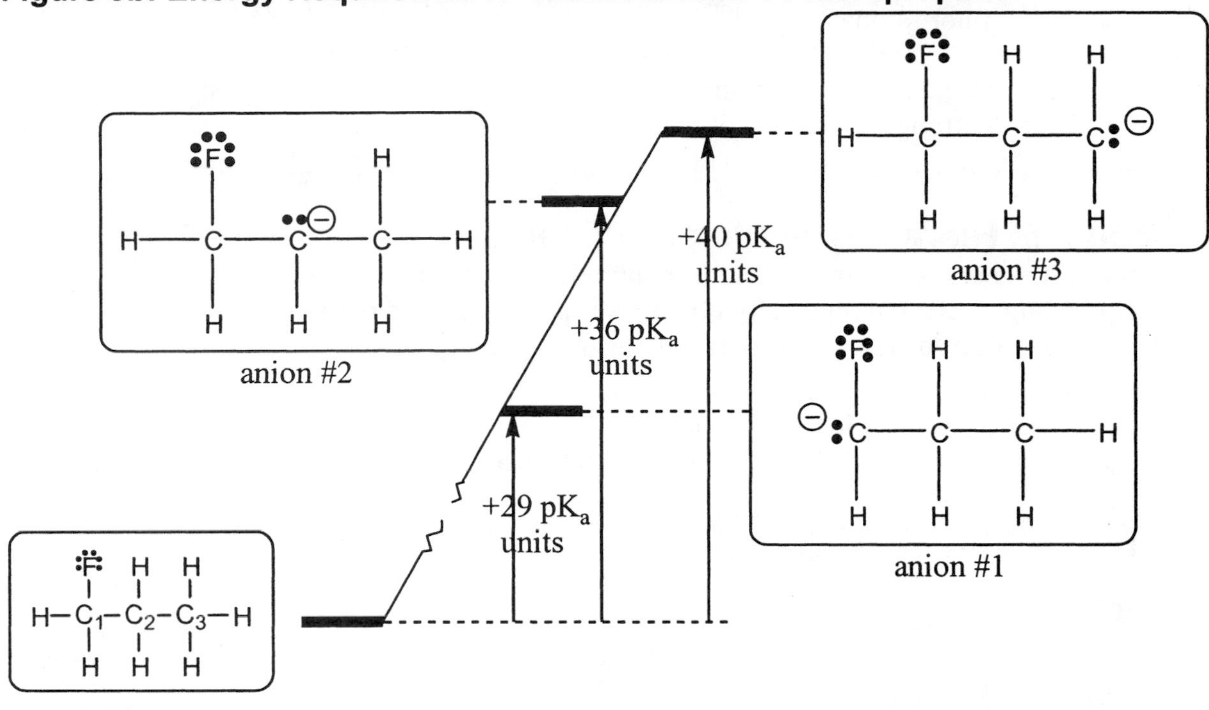

Information

There are two good ways of determining which is the most acidic H on a molecule such as 1-fluoropropane. (The second is generally more reliable.)

- Determine which H-C bond is most polarized toward C. That is, which H-C bond is already partially broken the most.
- Draw **all possible conjugate bases** (in this case there are three-*see* above), and determine which one is lowest in potential energy = happiest. The happiest one is most likely to form.

Critical Thinking Questions

10. There are three different **kinds** of H's on 1-fluoropropane. In the drawing below, they are labeled a, b and c. Explain why all the H's with a given letter are **chemically equivalent** and therefore indistinguishable from each other.

$$\begin{array}{ccccccc} & & :\ddot{F}: & & H_b & & H_c \\ & & | & & | & & | \\ H_a & - & C_1 & - & C_2 & - & C_3 - H_c \\ & & | & & | & & | \\ & & H_a & & H_b & & H_c \end{array}$$

a) Add polarization arrows to 1-fluoropropane so as to show the pull of electron density by the electronegative fluorine atom.

b) Construct an explanation for why it is easiest (least energy required) to remove an H^+ from C_1 of 1-fluoropropane.

c) Circle the most acidic H (or H's if more than one equivalent H) on 1-fluoropropane.

d) Which of the three anions shown in Figure 5b is the strongest base? (Strongest base = will release the most energy when combined with an H^+.)

Model 6: Hierarchy of Electron Effects

Strongest = **Formal charge effect** (= neutral happier than negative)
Very Strong = **Electronegativity effect** (= e^- pair on F happier than e^- pair on O>N>C)
Weak = **Inductive effect** (= long range electronegativity effect)

Critical Thinking Questions

11. For each compound below, circle the most acidic H (or H's if there is a tie).

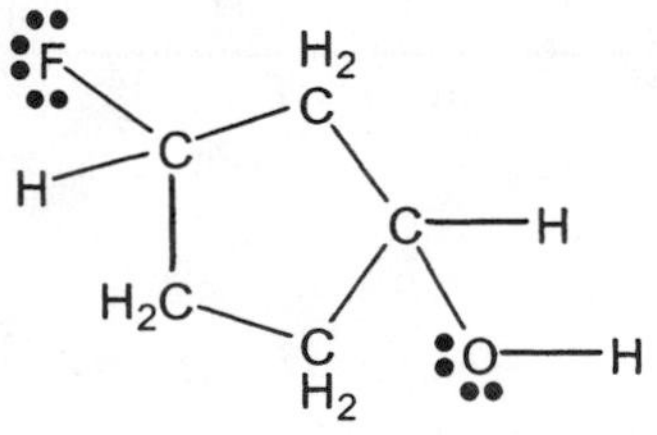

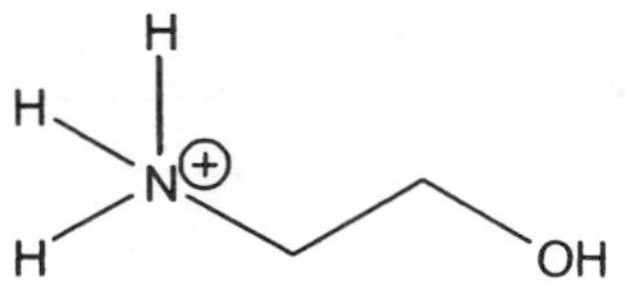

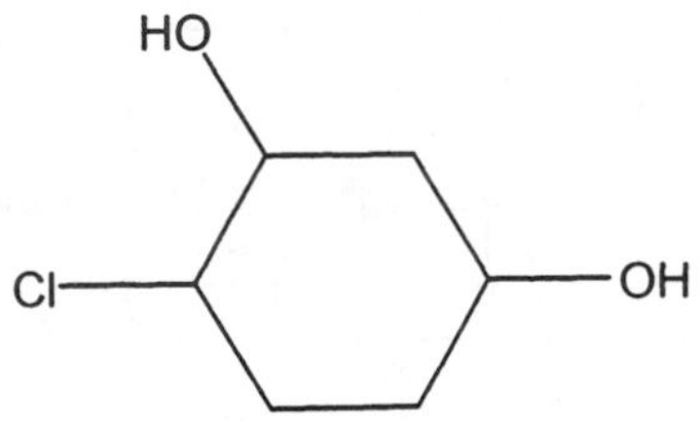

a) Explain how compound **I** demonstrates that the electronegativity effect is stronger than the inductive effect?

b) Explain how compound **II** demonstrates that the formal charge effect is stronger than the electronegativity effect?

c) Draw an isomer of compound **III** with the Cl in a position such that it has an equivalent inductive effect on both of the two OH groups. (This compound will have two equivalent most acidic H's.)

Exercises for Part A

1. Choose which species is more likely to act as the base and use the arrow formalism to illustrate the most likely acid-base reaction. Also draw the resulting products.

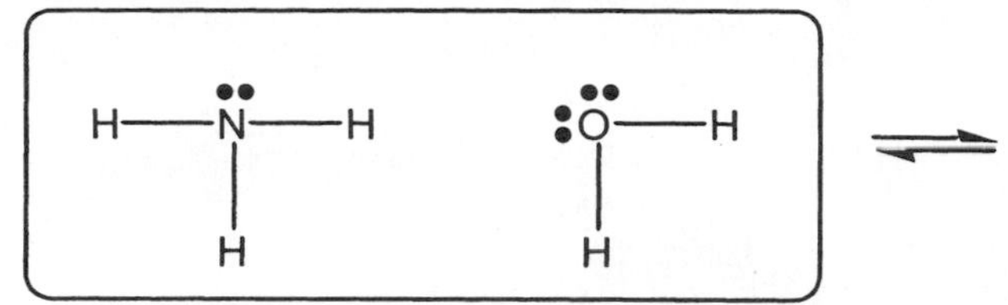

a) Which species is more likely to act as a base? Explain your choice.
b) Predict whether the reaction is up hill or down hill. Explain your reasoning.
c) What species would you look for on the table if you were looking for a pK_a value that would tell you the potential energy of the lone pair of electrons on NH_3?
d) Draw an energy diagram for the reaction, including an estimated value for ΔH in pK_a units.

Exercises for Part B

2. For each molecule below, draw in the **most acidic hydrogen/s**. If there is a tie, between 2 or more equivalent hydrogens—draw in all these H's.) Then circle the molecule in each row that is more acidic. The first one is done for you

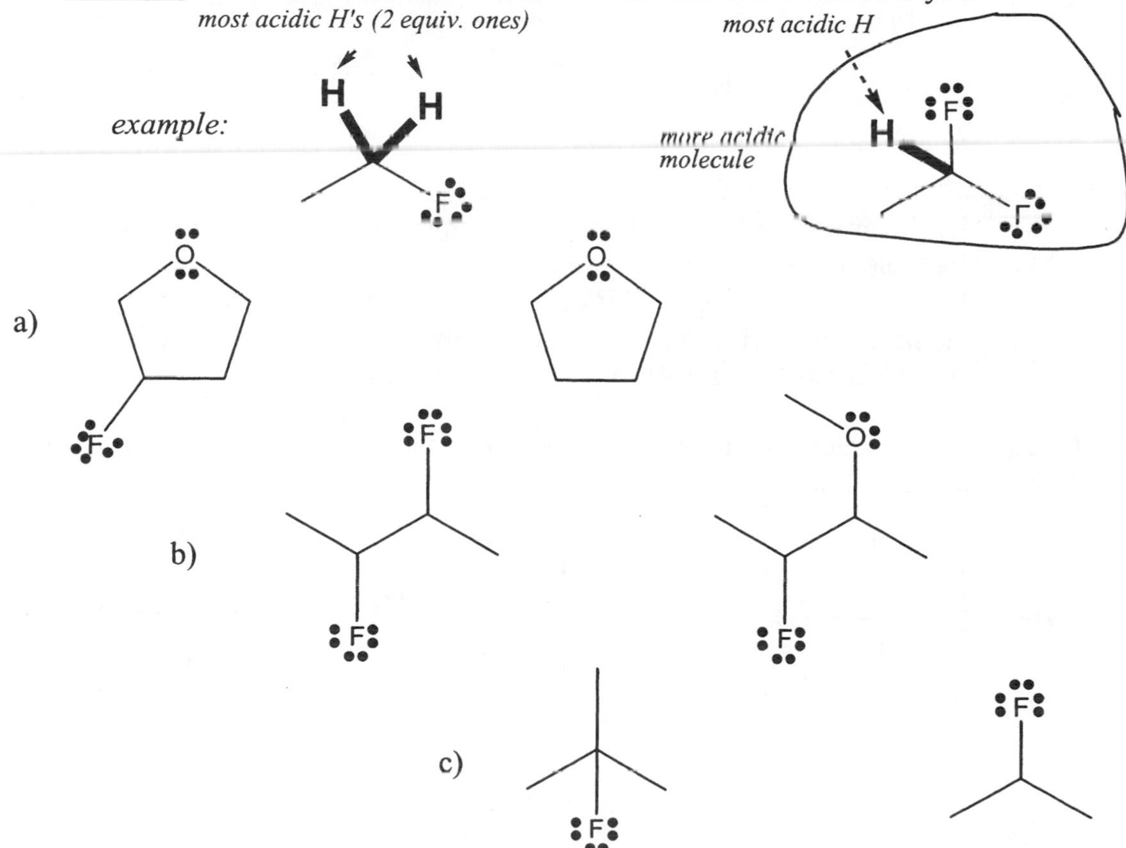

3. Rank the nitrogen bases below according to how much energy will be released by an acid-base reaction with H_3O^+. (**1** = most energy released...**3** = least energy released.) Assume no energy is released or consumed when breaking the O–H bond on H_3O^+.

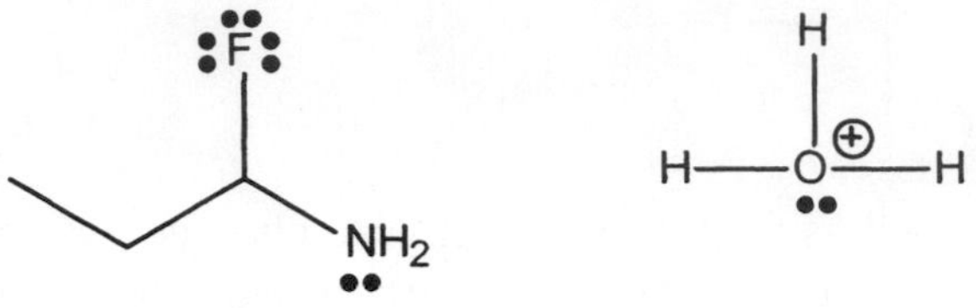

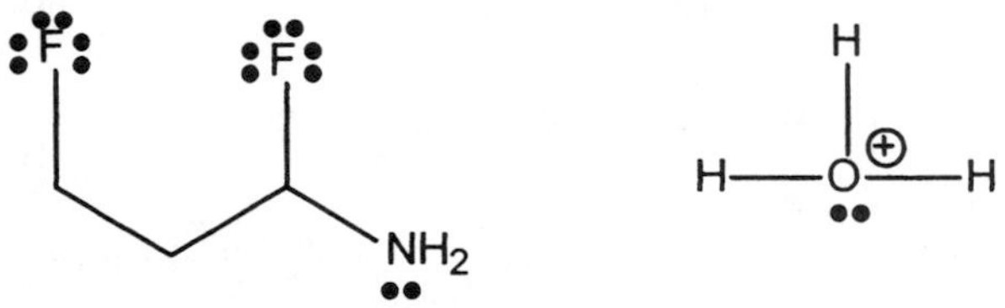

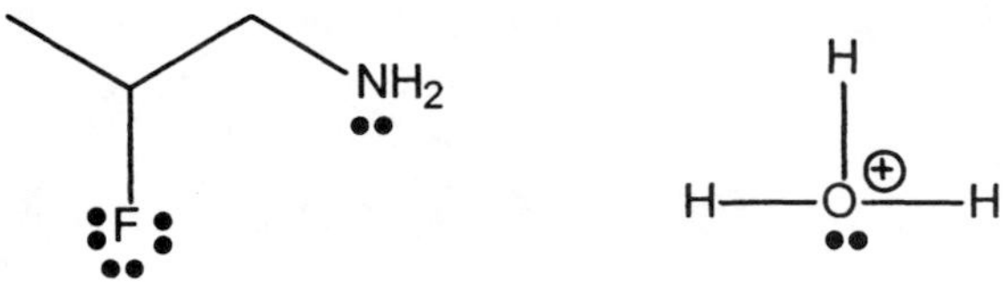

a) Use the arrow formalism to illustrate each reaction.
b) Draw the products of the reaction.
c) Draw the energy diagrams for the three reactions. Do not attempt to calculate exact ΔH_{rxn}, but your reaction diagrams should indicate which reaction is most (and least) downhill in terms of energy.

4. Circle the most acidic H on each structure below, then rank them from strongest acid (1) to weakest acid (3).

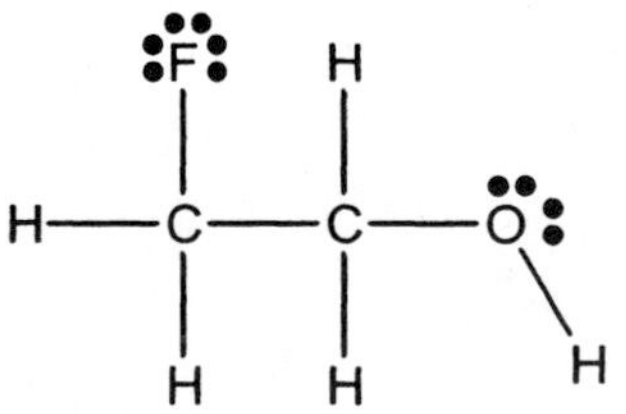

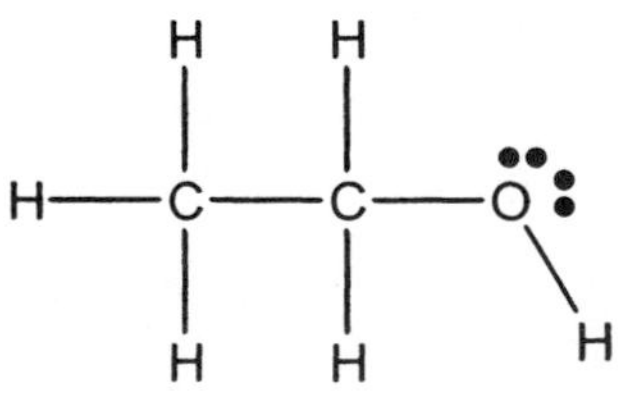

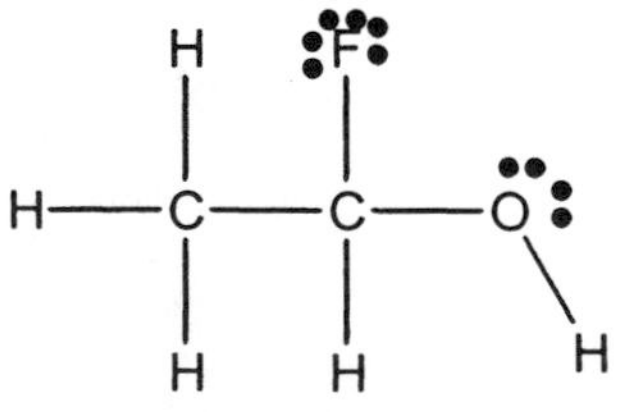

5. It is easier to remove an H^+ from **A** than **B** (*see* below). In general it is difficult to remove a 2nd H^+ to make a –2 anion. One explanation is that the anionic oxygen on **B** actually **donates** electron density, as shown by the squiggly arrows. Table 3 says that O is more electronegative than C. Explain why this oxygen on **B** donates electron density instead of withdrawing it from the carbon.

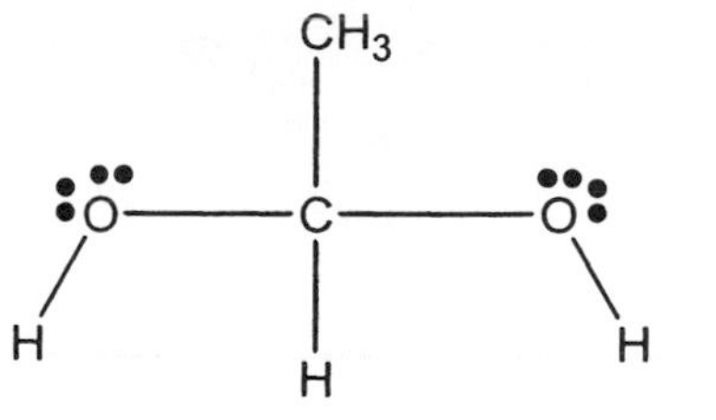

A 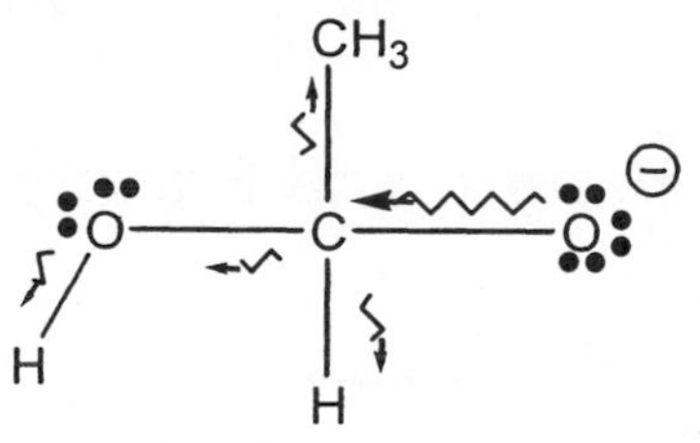 **B**

Hint: Consider a very absorbent sponge. The ability of this sponge to attract and absorb more water disappears as it becomes more water logged. In fact, it can eventually become a SOURCE of water.

6. Consider the following bases:

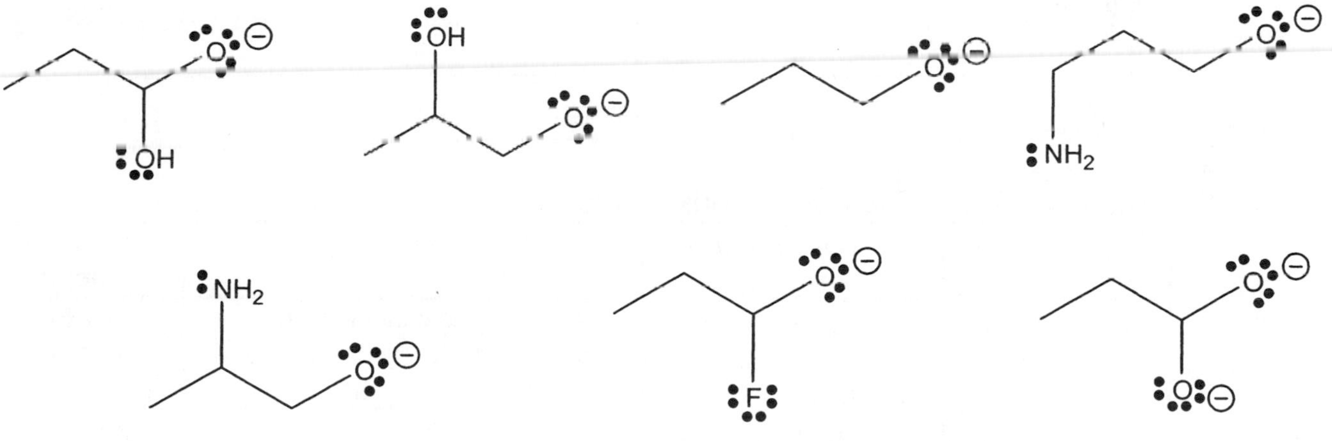

a) For each base above, circle the atom with the highest P.E. (most unhappy) lone pair of electrons. (Note that for one of them, there is a tie between two atoms.)
b) Put a **1** and the words **highest P.E.** next to the strongest base. Put a **7** and the words **lowest P.E.** next to the weakest base.
c) Number the others **2-6** according to base strength.

7. Read the assigned pages in the text and do the assigned problems.

ChemActivity 10

Resonance Effects

(How does the number of resonance structures affect the P.E. of a molecule or ion?)

Model 1: Wavelength and Energy

Figure 1a: Guitar Strings

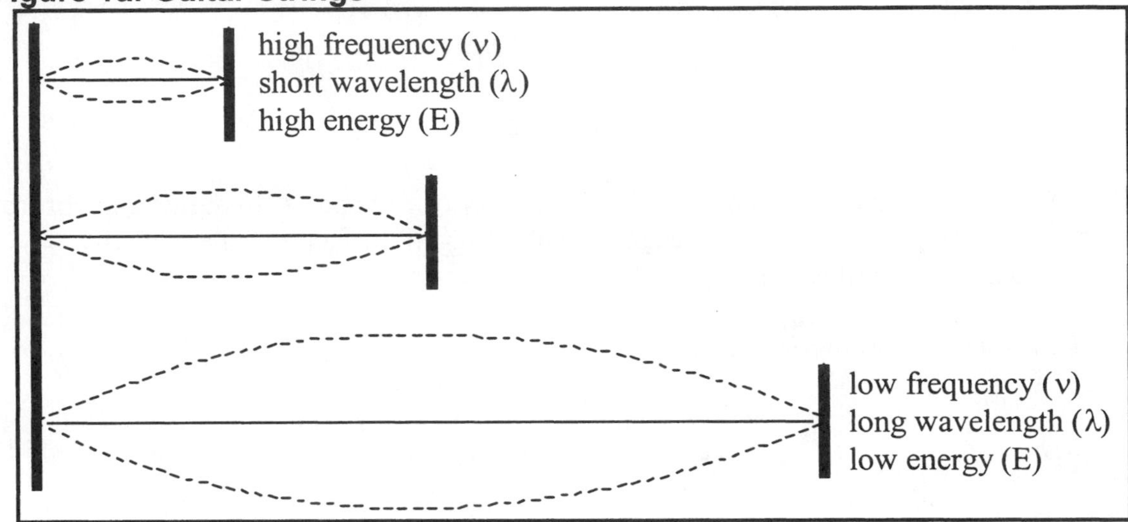

Table 1b: Properties of Various Kinds of Light

Type of Light	Wavelength (nm)	Effect on Humans	Relative Energy
X-ray light	$\lambda = 0.1$ to 200 nm	Will give you cancer if you are exposed for more than a few minutes over a lifetime.	**very high energy**
Ultraviolet light	$\lambda = 200$ to 400 nm	Gives you a sun burn and eventually can give you cancer.	**high energy**
Visible light	$\lambda = 400$ to 800 nm	Does not hurt you, but supplies energy to plants	**moderate energy**
Radio Waves	$\lambda = 0.2$ to 20 meters	no demonstrated effect on humans	**low energy**

There are 1 billion nanometers (nm) in 1 meter (1 m = 10^9 nm).

Critical Thinking Questions

1. According to Model 1:
   a) Light with a short wavelength is **low** or **high** [circle one] in energy.
   b) Light with a long wavelength is **low** or **high** [circle one] in energy.

Model 2: Electrons in a Box (a thought experiment)

- Imagine an electron confined to a box.
- The electron is constantly moving around in its box at nearly the speed of light. For this reason we say each electron is evenly spread out or **delocalized** over the entire box.. Chemists imagine the electron spread out into a mist they call an **electron cloud**.
- Electrons are so small that they behave like light and have a wavelength.
- The longer the box, the longer the wavelength of an electron in that box.

Figure 2: Electrons in Small and Large Boxes

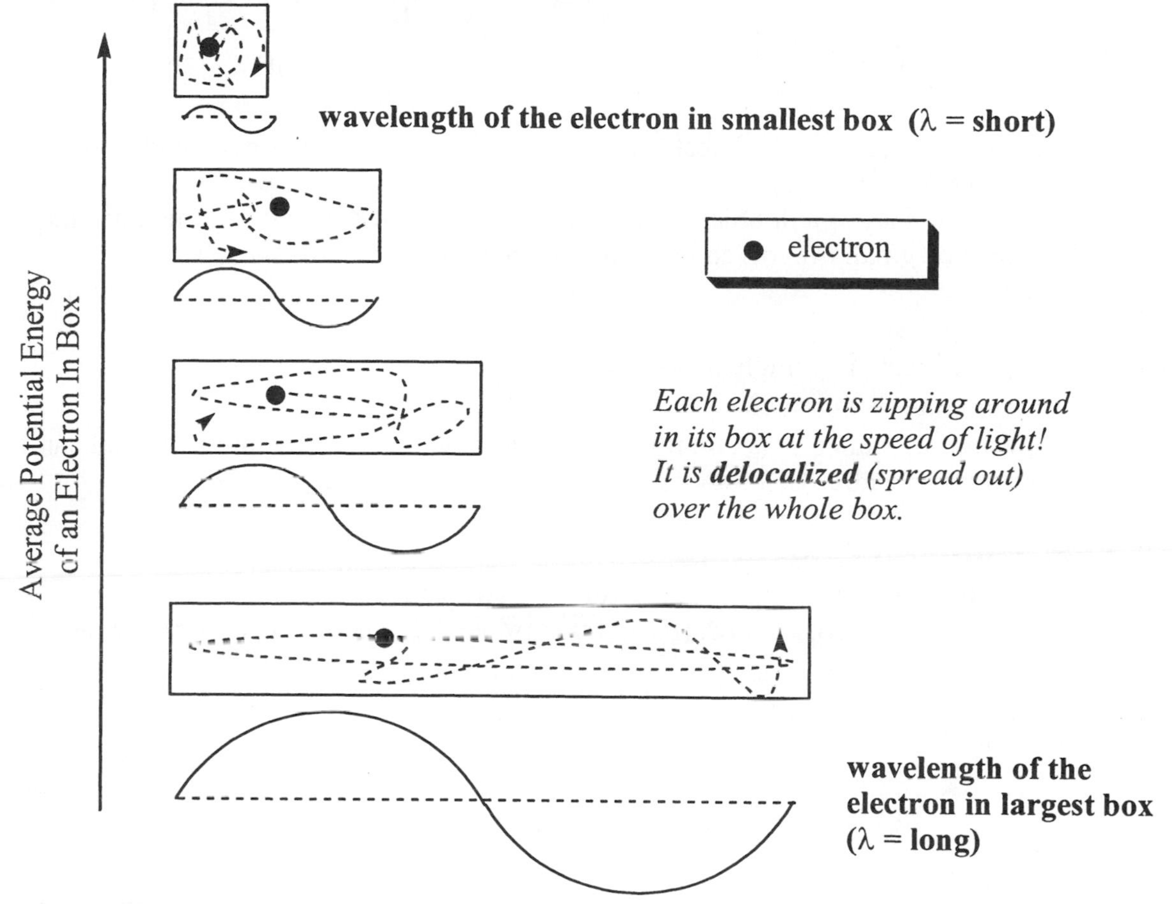

Critical Thinking Questions

2. Which electron in Figure 2 has the smallest wavelength?

3. Which electron in Figure 2 has the lowest energy? Explain your reasoning.

Review: Rules for Drawing Resonance Structures

The following are two of the three "legal" ways of moving electrons to generate a set of resonance structures. We will use the third way later on. Please study these two arrows carefully. They are the key to drawing resonance structures.

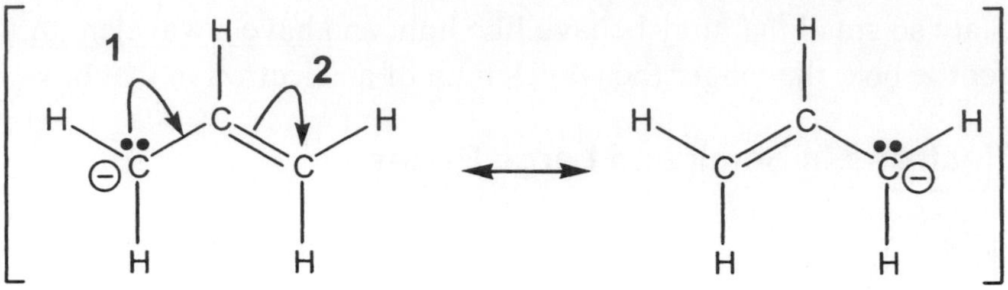

1. You may move a lone pair of electrons on an atom so as to form a new double bond to that atom (*see* **Arrow 1**).
2. You may break part of a double bond and make those two electrons a lone pair on one of the atoms originally involved in the double bond (*see* **Arrow 2**).

In addition:

- You may not break any single bonds.
- You may not move any nuclei.
- The resulting resonance structure must be a legitimate Lewis structure (e.g. you can't make 5 bonds to carbon)

Critical Thinking Questions

4. Draw all important resonance structures (if any) that go with the three Lewis structures (I, II & III) below. (Keep track of electrons using the curved arrow notation.)

I

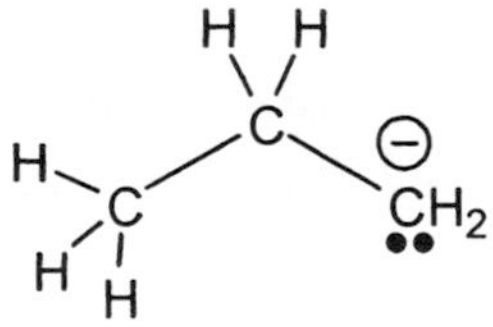

II

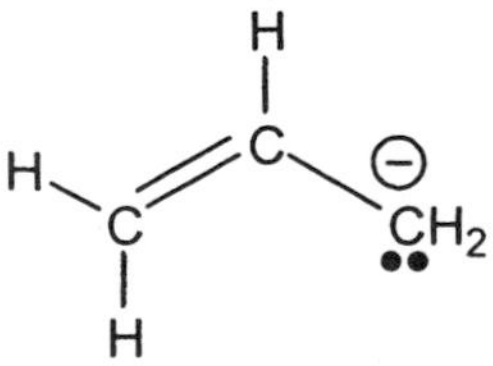

III

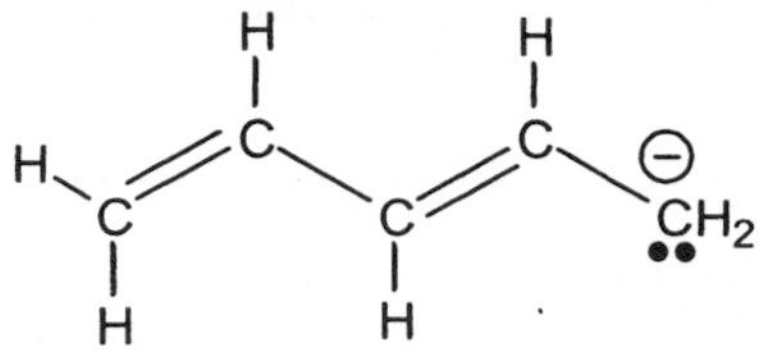

Model 3: More Resonance Structures = Larger "Box"

The electron pair on ion I (*previous page*) is stuck (localized) on C_3. However, the resonance structures you drew on the previous page tell you the electron pair on ion II is **delocalized** (spread out) over three carbons. It is therefore less confined, which is equivalent to saying it is in a "**larger box**." Recall from Model 2 that an electron in a large box has a long wavelength, and therefore a **low potential energy**.

The wavelength argument is not as important as the following end result called the resonance effect: The **Resonance Effect** states that "the **larger the # of important resonance structures** you can draw for a structure, the **lower the potential energy** of that ion or molecule."

Critical Thinking Questions

5. Explain why we can say "the electron pair on III is in the largest "**box**!"

6. Which ion (**I, II** or **III**) [circle one] do you expect to be lowest in potential energy? Explain your reasoning.

7. Fill in Columns 2 & 3 in Table 3a.

Table 3a: pKa Data for Various Conjugate Acid-Base Pairs

Conjugate Base	Total # Important R.S.'s	Energy to break C-H bond (or P.E. conj. Base)	Conjugate Acid	pK_a
I. H_3C–CH_2–$CH_2^{\ominus}$ (lone pair on CH_2)			H_3C–CH_2–CH_3	~ 50
II. H_2C=CH–$CH_2^{\ominus}$ (lone pair on CH_2) + other resonance structure			H_2C=CH–CH_3	43
III. H_2C=CH–CH=CH–$CH_2^{\ominus}$ (lone pair on CH_2) + other resonance structures			H_2C=CH–CH=CH–CH_3	33

8. Based on the pKa data in Table 3a, write "strongest base" next to one of the species in Table 3a and "weakest base" next to one of the species in Table 3a. Are your choices consistent with your answers to CTQ 7? Explain.

Review: Hierarchy of Electron Effects

Strongest = **Formal charge effect** = neutral happier than negative

Very Strong = **Electronegativity effect** = e^- pair on F happier than e^- pair on O>N>C

Strong = **Resonance effect** = the more "delocalized" an electron, the happier it is.

Weak = **Inductive effect** = long range electronegativity effect

Figure 3b: pK_a Data for Two Ions Demonstrating Resonance Effects

	Base	Conjugate Acid	pK_a
IV	[$H_2C{=}CH{-}CH_2^{\ominus}$ ⟷ $^{\ominus}H_2C{-}CH{=}CH_2$]	$H_2C{=}CH{-}CH_3$	43
V	[$O{=}CH{-}CH_2^{\ominus}$ ⟷ $^{\ominus}O{-}CH{=}CH_2$]	$O{=}CH{-}CH_3$	17

Critical Thinking Questions

9. Based on the pKa data, write **"weaker base"** next to the weaker of the two bases in Figure 3b (IV or V).

   a) Which conjugate base **IV** or **V** [circle one] is lower in potential energy?

   b) For Base IV, the two resonance structures are **equivalent**. For Base V the two resonance structures are **not equivalent**. One of the two resonance structures has an arrangement of electrons that is better/more happy/lower in potential energy. This better resonance structure is called the **more important resonance structure**. Circle this more important resonance structure for Base V.

   c) The two bases in Figure 3b have the same number of resonance structures. According to the resonance effect, they *should* have similar potential energies. **But they don't!** Explain why one base is much lower in potential energy than the other.

   d) Experiments show that Base V in Figure 3b has a carbon-oxygen bond order of 1.2 and a carbon-carbon bond order of 1.8. Is this information consistent with your answer to part c)? Explain.

Exercises

1. Consider the following acid-base reactions:

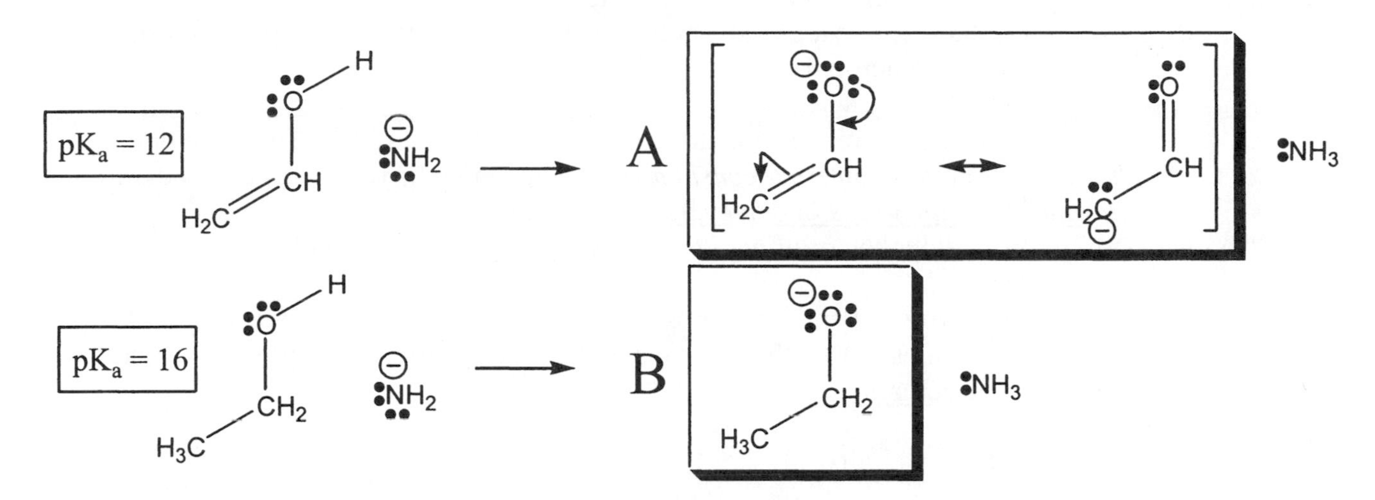

a) Use curved arrows to show each acid base reaction.
b) Based on the pKa data, which alcohol is a stronger acid?
c) Which conjugate base is lower in potential energy (a weaker base)?

Warning!! Common Sticking Point!! Many students think that conjugate base B should be lower in potential energy because A has an unfavorable negative carbon associated with it. **An additional resonance structure (even if it is not very favorable) can only "help" lower the potential energy of a species. It cannot hurt.** The following analogy helps illustrate this point.

Consider two teams of postal workers racing to postmark 1000 envelopes. Team A consists of a fast worker and a slow worker. Team B consists of a single fast worker. Assume the two fast workers work at the same pace, and the slow worker doesn't bother his fast teammate

This means Team A will postmark 1000 envelopes before Team B since the slow worker helps some, and doesn't hurt at all.

Explain how this story relates to the example at the top of the page.

2. A student made up the following analogy for the resonance effect: "*A kid lives on a block with very few fences separating the back yards. The larger the number of yards he and the other kids have access to, the happier they are.*"
   a) Explain what each of the underlined elements of the analogy represent in terms of chemistry.

   A second student thinks this analogy is incomplete. She adds: "*To figure out the happiness of a kid you must consider both the number of yards he has access to and the quality of the grass in each yard.*"
   b) What does *the quality of the grass in each yard* represent?

3. Use curved arrows to show the most likely acid base reaction between the following pairs of molecules. Draw the products of the reaction including any important resonance structures.

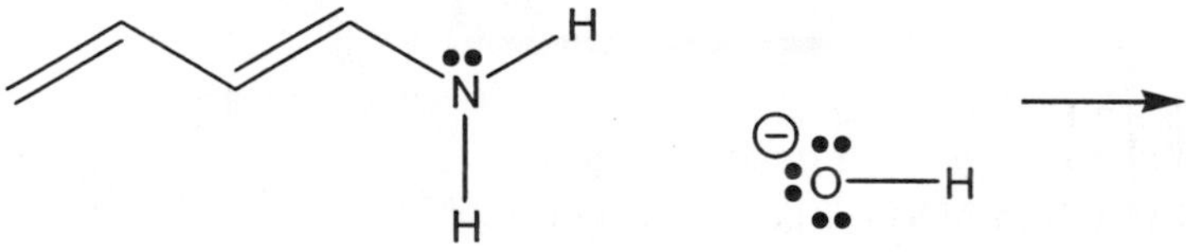

N, H, H, O, H → H_2O + ?

O, H, O, H → H_2O + ?

O, H, O, O, H → H_2O + ?

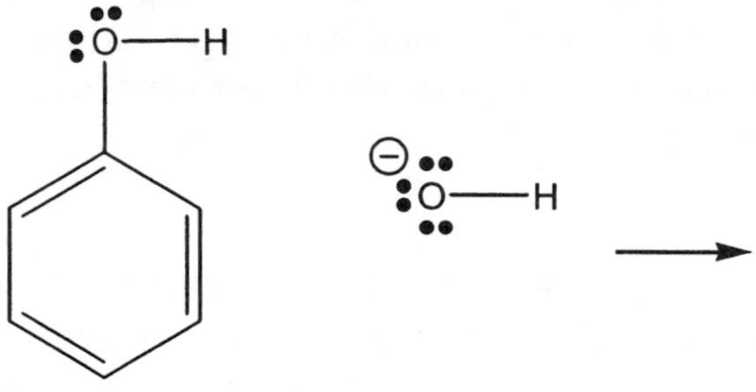

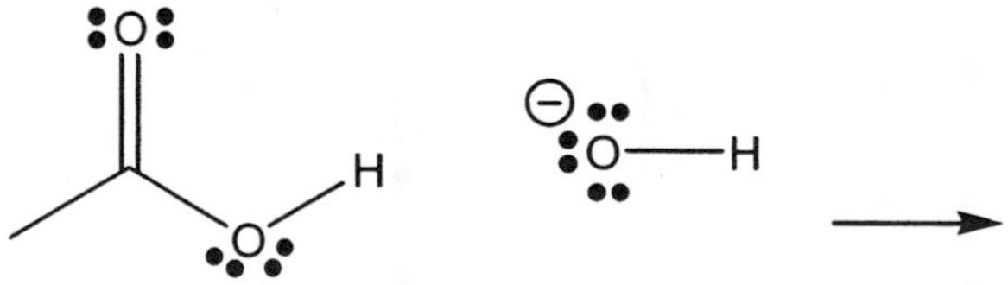

4. Construct an explanation for why sulfuric acid is such a strong acid. (Note that sulfur is in the third row of the periodic table and can have more than 8 electrons.)

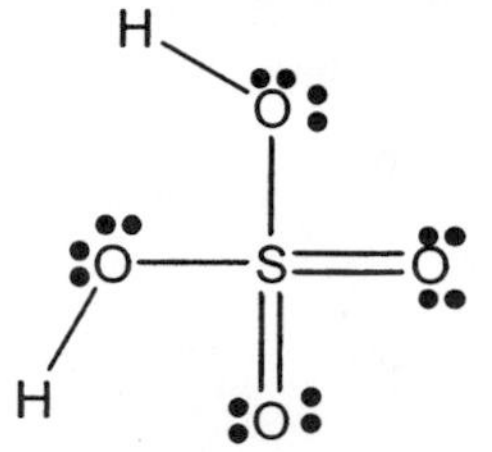

H, O, O, S, O, H, O

5. Consider the data in Table 3a.

- The pKa in row 3 tells you that it takes 33 pKa units of energy to remove an H^+ from the conjugate acid of III so as to create the new lone pair on conjugate base (III).
- Assume that all three conjugate acids in Table 3a have the same potential energy.
    a) Draw three energy diagrams showing the three acid base reactions depicted in Table 3a. (Show Conjugate Base + H^+ → Conjugate Acid)
    b) Construct an explanation for why the conjugate acid in Row 3 of Table 3a is the strongest acid listed in the table.

6. The pK_a of a molecule tells you the amount of energy it takes to remove the most acidic hydrogen. In most cases you are expected to know which H is most acidic. If you are not sure, sketch several conjugate bases and determine which one is lowest in potential energy.
    a) For each molecule below circle the most acidic hydrogen.
    b) For each pair, put a box around the molecule you expect to have a lower pK_a and explain your reasoning.

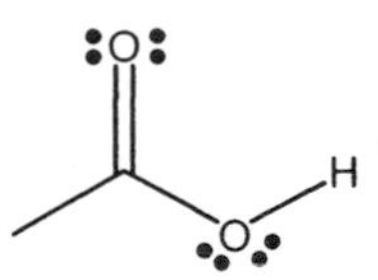

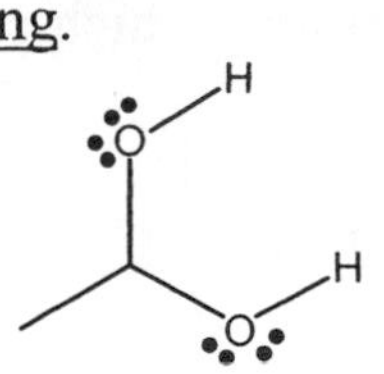

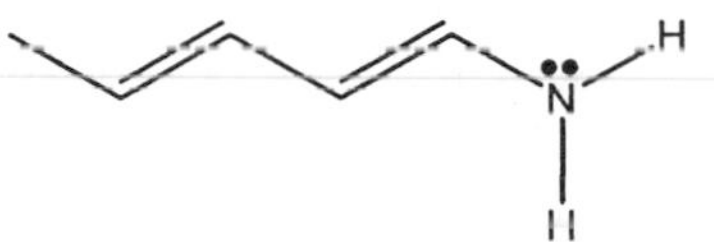

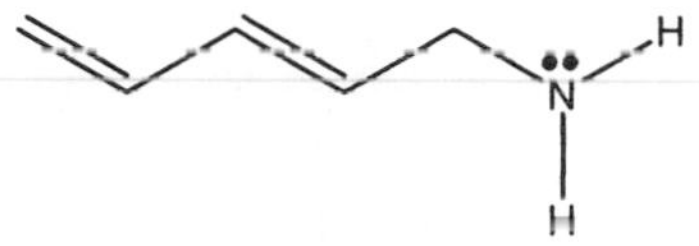

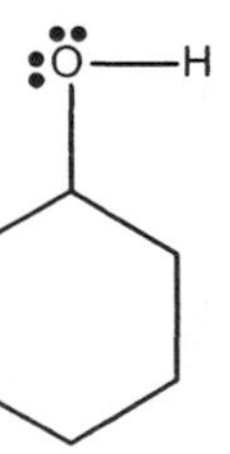

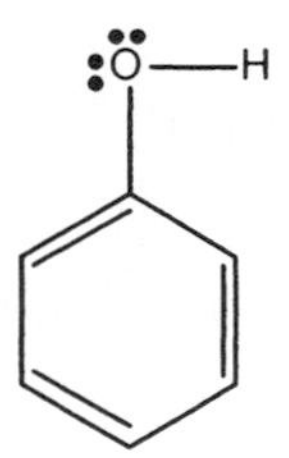

7. Read the assigned pages in your text and do the assigned problems.

Which H (proton) Is Most Acidic?

The following are equivalent ways of asking about the acidity of an H atom:

1. What is the "most acidic proton" on the molecule?
2. Which proton is associated with the published pK_a value?
3. Which proton on the molecule is easiest to remove?
4. Which proton on the molecule takes the least energy to remove?
5. For which H atom is removal least up-hill in energy?
6. Which bond to an H atom, when broken, results in the lowest P.E. lone pair of electrons?

The easiest way to answer question #1 is to ask yourself question #6.

To answer question #6, consider all the different H atoms in the molecule and draw the unique products (a.k.a. "conjugate bases") that result from each one's removal. Then consider the hierarchy of effects to sort out which conjugate base has the lowest P.E. new lone pair of electrons.

Formal Charge Effect > Electronegativity (Core Charge) Effect > Resonance Effect > Inductive Effect

Make a graphical sketch of the energies of all conjugate bases. For compound "S" below, there are 4 different kinds of unique H's, and thus 4 unique conjugate bases (I-IV).

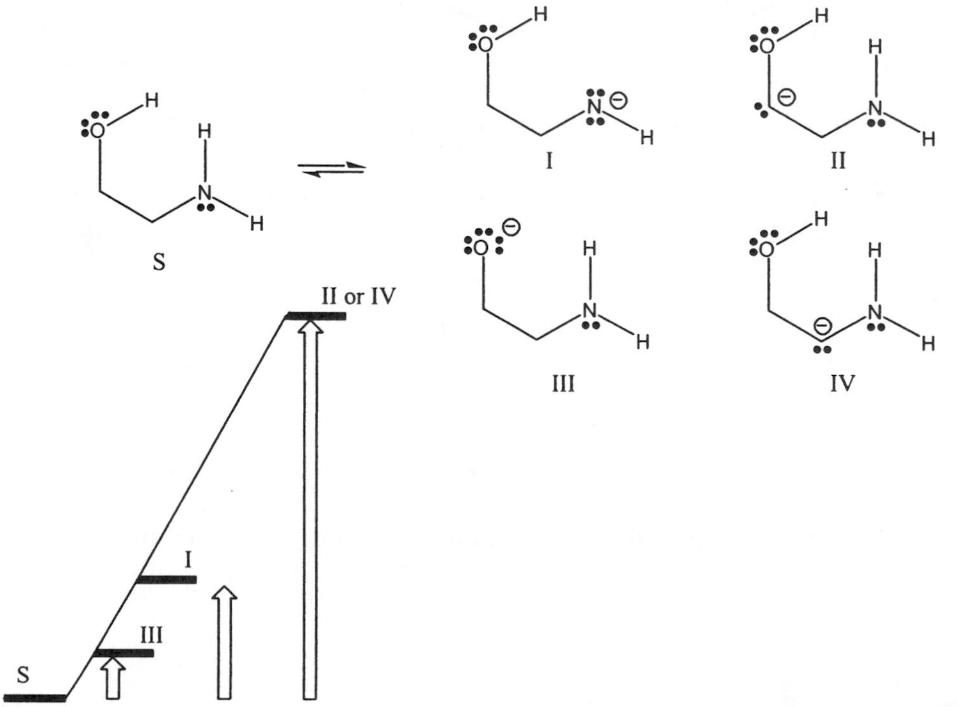

S ⇌ I, II, III, IV

Energy levels: S, III, I, II or IV

1. For the example above, answer the 6 questions in reverse order 6 ---> 1 and explain which is the most acidic hydrogen on the compound "S."

2. For each of the molecules below, draw ALL possible conjugate bases and
   a) Circle the conjugate base that would require the least amount of energy to make from the original molecule (conjugate acid).

   Note: There is a different pK_a associated with removal of each different H. A table would list the lowest pK_a of these. This would be the pK_a associated with the most acidic H and the lowest P.E. conjugate base.

   b) Circle the most acidic hydrogen on the original molecule (conjugate acid).

HO $\overset{\oplus}{N}H_3$

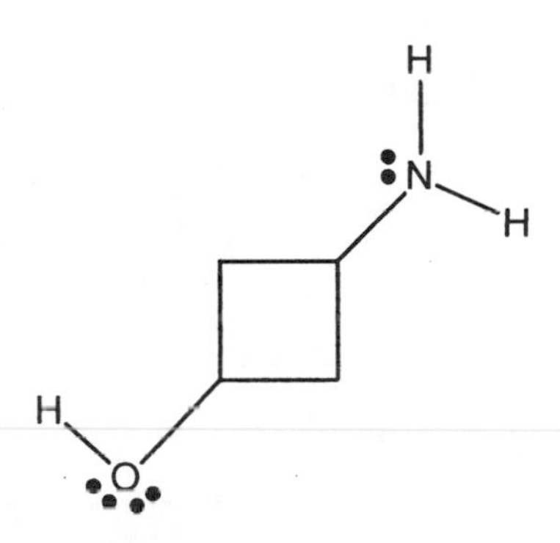

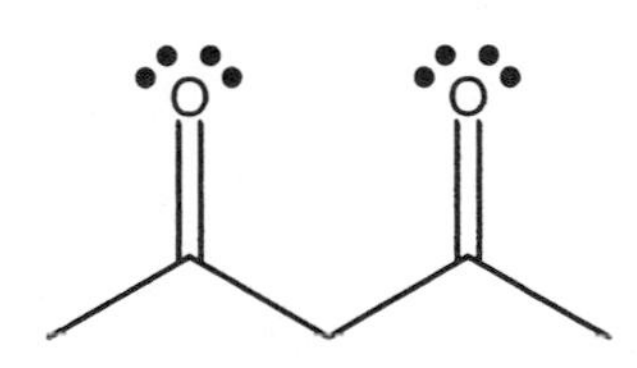

CF_3

F_3C

3. For each pair of compounds below, put a box around the molecule with the most acidic hydrogen (this is the molecule that would have a lower pK_a listed next to it on a table) and explain your reasoning.

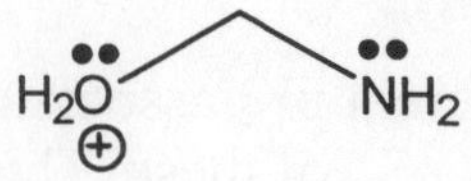

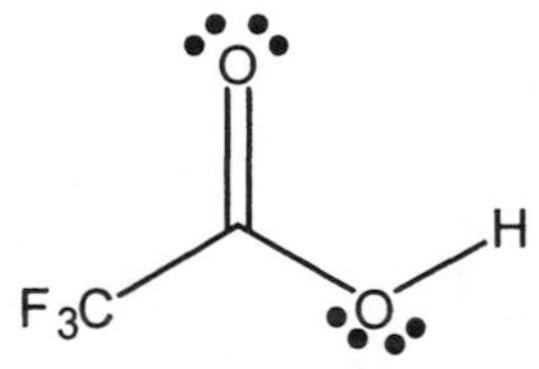

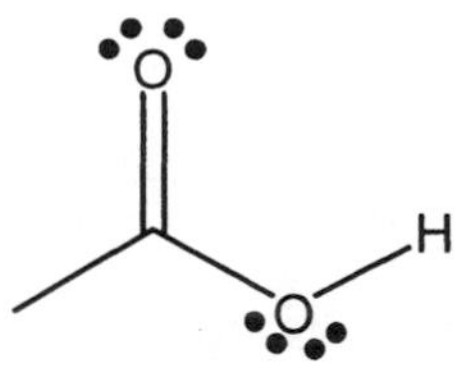

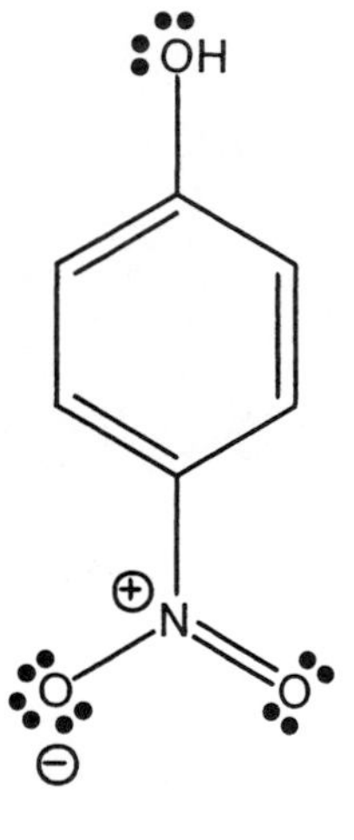

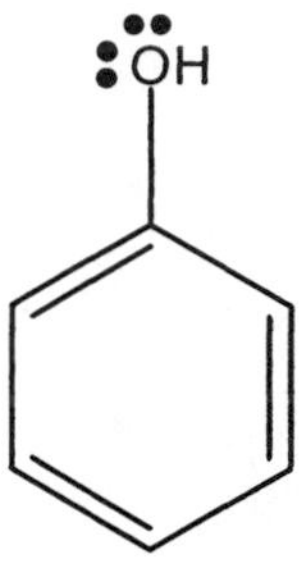

HO ⊕NH3   H2O⊕ NH2

F3C O H   O H

OH   OH

⊕N O⊖ O

ChemActivity 11

Part A: Orbitals and σ Bonds

(What exactly is an orbital?)

Model 1: Orbital vs. Orbit

So far in this course we have shown a hydrogen atom as a proton of +1 charge being orbited by an electron (-1 charge).

Figure 1a: Electron Orbit

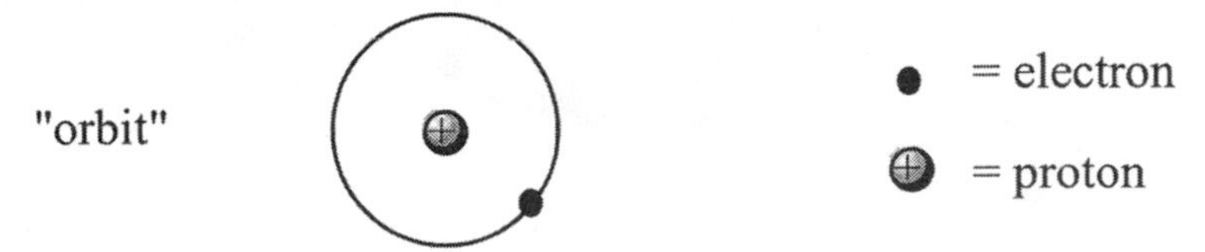

A more useful picture: the electron of hydrogen moves randomly at the speed of light in a 3D region of space near the nucleus. The region of space where the electron spends 90% of its time is designated as that electron's **orbital**.

A good way to think about an orbital is as **the home base territory of an electron** (or electron pair) or the area that contains most of the **electron cloud**.

Figure 1b: Electron Orbital

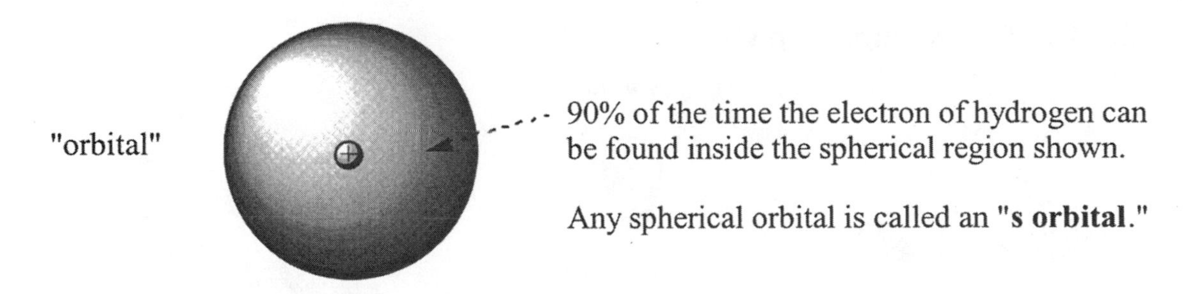

Critical Thinking Questions

1. Draw a circle on the diagram above indicating a spherical territory in which the electron could be found...

   a) 95% of the time.

   b) 85% of the time.

Model 2: *p* Orbitals

Most students find it easy to visualize a "s orbital" because it is spherical–but even s orbitals are strange. For example: Why doesn't the electron crash into the nucleus? Unfortunately, this and certain other questions will remain unanswered in this course.

Stranger than an s orbital is a "***p* orbital.**"

Figure 2a: Lobes of the *p* Orbital.

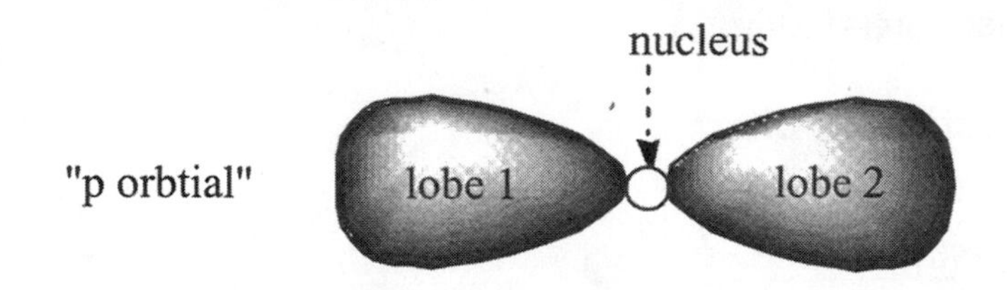

A *p* orbital is shaped like a dumbbell. Lobe 1 + lobe 2 make up a single *p* orbital. An electron whose territory is a *p* orbital can be found (90% of the time) in <u>one lobe or the other</u>, and can move between the two lobes even though the lobes do not touch. (This is allowed according to the strange laws of a branch of physics called quantum mechanics.)

There are three kinds of *p* orbitals: p_x, p_y and p_z. They are identical in size and shape. The only difference is their orientation in space, which is along the three perpendicular axes of 3D space: x, y and z.

Figure 2b: Three Kinds of *p* Orbitals.

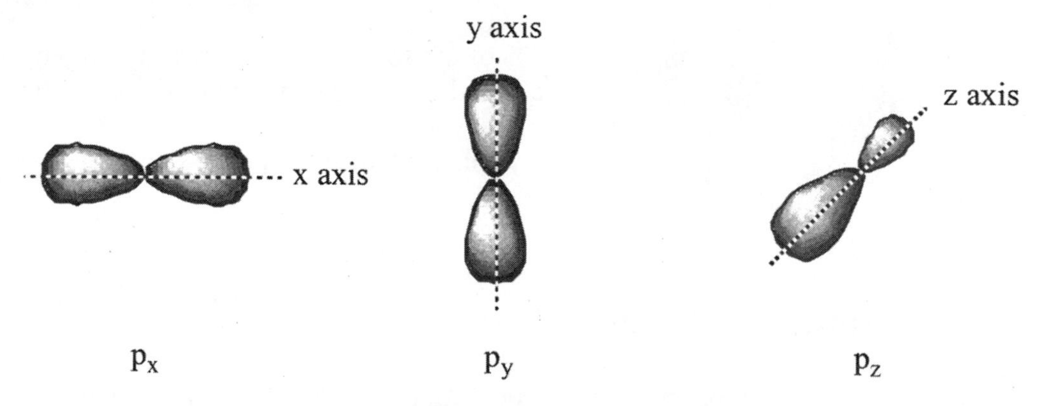

Critical Thinking Questions

2. Use an atom with 6 holes and the colored orbital lobes to construct a single model that has a p_x, a p_y and a p_z orbital (colored lobe = ½ of a *p* orbital).

Model 3: Orbital Representation of a Shell

Together with an *s* orbital p_x, p_y and p_z form a "shell." For example 2s, $2p_x$, $2p_y$ and $2p_z$ make up "shell #2." (We have previously represented shell #2 as an orbit that can hold a maximum of 8 electrons.)

Figure 3a: Orbital and Orbit Representations of Shell #2.

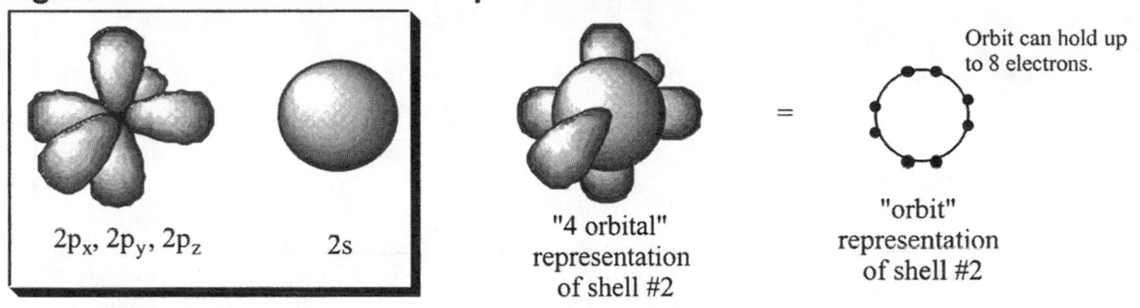

Critical Thinking Question

3. Based on information in Figure 3a, what is the maximum number of electrons that can fit in a single orbital?

Information

An electron in a 2s orbital is lower in potential energy than an electron in a 2p orbital because the territory of an *s* orbital is closer (on average) to the nucleus than that of a *p* orbital.

So far in this course we have ignored this energy difference and assumed that, all electrons in Shell #2 have the same potential energy.

Figure 3b: Potential Energies of Orbitals in Shell #2.

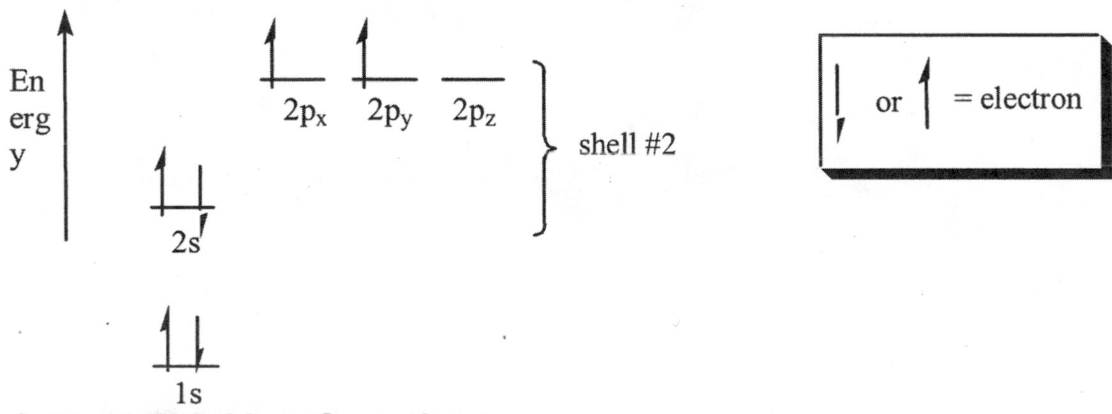

Critical Thinking Questions

4. What neutral atom is represented by the valence electron configuration shown in Figure 3b?

Model 4: Orbital Description of Bonding

Two overlapping, half-filled 1s orbitals form a σ **bond.** (All single bonds are σ bonds.)

Figure 4a: Four Representations of a σ ("sigma") Bond

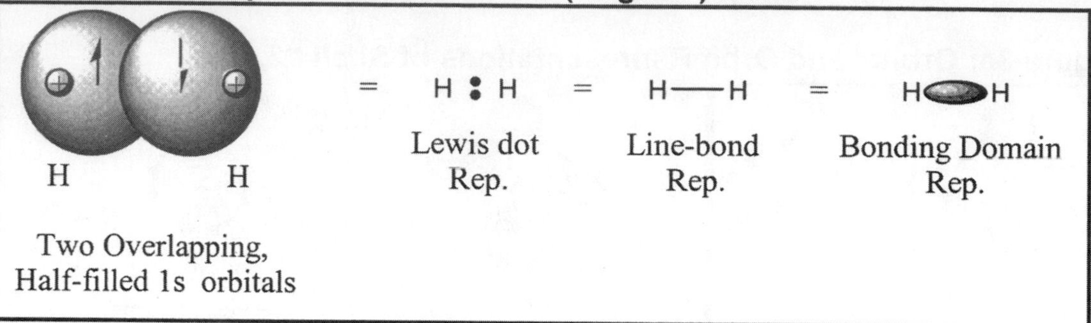

Note: σ ("sigma") is the Greek letter for "s."

Model 5: Other Ways To Make a σ Bond

Though the bond in H_2 is the classic example of a σ bond, in fact, a σ **bond** is formed when any two orbitals overlap to put an electron pair between two nuclei in a way that **does not restrict bond axis rotation**.

Figure 5: σ Bonds Involving *p* Orbitals

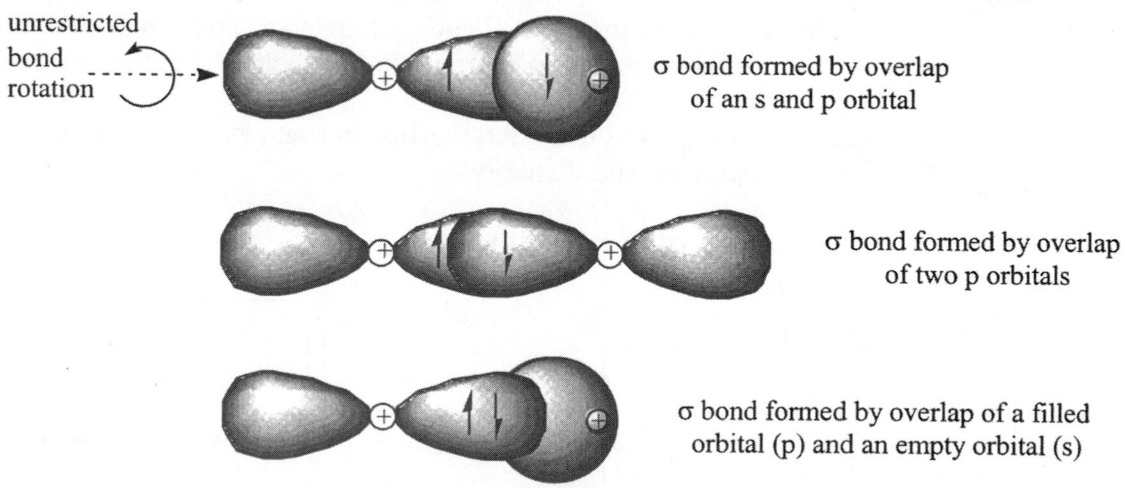

Model 6: Orbital Description of Bonding in Ammonia

A description of ammonia (NH_3) using the **standard orbital set: 2s, $2p_x$, $2p_y$ and $2p_z$** forces us to predict a HNH bond angle of 90° for this molecule (*see below*).

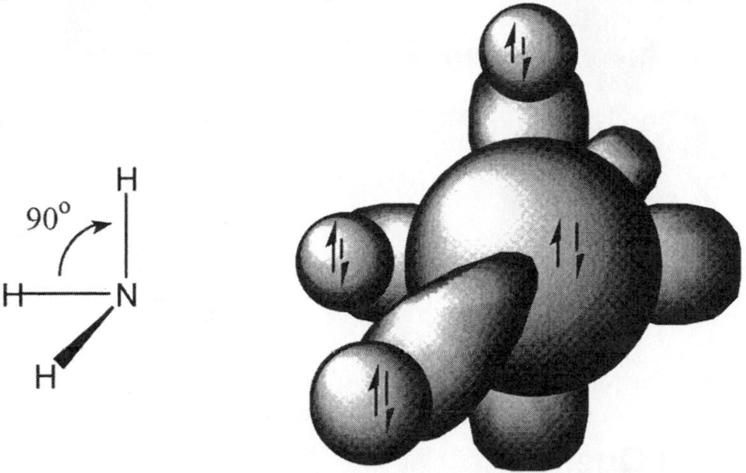

The experimentally determined HNH bond angle of NH_3 is about 109°. To explain this and other bond angles, chemists have invented alternate orbital sets called **hybrid orbital sets**.

Demonstration 1: Derivation of the 109.5° Hybrid Orbital Set

- Take four orbitals: 2s (red clay), $2p_x$, $2p_y$ and $2p_z$ (each green clay).
- Mix them together.
- Cut them into 4 equal pieces.

Each piece is a new orbital that is a **hybrid** (mix) of *s* and *p* orbitals. The four new orbitals are identical and are spread out equally around the center-point. **The angles between them are 109.5°.**

Critical Thinking Questions

5. Consider any one of the four identical hybrid orbitals in the 109.5° set.
   a) What fraction of the clay in this hybrid orbital was originally red (came from the 2s orbital)?

   b) What fraction of the clay in this hybrid orbital was originally green (came from any of the 2p orbitals)?

   c) Is "*s*(1/4)–*p*(3/4) hybrid orbital " a good name for this orbital? Explain.

   d) This hybrid orbital is actually called a sp^3 hybrid orbital. Explain the "sp^3".

Model 7: Hybrid Orbital Picture of Bonding in Ammonia

Figure 7a: sp^3 Hybridized Nitrogen

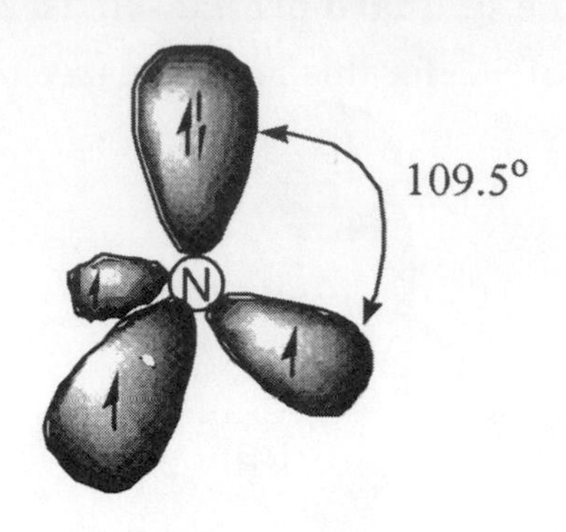

109.5° Hybrid Orbital Set for Nitrogen

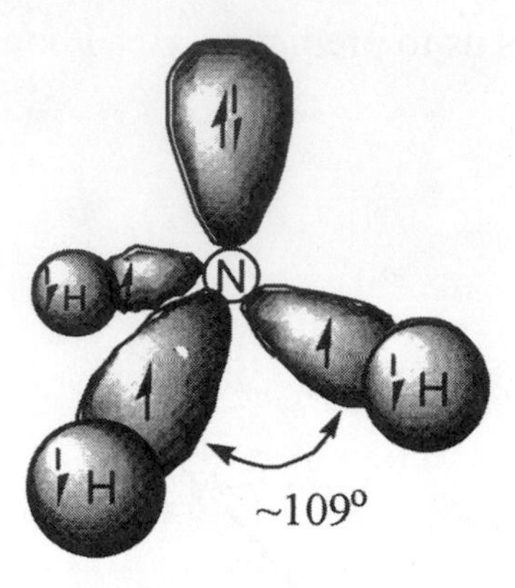

Ammonia Molecule

Critical Thinking Questions

6. Build an orbital model of nitrogen (like the one above, left) using the orbital lobes provided (1 colored lobe = the main part of a hybrid orbital).

   a) According to Model 7, what two orbitals are involved in each N–H sigma bond of NH_3?

   b) What is the **shape** of ammonia (assuming an sp^3 hybridized nitrogen)?

7. The diagram below left shows the energies and electron occupation of the valence orbitals of nitrogen using the standard set (2s, $2p_x$, $2p_y$ and $2p_z$). Complete the diagram below right by adding electrons to the energy diagram of sp^3 hybridized N.

Figure 7b: Electron Configuration of Nitrogen (standard and sp^3 hybridized)

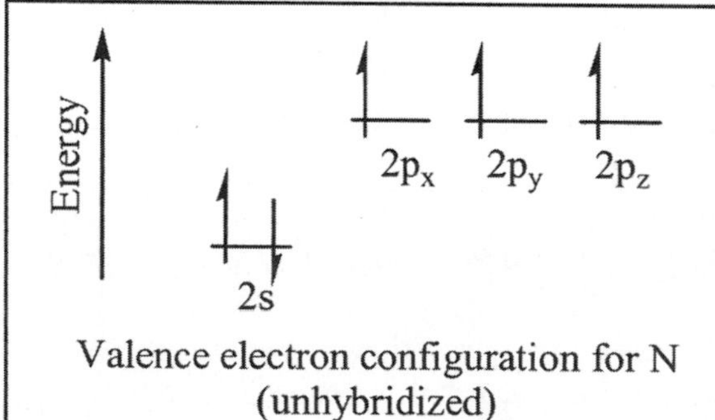

Valence electron configuration for N (unhybridized)

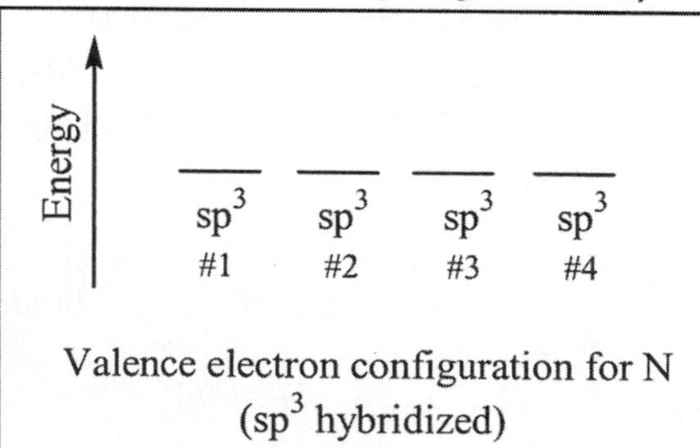

Valence electron configuration for N (sp^3 hybridized)

Demonstration 2: Derivation of the 120° Hybrid Orbital Set

A 109.5° hybrid orbital set was needed to explain the 109.5° bond angles of NH_3, H_2O, etc. Similarly, a 120° hybrid orbital set is needed to explain 120° bond angles such as formaldehyde (*shown below*).

formaldehyde

O, C, H, H, 120°, 120°

- Take **three** orbitals: 2s (red clay), $2p_x$ and $2p_z$ (each green clay).'
- Mix them together.
- **Leave $2p_y$ alone!!**
- Cut the mixed lump of clay into three equal pieces.

Each piece is a new hybrid orbital. The three new orbitals are spread out equally around the center-point. This leaves 120° between each one.

Figure 7c: Derivation of the 120° Hybrid Orbital Set

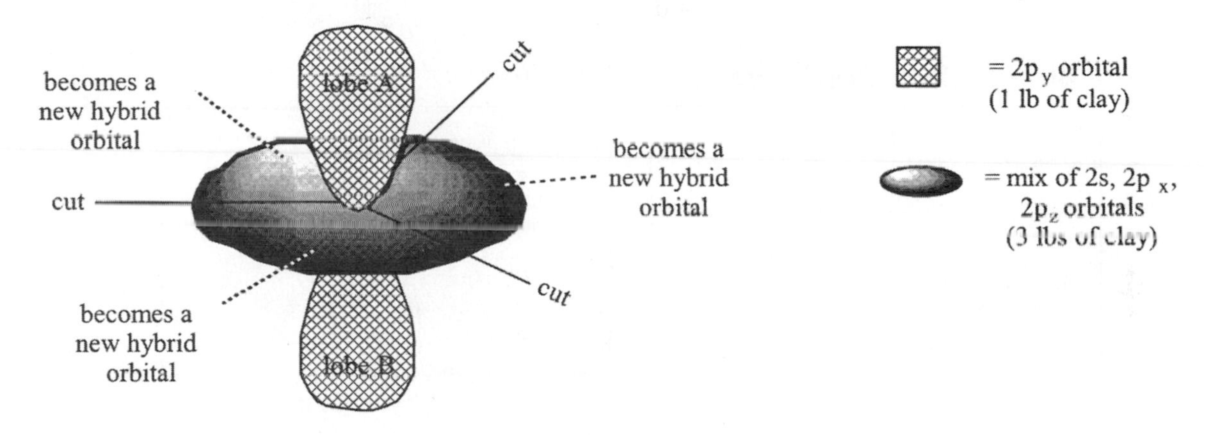

Critical Thinking Questions

8. Build an orbital model of the 120° hybrid orbital set using a 5-hole model piece and the orbital lobes provided. One colored lobe = ½ of a *p* orbital; or most of a hybrid orbital. (Note: a hybrid orbital has a second, very small lobe that we are ignoring.)

9. The 109.5° hybrid orbital set is made up of 4 identical sp^3 hybrid orbitals and zero *p* orbitals. An atom employing this orbital set for bonding is called "**sp^3 hybridized**." Complete the following sentence about the 120° hybrid orbital set:

The 120 ° hybrid orbital set is made up of ___ identical _____ hybrid orbitals and _____ *p* orbitals. An atom employing this orbital set for bonding is called "_____ **hybridized**."

Part B: Orbitals and π Bonds

(How can we explain double bonds using orbitals?)

Model 8: π ("pi") Bond

A π ("pi") bond is formed when two half-filled *p* orbitals are parallel to one another as shown below. Side-to-side overlap of the *p* orbital lobes forms one π **bond**.

Figure 8a: Overlap of Two Half-filled *p* Orbitals

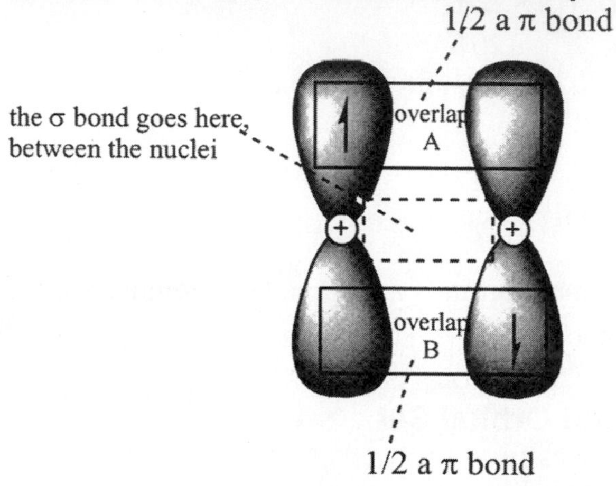

⊕ = nucleus

overlap A + overlap B = 1 π bond

Figure 8b: Electron Configuration of Oxygen

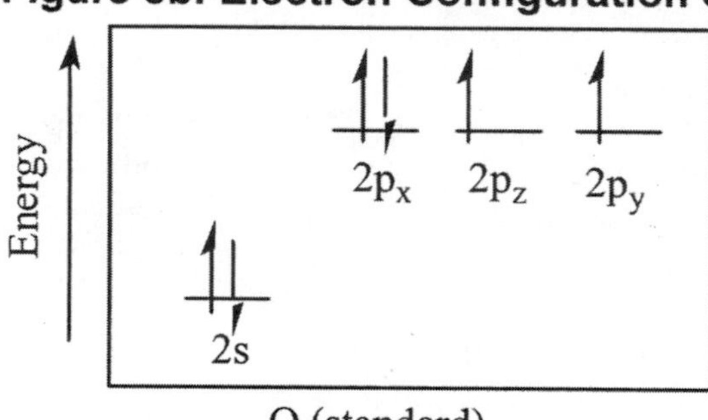

O (standard)

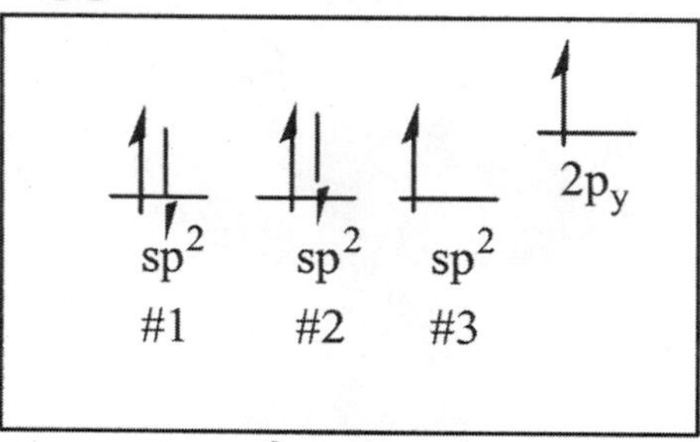

O (sp^2 hybridized)

Model 9: Orbital Description of a Double Bond

A double bond consists of one σ overlap and one π overlap.

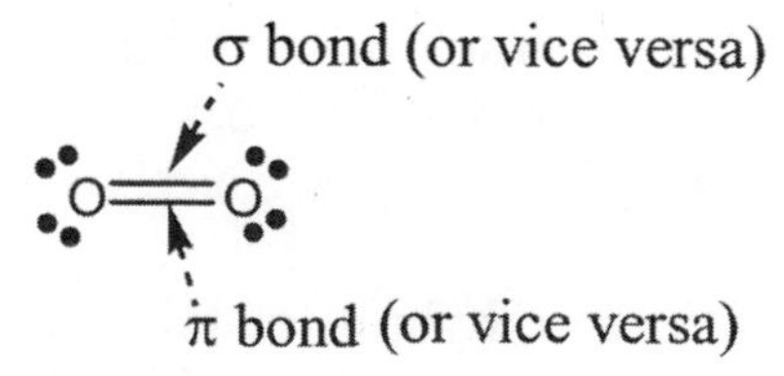

Lewis Drawing of O_2

Orbital Drawing of O_2

Critical Thinking Questions

10. In the drawing above, label **each** sp^2 hybrid orbital, *p* orbital, σ bond and π bond.
11. Label each orbital above as **full** (2 electrons); **half-full** (1 electron); or **empty**.

12. Below is an orbital drawing of ethyne H—C≡C—H.

Orbital Drawing of

a) A triple bond consists of one σ bond and two π bonds. Label the σ overlap and each of the two π overlaps holding the carbons together in the drawing above.

b) Label each C–H σ bond in the drawing above.

c) Label **each** hybrid orbital, *p* orbital, and 1s orbital in the drawing above.

d) How many *p* orbitals are there on a single carbon of ethyne?

e) How many hybrid orbitals are there on a single carbon of ethyne?

f) Which of the following is the best name for a hybrid orbital found in ethyne:

A. an *s* ¼ *p* ¾ hybrid orbital (a.k.a. an $\mathbf{sp^3}$ hybrid orbital)

B. an *s* 1/3 *p* 2/3 hybrid orbital (a.k.a. an $\mathbf{sp^2}$ hybrid orbital)

C. an *s* ½ *p* ½ hybrid orbital (a.k.a. an ***sp*** hybrid orbital)

13. Explain why the representation of ethene on the left is more accurate than the one on the right.

14. Explain why *cis* and *trans* 2-butene cannot inter-convert without breaking the π portion of the double bond.

cis-2-butene *trans*-2-butene

Exercises for Part B

1. Summarize how one determines the hybridization (also called **hybridization state**) of an atom in a molecule.

2. Explain what is wrong with each of the following statements.
   a) "A π bond is a double bond."

   b) "A π bond consists of four electrons, one in each of the four *p* orbitals involved in the π bond."

   c) "A π bond is twice as strong as a σ bond because it consists of two orbital overlaps instead of just one."

3. Explain why the representation of CH_2CCH_2 on the right is more accurate. Hint: assume a <u>*p* orbtial cannot be involved in two different π bonds at the same time</u>, and a hybrid orbital cannot be involved in a π bond and a σ bond.)

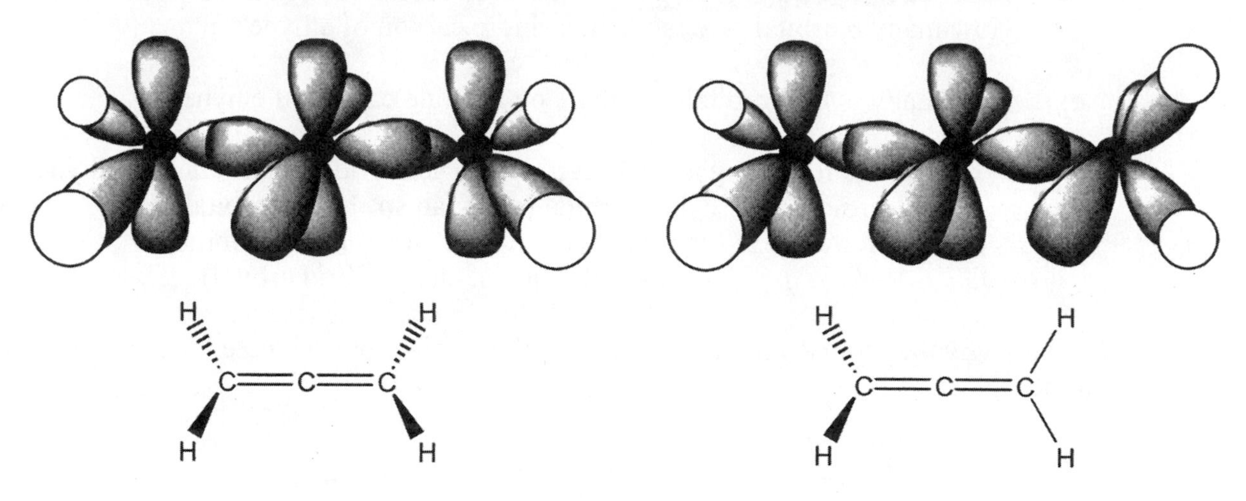

4. Use the play-dough ball analogy to explain how a *sp* hybrid orbital set can be made from the standard orbital set: 2s, $2p_x$, $2p_y$ and $2p_z$. (Assume each orbital is 1 pound of clay.)

5. Consider the pK_a's listed below:

H—C≡C—H pK_a 25

$H_2C=CH_2$ pK_a 44

$H_3C–CH_3$ pK_a 50

conjugate bases

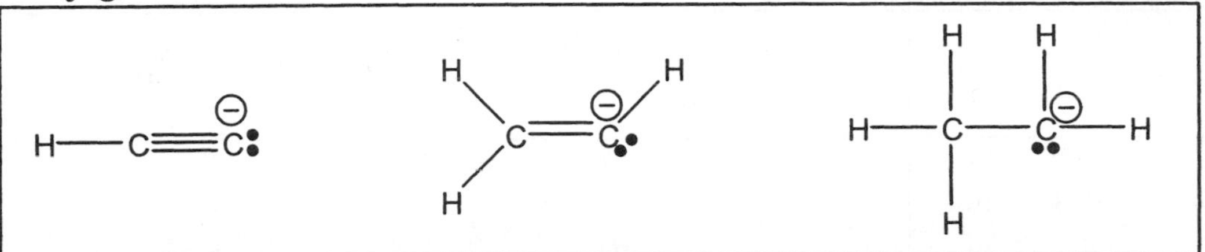

a) Based on the pK_a data, rank the conjugate bases according to the potential energy of their lone pair of electrons (label them **lowest P.E.**, **middle P.E.**, **highest P.E.**).

b) For each conjugate base, indicate the type of orbital that the lone pair of electrons resides in.

c) Indicate the % *s* character ("red clay") of each orbital holding a lone pair.

d) Which is lower in P.E., a *s* orbital or a *p* orbital.

e) Construct an explanation for the pattern in the pK_a data above.

6. For the following structures, indicate…

a) the hybridization state of each non-hydrogen atom (*sp*, sp^2 or sp^3)

b) the number of *p* orbitals on each (non-hydrogen) atom.

c) for each *p* orbital, if it is "**full**", "**empty**" or "**half-full and involved in a π bond**."

d) the type of orbital that each lone pair resides in.

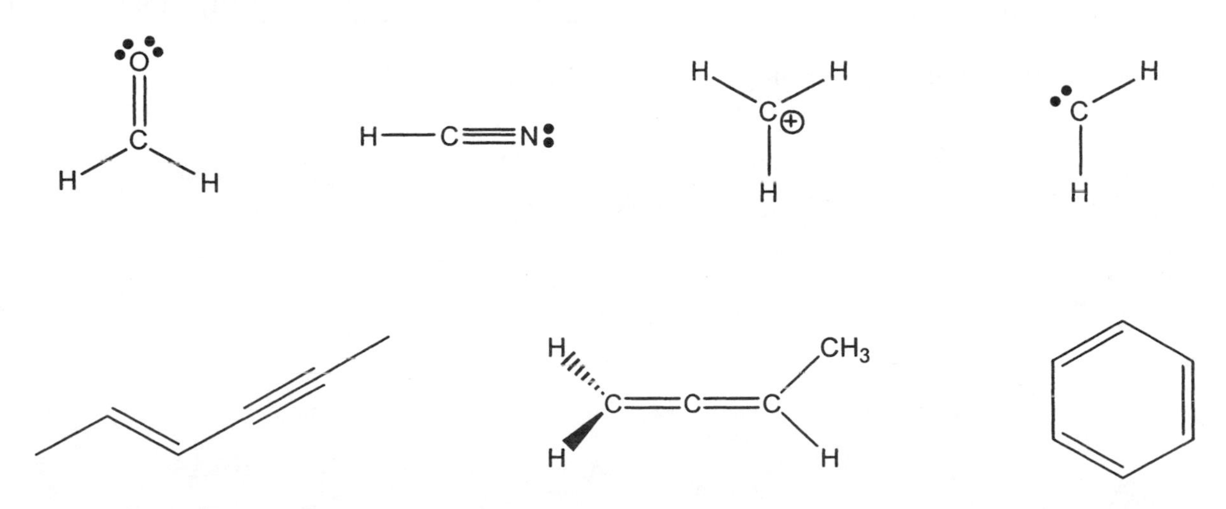

7. You should memorize the names of the three kinds of hybrid orbitals (sp^3, sp^2 and *sp*) and know all the component orbitals of the three hybrid orbital sets (109.5°, 120° and 180° hybrid orbital sets). Complete the following tables.

Table B: Syntax for Naming of Hybrid Orbitals

Old Name of Orbital	Hybrid Orbital is a Mixture of...	New Name of Orbital
s(1/2)–*p*(1/2) orbital	1 part *s* orbital; 1 part *p* orbital	"s^1p^1 orbital" or "*sp* orbital"
s(1/3)–*p*(2/3) orbital		or
s(1/4)–*p*(3/4) orbital		or

Table C: Summary Description of Hybrid Orbital Sets

Name of Hybrid Orbital Set	# of *p* Orbitals left untouched in Set	# of Hybrid Orbitals in Set	Angle Between Hybrid Orbitals in Set	Example of Atom Employing Set (hybridization state)
sp or 180° hybrid orbital set				C in HCCH or Be in BeH_2 (*sp* hybridized)
sp^2 or 120° hybrid orbital set				C in H_2CCH_2 or B in BH_3 (sp^2 hybridized)
sp^3 or 109.5° hybrid orbital set				C in CH_4, N in NH_3, O in H_2O (sp^3 hybridized)

8. Read the assigned pages in your text and do the assigned problems.

ChemActivity 12

Part A: One-Step Nucleophilic Substitution

(What are the characteristics of a good leaving group?)

Model 1: One-Step Nucleophilic Substitution Reaction

I. $I^- + CH_3-Br \rightleftharpoons I-CH_3 + Br^-$

II. $H-O^- + CH_3-O-SO_2-CH_3 \rightleftharpoons H-O-CH_3 + {}^-O-SO_2-CH_3$

Critical Thinking Questions

1. The two reactions in Model 1 are substitution reactions. In each, an "incoming group" displaces a "leaving group."
   a) Circle each "incoming group" among the reactants in Model 1.
   b) Put a box around each leaving group among the reactants in Model 1.
   c) Can leaving groups and incoming groups consist of more than one atom?

2. Add a δ+ and a δ- to the carbon containing reactants in Model 1 so as to indicate which way the C–LeavingGroup bond is polarized.

3. Use curved arrows to illustrate a <u>one-step</u> mechanism that will accomplish each substitution reaction in Model 1. (Hint: you will need to draw more than one arrow for each reaction.)

4. "Phile" = "lover of." (For example: an "anglophile" = a lover of anglo culture).
   a) The charge on a nucleus is (+ or –) [circle one].
   b) A **nucleophile** is an atom that loves to make a bond to another atom with excess (+ **charge** or – **charge**) [circle one].
   c) For each reaction in Model 1, write the word **nucleophile** below the atom among the reactants that is acting as a nucleophile.
   d) Label each atom among the reactants in Model 1 that is acting as an **electrophile** ("electron lover") and explain your reasoning.

Model 2: Transition State

The **transition state** (t.s.) of a reaction is the highest potential energy species between the reactant (R) and the product (P). The diagram below shows two possible transitions states for Reaction I in Model 1. Each is labeled with a "double dagger" (‡).

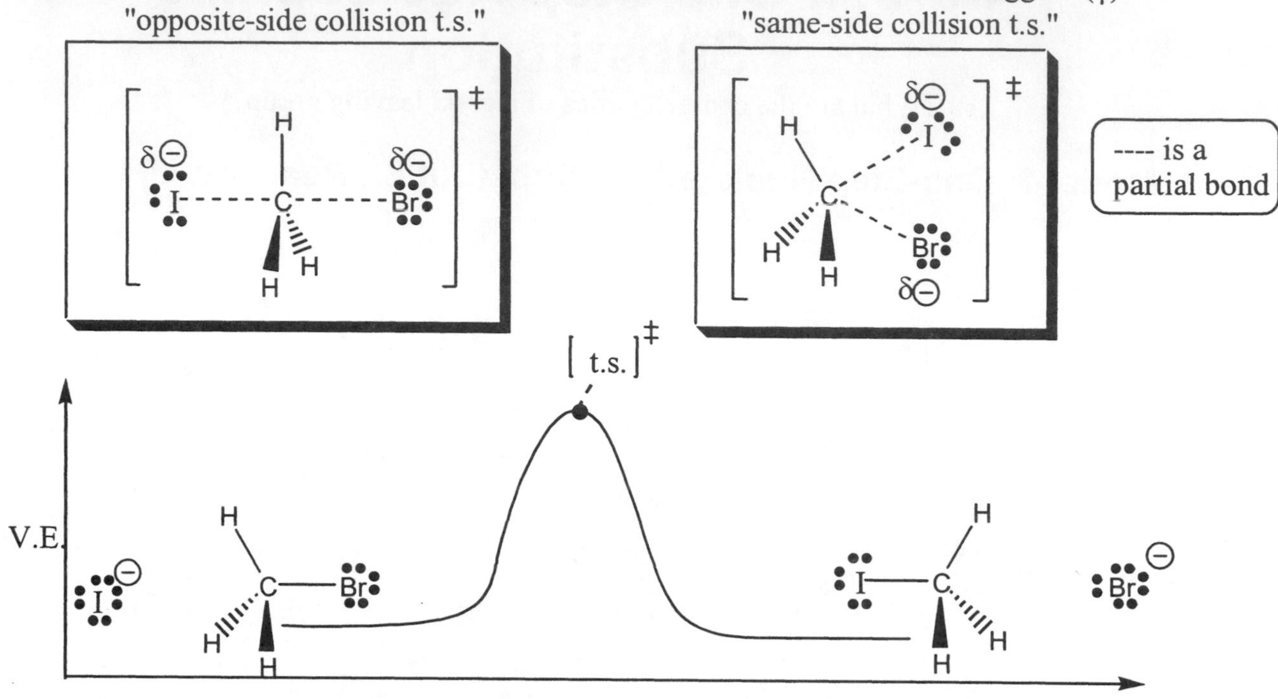

Reaction Progress

Critical Thinking Questions

5. Both the **nucleophile** and the **leaving group** have a partial negative charge in the transition state. Recall that like charges repel one another. Circle the transition state in Model 2 that minimizes the unfavorable repulsion between these two species.

6. You are having holiday dinner at your house, but your parents don't get along with your in-laws. You want to have one set of parents in the house at all times-otherwise it won't feel like a holiday-but never both sets! Which of the following scenarios minimizes unfavorable interactions between the two sets of parents?
   I. When the in-laws arrive at the front door, your parents exit out the back.
   II. When the in-laws arrive at the front door, your parents exit out the front door, passing the in-laws on the porch.

7. The stereochemistry of the product in the following reaction proves that the opposite-side-collision transition state is correct. Cross out the incorrect transition state in Model 2 and **explain how the product below proves which transition state is correct**.

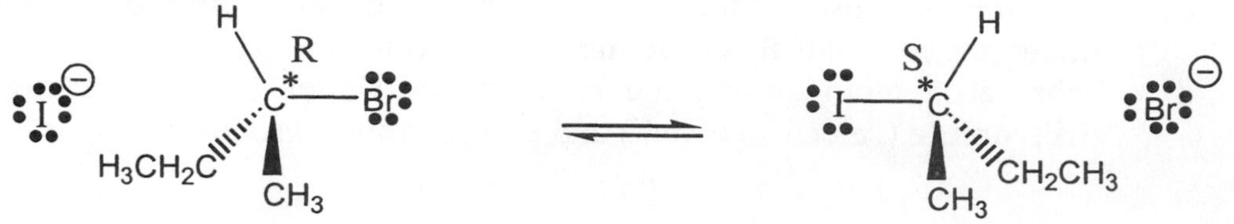

8. Explain how the changes at the electrophilic carbon in a one-step nucelophilic substitution reaction resemble the inverting of an umbrella in the wind.

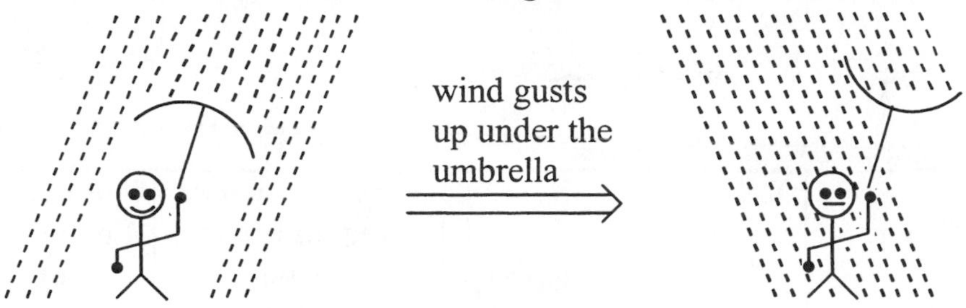

Model 3: Activation Energy (E_{act})

An acid-base reaction is very fast, partly because the only atom that needs to move is a tiny H^+. An H^+ can be transferred very quickly. Most other reactions including nucleophilic substitutions are much slower partly because elaborate atom movements are required to get to the transition state. (*For example:* the flipping of the bonds like an umbrella inverting in the wind). This takes energy and translates into high activation energies.

Reaction rate depends on activation energy (E_{act})

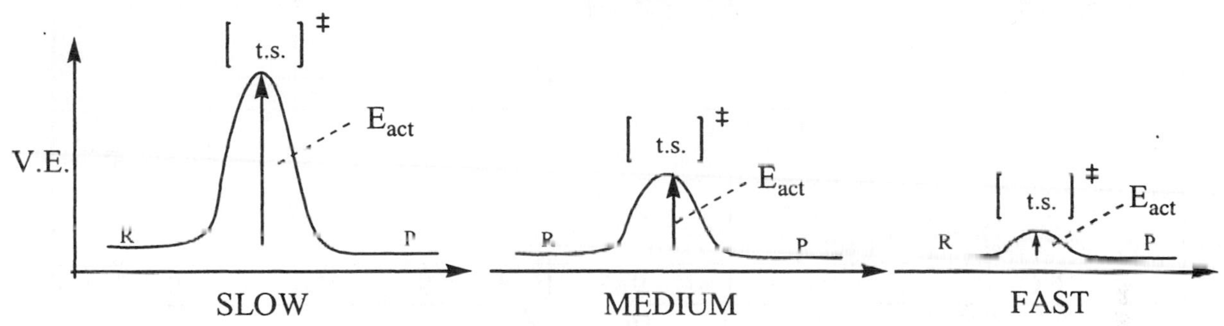

Each of the following lowers the activation barrier (lowers E_{act}) and therefore speeds up the rate of a substitution reaction.

- **Use of a good leaving group**
- **Use of a good nucleophile**
- **Use of a good electrophile**

(The remainder of this activity will be spent identifying the characteristics of each.)

Table 3: Rates of Substitution Reactions with Various Leaving Groups (X)

$$I^- + CH_3X \rightleftharpoons ICH_3 + X^-$$

Leaving Group (—X)	Relative Rate of Rxn	pK_a of Conj. Acid (H—X)	Approx. Energy Req. to Break C–X bond (in pKa units)	P.E. of New Lone Pair on X^- (in pKa units)
—I	10^7	-5	≈ 0	≈ 0
$R-S(=O)_2-O-$	10^6	-5	≈ 0	≈ 0
—Br	10^5	-5	≈ 0	≈ 0
—Cl	10^4	-2	≈ 0	≈ 0
$R(H)O^+-$ or H_2O^+-	10^4	-2	≈ 0	≈ 0
—F	1	3	3	3
$R-C(=O)-O-$	0.1	5	5	5
NC—	0.001	9	9	9
HO— or RO—	10^{-5} (v. slow)	16	16	16
H_2N—	No reaction	35	35	35
H_3C—	No reaction	50	50	50

Critical Thinking Questions

9. Add curved arrows to the reaction at the top of the page showing one-step nucleophilic substitution.
10. Label appropriate species in Table 3 with the words "**best leaving group**" and "**worst leaving group**" and explain your reasoning.
11. According to the information in Table 3…
    a) A good leaving group initially has **a strong bond** or **a weak bond** [circle one] to the electrophilic carbon.
    b) A good leaving group requires a **large** or **small** [circle one] amount of energy to break off from the electrophilic carbon.
    c) A good leaving group is **stable, happy & low P.E.** or **unstable, unhappy & high P.E.** [circle one] on its own after it leaves the electrophilic carbon.

Part B: One-Step Nucleophilic Substitution

(What are the characteristics of a good nucleophile and electrophile?)

Model 4: Effect of Nucleophile on Rate of Substitution

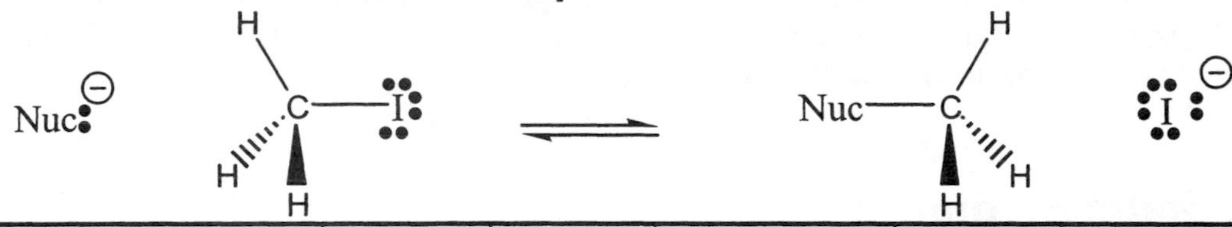

Nucleophile (Nuc)	Conj. Acid (H-Nuc)	Relative Rate of Rxn	pKa of Conj. Acid H-Nuc	Approx. P.E. of Nuc	Energy Released (ΔH_{rxn}) When Nuc–C Bond Forms
$H_3C:^-$		10^6	50	50	50
$H_2N:^-$		10^5	35	35	35
HO^- or RO^-		10^4	16	16	16
$R{-}C({=}O){-}O^-$		100	5	5	5
F^-		80	3	3	3
$H{-}O{-}R$ or $H{-}O{-}H$		1	-2	≈ 0	≈ 0
$O{=}S({=}O)(R){-}O^-$		0.0001	-5	≈ 0	≈ 0

Critical Thinking Questions

12. Fill in Column 2 in Model 4 with the conjugate acids (H-Nuc).

13. Add curved arrows to the reaction above the table. Label the electrophilic C with a "δ+" and circle and label the "**leaving group**" attached to this electrophilic C.

14. Based on the rate data, label appropriate species in Model 4 with the words "**best nucleophile**" and "**worst nucleophile**."

15. According to the data in Model 4, which ONE of the following statements is true?
    - I. A nucleophile must have a free pair of electrons which are available to make a new bond.
    - II. A nucleophile must have a negative charge and a free pair of electrons which are available to make a new bond.

16. According to the information in Model 4…
    a) A good nucleophile is **stable, happy & low P.E.** or **reactive, unhappy & high P.E.** [circle one] on its own before it makes a new bond to carbon.

    b) A good nucleophile releases **a large amount of energy** or **a small amount of energy** [circle one] when making a new bond to C.

    c) A good nucleophile makes a new bond to carbon that is **strong and hard to break** or **weak and easy to break** [circle one].

Model 5: Steric Hindrance

Figure 5a: Substitution Reactions with Different Electrophiles

Steric hindrance= unfavorable electron-electron repulsion that results when molecules or parts of molecules are forced too close together.

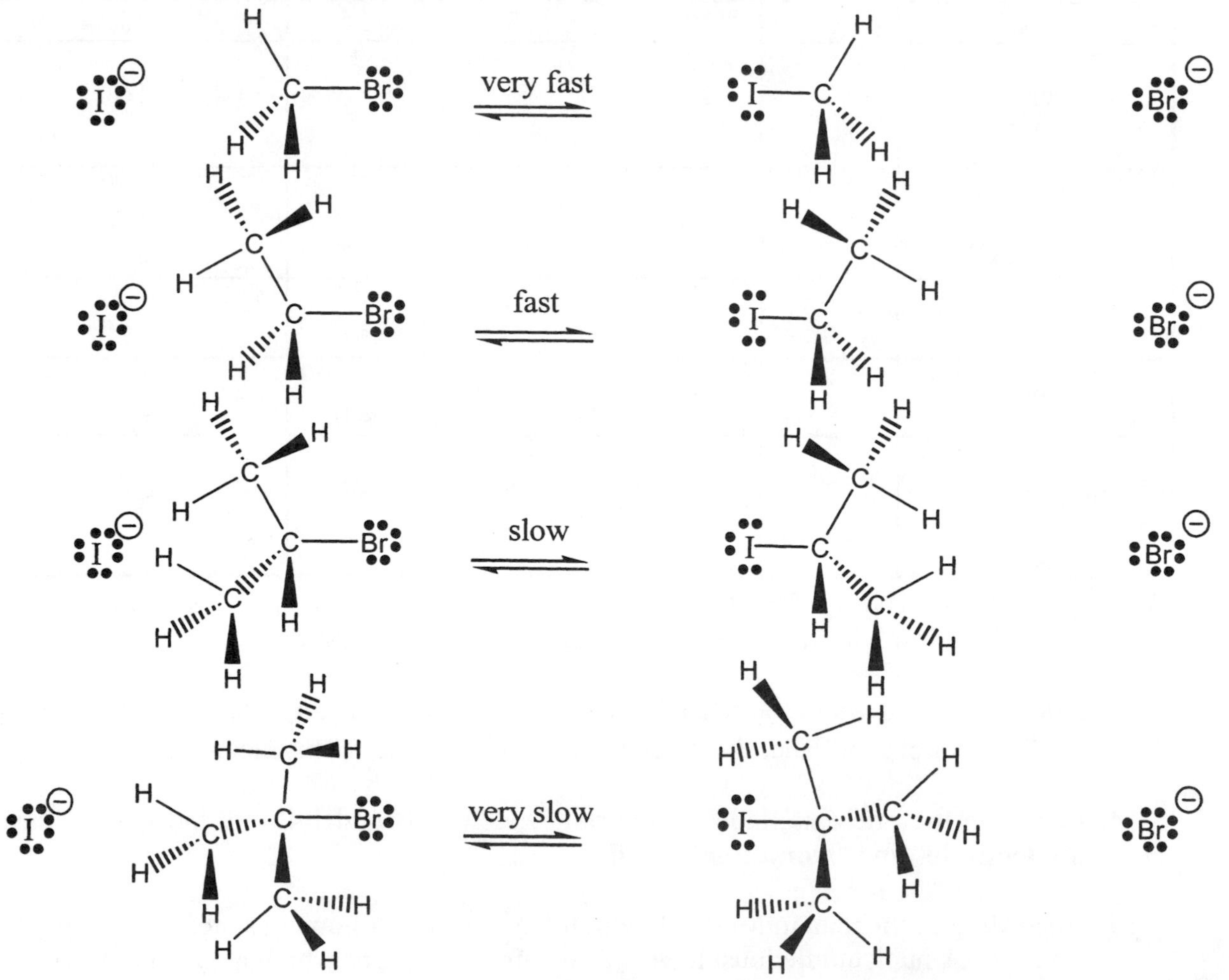

very fast

fast

slow

very slow

Figure 5b: Methyl, Primary, Secondary, Tertiary and Quaternary Carbons

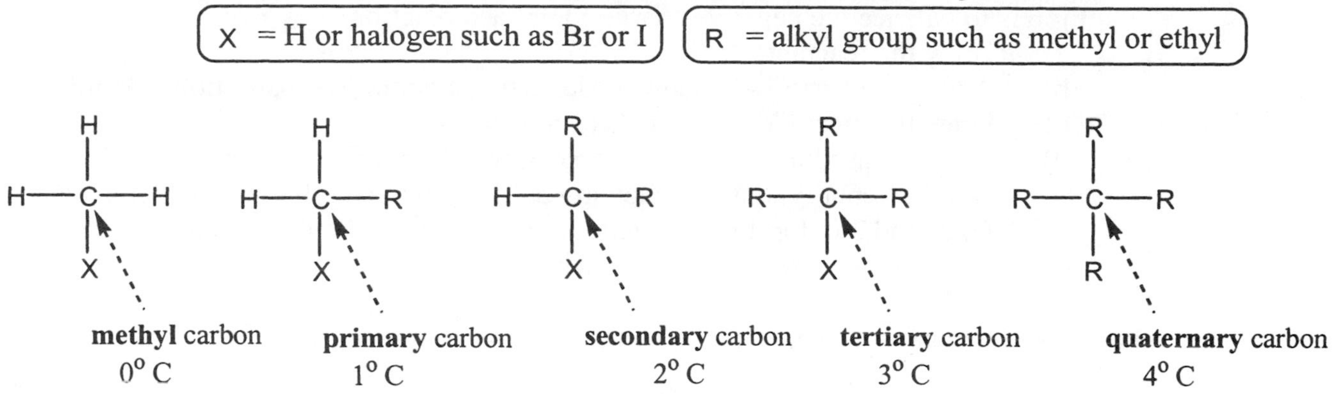

X = H or halogen such as Br or I | R = alkyl group such as methyl or ethyl

methyl carbon 0° C | **primary** carbon 1° C | **secondary** carbon 2° C | **tertiary** carbon 3° C | **quaternary** carbon 4° C

Critical Thinking Questions

17. What is the rule for determining whether a carbon is methyl (0°), primary (1°), secondary (2°), tertiary (3°), or quaternary (4°)?

18. Add curved arrows to illustrate each nucleophilic substitution reaction in Figure 5a.

19. Label each C in Figure 5a that is acting as an electrophile with a 0°, 1°, 2°, 3° or 4°.

20. Construct an explanation for the rate data that appears over each arrow in Figure 5a.

Exercises for Part A

1. Explain why the following are NOT the products you would predict for a one-step nucleophilic substitution reaction, and draw the correct product/s. (Note: Na^+ is never involved in the reaction. Its only job is to balance the negative charge.)

$Na^{\oplus}$ $HO^{\ominus}$ + H_3C, H, CH_2CH_3 C—I ⇌ H_3C, H, CH_2CH_3 C—OH + $I^{\ominus}$ $Na^{\oplus}$

2. The product shown above, (R)-2-butanol, can be made via reaction of (S)-2-iodobutane and NaOH. Draw an energy diagram for this reaction. Include on your energy diagram wedge-and-dash drawings of the reactants, transition state, and products. In your transition state drawing, use a dotted line to indicate a partial bond, and δ– to indicate partial negative charge on an atom.

3. Consider a reaction between NaOH and 2-iodopropane and a reaction between NaOH and 2-chloropropane. (Note: Na^+ is never involved in the reaction. Its only job is to balance the negative charge.) For each reaction…
   a) Draw the reactants.
   b) Add curved arrows to show a one-step nucleophilic substitution reaction.
   c) Draw the most likely products of the reaction.
   d) Predict, based on the data in Table 3, which reaction will be faster.
   e) Draw an energy diagram showing both reactions on the same set of axes. Draw the 2-iodopropane reaction profile with a dotted line and the 2-chloropropane reaction profile with a solid line.

4. Read the assigned pages in the text and do the assigned problems.

Exercises for Part B

5. Consider the following reactions.
   a) Construct an explanation for why Rxn A is slower than Rxn B.
   b) Construct an explanation for why Rxn A is slower than Rxn C.
   c) Construct an explanation for why Rxn A is slower than Rxn D.

A. Cl^- + $(CH_3)_2CHCl$ ⇌ $ClCH(CH_3)_2$ + Cl^- Rel. Rate = 1

B. Cl^- + $(CH_3)_2CHI$ ⇌ $ClCH(CH_3)_2$ + I^- Rel. Rate = 1000

C. HO^- + $(CH_3)_2CHCl$ ⇌ $HOCH(CH_3)_2$ + Cl^- Rel. Rate = 10

D. Cl^- + CH_3CH_2Cl ⇌ $ClCH_2CH_3$ + Cl^- Rel. Rate = 100

6. Each energy diagram below shows two reaction profiles. The dotted line shows the original reaction, the solid line show the reaction after a change has been made. Match each of the following changes with the correct energy diagram below.
   I. Change to a better nucleophile.
   II. Change from a secondary electrophilic C to a primary electrophilic C.
   III. Change to a better leaving group.

- - - - - - original reaction
———— changed reaction

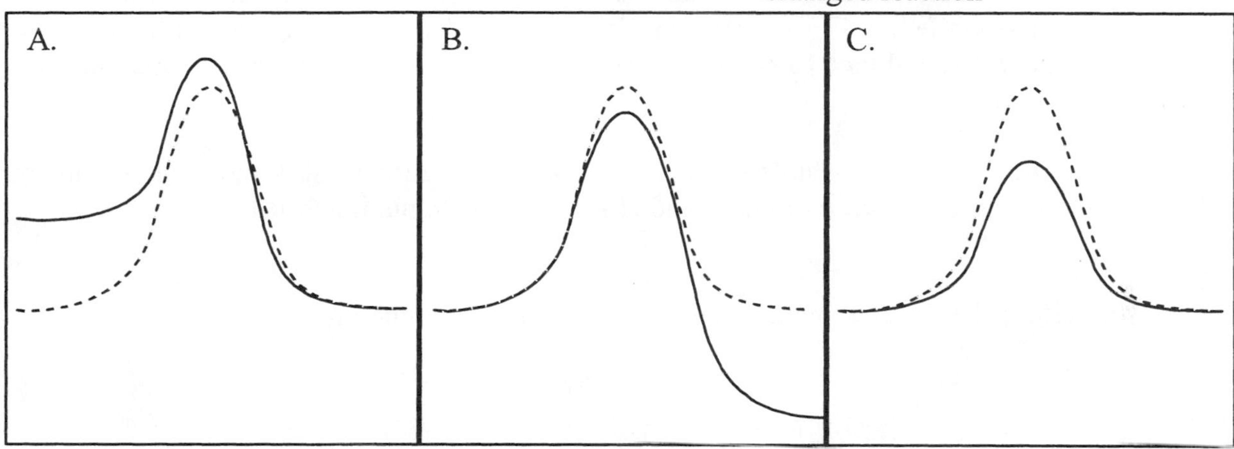

7. Draw the reaction shown in Model 4 using water as a nucleophile. (Hint: one of the products will have a +1 formal charge.)

8. In general a strong base (high P.E.) is a good nucleophile. Construct an explanation for the following exception. Both nucleophiles shown (hydroxide and *tert*-butoxide) are strong bases, but only one of them is good nucleophile.

HO^{-} + H_3C–CH(CH_3)–Cl $\xrightarrow{\text{fast}}$ HO–CH(CH_3)–CH_3 + Cl^{-}

$(H_3C)_3CO^{-}$ + H_3C–CH(CH_3)–Cl $\xrightarrow{\text{very slow}}$ $(H_3C)_3CO$–CH(CH_3)–CH_3 + Cl^{-}

9. There are also weak bases (low P.E.) that are very good nucleophiles. The explanation for this is mostly beyond the scope of this course. Please know that in each column of the periodic table, nucleophilicity increases as you go DOWN. This means that HS^- is a stronger nucleophile than HO^-, and that I^- is the strongest nucleophile among the halogens. (Interestingly, I^- is also the best leaving group among the halogens!)

   In brief, larger atoms have larger orbitals. Nucleophiles with large orbitals are called "soft" bases because they have big fluffy orbitals. Counterexample: A nucleophile like F^- is very small and is a "hard" base. Soft nucleophiles are malleable and tend to make lower energy transition states, thus fast substitution reaction rates.

   You are not responsible for this hard-soft base argument, it is supplied to help you remember the following: I^- and HS^- are the best nucleophiles.

10. Read the assigned pages in the text and do the assigned problems.

ChemActivity 13

Part A: Carbocations

(Which types of carbocations are most likely to form, and which types never form?)

Model 1: Carbocation Formation

- Polar solvents are very good at stabilizing + and – ions.
- In a polar solvent a good leaving group (X) can spontaneously leave, creating a C^+ ion and an X^- ion.
- Even in a very polar solvent such as water the vast majority of the C—X bonds remain intact because breaking the C—X bond is an uphill process.
- A carbon with a + charge is called a **carbocation**
- A carbocation is reactive and short lived because there is a C lacking an octet.

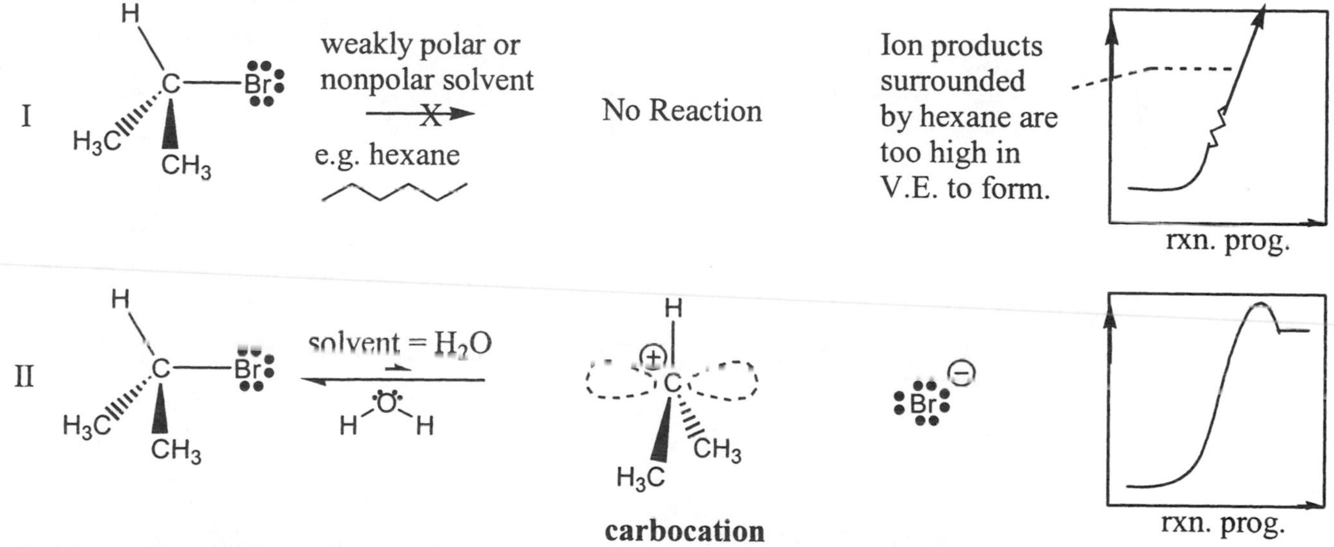

Critical Thinking Questions

1. Consider the carbocation in Model 1.
   a) Is the ∠CCC bond angle closest to: **109.5°**, **120°**, or **180°** [circle one]?

   b) What is the shape of the + charged carbon of a carbocation?

   c) What is the **hybridization state** of carbon in a carbocation?

   d) How many electrons occupy the *p* orbital? Label the *p* orbital shown in Model 1 with the words "**full**," "**empty**" or "**half-full**" as appropriate.

   e) Normally, we expect an atom with a + charge to be stable, unreactive and low in potential energy. A carbocation is short lived, reactive and "unhappy". What does a carbocation lack which might make it "unhappy?"

2. Based on the reaction arrows in Reaction II, there are more (**reactants** or **products**) [circle one] in the reaction mixture at equilibrium?

Model 2: Carbocation Stability

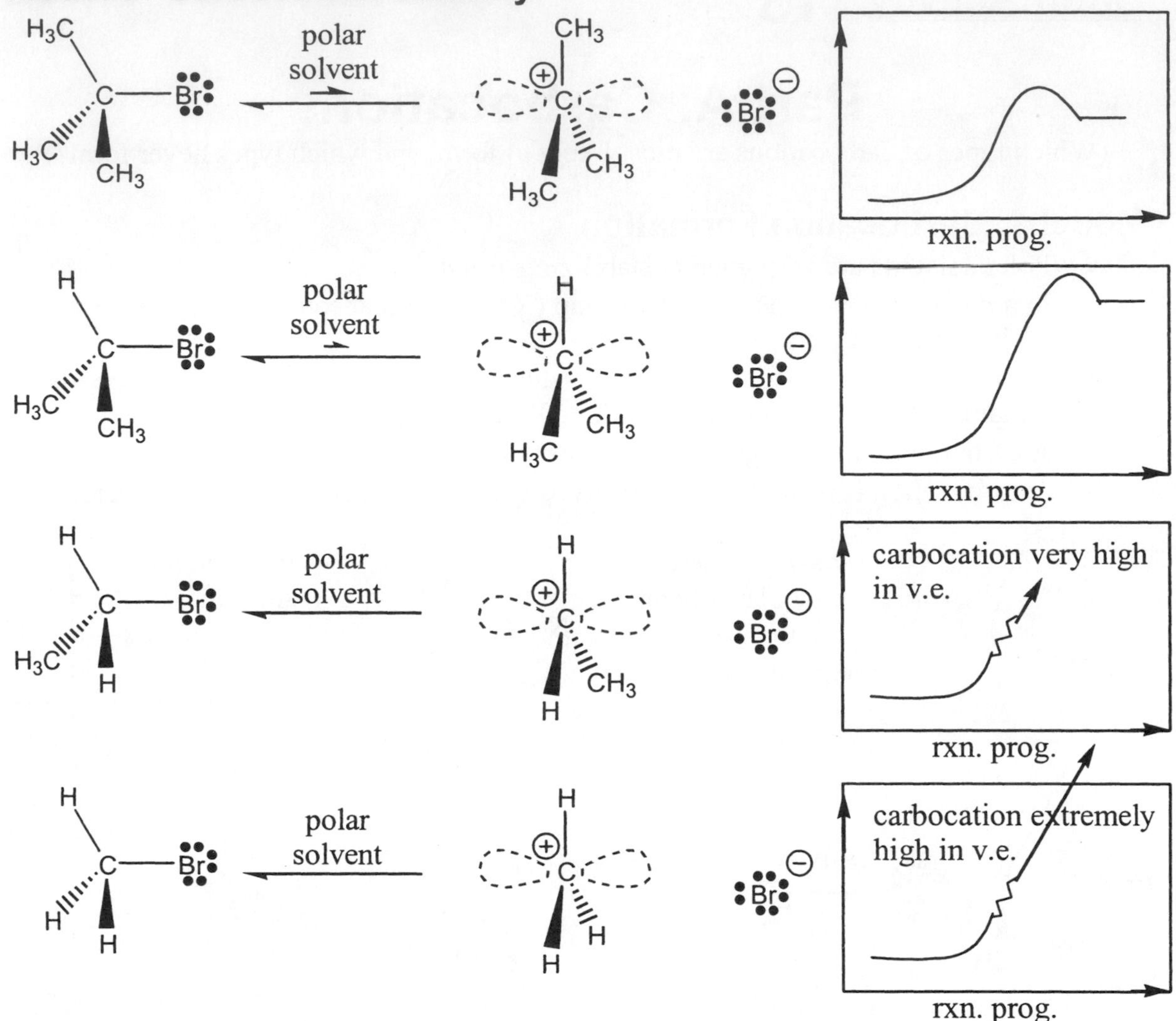

Critical Thinking Questions

3. Label each carbocation in Model 2 as 3°, 2°, 1° or methyl (there is one of each).

4. Based on the energy diagrams above, circle the carbocation in Model 2 that is easiest to form (requires the smallest input of energy to break the C–Br bond)?

5. Based on the reaction arrows above, put an X through the two carbocations in Model 2 that are least likely to form, even in water. (Little convincing evidence for these carbocations has been observed in a chemical reaction.)

6. Choose the statement which best explains why the carbocation you circled in CTQ 4 is closest to having an octet.
   I. Methyl groups on a carbocation inductively donate electrons to the + charged carbon (**methyl groups are electron donating groups**).
   II. Methyl groups on a carbocation inductively withdraw electron density from the + charged carbon (**methyl groups are electron withdrawing groups**).

   Explain your choice.

Model 3: Conjugation (Resonance Stabilized Carbocations)

- The ½ filled *p* orbital of a π bond can donate electron density into a neighboring empty *p* orbital of a carbocation (see below, LEFT).
- This strong orbital overlap is called **conjugation** and can be represented using resonance structures (see below, RIGHT).
- An **allyl carbocation** has a + charge next to a double bond.

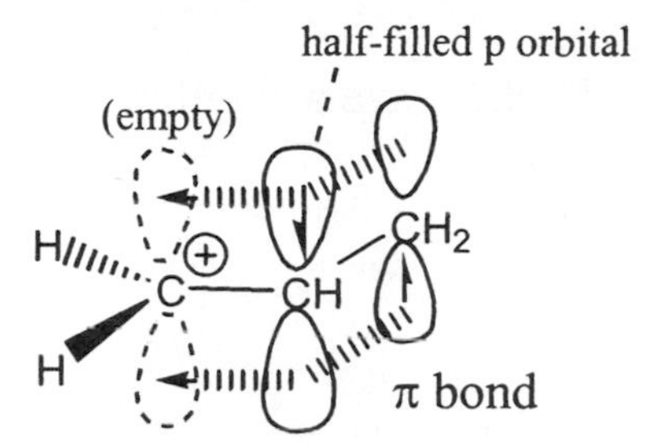

orbital description of allyl carbocation

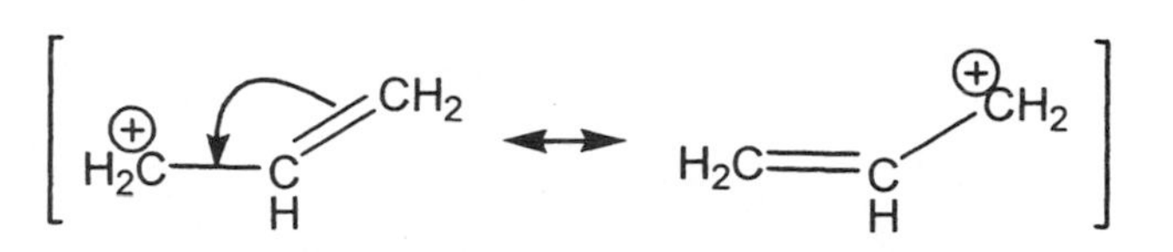

resonance description of allyl carbocation

Critical Thinking Questions

7. Consider the two carbocations below.

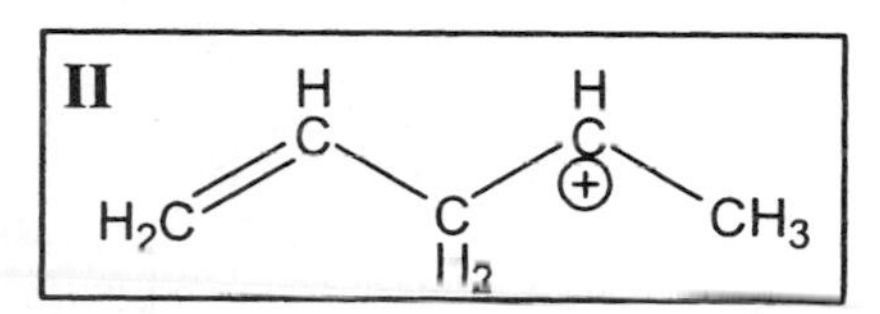

a) Use a curved arrow to generate a second resonance structure for I.

b) Are there other resonance structures for II?

c) In which carbocation (I or II) are the pi electrons more delocalized.

d) Is your answer above consistent with the fact that carbocation I is much lower in potential energy than carbocation II?

Part B: Two-Step Nucleophilic Substitution Reactions

(What molecules are likely to undergo two-step nucleophilic substitution reactions?)

Model 4: Rate Limiting Step

- The higher the concentrations of species involved in a step, the faster that step since the rate of collisions between the reactants is increased.
- A significant increase in the overall rate of reaction is only achieved if you speed up the slow step. (Also known as the "rate limiting step.")

Critical Thinking Questions

8. Consider the following example:
   - You have two buckets.
   - Each has a hole in the bottom.

   a) Which bucket is the "rate limiting step" in this overall process? Explain.

   b) Would increasing the size of the hole in the second bucket have a significant impact on the overall rate of water out?

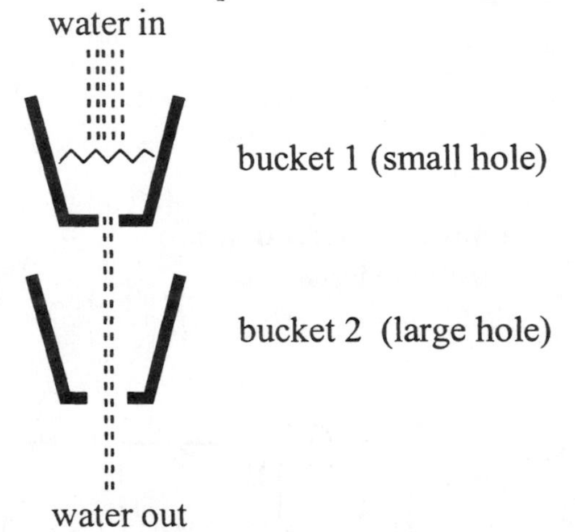

9. If a reaction has only one step, this one step must be the rate-limiting step (the slow step). The data in Table 4a confirms that both the nucleophile and electrophile are involved in this one (slow) step. Explain how the data tell you this.

solvent = mildly polar

1

only one step so it must be the slow step!

v.e.

rxn. prog.

Table 4a: Effect of Concentration on Rate for One-Step Substitution

$[I^-]$ (nucleophile)	[molecule 1] (electrophile&leaving group)	Relative Rxn. Rate
1 M	1 M	1
2 M	1 M	2
1 M	2 M	2
2 M	2 M	4

10. Consider a **two-step** nucleophilic substitution reaction:
    a) According to the reaction diagram below, which is the slow step in a two-step nucelophilic substitution reaction: **step 1** or **step 2** [circle one]? (Recall that the slow step is always the step with the higher activation barrier.)

    b) According to the data in Table 4b which species are involved in the slow step of this reaction? [circle all that apply]: **nucleophile ; electrophile-L.G.**

    c) To the box below, add a curved arrow to illustrate how the **S** stereoisomer product could be generated in step 2 of this reaction. Label this arrow "**S forming**."

    d) Draw a second arrow showing how the **R** stereoisomer is formed and label this arrow "**R forming**."

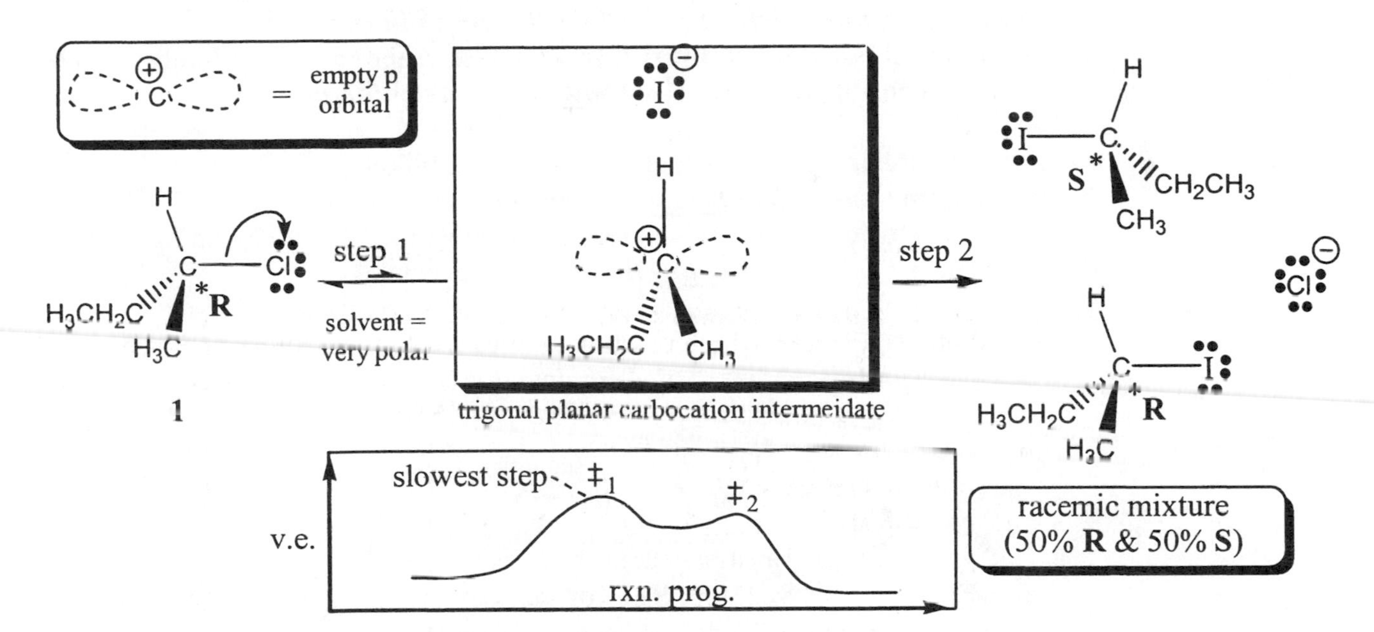

Table 4b: Effect of Concentration on Rate for Two-Step Substitution

$[I^-]$ (nucleophile)	[molecule 1] (electrophile&leaving group)	Relative Rxn. Rate
1 M	1 M	1
2 M	1 M	1
1 M	2 M	2
2 M	2 M	2

11. The book calls…
    a **one**-step substitution reaction a "S_N**2** reaction," and
    a **two**-step substitution reaction a "S_N**1** reaction."

    The "**S_N**" stands for "**substitution, nucleophilic**," but the 1 and 2 seem to be backwards. To what are the "**1**" and "**2**" referring? (Hint: how many species are involved in the slow step of each?)

12. A nucleophilic substitution reaction starts with pure R starting material and yields pure S product. Based on this information alone you can tell that the reaction was…
    one-step or **two-step** [circle one].

13. A nucleophilic substitution reaction starts with pure R starting material and yields a racemic mixture of R and S products. Based on this information alone you can tell that the reaction was…**one-step** or **two-step** [circle one].

14. In a polar solvent, one pair of reactants below…
    - goes to products exclusively by a S_N2 mechanism,
    - goes exclusively by a S_N1 mechanism,
    - goes by a mixture of S_N2 and S_N1 mechanisms.

    a) Based on **carbocation stability** (alone), which reaction is least likely to go by an S_N1 mechanism? [***circle S_N2 above this reaction arrow***]
    b) Based on **carbocation stability** (alone), which reaction is most likely to go by an S_N1 mechanism? [***circle S_N1 above this reaction arrow***]

    c) Based on **sterics** (alone), which reaction is least likely to go by an S_N1 mechanism? [***circle S_N2 below this reaction arrow***]
    d) Based on **sterics** (alone), which reaction is most likely to go by an S_N1 mechanism? [***circle S_N1 below this reaction arrow***]

    e) Based on each factor, which reaction is most likely to be a mix of S_N2 and S_N1 mechanisms? [***circle "mix" above/below this reaction arrow***]

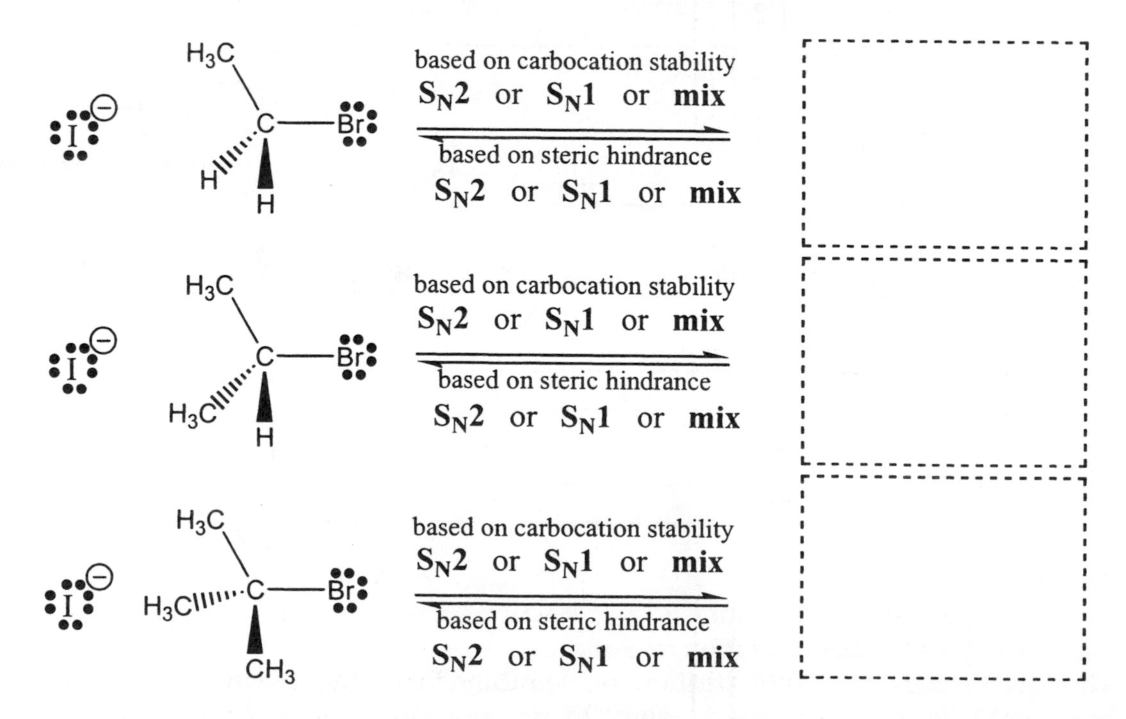

15. Are **carbocation stability factors** and **steric factors** in agreement, or are they in competition?

Exercises for Part A

1. Use curved arrows to generate all missing resonance structures for each of the carbocations below. The total number of resonance structure in the set (including the one shown) is given (x) below each structure.

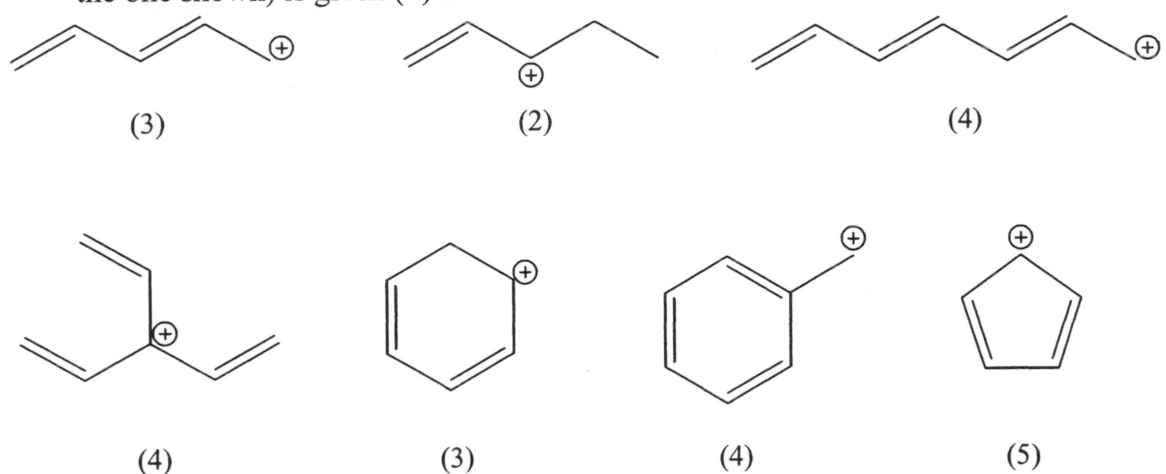

2. Explain why the following rule works: The more resonance structures you can draw for a carbocation, the lower its potential energy (the "happier" it is).

3. Which carbocation in Exercise 1 do you expect to be lowest in potential energy? Explain your reasoning.

4. A **benzene ring** is a six carbon ring with three double bonds. A + charge next to a **benzene ring** is a common occurrence so we give it a name: a **benzyl carbocation.** Circle the benzyl carbocation in Exercise 1.

5. Label each of the carbocations below the most appropriate of the following words: **methyl, 1°, 2°, 3°, allyl or benzyl.** (There is at least one of each type.)

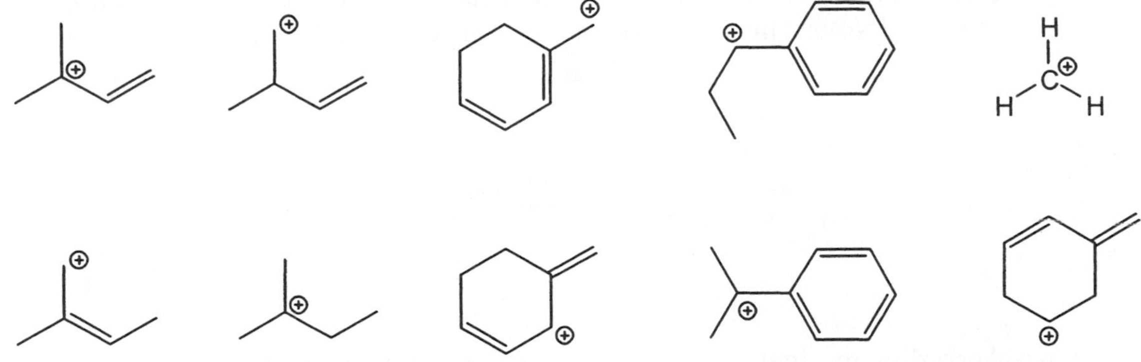

   a) Cross out the two carbocations in Exercise 5 that you expect do not exist.
   b) Circle together the two structures in Exercise 5that are really two resonance structures of the same carbocation.
   c) Construct an explanation for why the benzyl carbocation in the second row is lower in potential energy than the benzyl carbocation in the first row.
   d) Explain why the top left structure is lower in potential energy than the one directly below it. (Hint: draw the other resonance structure for each.)

6. Consider the following Figure.

Figure 6: Heats of Reaction for Nucleophilic Addition of Br⁻ to R⁺

Note: assume that all the neutral products have similar potential energies.

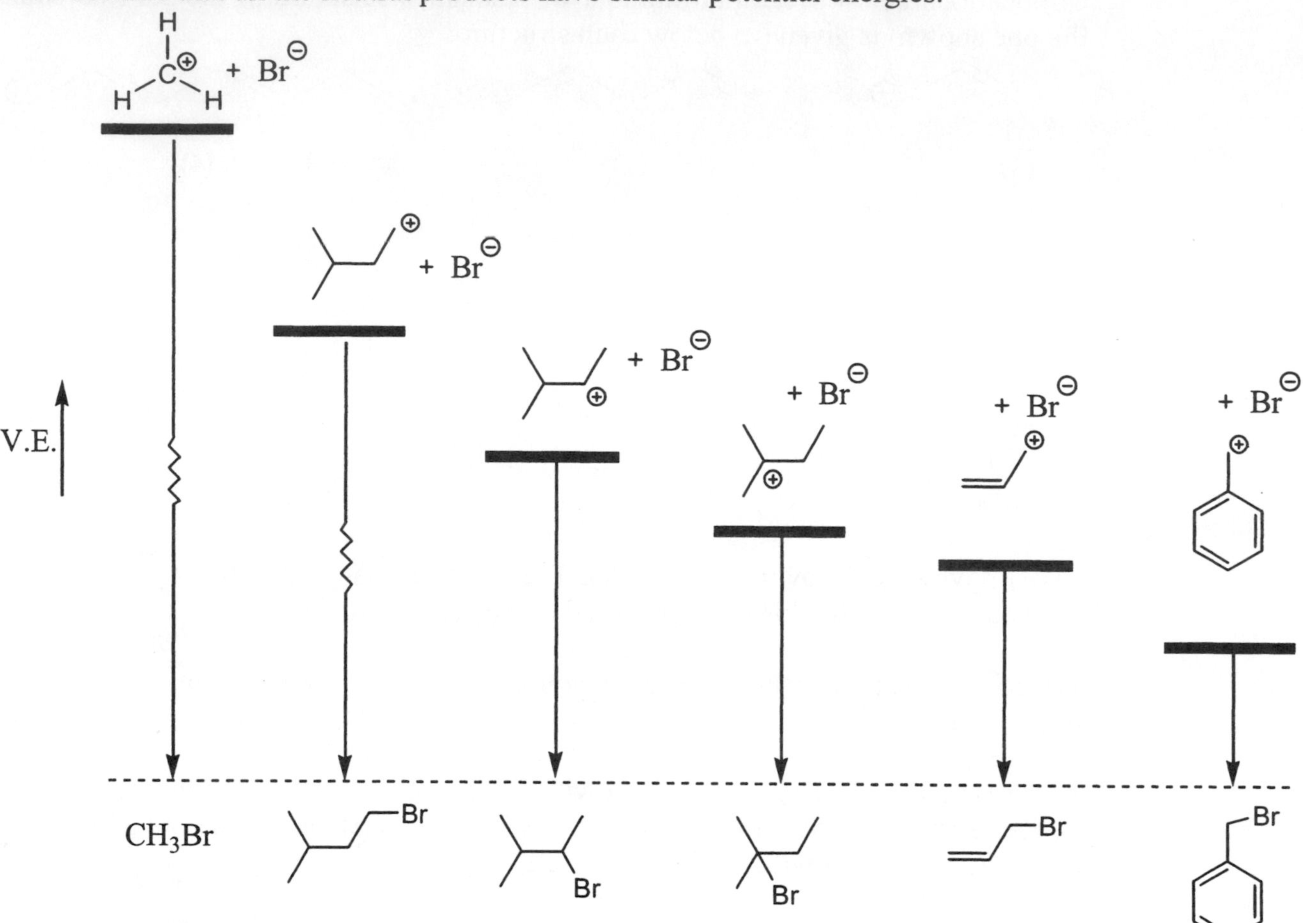

a) Use curved arrows to illustrate the reaction below, and draw the resulting product. (Note: This is one of the reactions from Figure 6)

H_3C CH_3
$\oplus$ C—CH_2 :Br:$^{\ominus}$ ⟶
H_3C

b) According to Figure 6, which is lower in potential energy: the allyl carbocation shown or the tertiary (3°) carbocation shown? Explain.

c) Based on the information in this chart, rank the carbocations in Exercise 5 from highest potential energy to lowest potential energy.

7. Read the assigned pages in the text and do the assigned problems.

Exercises for Part B

8. Place each of the following phrases in the correct box in Table 5.
   - Very polar solvent better, but weakly polar ok.
   - Very polar, protic solvent required to stabilize ion intermediates.
   - Must be 2°, 3°, allyl or benzyl
   - Methy or 1° preferred, 2° ok too.
   - Dependent on identity and concentration of both Nuc and Elec/L.G.
   - Dependent only on identity and concentration of Elect/L.G.
   - Inverted (switch from R to S or vice versa)
   - Racemic mixture produced

Tabel 5: Summary of Factors Leading to S_N2 vs. S_N1 Reactions.

Rxn Type	Solvent	Stereochemistry	Electrophile/L.G.	Rate
S_N2				
S_N1				

9. The following is an orbital explanation of a S_N1 reaction. Fill in the blanks with appropriate words or phrases and refer to the reaction in CTQ 11.

"In the first step, the ________ ________ leaves, taking its electrons with it. The resulting carbon changes hybridization state from sp^3 to ____ with one empty ___ __________ and bond angles of _____°. In the second step, the ____________ enters the picture for the first time and reacts with one or the other lobe of the ________ ___ ____________, filling it with two electrons that were formerly a ______ _______ on the nucleophile. This creates a new ___ bond. The electrons in this new bond are unhappy so close (_____° away) to the other bonds. To relieve this steric strain, the carbon atom changes hybridization states from ____ to ____. This moves the new σ bond to its proper location _______° away from the other bonds.

10. Hydroxide is a much better nucleophile than acetate ion so it is not surprising that Rxn I is much faster than Rxn II. (Assume solvent = water for reactions I-IV.)

I

HO^- + H_3C-Br ⇌ $H-O-CH_3$ + Br^- (Rel. Rate = 20,000)

hydroxide = good Nuc

pKa = 16 (and small in size)

II

$H_3C-C(=O)-O^-$ + H_3C-Br ⇌ $H_3C-C(=O)-O-CH_3$ + Br^- (Rel. Rate = 100)

acetate ion = fair Nuc

pKa = 5

a) Draw an energy diagram for reactions I and II. Be sure that your energy diagrams reflect the fact that Rxns I and II have different rates <u>and</u> different ΔH_{Rxns} (one of them is more down hill).

III

HO^- + R_3C-Br ⇌ $H-O-CR_3$ + Br^- (Rel. Rate = 1)

hydroxide = good Nuc

pKa = 16 (and small in size)

IV

$H_3C-C(=O)-O^-$ + R_3C-Br ⇌ $H_3C-C(=O)-O-CR_3$ + Br^- (Rel. Rate = 1)

acetate ion = fair Nuc

pKa = 5

b) Explain why Rxn III has exactly the same rate as Rxn IV, even though hydroxide is a much better nucleophile than acetate ion.

c) Draw an energy diagram for reactions III and IV. Be sure that your energy diagrams reflect the fact that the reactions have the same rate <u>but</u> different ΔH_{Rxns} (one of them is more down hill).

d) Draw a mechanism for each reaction.

11. Read the assigned pages in the text and do the assigned problems.

ChemActivity 14

Part A: Two-Step Elimination

(By what mechanism do carbocations easily become alkenes?)

Model 1: Hyperconjugation ("Weak Conjugation")

In the 3° carbocation *(below left)* **one** C—H σ bond is shown as an overlap of orbitals. The electron pair in this C—H bond donates some of its electron density into the empty *p* orbital, helping to complete the octet of the carbocation carbon. This weak orbital overlap is sometimes called **hyperconjugation**. For a 3° carbocation this effect is multiplied x9 (since all nine C—H bonds do this). This explains why alkyl groups are electron donating and tertiary carbocations are relatively stable.

The methyl carbocation *(below right)* has no such help completing its octet.

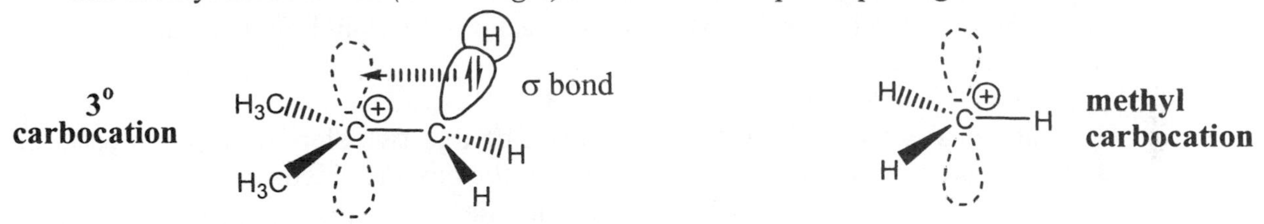

Critical Thinking Questions

1. Consider the C–H σ bond in Model 1 shown as an overlap of orbitals. Do you expect the presence of the empty *p* orbital to strengthen or weaken this C–H bond? Explain your reasoning.

2. Is your answer in CTQ 1 consistent with the following pK_a? Which H does this pK_a refer to?

$pK_a < 0$

3. A tertiary carbocation is such a strong acid that it will react with a very weak base such as sodium bicarbonate (baking soda).

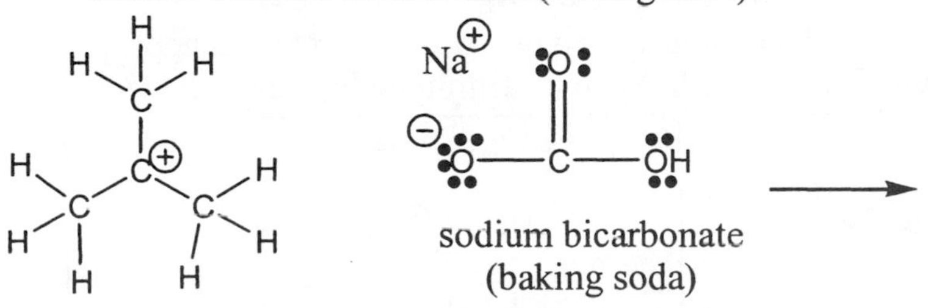

weak base & poor nucleophile

a) Show the mechanism and the products of this acid-base reaction.

b) Have you drawn the <u>most important</u> resonance structure of the C_4H_8 product? If not, draw a better representation of this product.

Model 2: Two-Step Elimination

Table 2: Concentration Dependence of Reaction Rate for Two-Step Elim.

[R—Br]	[HCO_3^-] (baking soda)	Rel. Rxn. Rate
1 M	1 M	1
2 M	1 M	2
1 M	2 M	1

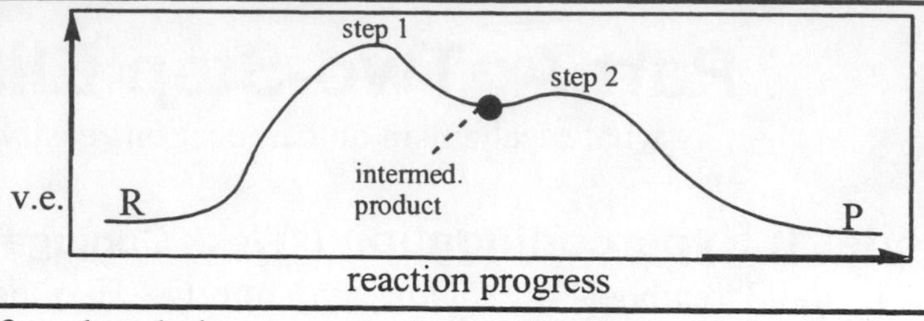

step 1
step 2
v.e.
R
intermed. product
P
reaction progress

Elimination = formation of a π bond via removal of a leaving group and proton.

Critical Thinking Questions

4. Based on the information in Table 2:
   a) Is baking soda involved in the slowest step of this reaction? Explain how you can tell.

   b) Which step is the slow step in this reaction? Explain how you can tell.

5. Use curved arrows to show the mechanism of this two-step elimination reaction. Draw the carbocation intermediate product in the box provided.

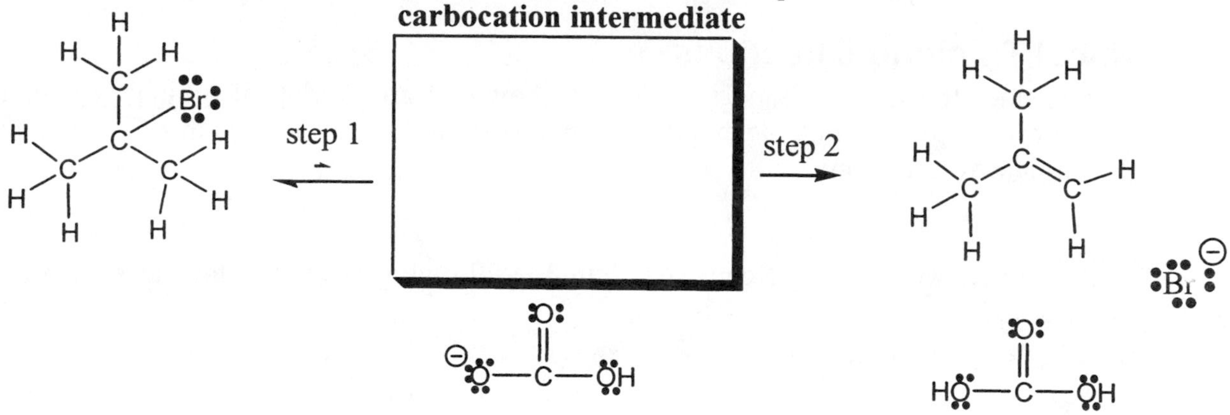

carbocation intermediate

step 1 step 2

6. Chemists use the Greek letters α,β,δ, etc. to designate how far away a C atom is from a functional group such as Br. H's attached to a C_α are called α–H's, H's attached to a C_β are called β–H's, etc. (**See the example below left**).
   a) Label the C's and H's in the example at right with α,β,γ, etc., as appropriate.
   b) What type of H (which Greek letter) can be eliminated by the base?

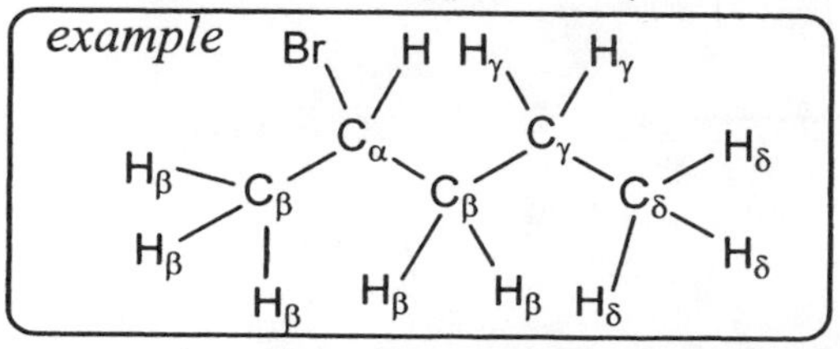

example

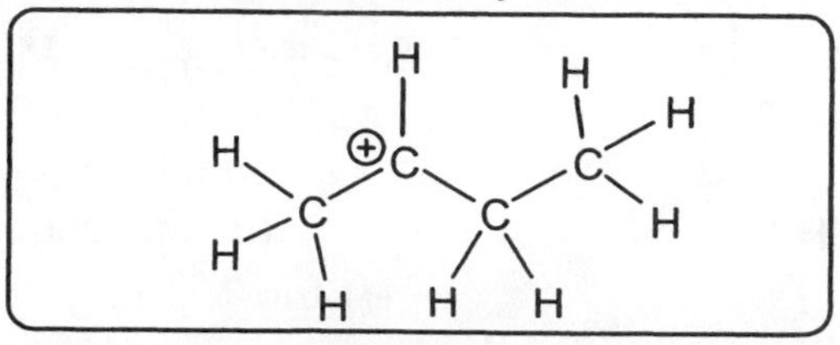

   c) Draw the <u>three</u> possible alkene products produced if a base reacts with the carbocation in the right hand box. (Hint: two of them are *cis/trans* isomers.)

7. Explain why the following molecule cannot undergo elimination, even though it readily forms the very stable benzyl carbocation shown. (Hint: review CTQ 6.)

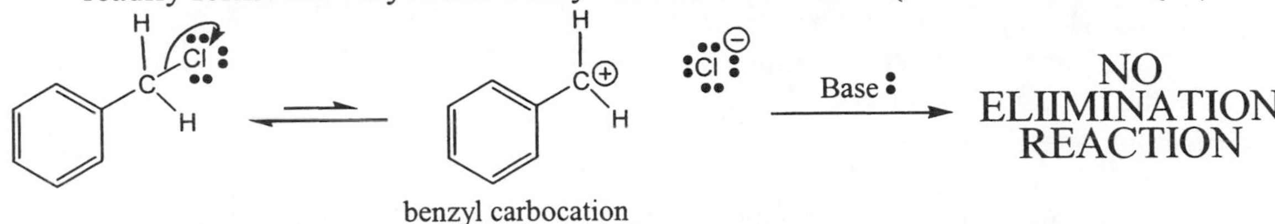

Part B: One-Step Elimination

(What reaction conditions favor elimination over substitution?)

Model 3: One-Step Elimination

Table 3a: Concentration Dependence of Reaction Rate for One-Step Elim.

[R—X]	*Tert*-butoxide (base)	Relative Rxn. Rate
1 M	1 M	1
2 M	1 M	2
1 M	2 M	2
2 M	2 M	4

v.e.
R
P
reaction progress

Critical Thinking Questions

8. Explain how the data in Model 3 show that this reaction happens in a single step.

9. Use curved arrows to show a mechanism of this reaction that is consistent with the kinetic (rate) data and the reaction energy diagram shown in Model 3.

10. Draw the transition state for this reaction. Represent each partial bond using a dotted line (---) and show partial charges as δ–.

11. The book calls…
    a **one**-step elimination reaction an "E2 reaction," and
    a **two**-step elimination reaction an "E1 reaction."

    The "**E**" stands for "**elimination**," but the 1 and 2 seem to be backwards. To what are the "**1**" and "**2**" referring?

12. Construct an explanation for why a weak base such as baking soda (Na^+ HCO_3^-) will not work as the base in an E2 reaction even though it is ok for an E1 reaction such as the one in CTQ 5.

13. If there is one thing that makes organic chemistry students go gray, it's the fact that many molecules will undergo more than one of the four reactions we have been discussing. In the laboratory, it is often impossible to avoid a mixture of substitution and elimination products. Fill in the following table to help sort out what molecules and conditions are likely to cleanly undergo only one of these mechanisms.

Table 3b: Conditions that Favor… (S_N1, S_N2, E1 or E2)

Favored Mech.	Base/Nuc	R—X	X (leaving group)	Rate dependent on concentration of	Solvent
S_N1					
S_N2					
E1					
E2					

Fill in the table above with the following phrases:

Col. 2: Strong base/poor nucleophile; Weak base/poor nucleophile; Weak base/good nucleophile (put this last phrase in two different boxes!)

Col. 3: 2° and 3° allyl or benzyl only; 2°, 3° allyl or benzyl only AND must have H_α to C^+, methyl and 1° best but 2° ok; must have acidic H_α to C-L.G., e.g. methyl will not do

Col. 4: excellent leaving group required (two boxes!); good leaving group (two boxes!)

Col. 5: R–X only (two boxes!); R–X and Base; R–X and Nucleophile

Col. 6: very polar, protic (two boxes!); polar is better but mildly polar ok (two boxes!)

Exercises for Part A

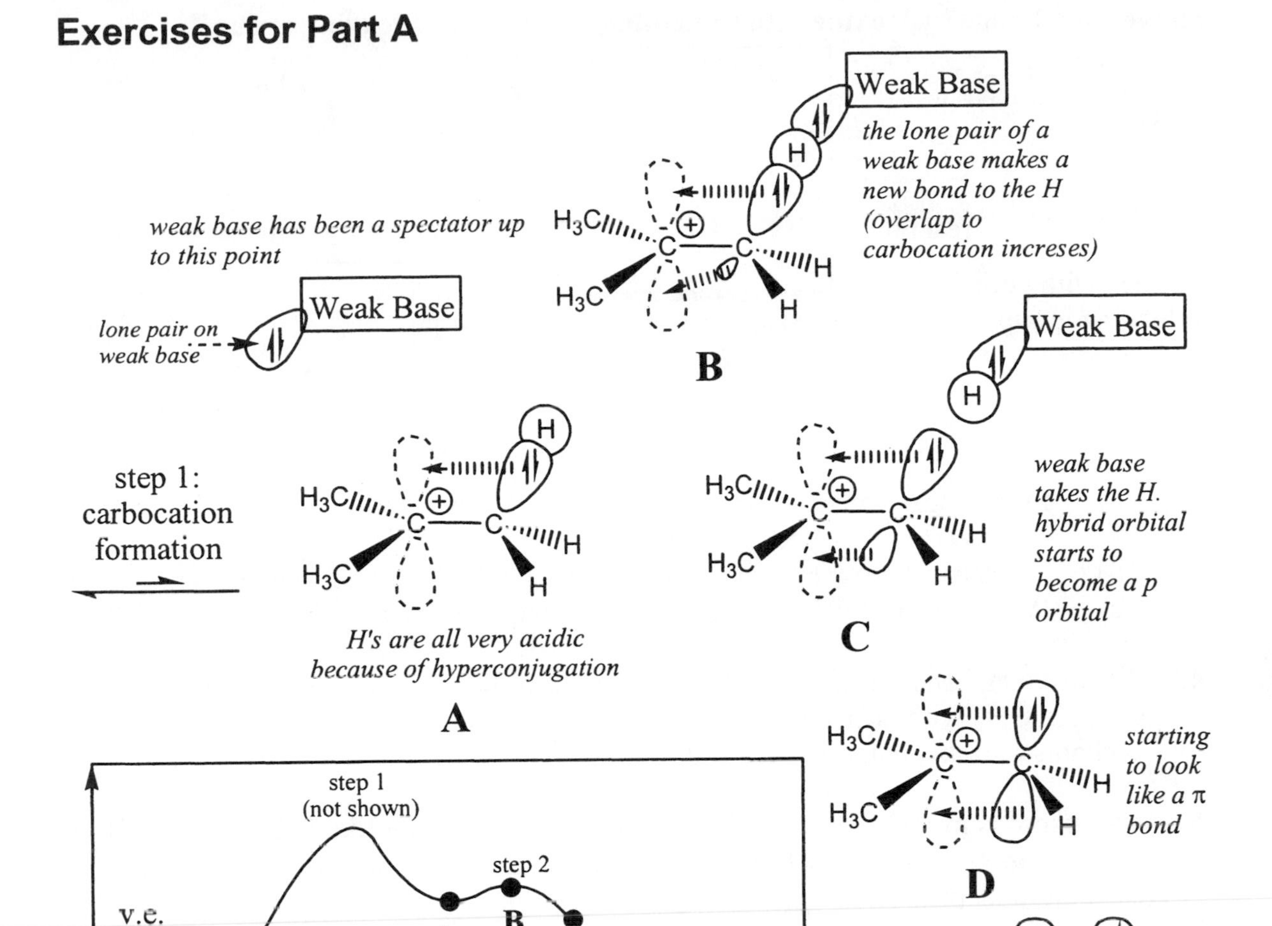

1. Which point (of A-E) is the transition state of step 2? Draw this transition state using dotted lines to represent partial bonds and δ's to represent partial charges.
2. One carbon involved in the reaction above changes hybridization state during the reaction. Label this C with its hybridization state in A (start) and in E (end).
3. Hydroxide (—OH) is a poor leaving group. This is because it takes so much energy (~16pK_a units) to break the C—OH bond and set HO^- free. (*See* Q's on next page.)

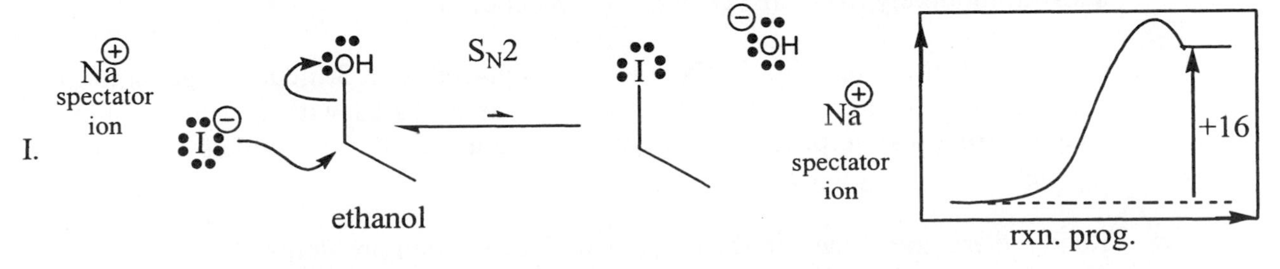

But we can change hydroxide into something that is a good leaving group!!

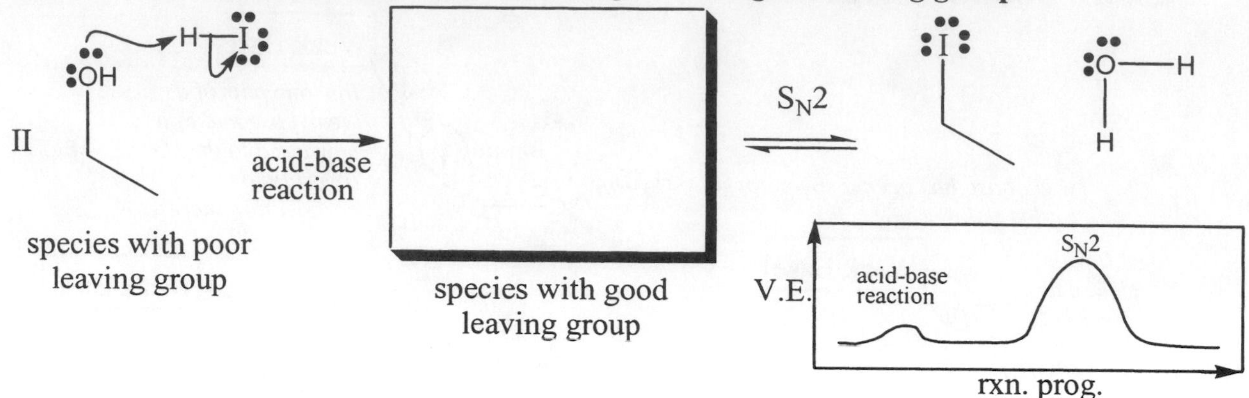

a) In the box above, draw the products of the reaction shown with curved arrows (step 1 of Reaction II).

b) Use curved arrows to show the S_N2 reaction in step 2 of Reaction II.

c) Why is water (OH_2) a better leaving group than hydroxide (HO^-)?

4. The products shown in Reaction II can also be achieved by mixing ethanol (CH_3CH_2OH) with sulfuric acid (H_2SO_4) and NaI. Use curved arrows to show the mechanism of this reaction. (Hint: the first step is an acid-base reaction.)

5. If t-butanol is mixed with sulfuric acid (and no nucleophile like I^- is introduced) the result is an acid-base reaction followed by a two-step elimination reaction. The reactants and products are shown below for this overall three step process. Use curved arrows to show the entire mechanism of this reaction and draw the two intermediate products. (Note that water is a stronger base than HSO_4^-)

t-butanol + sulfuric acid → three steps to → → → $CH_2{=}C(CH_3)_2$ + H_3O^+ + HSO_4^-

6. A lab technician accidentally spills some sodium iodide into the reaction mixture in Exercise 5. (NaI dissociates into Na^+ and I^-). As a result a substitution product is observed along with the elimination product. Use curved arrows to show the mechanism of this side reaction and draw the substitution product. (Is this substitution likely to be an S_N1 or an S_N2 reaction?)

7. If t-butanol in the reaction in Exercise 5 is replaced with 3-methyl-3-pentanol three different alkene products are produced. If it is replaced with 3-methyl-3-hexanol five different alkene products are produced. Draw all these products and show the mechanism leading to each one.

8. Read the assigned pages in the text and do the assigned problems.

Exercises for Part B

9. Most substitution or elimination reactions are competitions between two or more of the four reaction mechanisms we have discussed. Each of the following reactions go almost exclusively by the mechanism written over the arrow. For each reaction…
   a) Use curved arrows to show the mechanism and draw the product/s.
   b) Briefly explain why these starting materials can't or are unlikely to go by each of the other three reaction mechanisms.

H_3C—Br: $^{\ominus}$:OH $\xrightarrow{S_N2}$

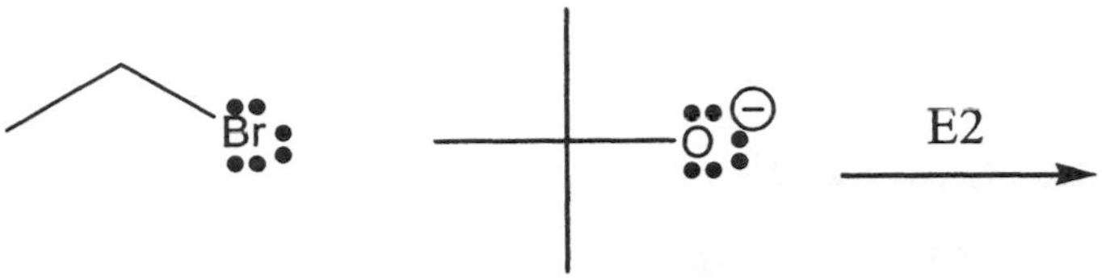

Br: O$^{\ominus}$ $\xrightarrow{E2}$

Br: NaI $\xrightarrow{S_N2}$

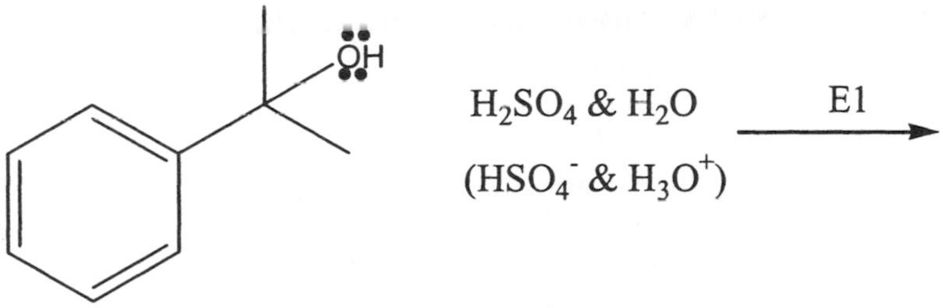

OH H_2SO_4 & H_2O (HSO_4^- & H_3O^+) $\xrightarrow{E1}$

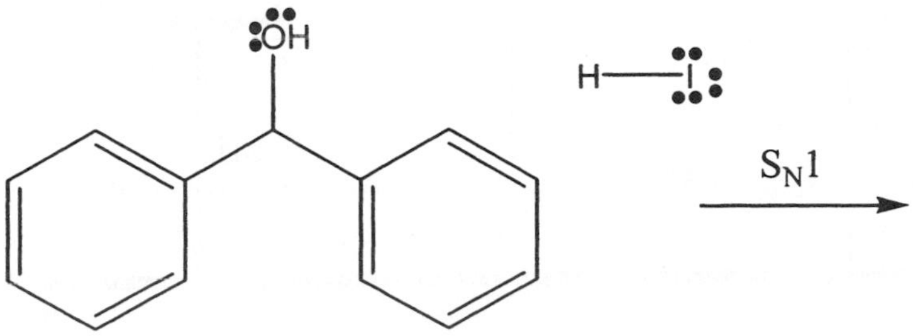

:OH   H—I: $\xrightarrow{S_N1}$

10. Read the assigned pages in the text and do the assigned problems.

ChemActivity 15

Part A: Alkene Potential Energy

(What are the relative potential energies of substituted and unsubstituted double bonds?)

Model 1: Heats of Hydrogenation ($\Delta H_{hydrogenation}$)

$\Delta\mathbf{H}_{\mathbf{hydrogenation}}$ = heat released when two H atoms are added to a double bond. You do not need to know the mechanism of this reaction (called **catalytic hydrogenation**). Know only that the metal, usually platinum (Pt) or palladium (Pd) delivers two H atoms simultaneously to the **same side** of the double bond, as in Figure 1a (this is called "**syn addition**").

Figure 1a: Catalytic Hydrogenation of 1,2-dimethylcyclopentene

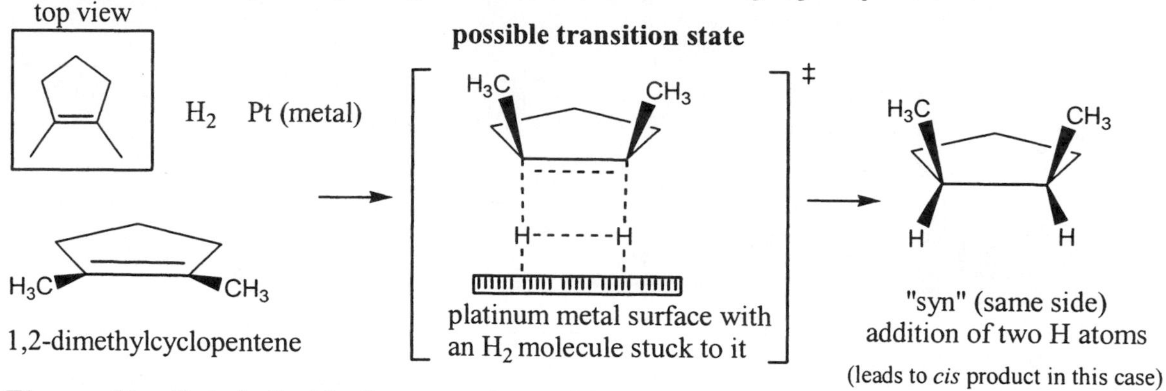

Figure 1b: Catalytic Hydrogenation of Various C_8H_{16} Constitutional Isomers

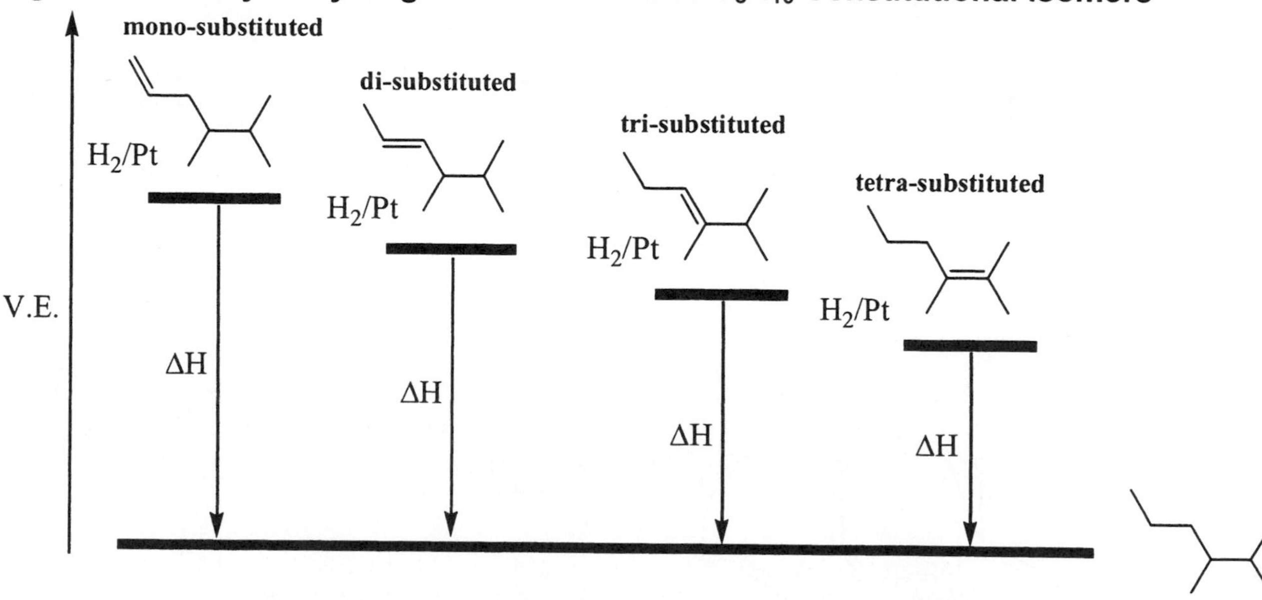

Critical Thinking Questions

1. Is catalytic hydrogenation an **exothermic** or an **endothermic** [circle one] reaction?

2. For the following catalytic hydrogenation reaction, circle the two products that do form, cross out the two that do not form, and explain your reasoning. (Hint: consider the stereochemistry of the catalytic hydrogenation reaction.)

H_2 / Pt metal

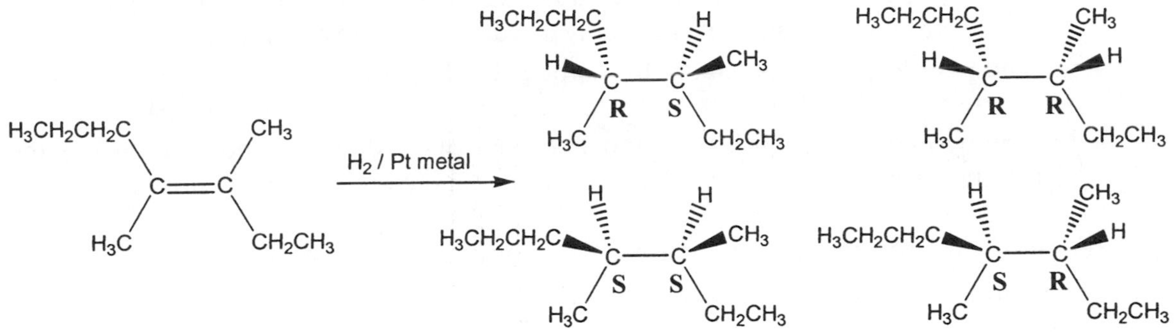

3. What do the labels **mono-substituted, di-substituted, tri-substituted and tetra-substituted** in Figure 1b refer to?

4. An **un-substituted** alkene has a molecular formula of C_2H_4. Draw the structure of this molecule and explain why it is called **un-substituted**. (This is the only "unsubstituted" alkene.)

5. According to the information in Figure 1b, what is the relationship between the level of substitution of a double bond and the potential energy of the alkene?

6. Consider the following elimination reaction:
   a) Label each alkene product with a word describing its level of substitution.

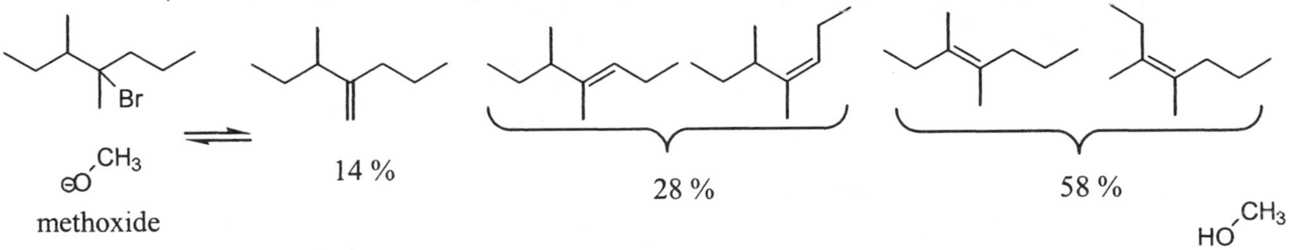

methoxide

14 %

28 %

58 %

   b) Is your conclusion in CTQ 5 consistent with the product distribution shown above? Explain.

   c) Explain why the words ***cis*** and ***trans*** are not adequate to describe the difference between the two tri-substituted products above.

Model 2: E vs. Z (more general than *cis* vs. *trans*)

As you saw on the previous page, some molecules are too complex to be called simply *cis* or *trans*. For these molecules, the E vs. Z system was invented.

- For the two groups attached to each sp^2 carbon, assign one as larger and the other as smaller using the same rules you learned for assigning R and S.
- If the two larger groups are on the same side of the double bond (*cis*-like), the structure is called **Z**, for the German word *zusammen* meaning "**together**."
- If the two larger groups are on opposite sides of the double bond (*trans*-like), the structure is called **E**, for the German word *entgegen* meaning "**opposite**."

There are several good ways to remember the definitions of E and Z:

- Think Z for "*zis*-like."
- Think Z for "zee zame zide" (say it in a German accent).
- Think consonant with consonant, vowel with vowel: **Z**/**T**ogether **E**/**O**pposite

Critical Thinking Questions

7. E vs. Z applies to four of the products in CTQ 6. Assign these structures E or Z.
8. E vs. Z does not apply to one of the alkene products in CTQ 6. Explain why.

Model 3: A Closer Look at E2 Reaction Product Distributions

- Based simply on alkene potential energies you would predict that the mono-substituted alkene is the minor product of the E2 reaction below.
- And, in general, *trans* or E molecules are lower in potential energy than *cis* or Z molecules, but this difference is not enough to explain the large preference for *trans* in this case. Indeed, something else is going on!

Br

2-bromobutane

CH_2CH_3

$\ominus$O

ethoxide ion

E2

60 %

23 %

17 %

CH_2CH_3

HO

Critical Thinking Questions

9. Why are *trans* or E molecules usually lower in potential energy than *cis* or Z molecules.

10. According to Model 3, is the energy difference between *trans*-2-butene and *cis*-2-butene large enough to account for the product distribution shown above?

Model 4: Leaving Group Position in an E2 Reaction

- Both reactions are run in a mildly polar solvent that does not support ions well.
- Reaction I, below, can ONLY go via a carbocation intermediate (E1 reaction).
- Reaction II can go via an E2 mechanism and is much faster in this solvent.

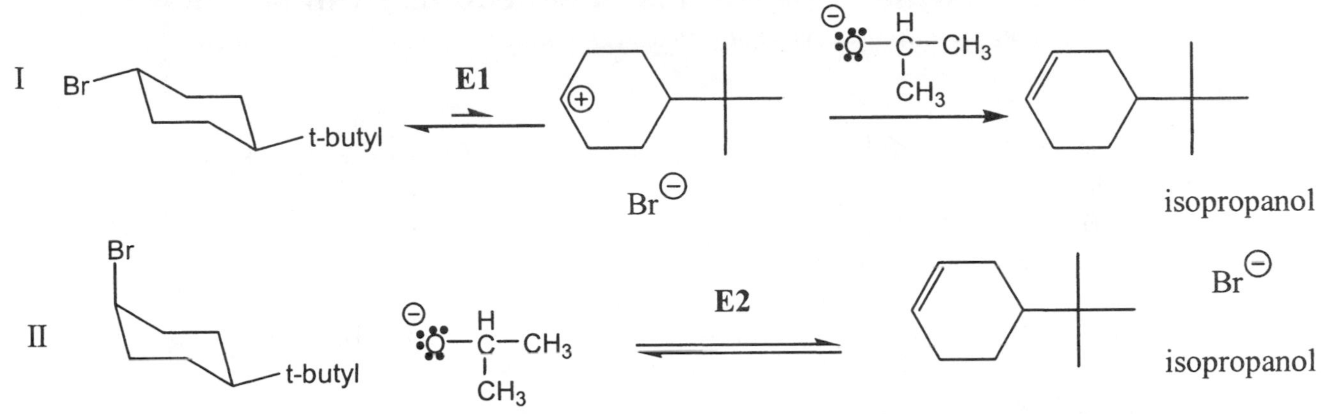

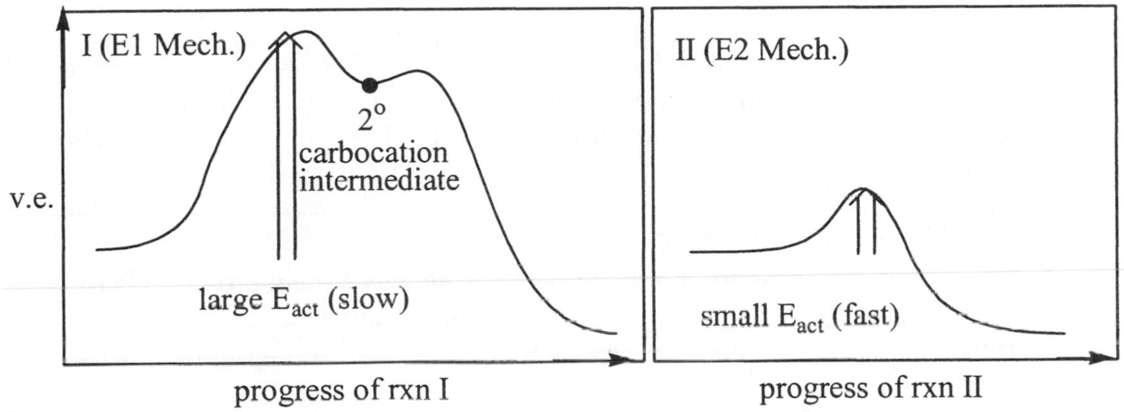

I Br — t-butyl ⇌ (E1) carbocation + Br^{-} → ($^{-}O-CH(CH_3)_2$) alkene + isopropanol

II Br — t-butyl + $^{-}O-CH(CH_3)_2$ ⇌ (E2) alkene + isopropanol + Br^{-}

I (E1 Mech.)

v.e.

2°
carbocation
intermediate

large E_{act} (slow)

progress of rxn I

II (E2 Mech.)

small E_{act} (fast)

progress of rxn II

Critical Thinking Questions

11. Why is an E1 mechanism much slower than an E2 mechanism in a mildly polar solvent?

12. Label one of the starting materials above *trans* and the other *cis*. (It may help to draw in the H's on C_1 and C_4 of the ring.

    a) Label each leaving group (Br) as being axial or equatorial in each . (Recall that the very large t-butyl group "demands" the more roomy equatorial position.)

    b) According to Model 4, an E2 mechanism is possible on a substituted cyclohexane ring when the leaving group is in an **axial position** or **equatorial position** [circle one].

Part B: Stereochemistry of E2 Reactions

(What conformation favors one-step elimination (E2)?)

Model 5: Newman Projections of Molecules from Model 4

(note that the t-butyl group takes an equatorial position in each case)

trans-1-bromo-4-*tert*-butylcyclohexane

cis-1-bromo-4-*tert*-butylcyclohexane

Review of Newman Projection Terminology

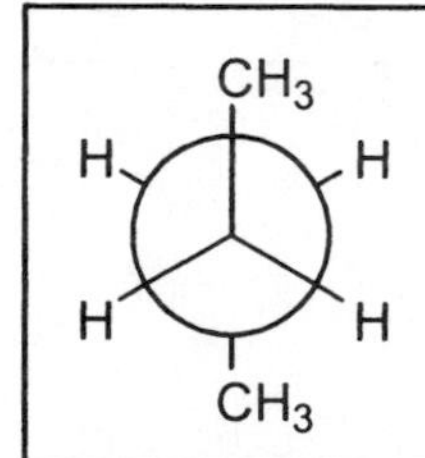

the two methyl groups are said to be **anti to one another**

the two methyl groups are said to be **gauche to one another**

Critical Thinking Questions

13. Based on the information on the previous page, write the words "**E2 favorable**" under one structure in Model 5 and "**E2 impossible-must go by E1**" under the other.

14. Circle the statement that is consistent with your conclusions above.
    I. For an E2 reaction to occur there mus. be an H **gauche** to the leaving group.
    II. For an E2 reaction to occur there must be an H **anti** to the leaving group.

15. The three possible staggered conformations of (S)-2-bromobutane are shown below.

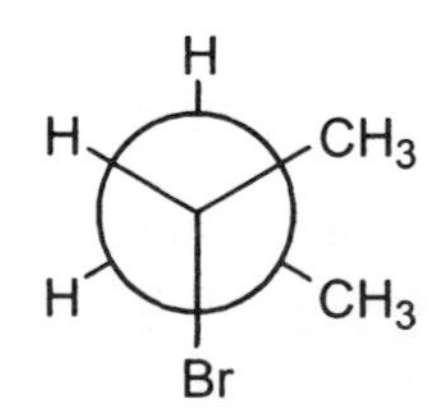

a) Based on your conclusions above, cross out the conformation that **cannot** lead to an E2 reaction.

b) Which of the remaining two conformations is more favorable, in terms of potential energy? Explain your reasoning.

16. Below are Newman and "sawhorse" representations of the two conformations of (S)-2-bromobutane that can lead to E2 reactions, along with the products of these E2 reactions.

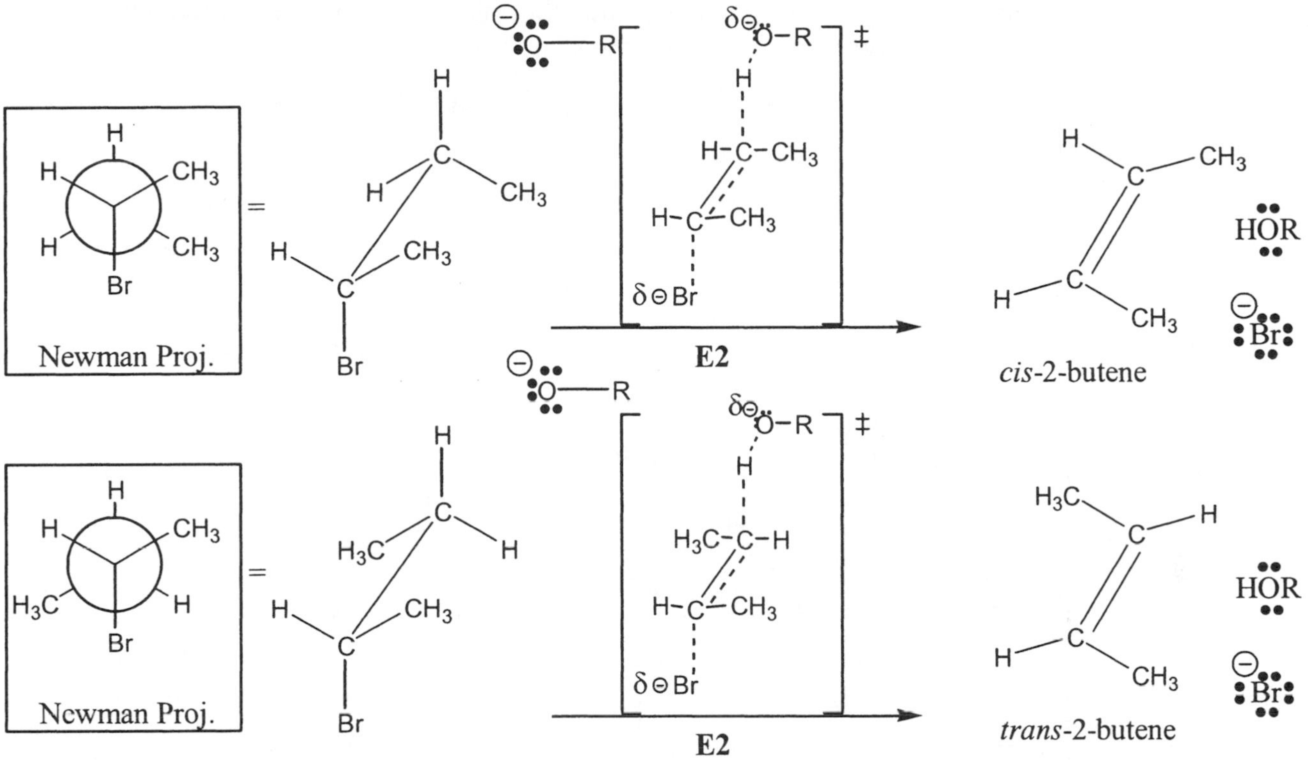

Newman Proj.

E2

cis-2-butene

Newman Proj.

E2

trans-2-butene

a) On the sawhorse representations of the reactants above, use curved arrows to show the flow of electrons during each E2 reaction.

b) Are your curved arrows consistent with the electron changes depicted in the transitions state for each reaction (shown in brackets above each reaction arrow)?

c) Relate the following statement to the example above: "The reactions above are E2 reactions so the changes happen all at once in one step. This '**traps**' the methyl groups: either **on the same side** of the newly forming double bond (*see* transition state leading to *cis* product), or **on opposite sides** of the newly forming double bond (*see* transition state leading to *trans* product)."

d) Label one of the Newman Projections above with the words "**lower P.E. – will spend more time in this conformation.**"

e) Based on the fact that each conformation gives exclusively the product shown, predict which will be more prevalent in the product mixture: ***trans*-2-butene** or ***cis*-2-butene** [circle one] and explain your reasoning.

f) Is your answer above consistent with the product distribution in Model 3?

17. For each conformation of the deuterated 3-bromohexane stereoisomer below, draw the alkene product that will form if the conformation shown undergoes an E2 reaction. (**Deuterated** means that one or more of the H atoms has been replaced with a "heavy" hydrogen called a deuterium atom (D). Assume D and H have the same chemical reactivity. (e.g. D_2O has the same chemistry as H_2O)

H
H_3CH_2C H
H_3CH_2C D
Br

D
H_3CH_2C H
H CH_2CH_3
Br

Model 6: Orbital Explanation for "Anti-Only" E2 Reaction

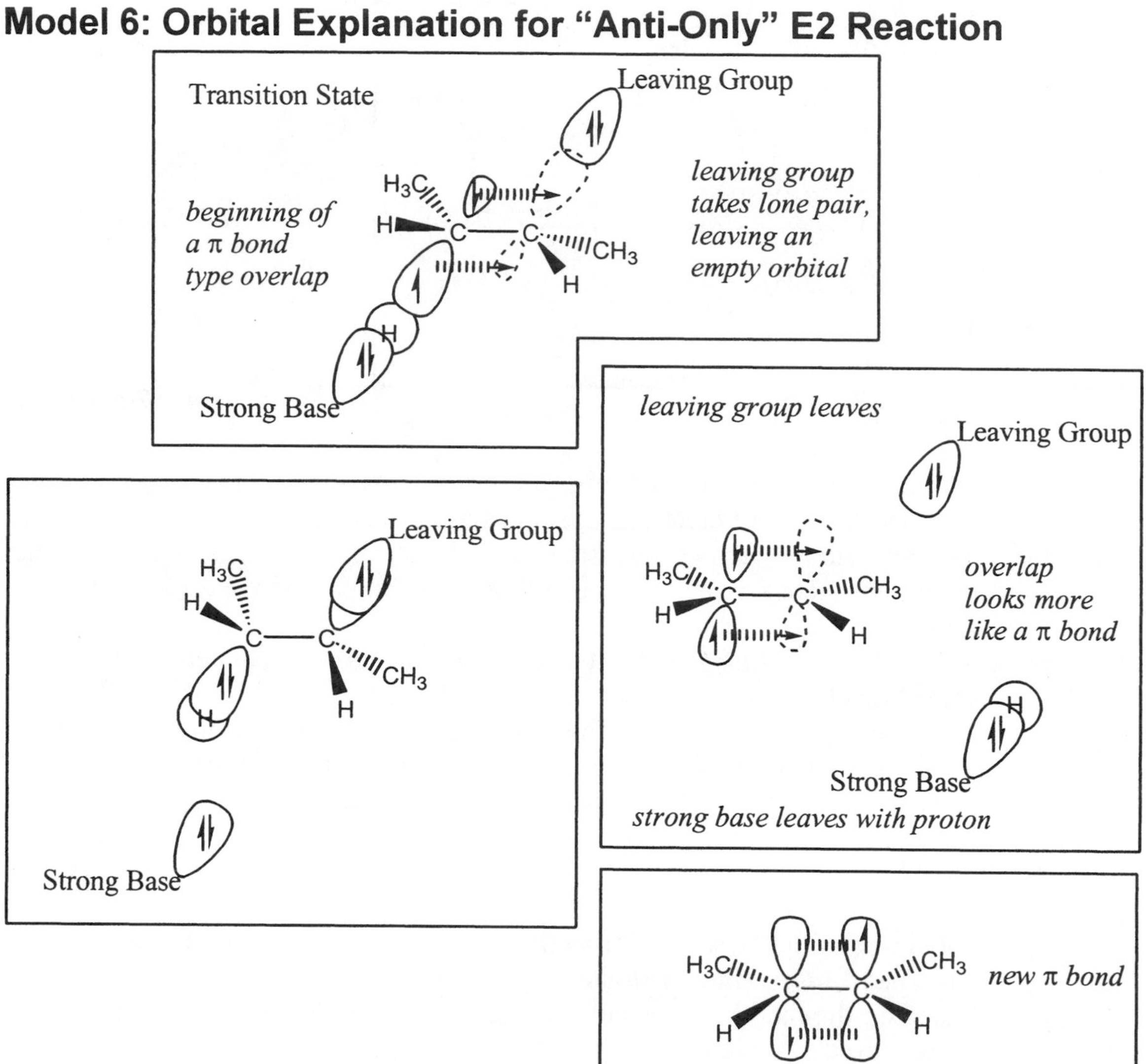

Critical Thinking Questions

18. Label the boxes above A-D according to the order in which they occur in an E2 rxn.

Exercises for Part A

1. Assign each double bond in the following molecules as E, Z, or *neither*.

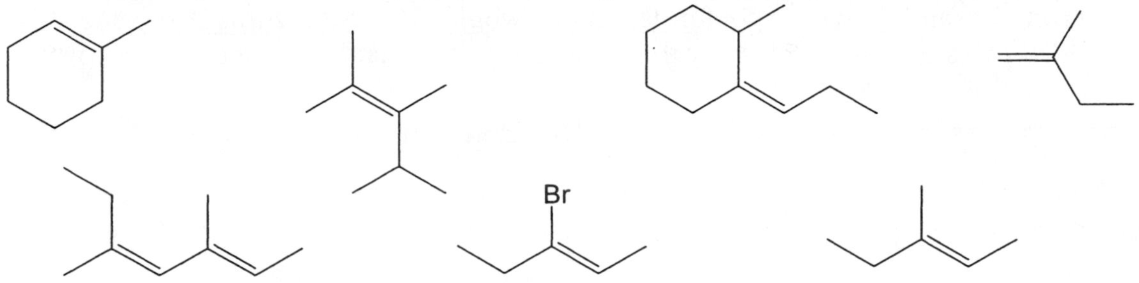

Br

2. For each of the following pairs of reactants
   a) Draw all possible E2 products.
   b) Use curved arrows to show the formation of each different product. Redraw the reactants for each new product so as to clearly show each set of arrows.
   c) Circle the lowest potential energy product in each case.

a) 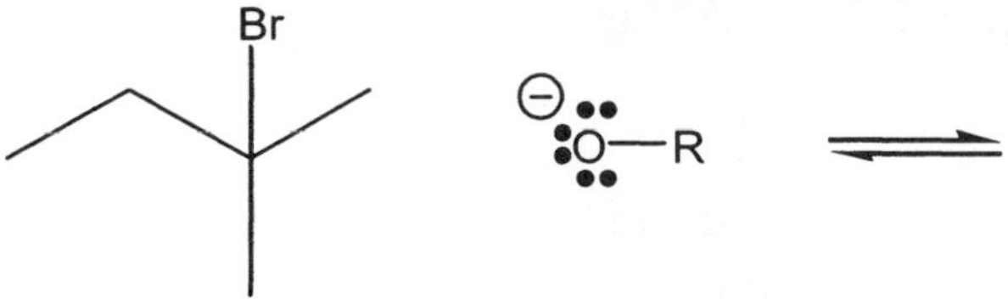 Br ⊖ O—R ⇌

b) 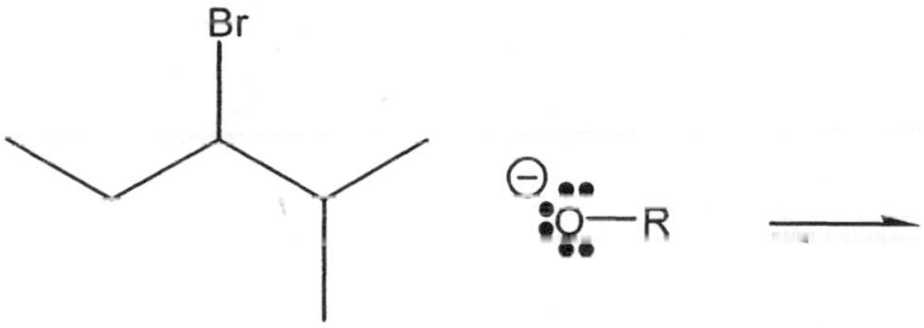 Br ⊖ O—R ⟶

c) 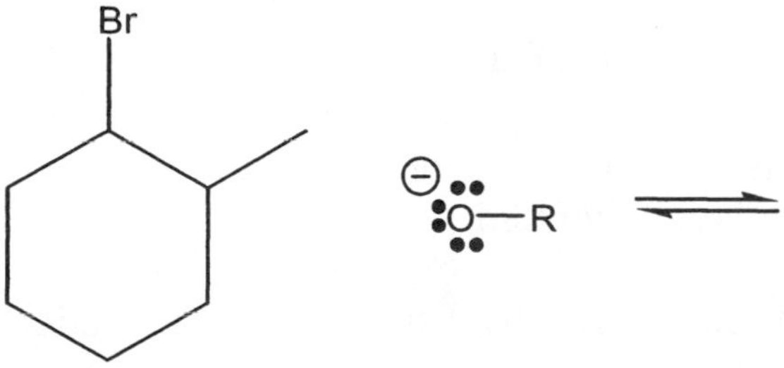 Br ⊖ O—R ⇌

3. Read the assigned pages in the text and do the assigned problems.

Exercises for Part B

4. Complete the following Newman and sawhorse representations of (S)-3-bromohexane showing a conformation that would give rise to ***trans*-3-hexene**. Also draw *trans*-3-hexene and show the mechanism of formation using curved arrows.

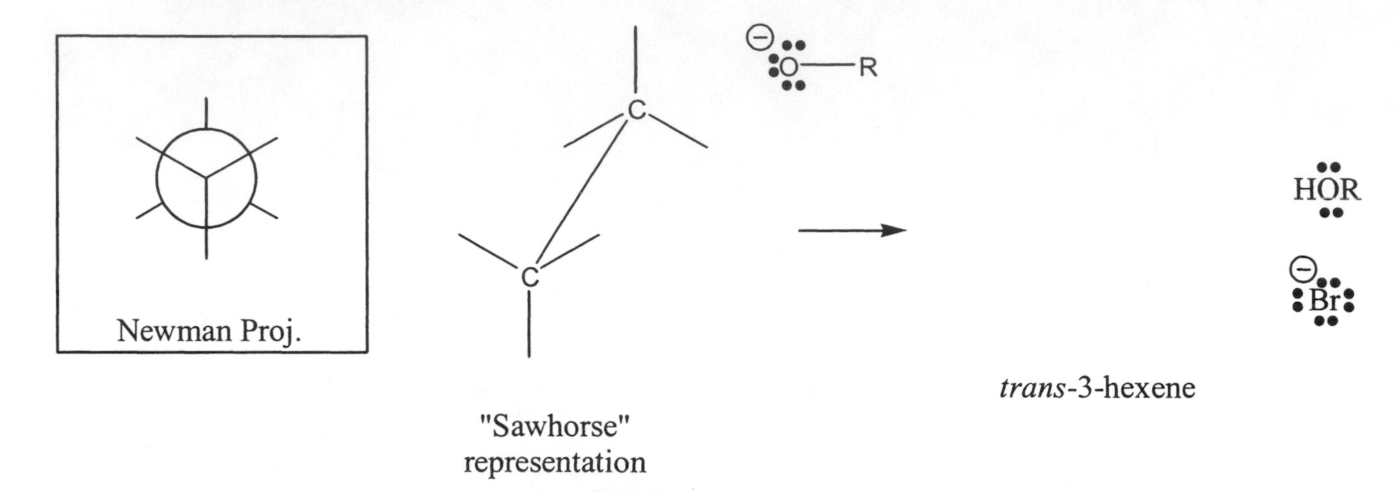

5. Even though the tri-substituted alkene product on the right is lower in potential energy, only the di-substituted product is observed.

Br 1 2 3 4 5 6 CH_3 = Br CH_3 ⊖O–t-butyl

CH_3 CH_3 ⊖Br HO–t-butyl

100% 0%

   a) Draw two Newman Projections of the reactant: one sighting down the C_1–C_2 bond and the other sighting down the C_1–C_6 bond.
   b) Use your Newman Projections as the basis for explaining why the di-substituted alkene is preferentially formed.

6. Mark TRUE each of the following statements that helps explain the product distribution in Model 3. (More than one may be true.)

   I. The conformation leading to the *cis* product is less favorable than the conformation leading to the *trans* product.
   II. The methyl groups of the *trans* product are farther apart than in the *cis* product. This leads to less steric hindrance and therefore lower P.E. for the *trans* product.
   III. A terminal double bond (one at the end of a chain) is higher in potential energy then an internal double bond (one in the middle of a chain).

7. Deuterium (D) has nearly identical reactivity as hydrogen (H). Consequently, the one-step elimination of Br from (S,S)-2-bromo-3-deuterobutane (shown below) yields two different alkene products.

(Note: D is given a higher priority than H when assigning absolute configuration.)

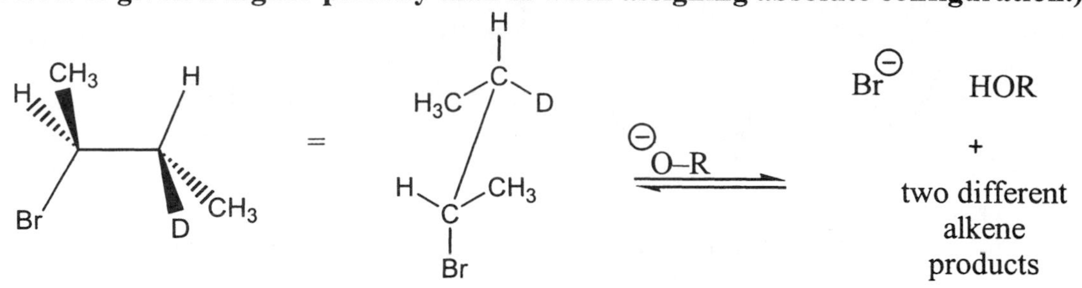

Consider the four possible di-substituted E2 products below...

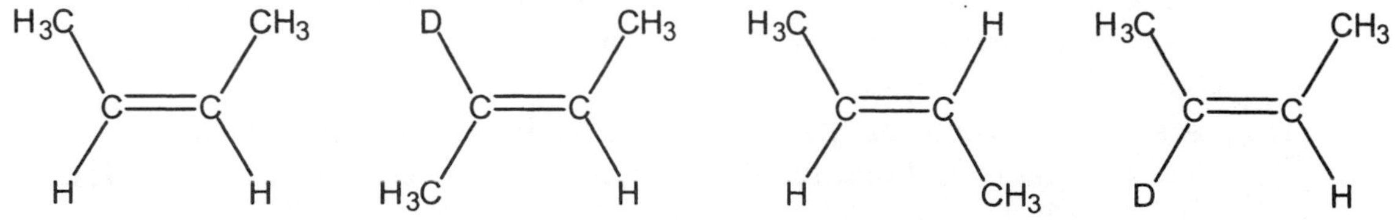

a) Circle the two that could be produced by E2 reaction of the (S,S) stereoisomer.
b) Mark one as the "major product" and one as the "minor product."
c) Draw a sawhorse representation of the conformation that would give rise to each of these two products.
d) Draw and name the enantiomer of (S,S)-2-bromo-3-deuterobutane.
e) Could E2 reaction with this stereoisomer give rise to either of the two uncircled alkene products? If so, show the sawhorse representation of the conformation giving rise to each product.
f) Draw AND NAME a diastereomer of 2-bromo-3-deuterobutane: either (S,R) or (R,S)-2-bromo-3-deuterobutane.
g) Can this stereoisomer give rise to the other two products? If so, show the sawhorse representation of the conformation giving rise to each product.

8. Draw an energy diagram for the reaction shown in Model 6 and put points on the diagram labeled A-D to correspond with orbital pictures A-D in Model 6.

9. In terms of orbitals, explain why an E2 reaction cannot occur via removal an H that is gauche to the leaving group—and why the H must be anti. Hint: sketch an orbital drawing of the transition state removing a H gauche to Br. Will the newly forming *p* orbitals be aligned in a way that can immediately form a double bond?

10. On the following drawing of (R)-2-bromobutane…

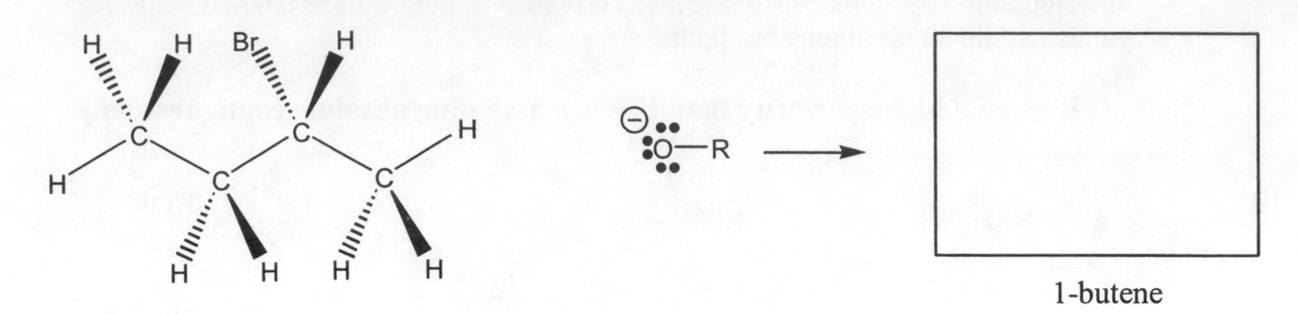

a) Circle any H that, if removed in an E2 reaction, would give rise to *cis*-2-butene, and label it "**gives *cis*.**"
b) Circle any H that, if removed in an E2 reaction, would give rise to *trans*-2-butene, and label it "**gives *trans*.**"
c) Draw 1-butene in the box provided.
d) Put a triangle around any H that, if removed, would give rise to 1-butene.
e) Use curved arrows to show the mechanism of an E2 reaction leading to 1-butene.
f) A student predicts that the ratio of *trans*-2-butene to 1-butene should be 1:3 in the final product mixture. Explain how she came up with the ratio 1:3, and explain the **flaw** in her reasoning.

11. Read the assigned pages in the text and do the assigned problems.

Worksheet for Degree of Unsaturation Calculation

(# of pi bonds) + (# of rings) = **degree of unsaturation**

It is important that you have a strategy for calculating degree of unsaturation from a molecular formula. Most books have a complicated formula that you can memorize.

Another method is as follows:

(e.g. for molecular formula C_6H_7N)

I. Draw a saturated structure (no rings or pi bonds) containing the right number of every element except hydrogen. (In this case, six carbons and one nitrogen.)

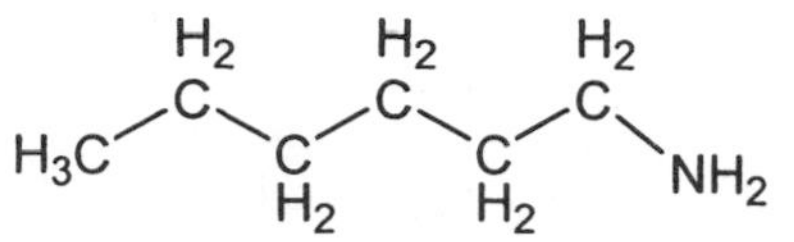

II. Count the # of H's on your saturated structure. (In this case = 15)

III. Subtract to calculate how many <u>extra</u> H's are contained in your saturated structure. (15-7 = 8)

IV. If you removed 2 H's from your saturated structure this would generate a ring or pi bond. Therefore, **each <u>pair</u> of extra H's = a degree of unsaturation**. (In this case you have 4 pairs of extra H's, so the unknown structure must have four degrees of unsaturation.)

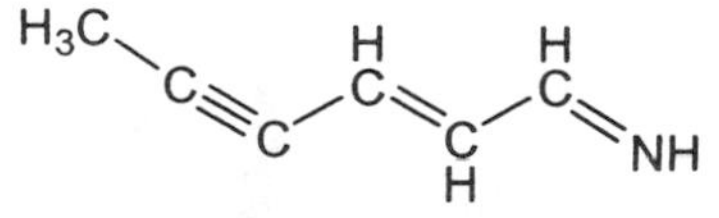

or 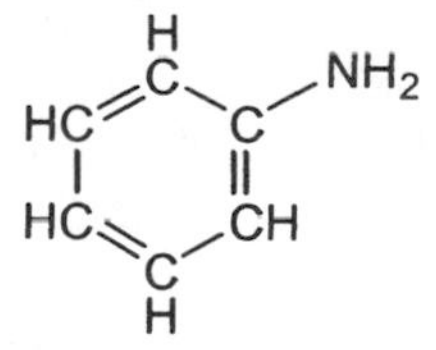   or 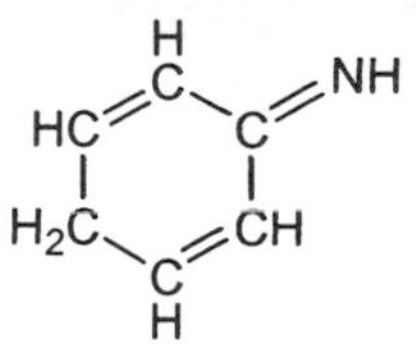   or others

ChemActivity 16

Spectroscopy Part A: IR

(What can an IR spectrum tell you about the structure of a molecule?)

Model 1: Resonant Frequency

Consider a parent pushing a child on a swing. If the parent pushes at the exact right intervals energy will be absorbed by the swinging child and she will go higher. If the parent pushes at the wrong moments the swinging child will not absorb energy from the parent and the swing will not go higher.

Pushing in **Resonance** with Swinging of Swing Pushing **not in Resonance** with Swinging of Swing

Frequency = cycles per unit time (minute or second). If the frequency of pushing matches the frequency of swinging, then the two are in resonance with each other.

Demonstration: A Weight Bouncing on a Spring

Soon you will see a demonstration in which a weight, suspended from a spring, will be dropped and allowed to bounce. Before the demonstration begins, **predict** which case below will have the **fastest frequency** (# of bounces per unit time) and which will have the **slowest frequency**.

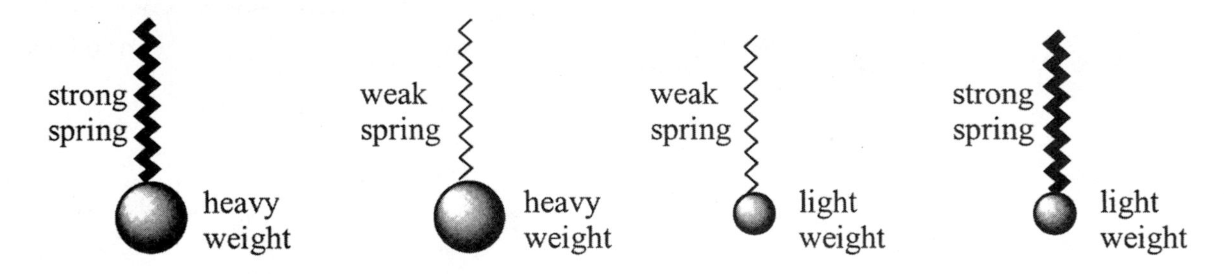

Critical Thinking Questions

1. Circle each of the following statements that fits with what you observed in the demonstration. **Cross out** each one that **does not fit** with what you observed.

   I. The stronger the spring, the higher (faster) the frequency of bouncing.
   II. The weaker the spring, the higher (faster) the frequency of bouncing.
   III. The heavier the weight, the higher (faster) the frequency of bouncing.
   IV. The lighter the weight, the higher (faster) the frequency of bouncing.

Figure 1: Some of the Stretching and Bending Motions of Water

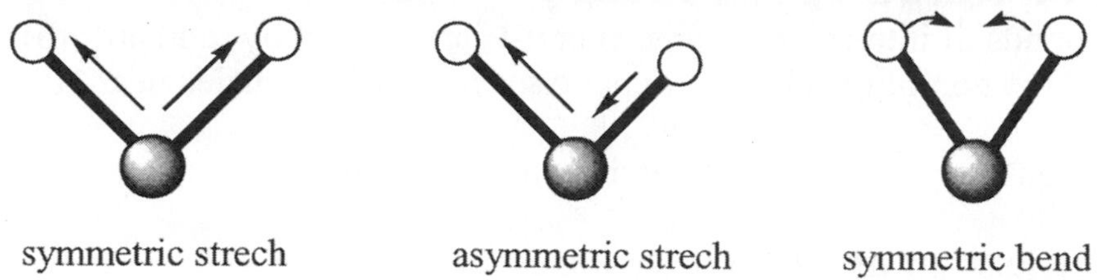

Bonds stretch, bend and wag as if they were balls attached by springs. A light wave of the correct frequency will transfer its energy to these motions, like the parent pushing the swinging child. We say the light is **absorbed** by the molecule. For example: microwaves (frequency near 10^{12} cycles per second) resonate with the above motions of water molecules. This is how your microwave oven adds energy to (heats) water.

Critical Thinking Questions

2. Pretend that your hands are the H atoms of water and your torso is the oxygen of water. Do the water molecule dance with your group, demonstrating the **symmetric stretch**, the **asymmetric stretch** and the **symmetric bend**.

3. Water molecules do all these at once (try that!), plus some other motions not shown. Each unique motion absorbs a unique frequency of light in the microwave range. Answer the following questions relating to the cartoon experiment below:
   a) What happened to the light not **transmitted** through the water?

   b) Why are some frequencies of microwave light not **absorbed** by the water?

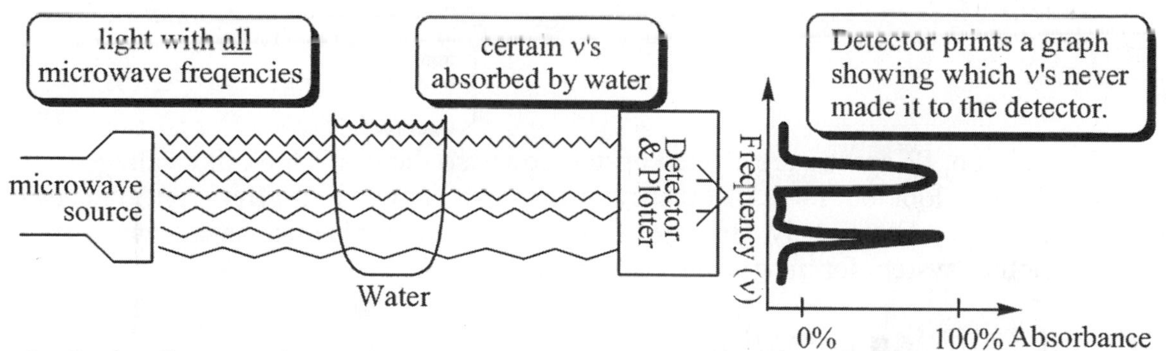

4. In the diagram above, does a peak on the graph represent frequencies that were **absorbed** by the water or frequencies that were **transmitted** through the water?

5. Which of the following carbon-carbon bonds is stronger?

$H_2C{=}CH_2$	$HC{\equiv}CH$
Most carbon-carbon double bonds absorb light of ν = A	Most carbon-carbon tripple bonds absorb light of ν = B

6. The Greek letter "nu" (ν) means **frequency.** Label light ν=A and light ν=B, above, with "**high** ν" or "**low** ν" based on the spring demonstration.

Model 2: Infrared Spectroscopy

Most bonds of interest to organic chemists are excited by (and absorb) light in the infrared (IR) region of the electromagnetic spectrum (frequencies near 10^{13} cycles per second).

Measurements are made using an infrared spectrometer (see Figure 2).

Figure 2: Diagram of Infrared Spectrometer

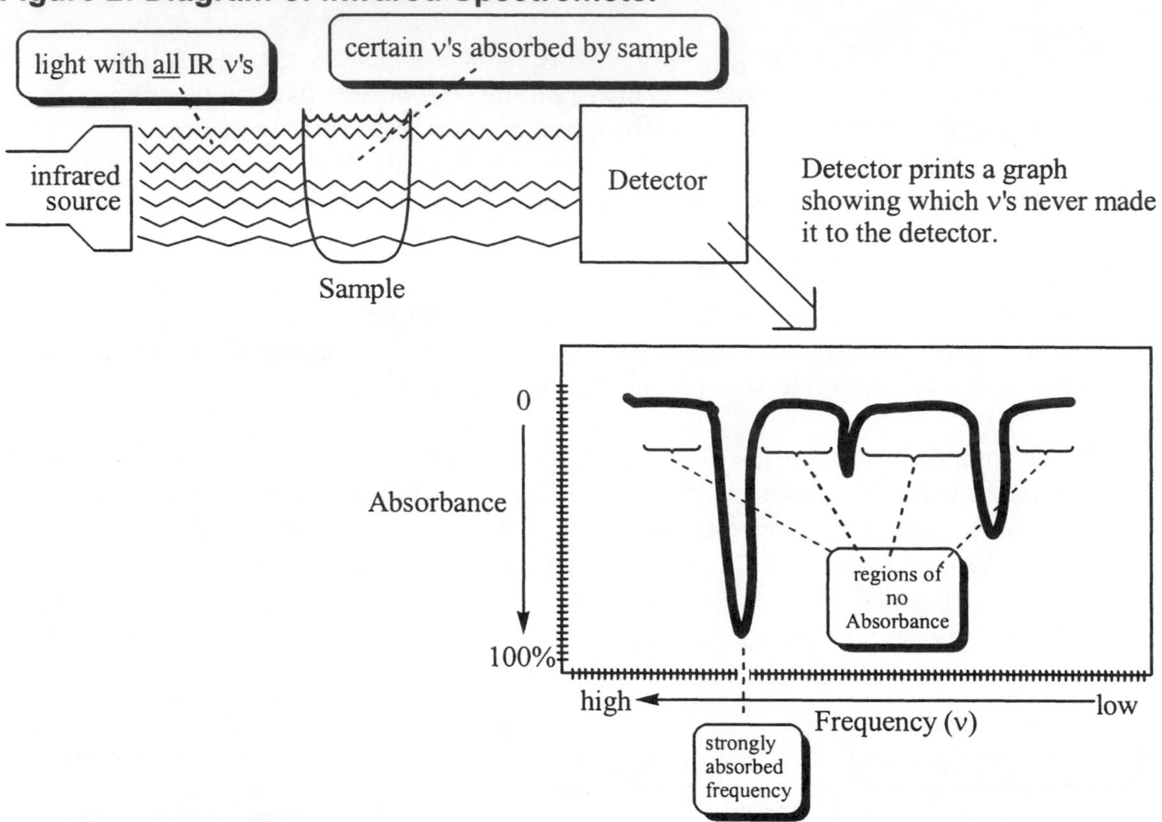

By tradition, IR spectra are printed upside-down, so the peaks look like valleys. Another thing to look out for is that the x axis is marked in **wavenumbers** (cm^{-1}) instead of frequency (Hz). Fortunately, cm^{-1} are proportional to Hz. This means wavenumbers are just another system for measuring frequency.

Critical Thinking Questions

7. What happened to the frequencies of light that never made it to the detector?

8. Light hitting the detector in Figure 2 is missing light of frequency A, but is not missing light of frequency B (see CTQs 5 & 6). What can you conclude about the structure of the molecule in the sample tube? That is, what structural features does the molecule likely have and not have?

Table 2b: Characteristic IR Absorption Frequencies of Organic Molecules

Type of Bond (functional group)	Absorption Range (wavenumbers = cm^{-1})	Type of Bond (functional group)	Absorption Range (wavenumbers = cm^{-1})
R—O—H (O-H of alcohol)	3200-3650	C≡C bond (of alkyne)	2100-2260
R—C≡C—H (C-H of alkyne) C_{sp}–H	3260-3330	C=O (of carbonyl)	1690-1760
$R_2C=C(R)H$ (C-H of alkene) C_{sp2}–H	3050-3150	C=C bond (of alkene)	1620-1680
R_3C—H (C-H of alkane) C_{sp3}–H	2840-3000	C-O bond (of alcohol)	1000-1260

9. For the IR spectrum below, use the data in Table 2b to match each major absorption peak outside the fingerprint region with a bond in the structure. (Note: each "peak" looks like a valley.)

Note: The "fingerprint" region of an IR spectrum is so crowded with peaks that it is almost impossible to make assignments there. However, this region can be used like a fingerprint to identify a compound. This is done by comparing an IR spectrum that you obtain in the laboratory with published libraries of IR spectra until you find a positive match.

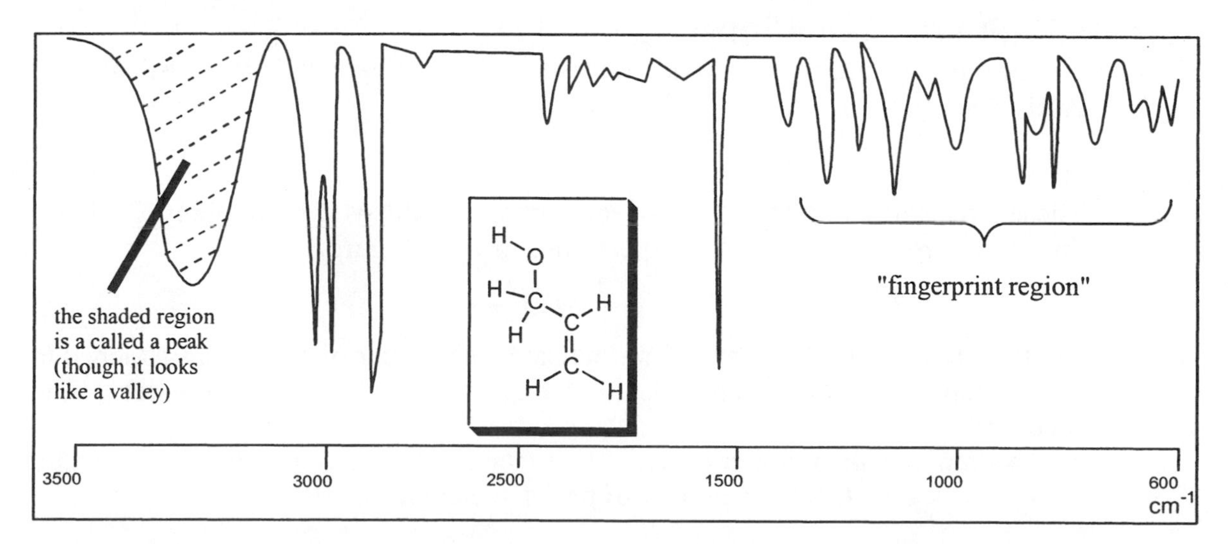

Spectroscopy Part B: ^{13}C NMR

(What can a ^{13}C NMR spectrum tell you about the structure of a molecule?)

NMR (nuclear magnetic resonance) spectroscopy is the most powerful tool that scientists have for looking at the structure of organic molecules. Medicine makes extensive use of this technique–though they drop the word "nuclear" and call it MRI (magnetic resonance imaging). The physics behind NMR is very complex, although it is similar to IR—except that radio frequency light is used instead of infrared light. The very low energy radio waves excite a property called **nuclear spin**. In the rest of this ChemActivity we will focus on how to interpret data from an NMR spectrum.

Model 3: ^{13}C NMR ("Carbon-13 NMR")

Figure 3: ^{13}C NMR Spectrum of 1-bromo-2,3,3-trimethylbutane

Each carbon in the structure below is labeled with a letter. This letter is used to associate the carbon with a cluster of lines (peak cluster) on the spectrum (e.g. peak cluster **a** goes with $\mathbf{C_a}$).

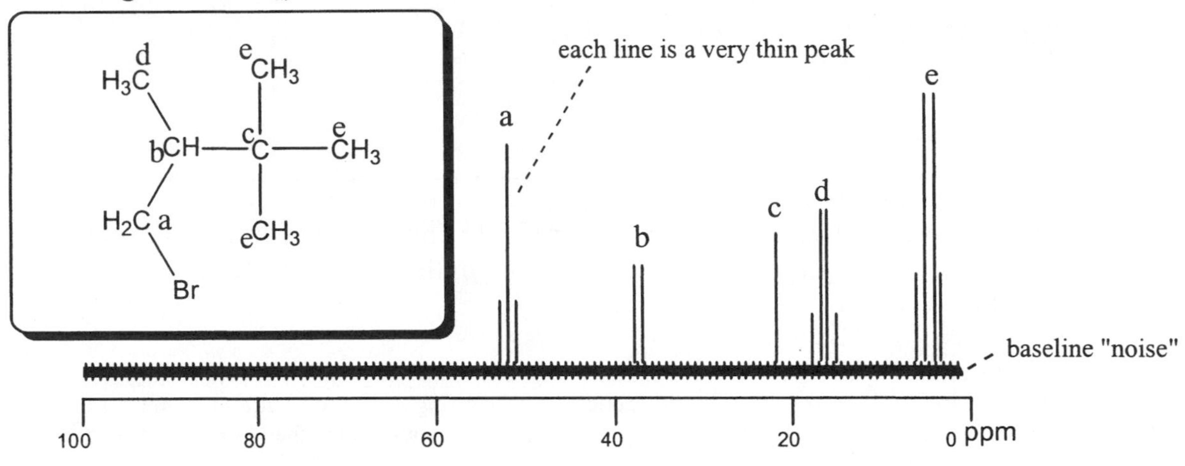

Critical Thinking Questions

10. The boxed structure the three carbon atoms labeled "e" are called "**chemically equivalent**." Explain why.

11. These three carbons labeled "e" are sometimes called **NMR equivalent**. Explain this term in relation to the NMR spectrum shown in Figure 3.

12. Next to each carbon in the structure, indicate the number of hydrogens attached to that carbon. Also write this number next to the corresponding **peak cluster** on the spectrum.
    a) Explain how the number of hydrogens attached to a given carbon is related to the **peak multiplicity** (number of peaks in a cluster).

b) Chemists say that "the two H atoms attached to C_a (we will call them H_a) '**split**' the signal of C_a into three peaks." This is illustrated in the diagram below for signal a and signal b. The dotted line illustrates what the peaks would look like without any splitting. Draw a similar diagram for signal d.

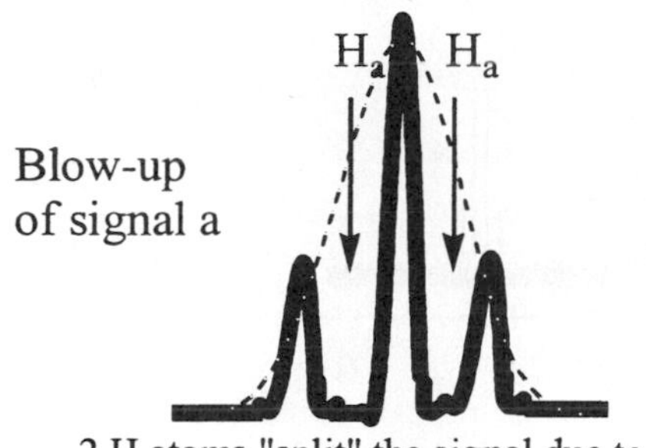

2 H atoms "split" the signal due to C_a

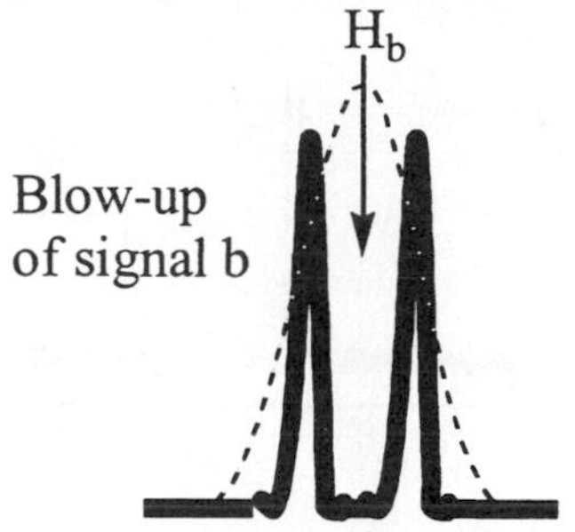

1 H atom "splits" the signal due to C_b

c) A peak cluster with…
only **one** peak is called a **singlet**,
two peaks is called a **doublet,**
three peaks is called a **triplet,**
four peaks is called a **quartet**
with **five or more** peaks is called a **multiplet**
Label each peak cluster in Figure 3 with one of these words.

13. Consider the placement of the peak clusters along the x axis of Figure 3.
a) Which of the following best explains the placement of peak clusters along the x axis?
   I. The number of hydrogens attached to that carbon atom.
   II The total number of bonds to that carbon atom.
   III. Distance from the Br atom. (# of bonds away from Br atom).

b) The property measured by the x axis is called **chemical shift** (think of it as frequency if you like). It is measured in strange units called ppm. Basically, the chemical shift of an atom is proportional to the **density of the electron cloud around that atom**. Based on the summary below, where do you expect the peak cluster due to a C attached to an electronegative F: at **high ppm** or at **low ppm**? Explain your reasoning.

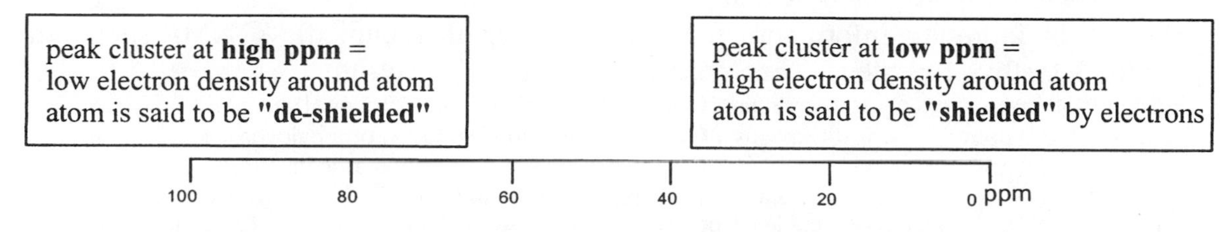

c) Is your answer in b) consistent with the peak cluster placements along the x axis in Figure 3? (Note that F, Br and all halogens are "electron hogs.")

14. The following is a proton coupled ^{13}C NMR spectrum of 2-bromobutane. Draw the structure of 2-bromobutane and match the peak clusters in the spectrum to the carbon atoms in your structure.

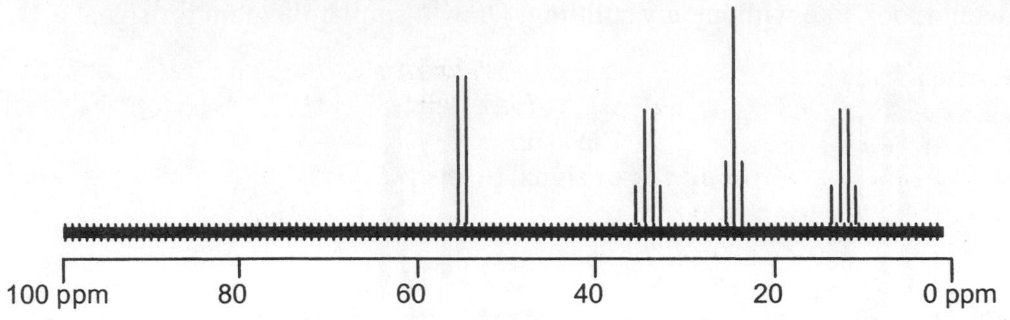

Model 4: Proton De-Coupled ^{13}C NMR Spectra

If you have a complex molecule with more than five different types of carbon atoms the peak clusters generally overlap. This makes interpretation very difficult. To solve this problem, chemists "**decouple**" the H's from the C's. A decoupled ^{13}C NMR spectrum of 2-bromobutane is shown below.

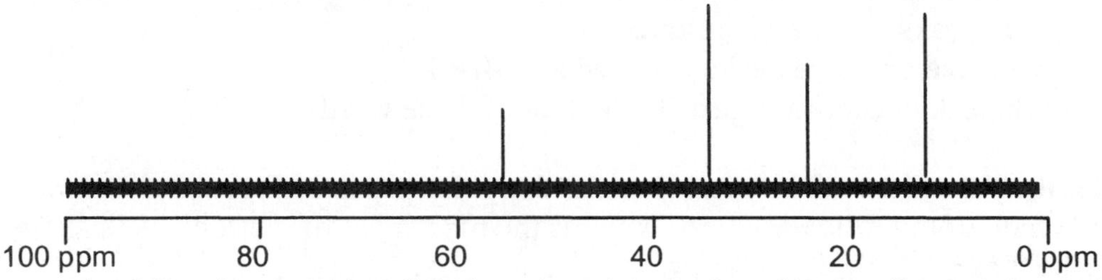

The following is a simplified explanation for how decoupling works. Think of each C–H bond as a phone link. In the coupled spectrum, each H is "splitting" the carbon signal through this "phone line." To make a decoupled spectrum, we produce lots of static on the "phone line" so each carbon cannot "hear" the H's attached to it. Consequently, each carbon shows up as a **singlet** (1 peak).

Critical Thinking Questions

15. What structural information is lost when we decouple the C's from the H's?

16. What basic structural information does a decoupled ^{13}C NMR spectrum still convey?

17. T or F: In ^{13}C NMR the height of a peak tells you exactly how many equivalent carbons are represented by that peak.
18. Use the following information to sketch a reasonable (coupled) ^{13}C NMR spectrum for 3-methyl-1-pentene. Then sketch the decoupled spectrum.

Peak at 0-50 ppm	means a **C** with only single bonds to C's or H's
Peak at 50-100 ppm	means a **C** with a single bond to O, N, or a halogen
Peak at 80-150 ppm	means a **C** with a double or triple bond attached to it
Peak at 150-200 ppm	means a **C** double bound to an O (C=O "carbonyl group")

Coupled ^{13}C NMR Spectrum of 3-methyl-1-pentene **Decoupled ^{13}C NMR Spectrum of 3-methyl-1-pentene**

Spectroscopy Part C: ^{1}H NMR

(What can a ^{1}H NMR spectrum tell you about the structure of a molecule?)

Model 5: ^{1}H NMR ("Proton NMR")

In ^{13}C NMR the signals are generated by carbon nuclei. ^{1}H NMR signals are generated by **hydrogen nuclei.** (Note that most ^{1}H NMR peaks appear in the 0-10 ppm range, while most ^{13}C NMR peaks appear in the 0-220 ppm range.)

Figure 5: ^{1}H NMR Spectrum

The peak at 0 ppm is a **reference peak**. It is not associated with any H on the structure shown.

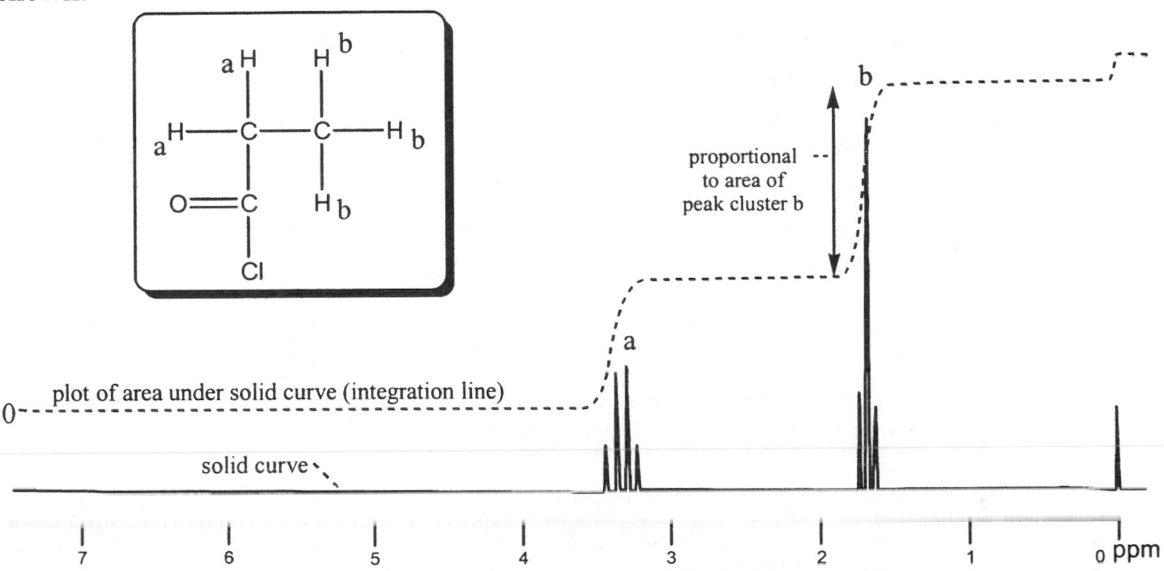

Critical Thinking Questions

19. How many chemically distinct H's and C's are there on the structure in Figure 5?
20. How many different peak clusters are associated with the structure in Figure 5?
21. Complete the following sentence: In a proton NMR spectrum there is a peak cluster for each chemically distinct **H** or **C** [circle one].
22. On a proton NMR spectrum, the chemical shift (ppm/placement along the x axis) conveys similar information as in ^{13}C NMR. Construct an explanation for why **peak cluster a** is farther to the left (higher ppm) than **peak cluster b**.

23. The dotted line stepping up from Area = 0 at the left is a cumulative measure of the integrated **area under the solid curve**. This is called the **integration line**.
    a) Draw a double-headed arrow with a length proportional to the area of **peak cluster a**.
    b) Which peak cluster **a** or **b** [circle one] has a larger area? How you can tell?
    c) What is the closest integer ratio **1:1, 1:2, 1:3, 2:3, 3:4** [circle one] that describes the **(area of peak cluster a) : (area of peak cluster b)**?

d) If H's could sing, which would be louder…two H's singing a tune together or 3 H's singing a tune together? (Assume each H's sings at a similar volume.)

e) What structural information is conveyed by peak cluster areas in a ^{1}H NMR spectrum?

24. Explain the chemical shift (ppm values), and peak cluster areas (integration line) for the following proton NMR spectrum. Assign a letter to each peak cluster and match this to appropriate H/H's on the structure. (The peak at 0 ppm is a reference.) For now, don't worry about the number of peaks in each peak cluster.

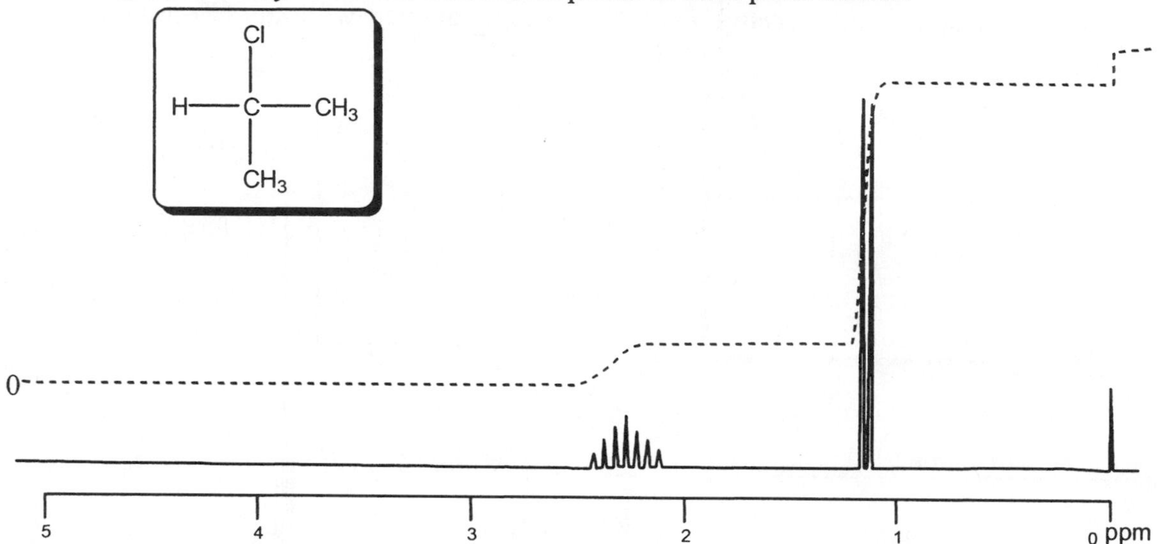

25. The easy way to determine that two H's are chemically equivalent (NMR equivalent) is to either find that …
- the H's in question are linked by an internal mirror plane.
- that rotation of single bonds allows you to switch the positions of H's in question.
  - a) Find the mirror plane between the 2 equivalent CH_3 groups in CTQ 24.
  - b) Explain why the 3 H's of a methyl group are NMR equivalent.

26. Interestingly, you can prove that H's on different molecules are NMR equivalent by finding an external mirror plane that links them. This means that **a molecule and its enantiomer will produce the exact same NMR spectrum.**

external mirror plane

CH_2CH_3 S C H H_3C Br | CH_2CH_3 R C H CH_3 Br

Imagine you have two bottles: one with (R) the other with (S)-2-bromobutane. Unfortunately, your lab partner messed up and labeled both bottles simply "2-bromobutane". Can you use NMR to sort out this problem? Explain why or why not.

27. In a ^{13}C NMR spectrum attached H's **"split"** the signal of a given carbon atom. (review CTQ 12b). In ^{1}H NMR nearby H's **"split"** the signal of a given H atom. Figure 5 is reproduced below:

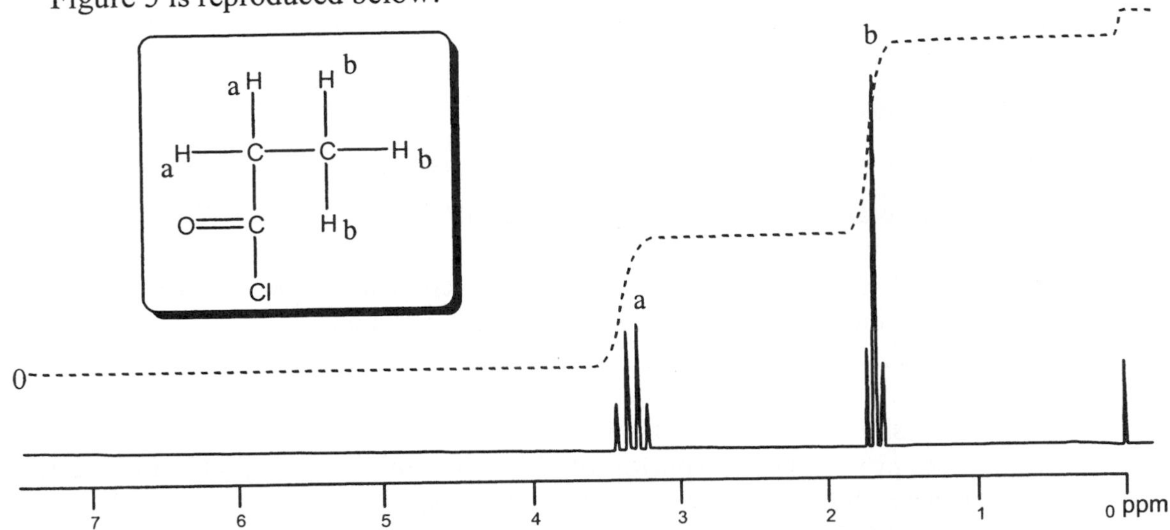

a) Label each peak cluster in the spectrum above as a singlet, doublet, triplet, quartet or multiplet, as appropriate.
b) How many H's are splitting signal a?
c) How many H's are splitting signal b?

Summary of Rules for Splitting in Proton NMR

- H's split C signals in ^{13}C NMR. (C's do not split other C signals.)
- H's split other H signals in ^{1}H NMR. (C's do not split H signals.)
- The NMR signal of an H atom is **split by any (nonequivalent) H within 3 bonds**.
- Chemically equivalent H's **do not** split each other.

d) How many bonds separate a representative H_a from a representative H_b?
e) According to the rules above, should H_a split H_b and vice versa? Explain.

f) What rule above tells you that no H_b will split signal b, nor will any H_a split signal a?
g) Blow-ups of peak clusters a & b are provided below. Replace each "?" with an **a** or **b** so as to show which H's split each signal.

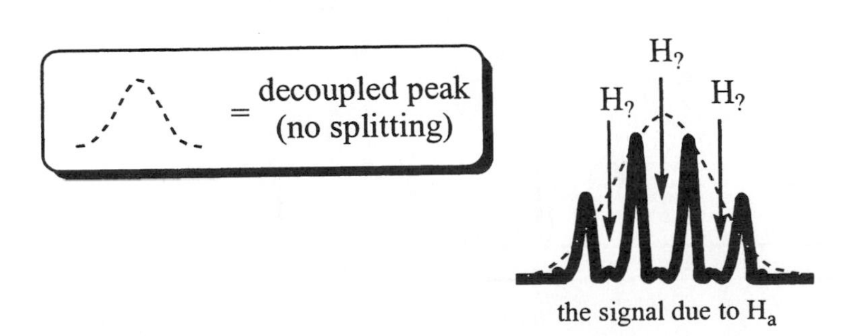

28. Consider the structure of 2-bromopropane.

Br
H—C—CH_3
CH_3

a) How many types of chemically distinct H's are there in 2-bromopropane? Label them a, b, …etc.

b) Predict the multiplicity (# of peaks in each peak cluster) for the peak clusters a, b,…etc. in the ^{1}H NMR spectrum of 2-bromopropane.

c) Sketch these peak clusters in the space to the right of the structure. Be sure to put the peak clusters in the correct order. (The one you expect at higher ppm should be on the left.)

d) The proton NMR spectra of 2-bromopropane and 2-chloropropane are nearly identical (except for slight differences in the ppm values.) Are your answers above consistent with the spectrum of 2-chloropropane in CTQ 24?

29. Proton NMR can be useful in determining the purity of your sample. If there are only peaks associated with your molecule then your sample is pure. However, it is often the case that your sample is contaminated by a second substance. Peak cluster areas can be used to estimate the % of impurity in your sample. A sample of fluoroethane is contaminated by a sample of 2-bromopropane. The proton NMR spectrum of this mixture is shown below. Estimate the ratio of [2-bromopropane]:[fluoroethane] in the sample tube.

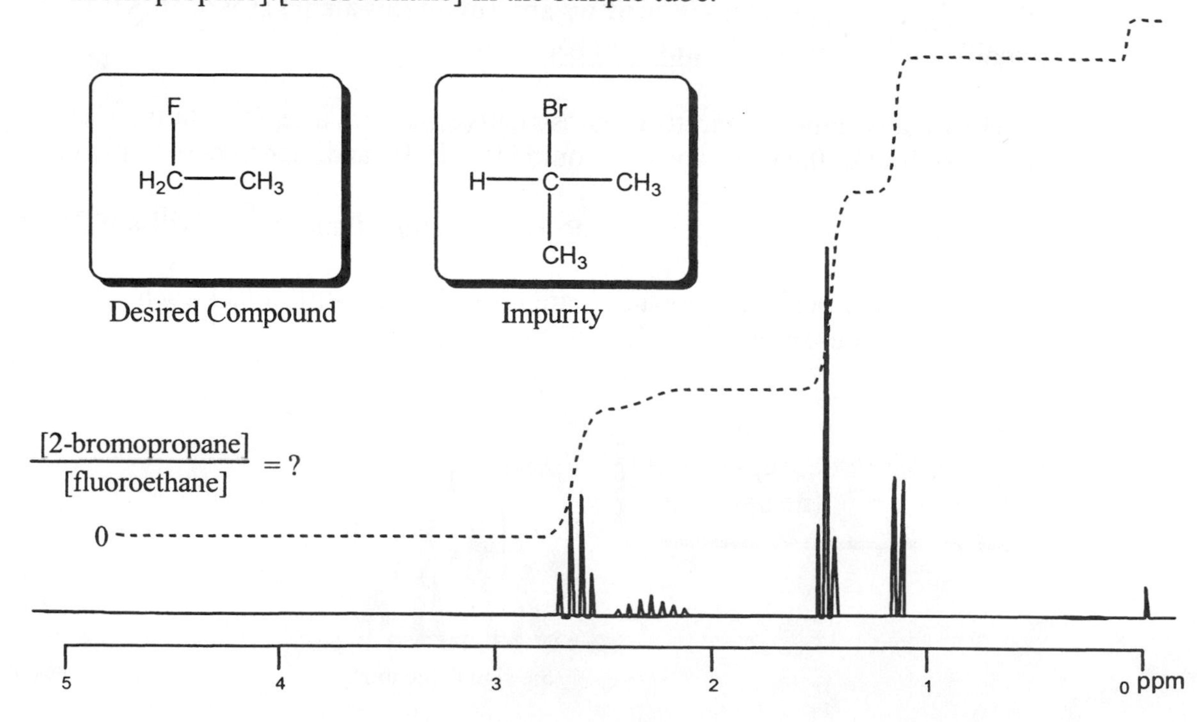

Exercises for Part A

1. Match each IR spectrum to a molecule from the pool of structures below each one.

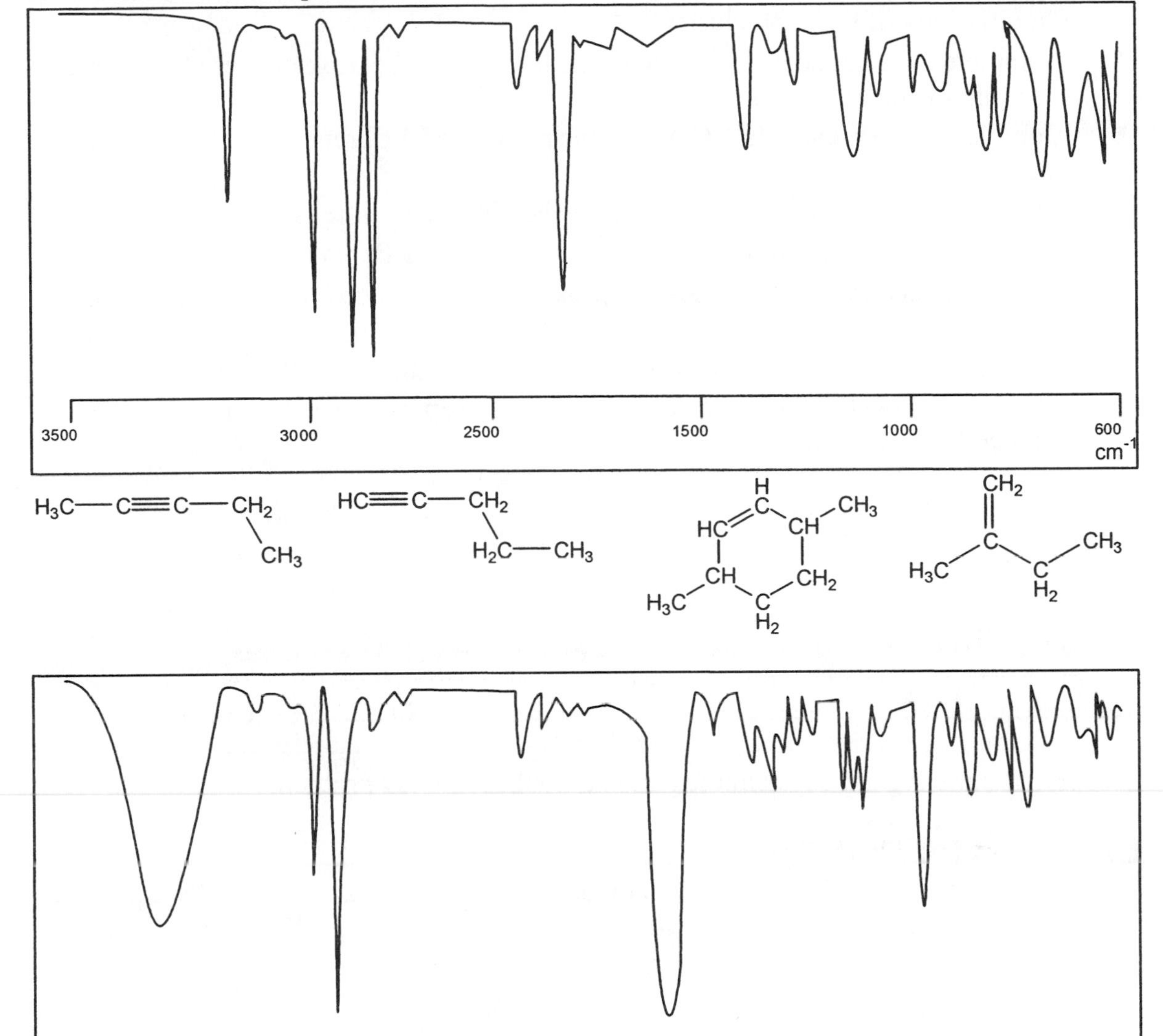

H2C=CH
OH
H2C—CH2
O
C
CH3
H3C
C
H2
HC≡C—CH2
C—CH3
O
CH3
C
OH
O
C
H2

2. Read the section in your text on **mass spectrometry**.

3. Read the assigned pages in the text and do the assigned problems.

Exercises for Part B

4. Complete the sentence: In coupled ^{13}C NMR, the number of peaks in a peak cluster ("multiplicity of the peak cluster") tells you…
5. Complete the sentence: In ^{13}C NMR, the location of the peak cluster along the x axis (ppm value) tells you…
6. Complete the sentence: In ^{13}C NMR, the number of different peak clusters tells you…
7. Complete the sentence: In ^{13}C NMR, the height of a peak tells you…
8. Chemists rarely use proton coupled ^{13}C NMR spectra Explain why.
9. T or F: In decoupled ^{13}C NMR each peak cluster is reduced to a singlet (a single peak).
10. The decoupled ^{13}C NMR spectrum of a molecule with the molecular formula C_6H_{12} is shown below. On the basis of this spectrum, propose a structure for this molecule.

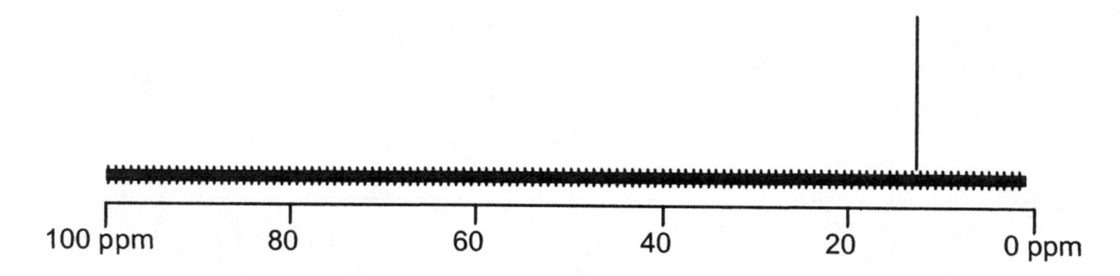

11. Read the assigned pages in the text and do the assigned problems.

Exercises for Part C

12. On the structure below, use letters (a, b, c, ...etc.) to indicate chemically equivalent and distinct hydrogens. (Also called NMR equivalent and NMR distinct H's.)

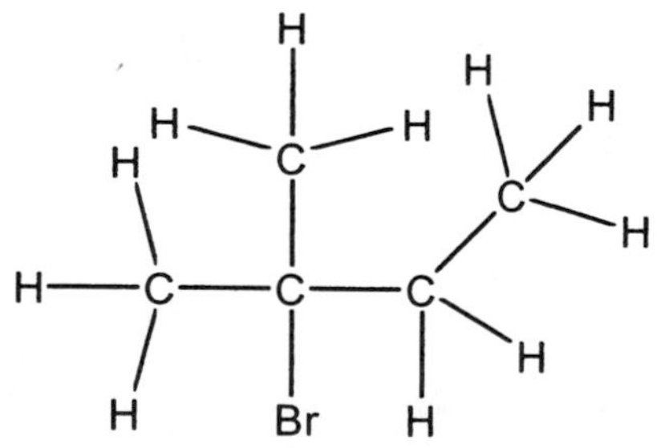

    a) How many peak clusters to you expect in the ^{1}H NMR spectrum of this molecule?
    b) Predict the ratio of areas of the peak clusters (a:b:c).
    c) Predict the **multiplicity** (= the number of peaks in a cluster) of each peak cluster and EXPLAIN your reasoning.

13. The proton NMR data for 1-bromopropane is summarized below on the far right. The chemically distinct hydrogens are labeled a, b, c...etc. and listed with their multiplicity and integration.

Example:

1 peak	**singlet**	s
2 peaks	**doublet**	d
3 peaks	**triplet**	t
4 peaks	**quartet**	q
5 or more	**multiplet**	m

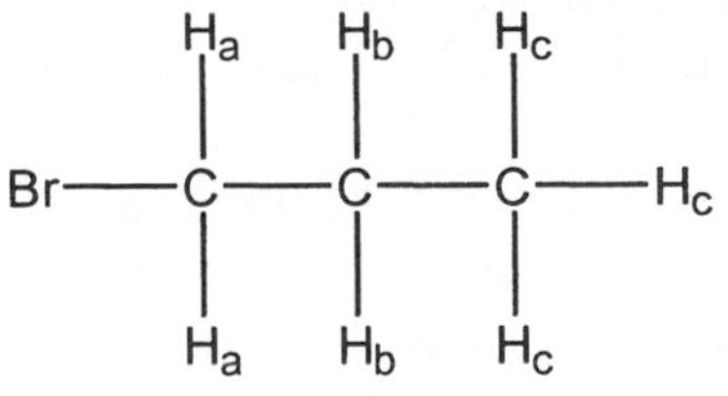

H_a : t, Int.= 2H

H_b : m, Int. = 2H

H_c : t, Int. = 3H

a) Through how many bonds can a hydrogen split another hydrogen?

b) According to this splitting rule, does H_a split H_c?

c) Is your answer in part a) consistent with the multiplicity listed for peak clusters a and c?

d) How many hydrogens split H_b?

e) Upon very close inspection of the proton NMR spectrum of 1-bromopropane, you would find that peak cluster b has 6 peaks. Is this consistent with your answer in part d)?

f) Speculate as to why any peak cluster with more than five peaks is listed simply as a "multiplet."

14. For each molecule below, label the chemically distinct hydrogens a, b, c...etc. and predict the associated multiplicity and integration for each set of hydrogens (*see* the example, above).

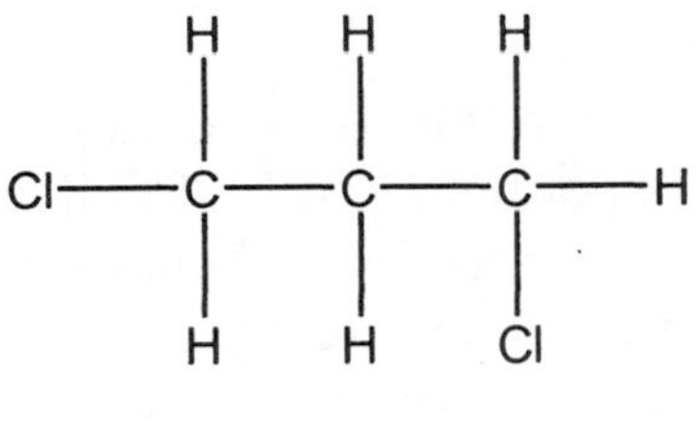

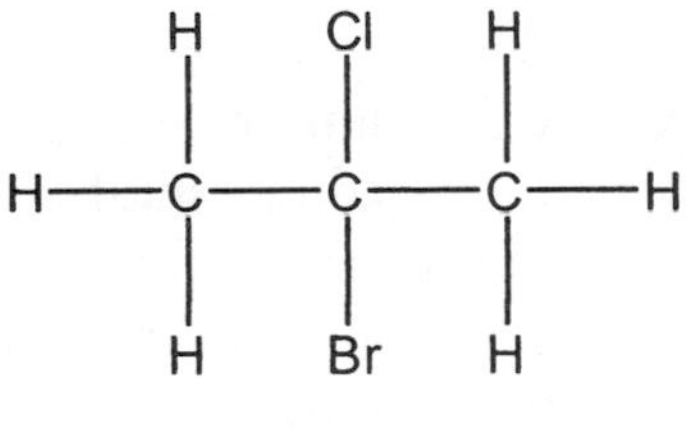

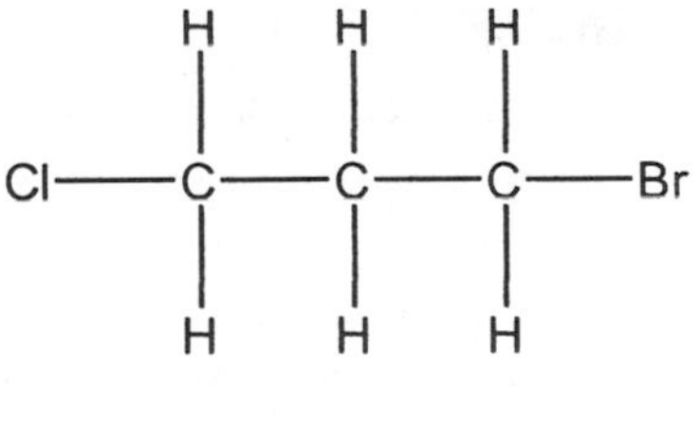

15. For reasons we did not discuss, the H of an alcohol functional group usually shows up on a proton NMR spectrum as a singlet. For each alcohol below, label the chemically distinct hydrogens a, b, c...etc. and predict the associated multiplicity and integration for each set of hydrogens.

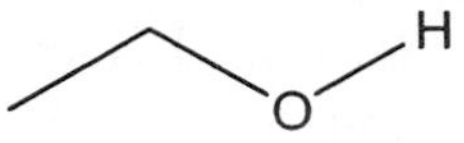

ethanol 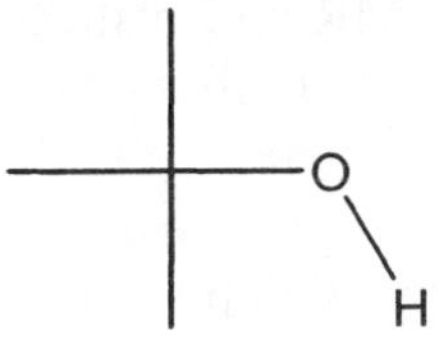 t-butanol 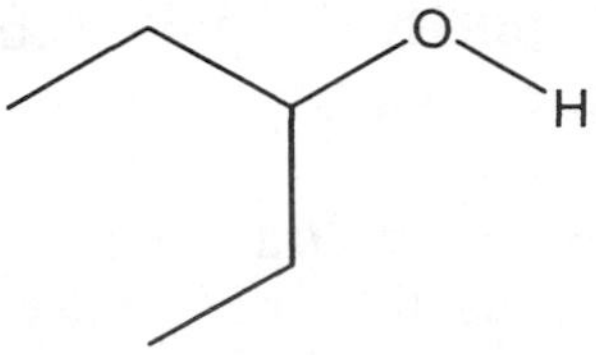 3-pentanol

16. Read the assigned pages in the text and do the assigned problems.

Summary: Interpretation of NMR Spectra

This ChemActivity is designed to help you remember the important aspects of interpreting NMR spectra. The following is a summary of important points:

^{1}H NMR

- **ppm** or **chemical shift** (*given by location along the x axis*) tells you the amount of electron density around an H. The closer the H is to an electronegative element, the more "deshielded" it is and therefore the higher the ppm number of its peak cluster (farther left on the spectrum). Multiple bonds also cause the signal of nearby H's to be shifted to the left. Look in the index of your text for "chemical shifts" to find a list of common functional groups with their chemical shifts. Each chemically distinct H should have a unique chemical shift, though in practice different peak clusters sometimes overlap just by coincidence. This can make spectrum interpretation very difficult. Proton NMR peaks usually appear in the 0-10 ppm range.
- **Integration** or **peak area** *(given by integration line)* tells you the relative size of each peak and therefore the relative number of equivalent H's represented by each peak. Note that integration only gives you a ratio of peak areas, making it impossible to tell the difference between a 1H to 3H ratio and a 2H to 6H ratio.
- **Multiplicity** is the number of peaks in a peak cluster (also called **splitting** or **proton-proton coupling**). It tells you the number of nonequivalent neighbor H's within three bonds. For example, a doublet (two peaks) tells you there is exactly 1 non-equivalent H within three bonds of the H responsible for this signal. Equivalent H's don't split each other.
- **NMR Equivalent Hydrogens** = hydrogens that can be associated via an internal mirror plane or are the same distance away (through bonds AND through space) from every other atom or feature in the molecule. This means that a molecule and it's enantiomer will produce the exact same NMR spectra. (Enantiomers are not distinguishable by NMR!!)

^{13}C NMR

- **ppm** or **chemical shift** in ^{13}C NMR is similar to that of proton NMR. If a C is close to an electronegative element or involved in a multiple bond, or both, you will find the corresponding peak at higher ppm (farther left on the spectrum). Each chemically distinct C should have a unique chemical shift, though in practice different peak clusters sometimes overlap just by coincidence. This can make interpretation of coupled spectra very difficult. This is why decoupled spectra are usually taken. C-13 NMR peaks appear in the 0-220 ppm range, though most peaks are found < 100ppm.
- **Peak area** has LITTLE MEANING in C-13 NMR. Peak area is a function of many things, one being whether the C is 1°, 2°, 3° or 4°. You will not be asked to interpret the meaning of peak area in C-13 NMR.
- **Multiplicity** in a proton coupled ^{13}C NMR spectrum tells you the number of H's attached to a given carbon. For example, a C that produces a doublet (two peaks) must have exactly 1 H attached to it. Very often chemists record "proton decoupled" ^{13}C NMR spectra. Such spectra have a singlet for each chemically unique carbon. This is useful for complex molecules for which the peak clusters would overlap. A decoupled ^{13}C NMR spectrum tells you the number of different carbon atoms present in the sample.

ChemActivity 17

Part A: Electrophilic Addition

(What is the mechanism of electrophilic addition?)

Model 1: Electrophilic Addition

Figure 1: Addition of H—Br to an Alkene

$$R_2C{=}CR_2 + H{-}Br \xrightleftharpoons{\text{solvent = diethyl ether}} H{-}CR_2{-}CR_2{-}Br$$

Table 1: Effect of Solvent and Substrate on Rate of Electrophilic Addition

	$H_2C{=}CH_2$	$H_2C{=}CH(CH_3)$	$H_3C(H)C{=}C(H)CH_3$	$H_2C{=}C(CH_3)_2$	$(H_3C)_2C{=}C(CH_3)_2$	$C_6H_5CH{=}CH_2$
Solvent = Non-Polar (hexane)	Rate = NO Rxn	Rate = 0.0001	Rate = 0.0001	Rate = 0.01	Rate = 0.01	Rate = 0.1
Solvent = Polar (Diethyl ether)	Rate = NO Rxn	Rate = 1	Rate = 1	Rate = 100	Rate = 100	Rate = 1000

Critical Thinking Questions

1. According to the rate data in Table 1, which solvent best facilitates this type of reaction: a **non-polar solvent** or **a polar solvent** [circle one]?
2. Is your conclusion above consistent with a hypothesis that this reaction involves a **carbocation intermediate**? Explain.

3. Addition of sulfuric acid = H^+ (HSO_4^-) speeds up the reaction in Figure 1. Addition of sodium bromide = (Na^+) Br^- has no effect on the rate of this reaction. (Inert counter-ions are in parenthesis.) Based on this information:
   a) Is H^+ involved in the slow step of this reaction?
   b) Is Br^- involved in the slow step of this reaction?

Information

In the game of pool it is almost impossible for three balls, all in motion, to collide at the exact same instant. Similarly, in organic chemistry three separate molecules are unlikely to collide in a single instant. Practically, this means that we cannot propose a reasonable mechanism in which three separate molecules are involved in a single step. Also, any reaction that involves **three or more reactants must go by a multi-step mechanism.**

4. Use curved arrows to propose a two step mechanism for the reaction in Figure 1, which is redrawn for you below. Assume the first step is the slower step and make sure your mechanism is consistent with your answer to CTQ 2.

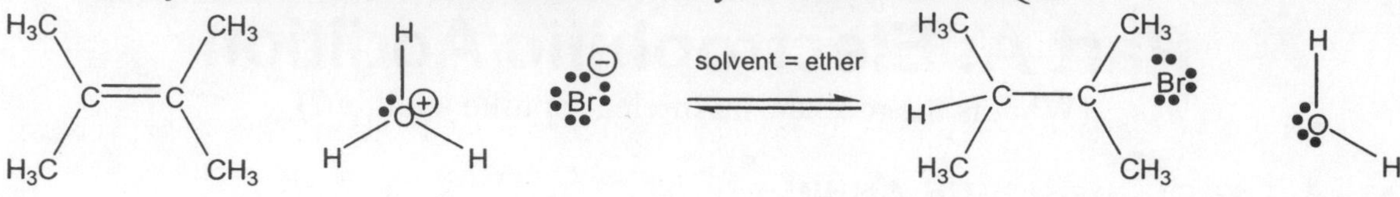

5. Use curved arrows to show step 1 of each electrophilic addition below and draw the intermediate carbocation that results in each case.

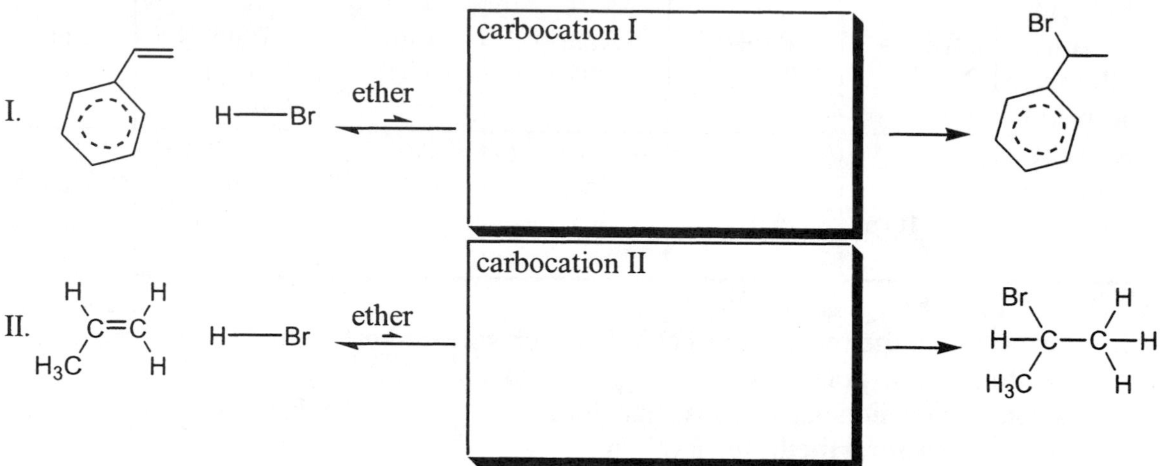

6. Consider the following energy diagram for the two reactions in above:

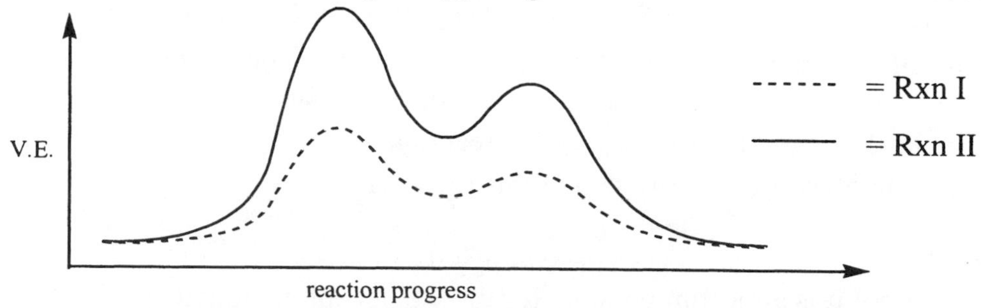

a) Mark the points on the energy diagram that correspond to carbocations I & II.
b) Construct an explanation for why carbocation I is lower in potential energy than carbocation II.

Model 2: Energy Diagrams, Reaction Rates and the Hammond Postulate

- **The height of an activation barrier (E_{act}) determines the rate of a reaction.** The higher the potential energy of the transition state, the higher the barrier, and the slower the reaction.
- It is very difficult to predict the potential energy of a transition state, but it is often easy to predict the potential energy of an intermediate.
- One application of the **Hammond Postulate** is "**For two similar reactions, the one with a higher potential energy intermediate will also have higher potential energy transition state.**"

Critical Thinking Questions

7. Based on the information in Model 2 and the energy diagram on the previous page, which reaction in CTQ 5 is faster: **I** or **II** [circle one]? Explain your reasoning.

8. Is the Hammond Postulate true for the two reactions in CTQ 5? Explain.

9. According to Table 1, the following reaction is so slow as to not occur at all. Use curved arrows to show the mechanism of Step 1 of this reaction. Draw the carbocation intermediate and explain why ethene doesn't react with HBr.

H₂C=CH₂   H—Br   (ether) ⇌ [carbocation] → CH_3CH_2Br

this product is not observed!!

10. Consider the following alternative product for Rxn I in CTQ 5.
    a) Draw the carbocation intermediate that would give rise to this product.

b) Explain why the product shown in CTQ 5 is formed instead of this product.

11. Draw the complete mechanism including the intermediate and most likely product for each of the following addition reactions.

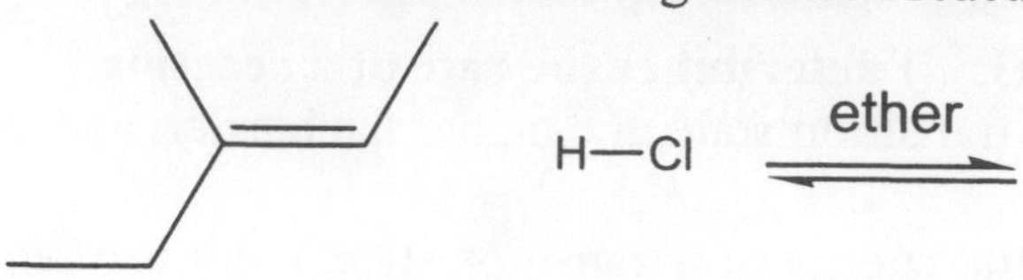

H—Cl ether

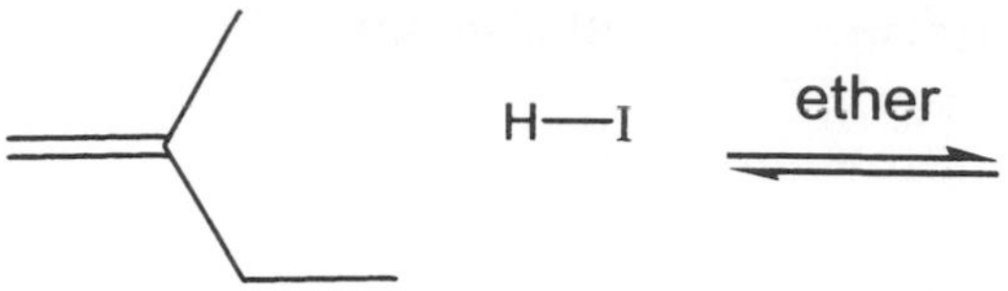

H—I ether

Part B: Carbocation Rearrangements

(How do we explain certain unexpected electrophilic addition products?)

Model 3: Markovnikov and Anti-Markovnikov Addition

In 1869, Russian Chemist, Vladimir Markovnikov (1838-1904) formulated the following rule: "Electrophilic addition involving a double bond that is not equally substituted at both ends results in the nucleophile attaching itself to the more substituted carbon."

- This is now known as **Markovnikov's Rule**
- An addition product obeying this rule is called a **Markovnikov Product.**
- One that does not obey this rule is called an **anti-Markovnikov Product.**

Critical Thinking Questions

12. Label each of the products you drew in CTQ 10 as "Markov. Products."
    a) Draw the anti-Markovnikov products that DID NOT FORM in the reactions.
    b) Label these products "anti-Markov" and note that they **"do not form"**.

Model 4: Non-Reactivity of Benzene Ring

The three π bonds of the benzene ring are **less reactive** than normal π bonds. For now think of them as "**resonance stabilized**" making them harder to break. Later we will discuss this phenomenon, known as **aromaticity**, in more detail.

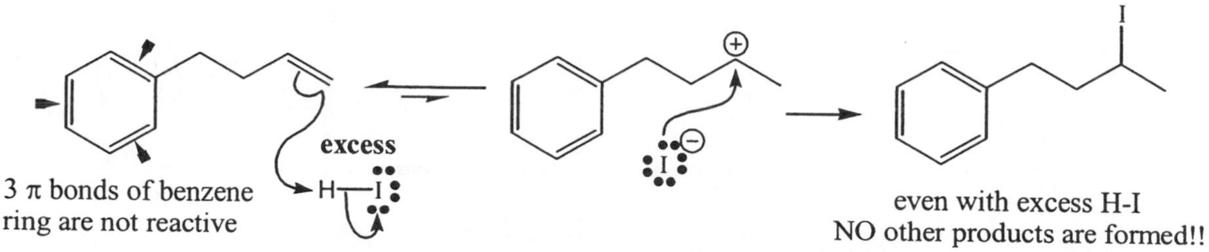

excess

H—I

3 π bonds of benzene ring are not reactive

I

even with excess H-I
NO other products are formed!!

Note the way the curved arrow "bounces" to one end of the π bond. This is a way of saying that the electrons form a bond from the terminal C to the H^+, leaving a 2° C^+.

Critical Thinking Questions

13. Use curved arrows to show the most likely electrophilic addition reaction and...
    a) Complete the drawing of the carbocation intermediate in Box 1.
    b) Draw the resulting Markovnikov Br addition product.

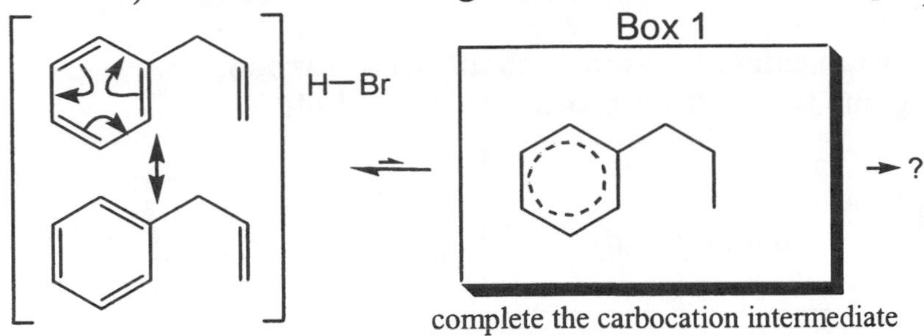

14. An unexpected product is formed from the reaction in the previous question. The **actual product formed** is shown *below, on the far right.*
    a) In Box 2 (thinking backwards from the **Actual Product Formed**) draw the carbocation that would most easily give rise to the product shown.
    b) Draw all important resonance structures of the carbocation you drew in Box 2.

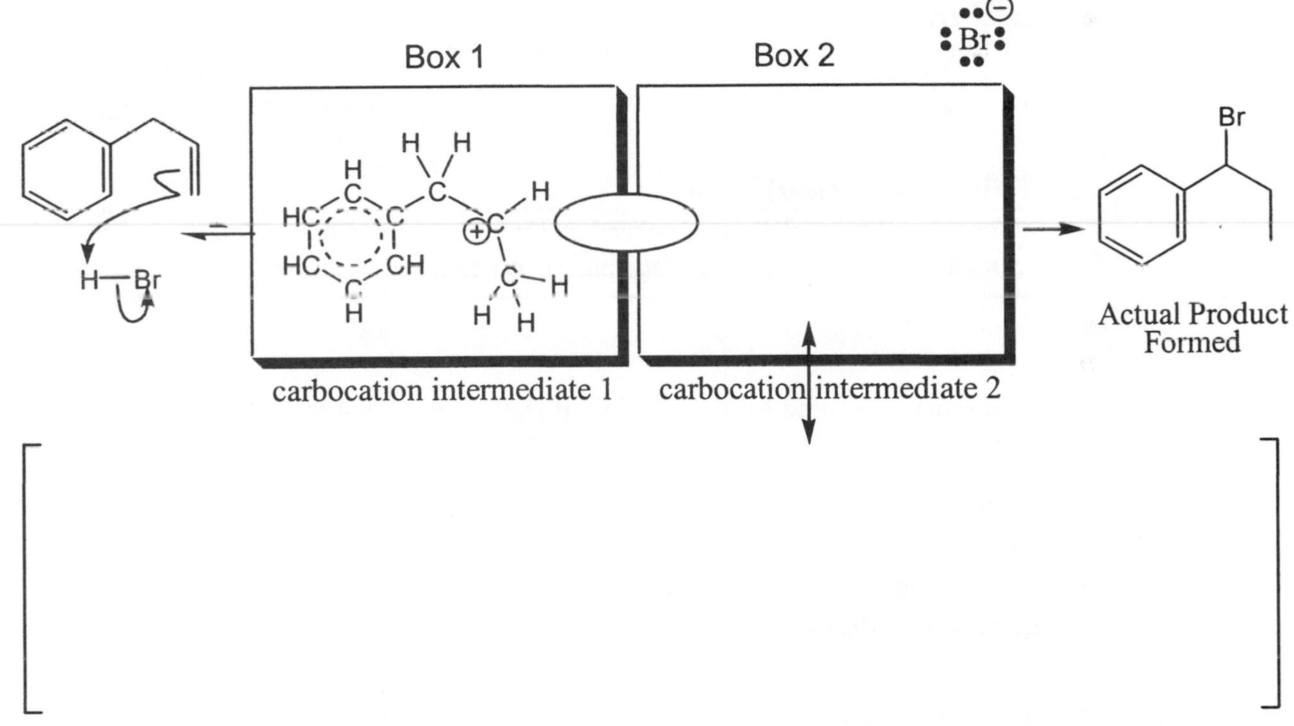

Important Resonance Structures of Carbocation 2

c) Can the + charge on carbocation 1 be spread out at all? Explain.

d) Which carbocation (in **Box 1** or in **Box 2**) [circle one] is lower in potential energy? Draw appropriate equilibrium arrows in the oval between the boxes to indicate which carbocation is favored at equilibrium.

Model 5: Carbocation Rearrangement

A **carbocation rearrangement** will occur when there is a driving force (an energy incentive) to do so. There are two kinds of carbocation rearrangements.

I. Hydride shift.

II. Alkyl shift.

Note: A hydride or methyl group can only jump to an adjacent carbon.
(In each example below, the group that shifts/migrates is shown in **bold**.)

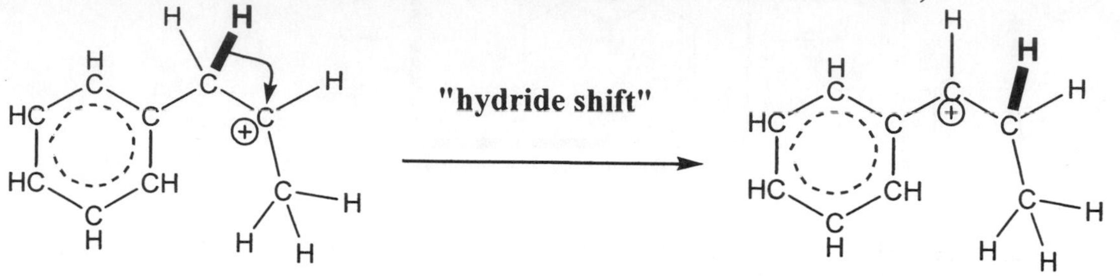

"hydride shift"

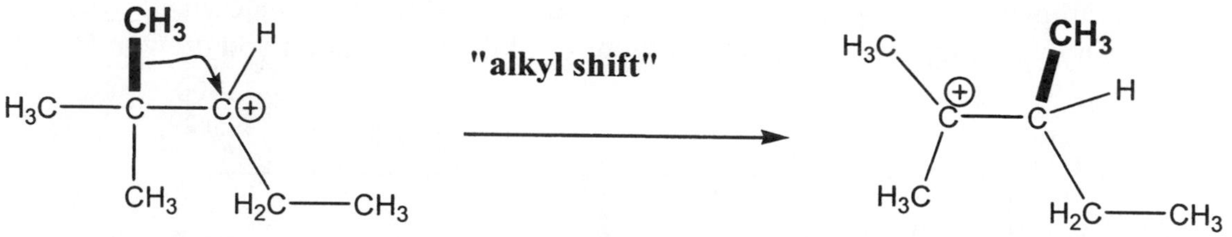

"alkyl shift"

Critical Thinking Questions

15. The top example in Model 5 is a curved arrow depiction of how carbocation 1 turns into carbocation 2 (from the previous page). Write in words what this curved arrow "says."

16. What is the driving force for the alkyl shift shown in Model 5?

17. Use curved arrows to depict the hydride shift that will likely occur for the carbocation below and **explain why the new carbocation is lower in potential energy** than the original. (Hint: Draw any important resonance structures of each!)

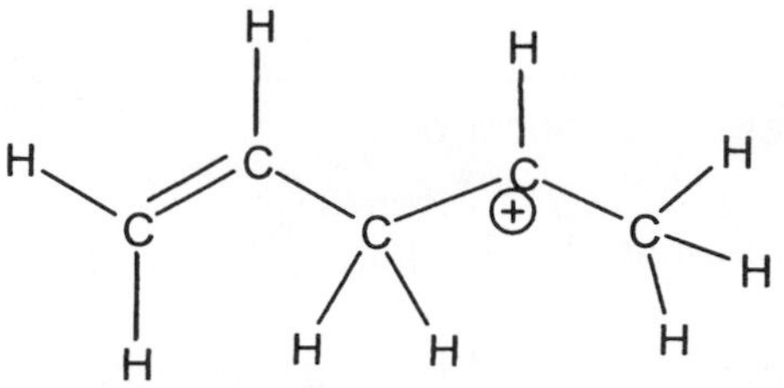

Part C: Acid Catalyzed Electrophilic Addition

(What is does it mean to say a participant in a reaction is a catalyst?)

Model 6: Effect of a Catalyst on a Reaction Profile

A species is considered a catalyst in a reaction if both of the following are true:

- The catalyst speeds the rate of the reaction.
- The catalyst is neither consumed nor produced, on net, by the reaction. (If it is consumed it is a reactant, if it is produced it is a product.)

A typical effect of a catalyst on the progress of a reaction is shown below:

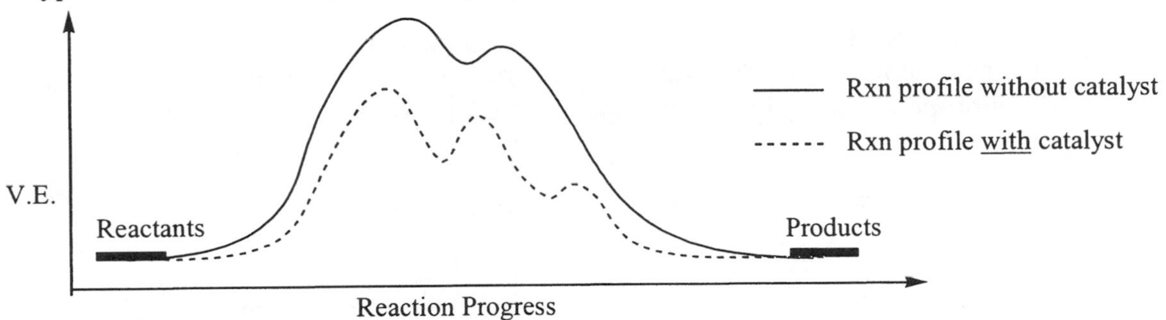

Critical Thinking Questions

18. Explain how you can tell from the energy diagram that the reaction with the catalyst is faster than the reaction without the catalyst.

19. **True** or **False**: In the diagram above, the catalyst changes the energy of the intermediates and transition states, but not the energy of the reactants or products.

20. How many steps are involved in each reaction (with and without the catalyst)? Note: a **reaction step** is defined as a change from one valley ("local energy minimum")-through a transition state-to another valley.

21. The following electrophilic addition reaction does not occur.

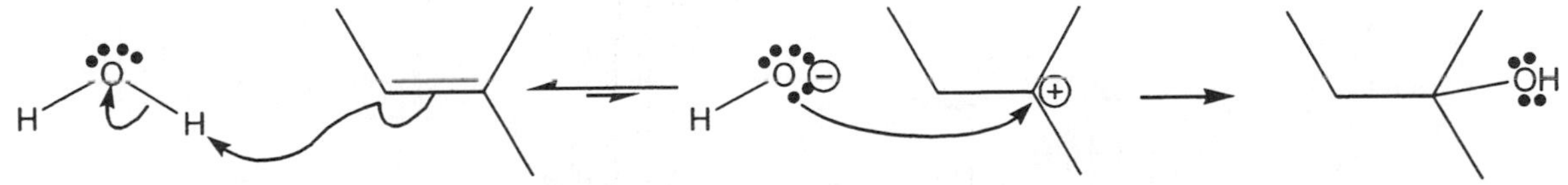

   a) Circle the highest potential energy species involved in the reaction.
   b) Which step in this reaction mechanism is very unfavorable. Explain your reasoning.

   c) Which reaction profile in Model 6 matches this mechanism: **solid** or **dotted** [circle one]?

Model 7: Choosing the Best Mechanism

- For a given reaction we can think of different competing mechanisms.
- The mechanism that is **most likely** is the one that is **FASTEST**.
- A good trick for finding the fastest mechanism: **Look for the mechanism with the lowest potential energy intermediates!**
- **Rule of Thumb**: Try always to think of a mechanism that does not involve any intermediates that contain a C, N, or O with a -1 formal charge.

Critical Thinking Questions

22. Explain why the "Rule of Thumb" in Model 7 works. That is, why are mechanisms involving an intermediate with a negatively charged C, N or O usually very slow? Hint: see the solid energy diagram in Model 6.

23. Use curved arrows to devise the most likely mechanism for the following reaction. Hint: The reaction takes three steps and it is possible to complete this mechanism without proposing any high energy species such as negatively charged C, N or O.

H O H + (alkene) + H_3O^+ ⇌ (alcohol) OH + H_3O^+

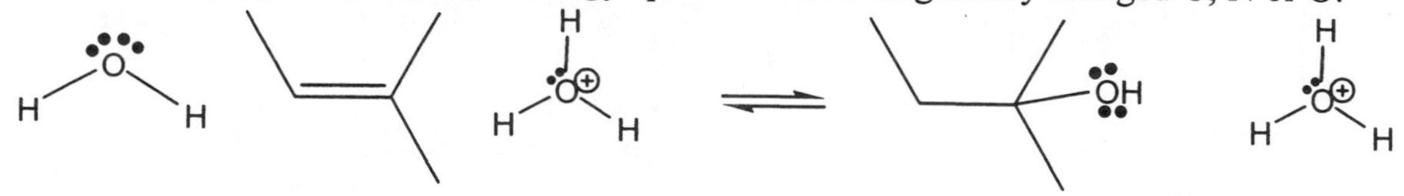

24. Does your mechanism in CTQ 23 match one of the reaction profiles in Model 6?
25. Explain why H_3O^+ is considered a **catalyst** in the reaction in CTQ 23. (See the definition of catalyst in Model 6.)

26. A student proposes the following reaction mechanism for the reaction in CTQ 23.
    a) Is H_3O^+ acting as a catalyst in this reaction? Explain.
    b) Which step in this mechanism is NOT reasonable? Explain your reasoning.

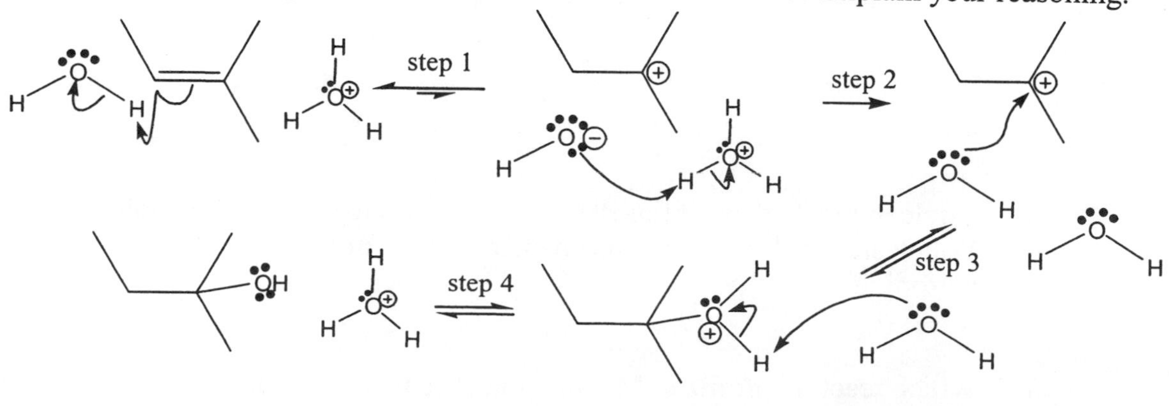

step 1

step 2

step 3

step 4

27. Confirm that without an acid catalyst (H_3O^+) there is no alternate mechanism for the reaction in CTQ 23 that avoids the high energy oxygen with a -1 formal charge.

28. Sketch an energy diagram for a reaction that has…
    a) Low energy reactants, slightly higher energy intermediate, and low energy products.
    b) Low energy reactants, high energy intermediate, and low energy products.
    c) High energy reactants, and high energy products.
    d) High energy reactants, and low energy products.

29. The reaction below is an example of a step that produces a high energy species (hydroxide, HO^-), but is still favorable.

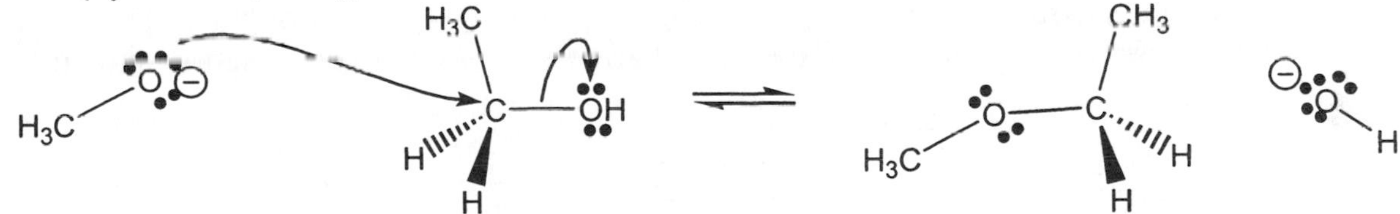

a) Explain how you know that the reactants and products in this step have about the same potential energy.

b) Which energy diagram that you drew in CTQ 28 fits this reaction?

c) Explain why the following addendum to the Rule of Thumb in Model 7 works: **It is ok to propose a high energy product/intermediate in a step if you start with a high energy reactant.**

Part D: Electrophilic Addition to a Diene

(How do multiple products arise during elelctrophilic addition to a diene?)

Model 8: Electrophilic Addition to a Diene

Figure 8: Addition of H–Br to 1,3-Butadiene

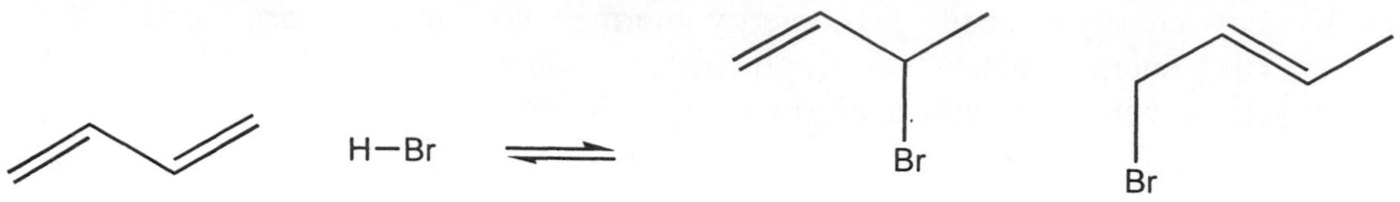

H—Br

Br

Br

Product I
(major product at -20°C)

Product II
(major product at 40°C)

Critical Thinking Questions

30. Draw a mechanism for the reaction in Figure 8 that gives rise to **Product 1**. ("Draw mechanism" always means draw curved arrows and show all intermediates.)

31. Thinking backwards from Product 2, propose a carbocation intermediate that would easily give rise to Product 2. (Use curved arrows to illustrate the reaction shown.)

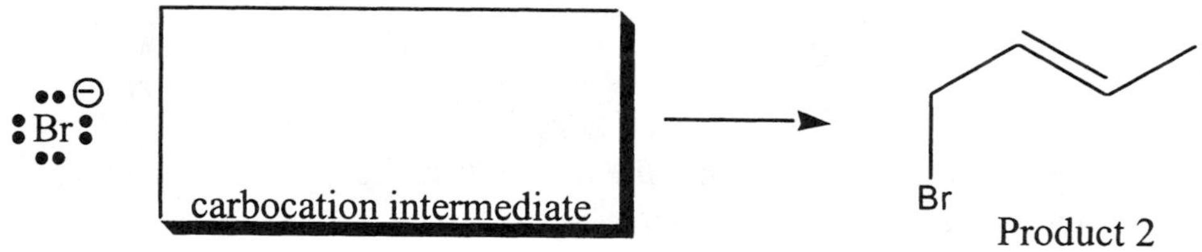

$:\ddot{\underset{..}{Br}}:^{\ominus}$

carbocation intermediate

Br

Product 2

32. What is the relationship between the intermediate that gives rise to Product 1 and the intermediate that gives rise to Product 2?

33. Show a mechanism starting from the reactants in Figure 1 that gives **Product 2**.

Model 9: Kinetic vs. Thermodynamic Products

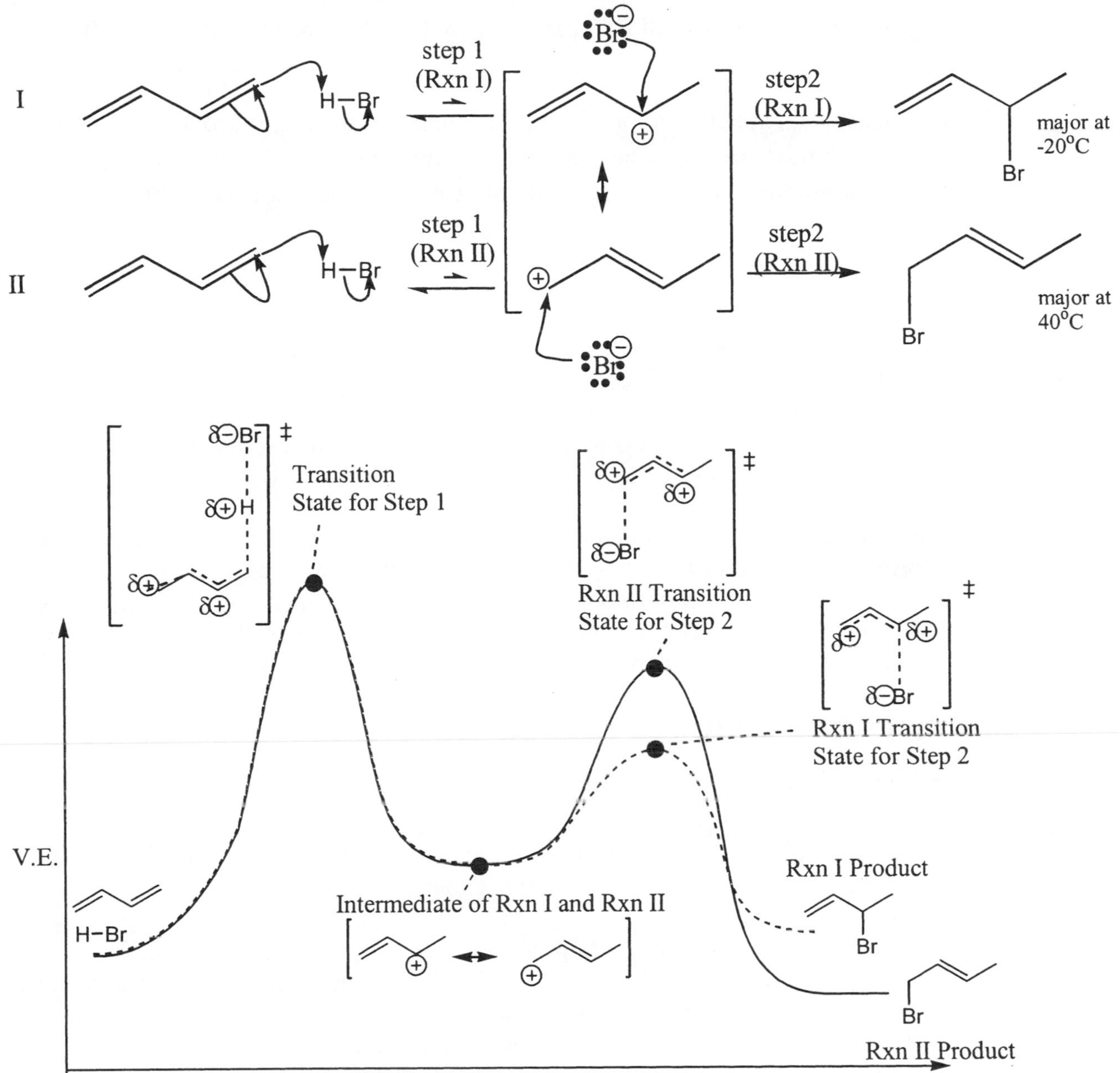

Critical Thinking Questions

34. Identify the drawing of the transition state for step 1 on the energy diagram above and mark each bond that is in the process of breaking as "bond breaking" and each bond that is in the process of forming as "bond forming."
35. Based on the energy diagram above, circle either True or False for each of the following:
    a) **True or False:** The first step, including the structure of t.s.#1 (= "transition state of step #1") is the same for Rxn I and Rxn II.
    b) **True or False:** The intermediate products of Reactions I and II are identical.
    c) **True of False:** t.s.#2 of Rxn I is identical to t.s.#2 of Rxn II.

36. Which is higher in P.E.: **t.s.#2 of Rxn I**. or: **t.s.#2 of Rxn II**?

37. Explain why the product of Rxn I is higher in P.E. than the product of Rxn II.

38. Construct and explanation for why Rxn I Product is the major product when there is a very limited amount of heat energy available (e.g. -20°C), but Rxn II Product is the major product when there is plenty of heat energy available (e.g. 40°C).

39. The product produced by the faster reaction (lower activation energy) is called the "**kinetically favored product**." This product is favored at lower temperatures when getting over the higher activation barrier is very difficult-making the reaction rate the deciding factor. Write this term next to the appropriate product on the energy diagram on the previous page.

40. The product at lower potential energy is called the "**thermodynamically favored product**." This product is favored when there is plenty of energy so either activation barrier can be surmounted-making the potential energy of the products the deciding factor. Write this term next to the appropriate product on the energy diagram on the previous page.

Exercises for Part A

1. Look back at Table 1 on the first page of this ChemActivity. Explain the rate data in this table. In particular, explain why one disubstituted alkene in the table (2-methylpropene) undergoes electrophilic addition 100 times faster than the other disubstituted alkene in the table (*trans*-2-butene). (Note that *trans*-2-butene and *cis*-2-butene undergo electrophilic addition at about the same rate.)

2. For each alkene draw all the likely products of electrophilic addition of one equivalent of strong acid, HX (where X = Cl, Br or I) and use curved arrows to show the mechanims of each reaction. Note that in cases where a one new chiral center is generated, a racemic mixture (50% R and 50% S is created). In cases where two chiral centers are generated, four stereoisomers are generated.

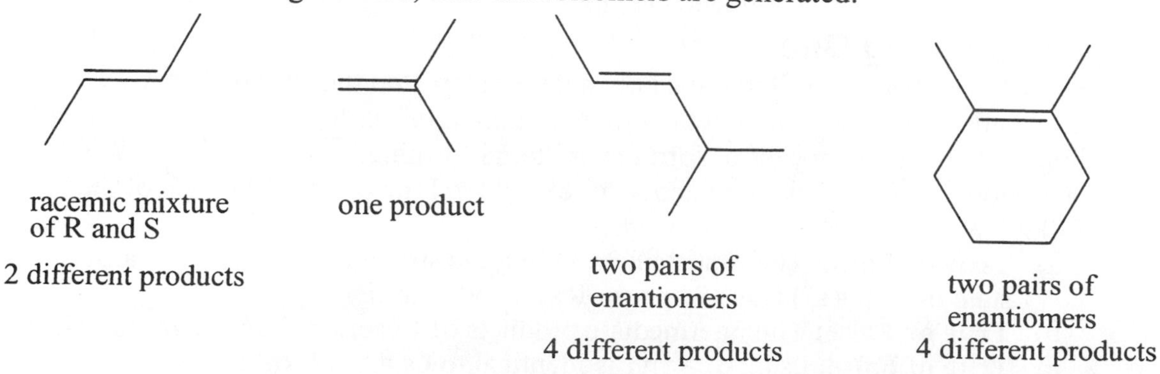

3. Explain why ethene does not react with HX (X = Cl, Br or I).

4. Which of the following pairs of reaction profiles is a violation of the Hammond Postulate? Explain your reasoning. (Assume the two reactions depicted on a diagram are similar.)

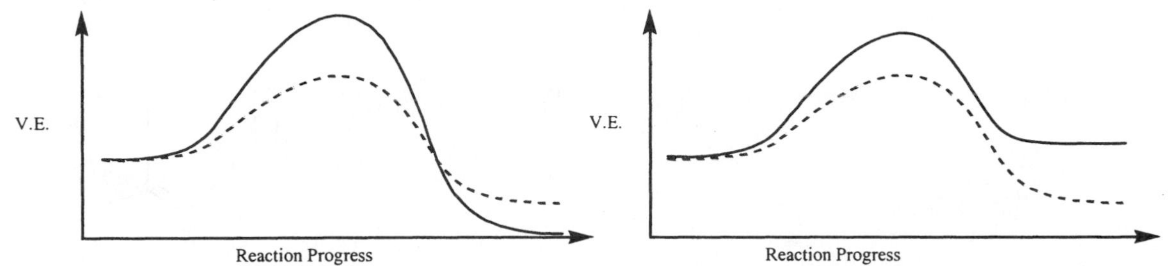

5. Consider the energy diagram above, LEFT. At high temperature there is an excess of thermal energy available so either energy barrier (E_{act}) can be surmounted. At a low enough temperature, only the lower of the two activation energies can be achieved. Based on these facts, explain why the reaction shown with a solid line is favored at high temperature and the reaction shown with a dotted line is favored at low temperature.

6. Read the assigned pages in the text and do the assigned problems.

Exercises for Part B

7. Consider the following reaction:

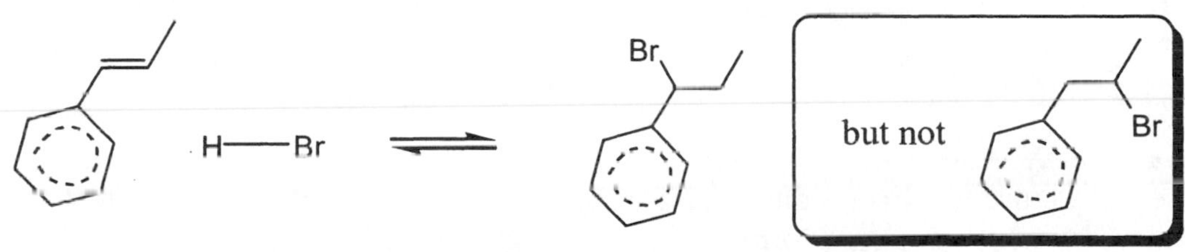

a) Label the products above with the words Markovnikov and anti-Markovnikov, as appropriate.

b) Use curved arrows to show the complete mechanism of this reaction including any intermediate products.

c) Draw all five resonance structures needed to properly represent the electron distribution of this intermediate.

d) Explain why the product in the box is not formed.

Note: addition of the bromine does not take place at the 2, 4 or 6 position on the ring even though (according to the resonance structures) there is partial positive charge at these positions. This is because of the special stability of the benzene ring known as **aromaticity**. You don't want to break up the ring if you don't have to.

8. Show the carbocation rearrangement (if any) that will occur on each of the following ions. Draw the new carbocation and briefly describe the driving force for the rearrangement.

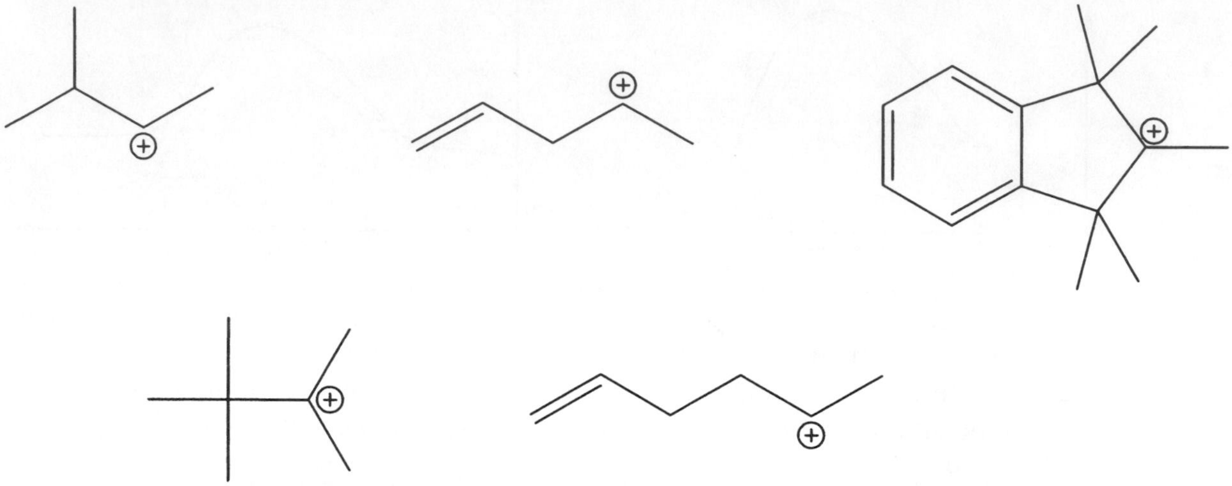

9. The product of the following electrophilic addition reaction has an NMR spectrum that consists of only two peak clusters.

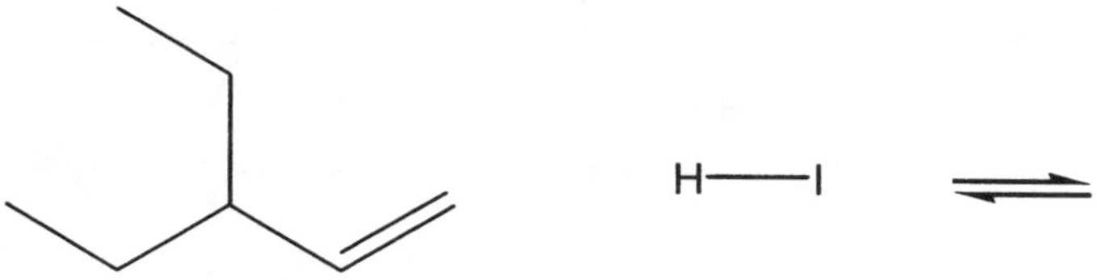

H—I

10. Propose a structure for this product and use curved arrows to show a reasonable mechanism by which it would form.
11. Sketch the ^{1}H NMR spectrum of the product showing the multiplicity and integration of each of the two peak clusters, as well as their relative positions along the x axis of the spectrum.
12. Read the assigned pages in the text and do the assigned problems.

Exercises for Part C

13. Use curved arrows to show the best mechanism of the following reaction. Include all intermediate products. Note that because the catalyst is neither a reactant or a product it is often placed above the equilibrium arrows, between the reactants and products (as shown below).

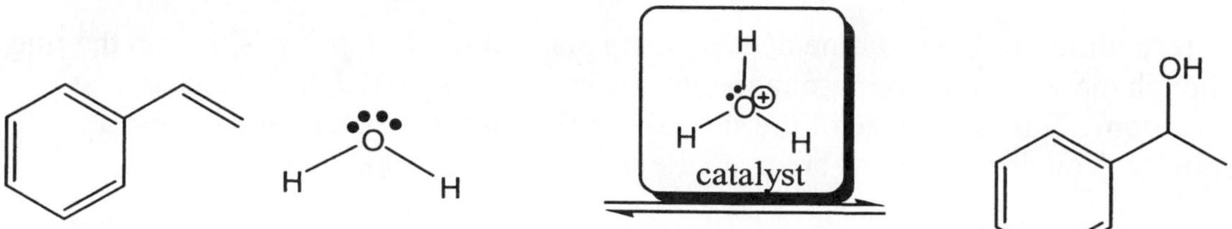

H O H

H O H H catalyst

OH

14. In water, electrophilic addition of H–Br (or H–Cl or H—I) competes with an electrophilic addition side reaction. (The side product shown occurs in water, and a different side product forms in methanol. See part c) of this question.)

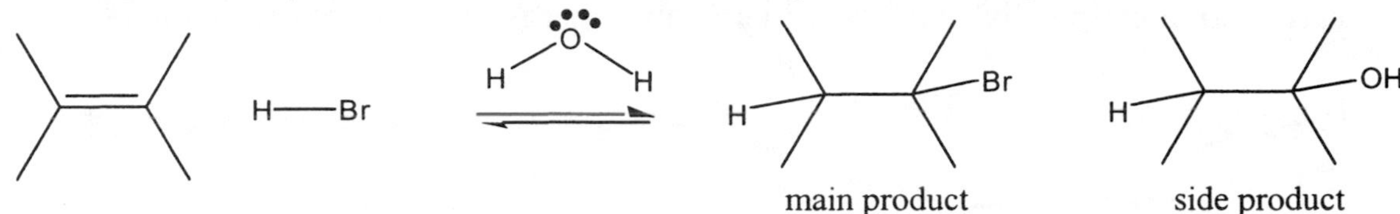

a) Devise a mechanism for formation of the side product.
b) The side product accounts for less than 5% of the product mixture. What does this say about the nucleophilicity of bromide vs. water.
c) Draw the side product you would expect if the same reaction were run using methanol as the solvent.

15. Read the assigned pages in the text and do the assigned problems.

Exercises for Part D

16. For the reaction mechanism shown below:

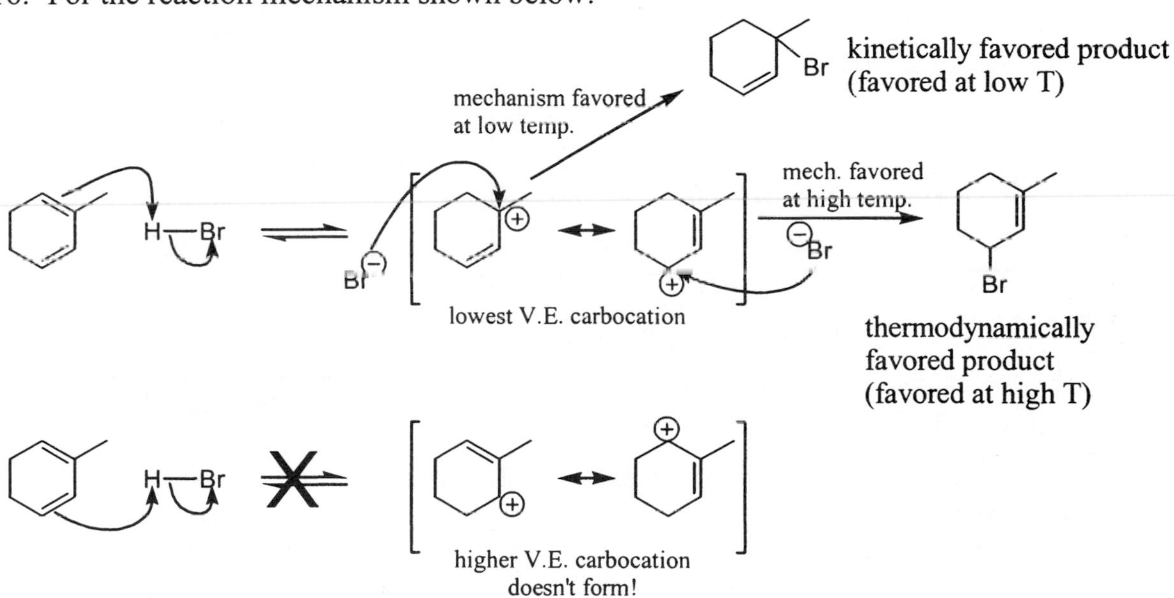

a) Explain why the carbocation labeled "lowest P.E. carbocation" is lower in P.E. than the other one (which doesn't form because it is higher energy).
b) Draw an energy diagram showing the formation of both the kinetically favored product and the thermodynamically favored product. Show the profile of the kinetically favored product using a dotted line.
c) Draw each unique transition state and intermediate product, as in the energy diagram on the previous page.

17. Notice that in the example in the previous question, electrophilic addition to each double bond results in a different carbocation intermediate, and that one of these carbocation intermediates is lower in potential energy. For the following diene…

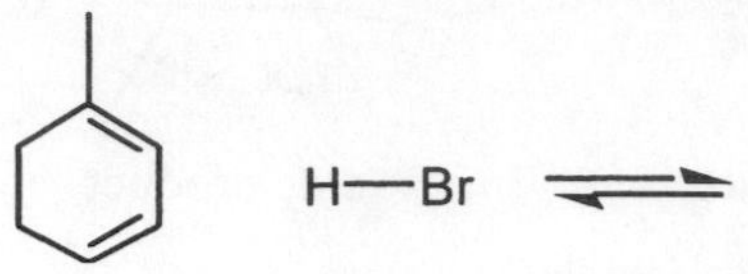

H—Br

a) Draw the two carbocation intermediates that result from electrophilic addition to each of the two different double bonds by H-Br.
b) Determine which of the two carbocations is lower in potential energy.
c) Draw the two different products that result from the nucleophile (Br^-) colliding with this carbocation intermediate. (One of these is the kinetically favored product. The other is the thermodynamically favored product.)

18. Consider the reaction of the triene shown below and HBr. Because of a carbocation rearrangement, Product I is not observed. Instead two other products are observed (Products II and III).

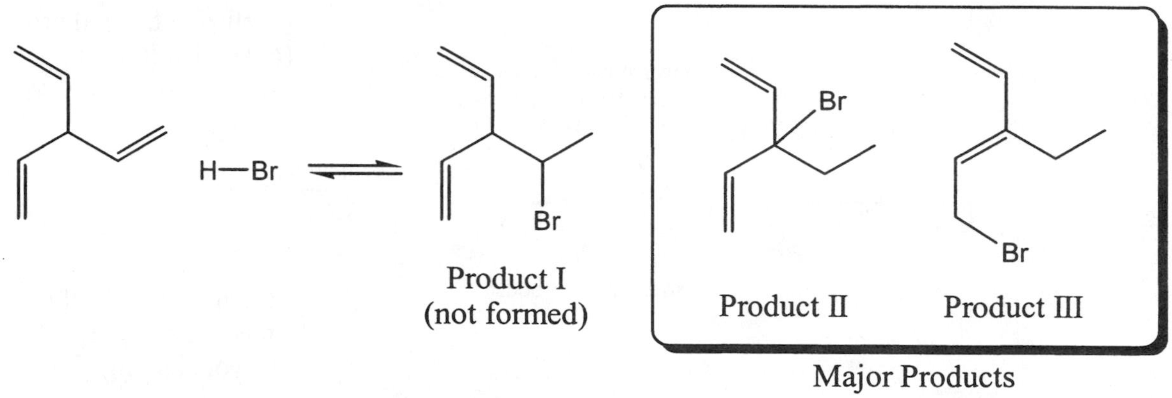

H—Br

Br

Product I
(not formed)

Br

Br

Product II

Product III

Major Products

a) Show the mechanism of formation of Products II and III including the carbocation rearrangement.
b) Explain what the driving force for this carbocation rearrangement is.

19. Read the assigned pages in the text and do the assigned problems.

ChemActivity 18

Epoxide & Bromonium Ring Opening

(How can we explain the products produced by some electrophilic additions?)

Model 1: Epoxides

A three-member ring containing 1 oxygen and 2 carbons is called an **epoxide**.

Figure 1: Nucleophile Colliding with an Alcohol vs. an Epoxide

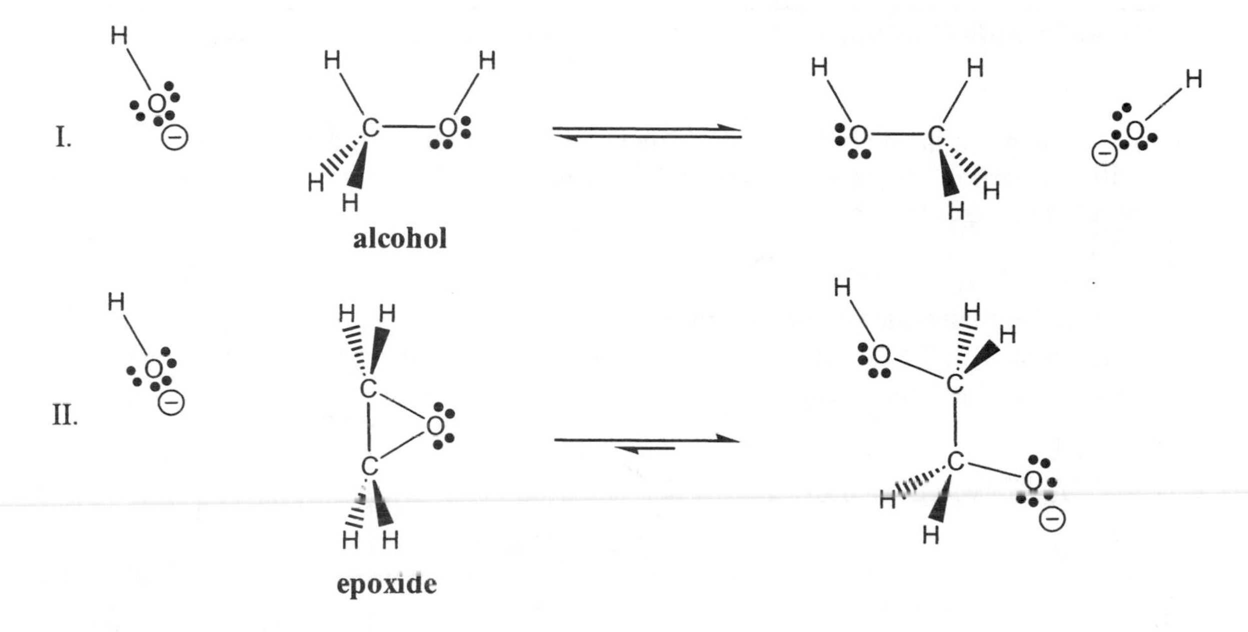

Rxn I is thermo-neutral (neither up-hill nor down-hill), as indicated by the reaction arrows. Although hydroxide is a poor leaving group, it is also a very good nucleophile. These balance each other and make Rxn I neither up-hill nor down-hill. Rxn II is downhill.

Critical Thinking Questions

1. Draw curved arrows to illustrate each reaction mechanism in Figure 1. Both are one-step reactions.

2. Both reactions in Figure 1 are examples of S_N2 type mechanisms. Identify the **nucleophile, electrophilic atom** (carbon), and **leaving group** in each.

3. Explain the following statement. The leaving group in Rxn II doesn't actually depart like a normal leaving group because it is attached with two bonds, and only one of them breaks.

4. Consider the following mechanisms of epoxide ring opening. Construct an explanation for why "**opposite-side collision**" is favored over "**same-side collision**."

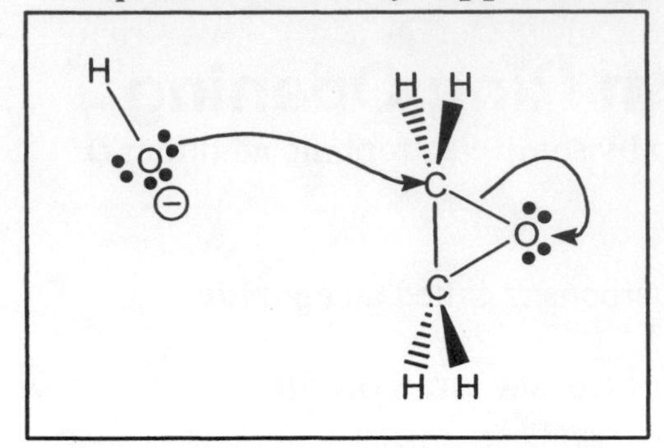

Opposite-Side Collision by HO⁻

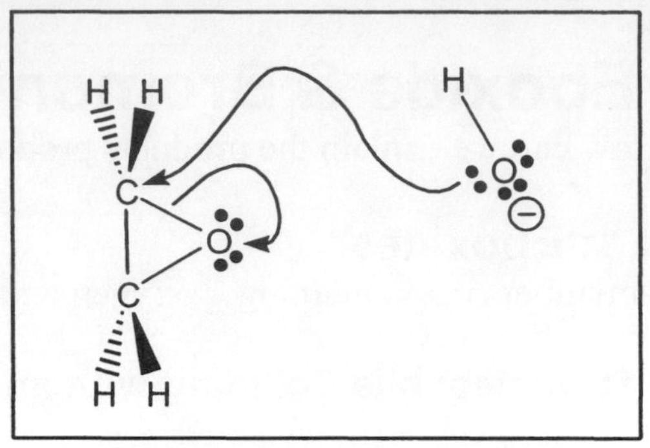

Same-Side Collision by HO⁻

5. In the diagram labeled "**Opposite-Side Collision by HO⁻**," the hydroxide collides with the carbon on the side opposite from what? (That is, why is this called opposite-side collision?)

6. Is the "**Opposite-Side Collision by HO⁻**" mechanism at the top of the page consistent with the fact that epoxide ring openings (like the reaction below) yield ONLY *trans* product? Explain.

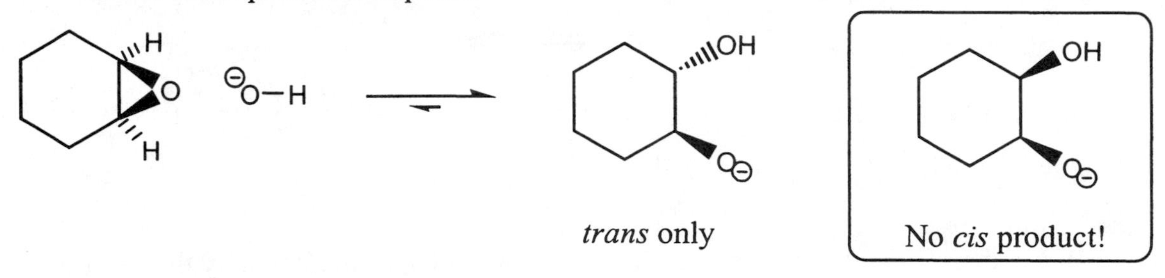

7. Upon neutralization with dilute HCl, the *trans* product in CTQ 6 becomes a di-ol (di-alcohol) as shown below.

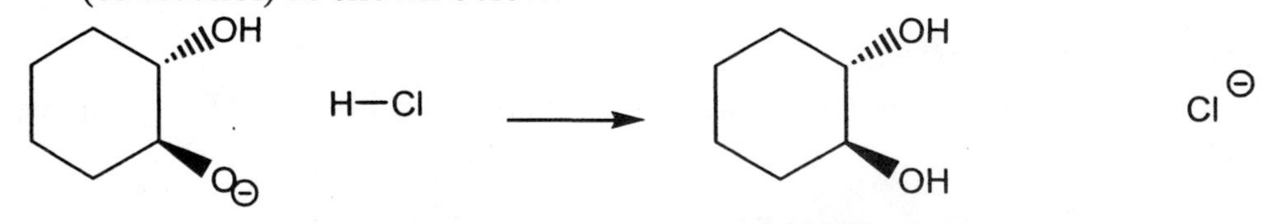

***trans* di-ol**

a) Use curved arrows to show the mechanism of this acid-base neutralization reaction.

b) A second *trans* di-ol (not shown) is produced when the epoxide in CTQ 6 is treated with hydroxide followed by dilute HCl. Draw this second *trans* di-ol. (Hint: it is a stereoisomer of the *trans* di-ol shown above.)

Model 2: Bromonium Ion Intermediate (vs. Carbocation)

Br_2 is called "polarizable" because electrons in the large orbitals that make up the Br—Br bond can slosh back and forth. Think of waves in a bathtub. At a given moment more water can be at one end of the tub than the other.

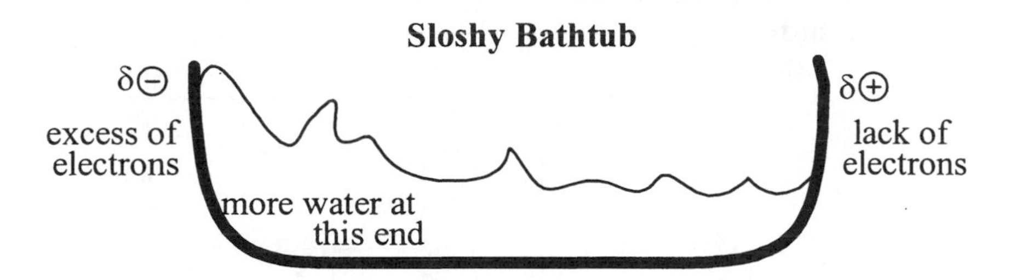

In Figure 2a, the electrons in Br—Br have "sloshed" to the left, leaving a partial + charge on one Br atom. At this moment, the π electrons of an alkene, acting as a nucleophile, react with the electrophilic Br atom.

Figure 2a: Polarized Br—Br as Electrophile and Nucleophile

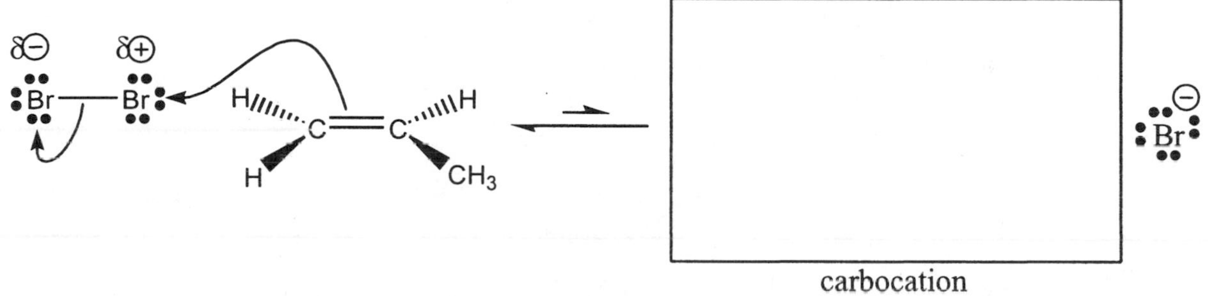

Critical Thinking Questions

8. In the box above, draw a carbocation that might result from this reaction. Be sure to put the + charge on the correct carbon.

Figure 2b: Carbocation vs. Bromonium Ion

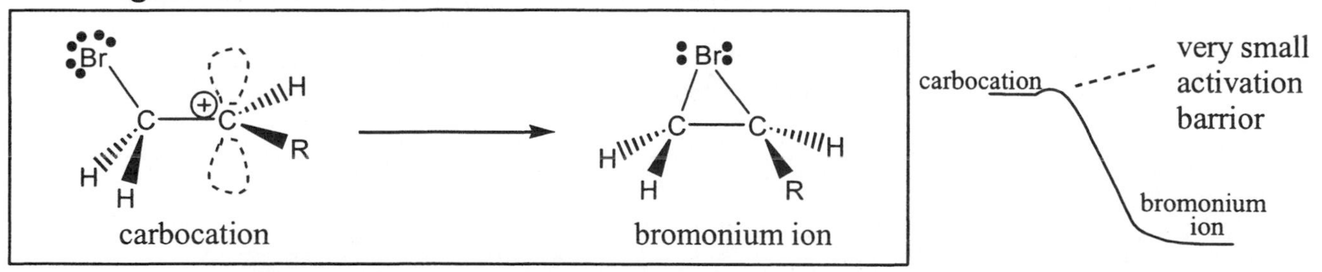

9. The carbocation in Figure 2b is not stable because there is no activation barrier preventing it from changing into a much lower potential energy **bromonium ion**.
   a) Add the missing formal charge to the Br of the bromonium ion in Figure 2b.
   b) A carbocation intermediate is not observed in this reaction, it may be that it forms but immediately changes into the bromonium ion shown. Add curved arrows to show a mechanism for changing the carbocation in Figure 2b into a bromonium ion.

10. Bromonium ion ring opening is very similar to epoxide ring opening.

a) Label one of the arrows going from Br^- on the diagram below "**same-side collision**" and label the other one "**opposite-side collision**," as appropriate.

b) Predict which type of collision (**same-side collision** or **opposite-side collision**) [circle one] is more likely in a bromonium ring opening reaction and explain your reasoning.

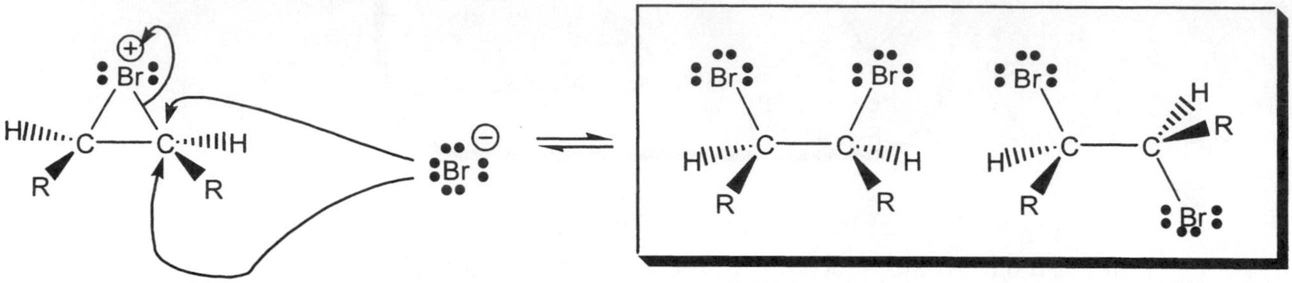

ONLY ONE of these products forms [circle one]

c) Circle the more likely of the two products shown above.

11. Consider the following reaction, which yields the product shown below, right.

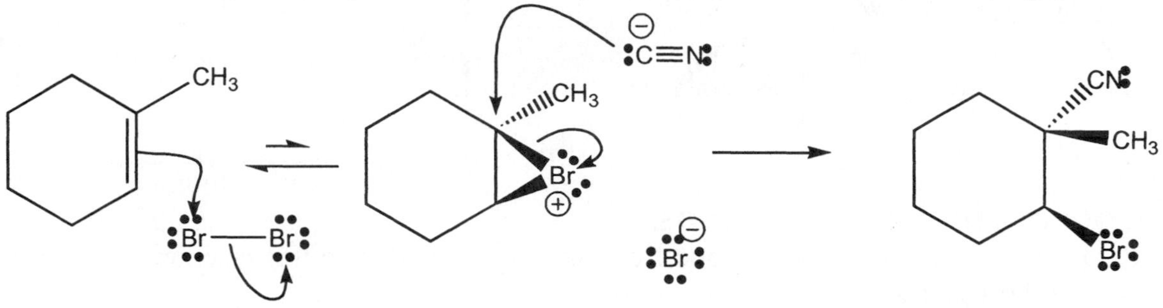

a) According to the product shown, which is a stronger nucleophile: bromide or cyanide (^-CN)?

b) Is the reaction of the nucleophile with the bromonium ion Markovnikov or Anti-Markovnikov? Label the product "**Markov.**" or "**anti-Markov.**"

Exercises

1. In CTQ 11 two enantiomeric bromonium ions are formed (shown below, center), leading to two enantiomeric products (shown below, right). This question asks you to draw products that do **NOT** form in the this reaction.

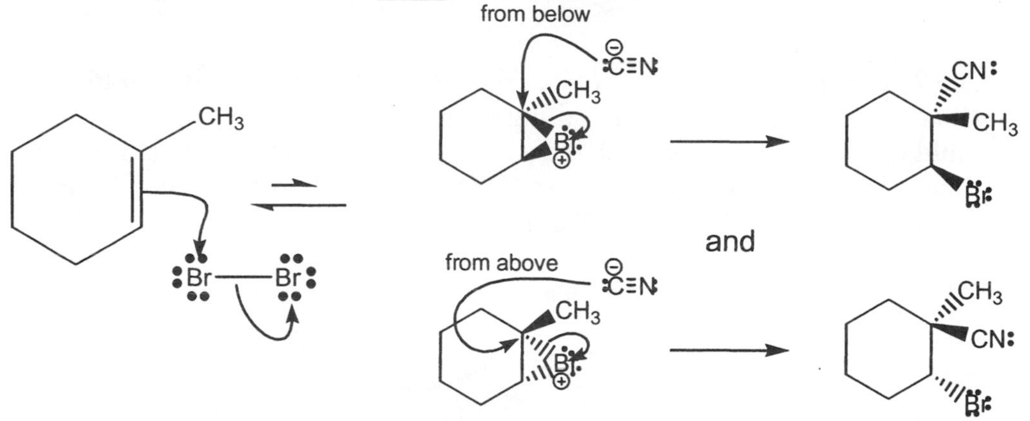

from below

and

from above

   a) Draw a product that does NOT form which has the opposite **regiochemistry** but the correct stereochemistry, as compared to the products shown above (there are two possibilities).

   The regiochemistry of the product is determined by which carbon collides with the nucleophile (the Markovnikov C or the anti-Markovnikov C). Label your does-not-form product "Markov." or "Anti-Markov.," as appropriate.

   b) Draw a product that does NOT form which has the opposite **stereochemistry**, but the correct regiochemistry, as compared to the products shown above (there are two possibilities).

   The stereochemistry of the product is determined by which side of the molecule collides with the nucleophile. Label your does-not-form product "syn addition-leading to cis product" or "anti addition-leading to trans product."

2. Construct an explanation for why the products of Rxn II in Model 1 are lower in potential energy than the reactants. (That is, explain why Rxn II is down-hill.)

3. Show a mechanism for the following reaction via a bromonium ion intermediate and explain why the *cis* di-bromide product is NOT formed.

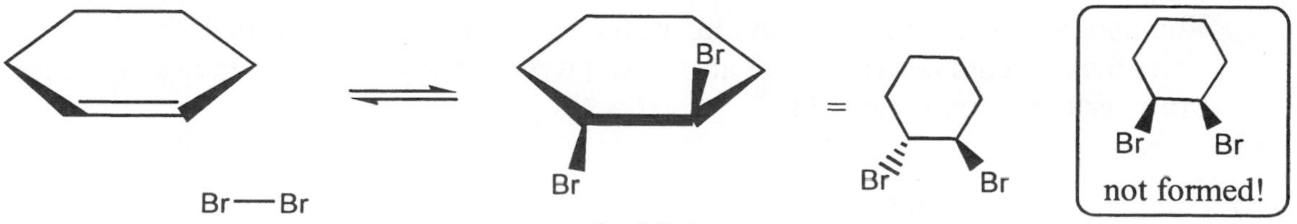

Br—Br

anti addition
(*trans* product)

not formed!

syn addition
(*cis* product)

4. Circle the correct product or products for the following epoxide ring opening.

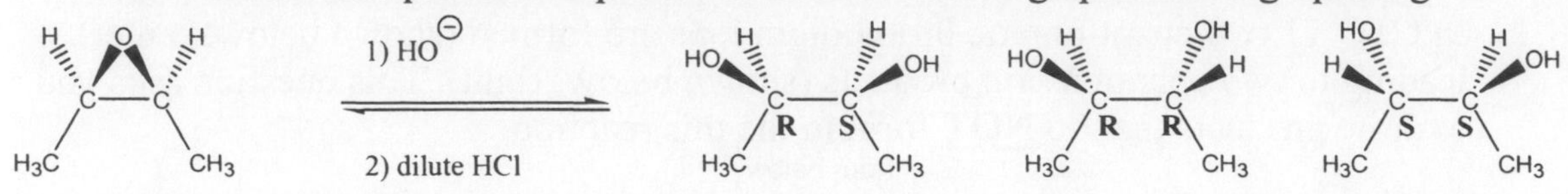

5. The 1) and 2) next to the reagents above designate that the epoxide is first treated with hydroxide, and then, only after waiting for HO^- to react, is the acid added. What would go wrong if we add both reagents, HO^- & HCl, at the same time?
6. Careful analysis of the reaction mixture in CTQ 11 shows two different dibromide products in equal amounts! (Hint: neither is a *cis*-dibromide)

   a) Propose a structure for this second *trans*-dibromide product and show the mechanism by which it is formed.
   b) Explain why the two products are found in exactly equal amounts.
   c) Explain why mass spectrometry would not be useful for showing that there are two different products.
   d) Explain why NMR would also not be useful for proving there are 2 different products.
   e) Which of the following analytical methods could you use to prove there are two different trans di-bromide products in this reaction:
   IR, TLC, or **chiral stationary phase GC?**

7. Assume hydroxide ion and bromide ion are comparably strong nucleophiles. Show the complete mechanism for formation of a trans-dibromide product and an alcohol product. (Note: sodium ion is a spectator ion and is not involved in the reaction.)

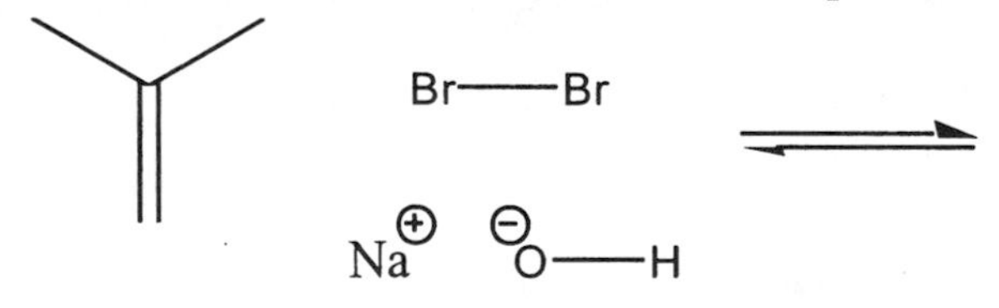

8. Design a synthesis of each of the following target molecules starting from 3-methyl-2-butanol and using any combination of the reactions we have learned thus far. Write the reagents above the rxn arrow, and write the type of reaction below the rxn arrow, as in the example below.

Example:

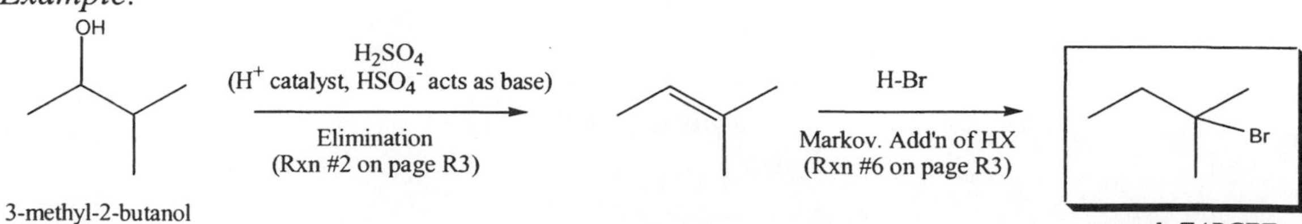

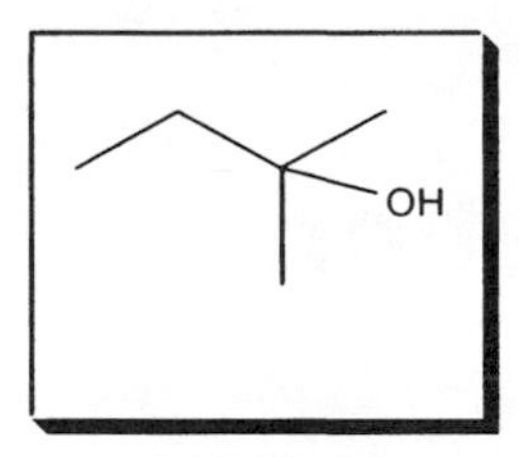

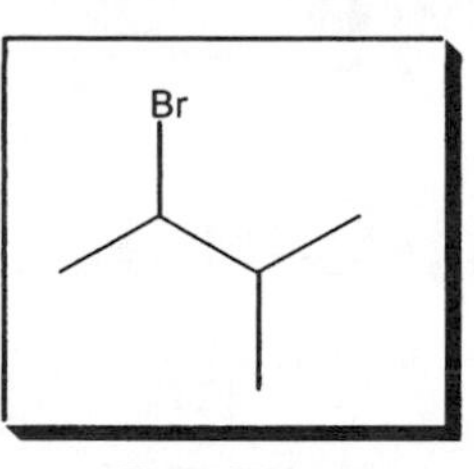

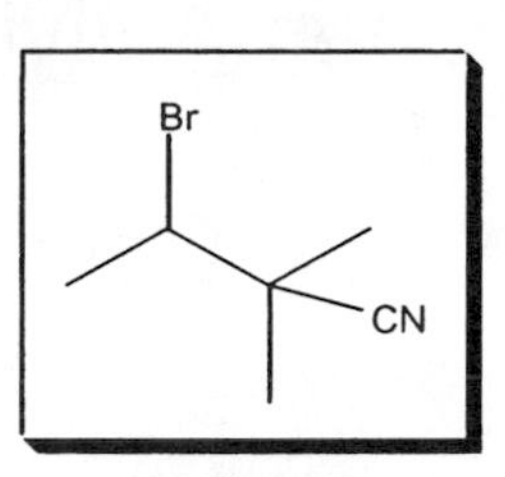

TARGET A **TARGET B** **TARGET C** **TARGET D**

9. Consider the following reaction in which all three product shown are observed.

Br—Br

A **B** **C**

a) Is the stereochemistry of the products more consistent with a bromonium ion intermediate or a carbocation intermediate? Draw both possible intermediates and explain your reasoning.

b) Show the complete mechanism of the reaction including formation of each product shown.

c) Construct an explanation for why a carbocation intermediate might be favored over a bromonium ion with this unusual alkene. (Draw the carbocation!!)

d) What word best describes the relationship between…
   Product A&B ?
   Product A&C ?
   Product B&C ?

e) Assign absolute configuration (R or S) to each chiral center in Products A-C.

f) Which products (A-C) are **chiral** (be sure the <u>molecule as a whole</u> is chiral).

10. Read the assigned pages in the text and do the assigned problems.

ChemActivity 19

Carbon-Carbon Bond Formation

(How can we make a new carbon-carbon σ bond?)

Model 1: Introduction to Organic Synthesis

Organic synthesis is the art of constructing complex target molecules from smaller molecules. The reactions we will use can be put in two categories.

1. **Functional group transformations** (changing one functional group into another)
2. **Carbon backbone assembly** (generation of the carbon backbone of the molecule)

So far we have learned several functional group transformations (e.g. –Br to –OH, alkene to dibromide, alkene to alkane etc.). Now we will learn a few ways to form carbon-carbon single bonds.

Critical Thinking Questions

1. The target molecule shown below can be made in several steps from the starting materials, but we have not yet learned all of the necessary reactions.

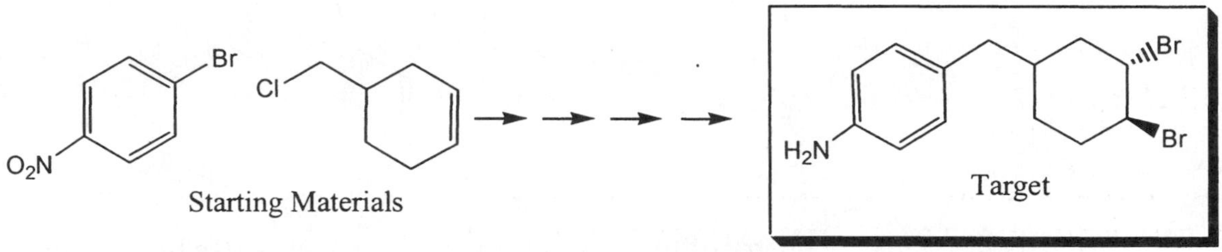

Br Cl O_2N
Starting Materials
H_2N Br Br
Target

a) Label the "**new C-C single bond**" on the target molecule backbone.

b) Identify the two functional group transformations involved in this synthesis by completing the following sentences. (An $-NO_2$ group is called a "nitro" group,)

i. The **nitro group** was turned into an ________________ group.

ii. The **alkene/π bond** was turned into a di-________________ group.

Model 2: Lithium and Grignard Reagents

We have learned so far in this course that C^- is very high in potential energy. In fact it is so high in potential energy that carbon salts with common counterions (Na^+ & K^+) are not stable. Instead, lithium and magnesium salts are used. These reagents are prepared from alkyl halides by a mechanism that we will not discuss. Assume that these reagents act like a carbon anions, that is, they are excellent nucleophiles and extremely strong bases (release about 50 pKa units of energy upon forming a new bond).

Figure 2: Preparation of Grignard and Lithium Reagents

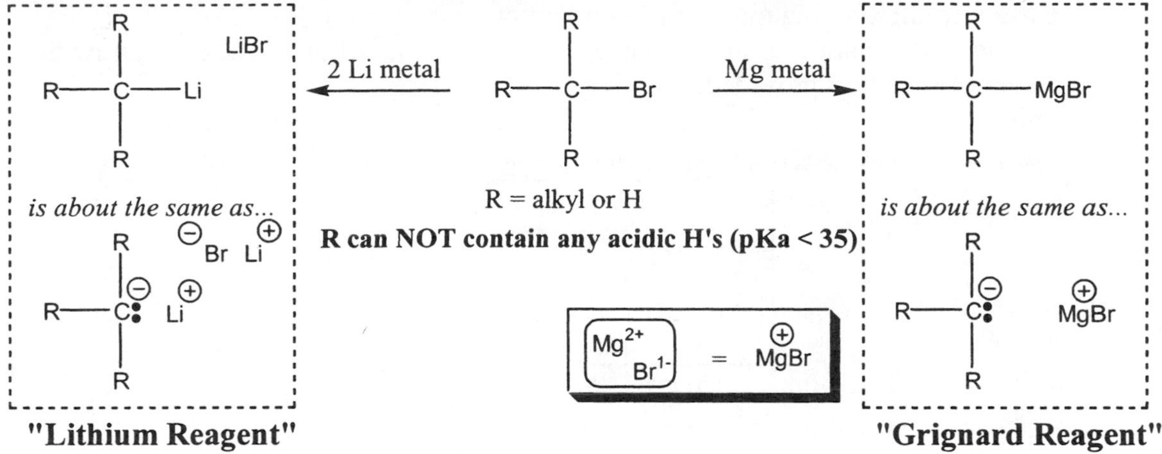

Critical Thinking Questions

2. Consider the "Grignard Reagent," named after French chemist Francois Grignard (1871-1935).
   a) What is the charge on a Mg ion?
   b) What is the charge on a Br ion?
   c) What is the total charge on a MgBr ion?

3. There are two representations of a Grignard reagent in Model 2. One of them shows an intact bond between C and MgBr, but this bond is very very polarized-almost to the point of being broken.
   a) Add a **LARGE** $\delta-$ **and** $\delta+$ to this structure to show the super-polarization of the C–MgBr bond.
   b) Add a curved arrow to this structure to show how it could be converted into the pair of ions drawn below it.

4. Alkyl bromides are the most common starting materials for preparing Grignard and lithium reagents, but Cl and I also work. Draw the products of each of the following reactions and **show the products as ions** to emphasize the polarization of the C–M bond (M = metal).

← 2 Li metal — $H_3C-CH_2-CH_2-Cl$ — Mg metal →

← 2 Li metal — C_6H_5I — Mg metal →

5. Synthesis of 1/2 mole of the following target molecule can be accomplished using one mole of the starting material shown below.
   a) 1/2 mole of the starting material is used to prepare a Grignard reagent. Draw this reagent in the box provided, showing a full -1 charge on C.
   b) This Grignard reagent is added back to the rest of the starting material. Use curved arrows to show the S_N2 reaction that results to give the target product.

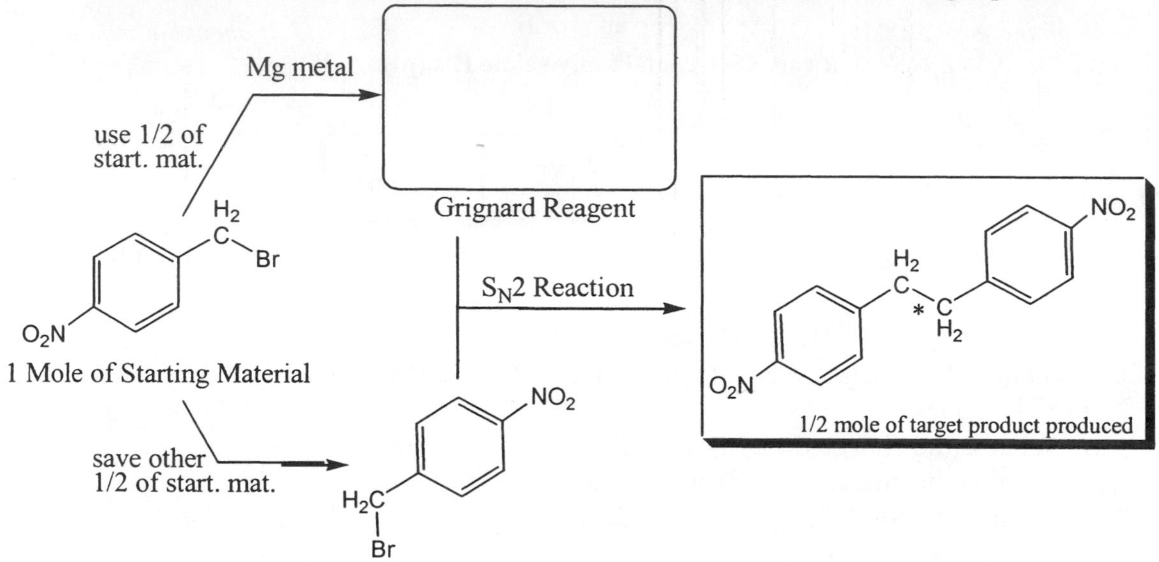

Mg metal

use 1/2 of start. mat.

H_2 C Br

O_2N

1 Mole of Starting Material

save other 1/2 of start. mat.

Grignard Reagent

S_N2 Reaction

NO_2 H_2C Br

NO_2 H_2 C * C H_2 O_2N

1/2 mole of target product produced

*Note: The C-C bond in this synthesis would best be formed using a copper coupling, but this reaction is left for an advanced course.

6. Grignard and lithium reagents are extremely strong bases. Upon making a bond to H^+ they release a large amount of energy (you may assume about 50 pKa units, but actually the value is closer to 35-40 pKa units of energy). Recall that the pKa of an alcohol is about 16. (Requires 16 pKa units of energy to remove an H^+.)

   a) In the reaction below, the moment a molecule of Grignard reagent is formed, it immediately undergoes an acid-base reaction with a second molecule of starting material. Use curved arrows to show this unwanted side reaction, and draw the products.

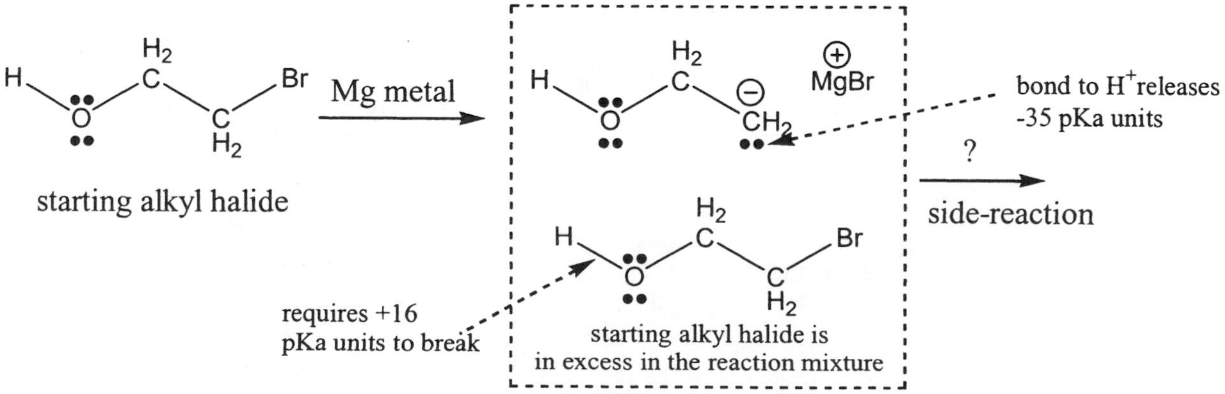

H O H_2 C C H_2 Br

starting alkyl halide

Mg metal

H O H_2 C ⊖ CH_2 ⊕ MgBr

bond to H^+ releases -35 pKa units

?

side-reaction

H O H_2 C C H_2 Br

requires +16 pKa units to break

starting alkyl halide is in excess in the reaction mixture

   b) Construct an explanation for why preparation of a Grignard or lithium reagent **cannot** be performed with an alkyl halide containing an acidic H (pKa < 35).

Model 3: Acidity of Terminal Alkynes (RC≡CH)

Recall the **hybridization effect**: an electron pair in an *sp* hybrid orbital is lower in potential energy than an electron pair in an sp^2 hybrid orbital. This is because the *sp* orbital has more **"*s* character"** (50% vs 33%), and ***s* orbitals are lower in energy**.

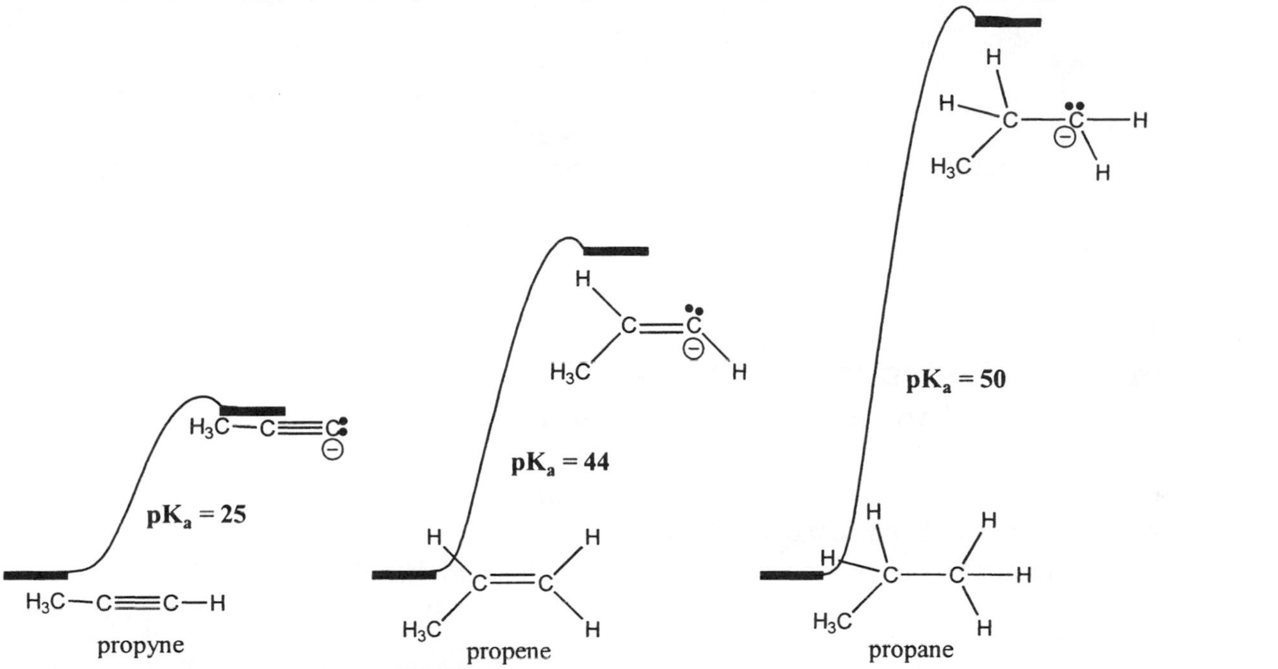

$H_3C{-}C{\equiv}C{-}H$

propyne

$\mathbf{pK_a = 25}$

$H_3C{-}C{\equiv}C:^{\ominus}$

propene

$\mathbf{pK_a = 44}$

propane

$\mathbf{pK_a = 50}$

Critical Thinking Questions

7. Label each lone pair in Model 3 with the name of the hybrid orbital in which it resides?
   a) Which is most acidic: **propyne, propene** or **propane** [circle one]?
   b) Why is it more acidic?

8. Alkynes such as propyne are acidic enough that a strong base (Na^+ H_2N^-) can pull off the terminal H (**Rxn II**). This same base cannot pull an H off of propene or propane (**Rxn I**). The resulting conjugate base is another type of carbon nucleophile.
   a) Use curved arrows to show the mechanism of **Rxn III**, an S_N2 reaction.
   b) Label the **new C-C σ bond** in the product, 2-butyne.

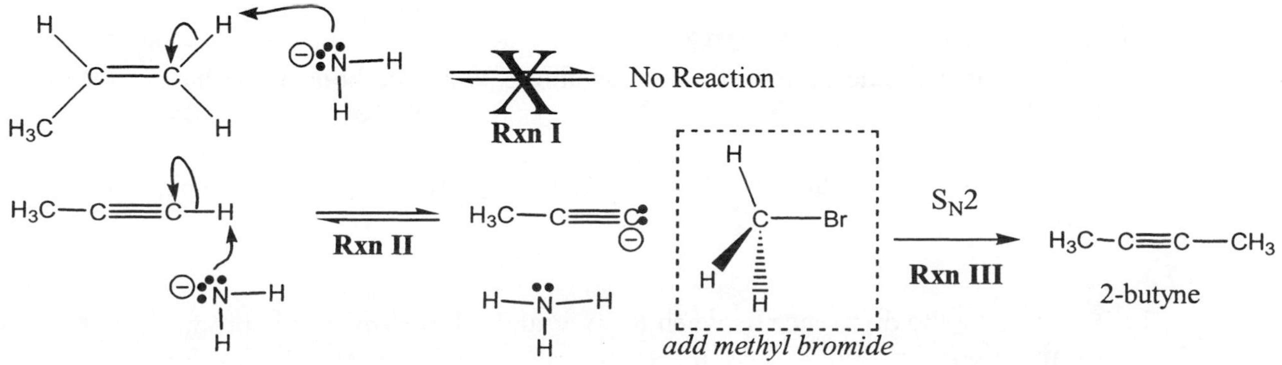

No Reaction

Rxn I

Rxn II

add methyl bromide

S_N2

Rxn III

$H_3C{-}C{\equiv}C{-}CH_3$

2-butyne

9. Nucleophilic substitution involving an alkyl halide (such as in Rxn III on previous page) only gives a reasonable yield of product when a 1° (or methyl) alkyl halide is used. Based on this, cross out each nucleophilic substitution below that will **not** work well.

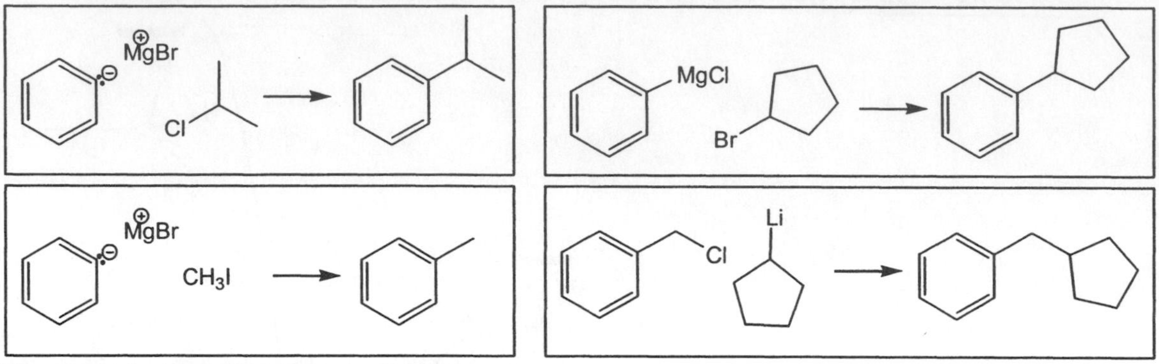

Model 4: Reduction of Alkynes

- Reduction of a molecule is the addition of H atoms (not H^+) or the removal of O atoms.
- You do not need to know the mechanisms of these three reactions.
- H's that have been added are shown in **bold**.
- It is easier to reduce the first π bond of a triple bond than the second π bond.

IV. $H_3C—C≡C—CH_3$ —(H_2 / Pt metal)→ $H_3C—CH_2—CH_2—CH_3$

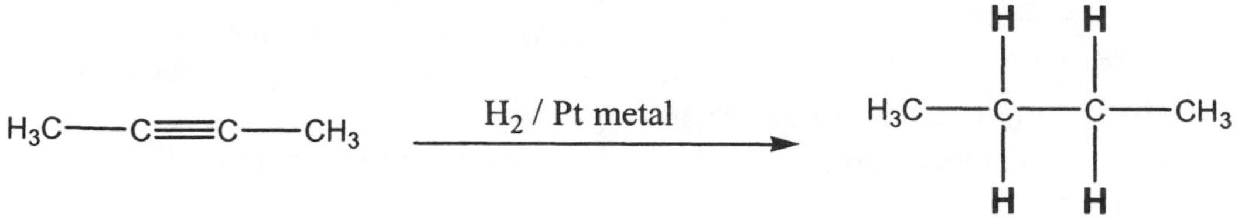

V. $H_3C—C≡C—CH_3$ —(H_2 / Lindlar Catalyst)→ cis-$H_3C(H)C=C(H)CH_3$

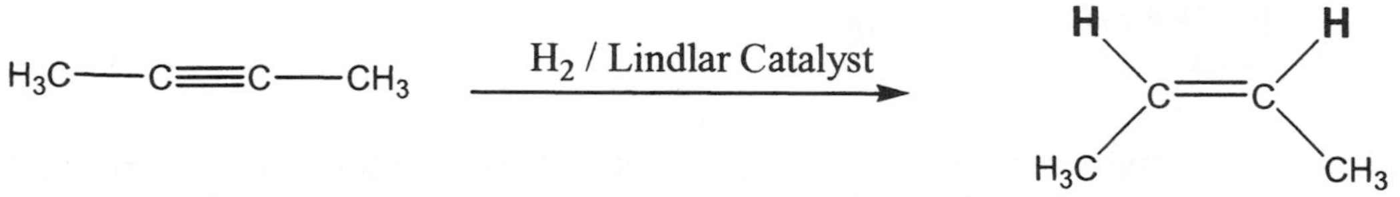

VI. $H_3C—C≡C—CH_3$ —(1) Na metal/ liquid NH_3; 2) H_2O)→ trans-$H_3C(H)C=C(H)CH_3$

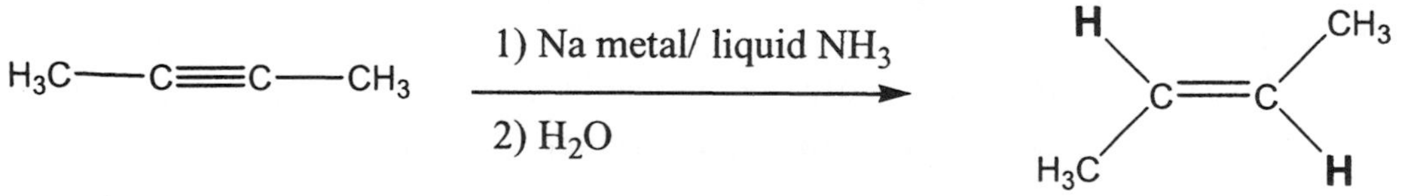

Critical Thinking Questions

9. According to the definition of reduction above, which carbons are reduced in Rxns IV-VI.

10. Which **reducing agent** in Model 4 is the strongest reducing agent? Explain your reasoning.

11. Summarize the difference between Rxns V and VI in terms of the products that form.

Model 5: Functional Group Approach

Most textbooks and courses use what is called the "functional group approach." In these courses the study of organic chemistry is organized by functional group. For example, a chapter in the text is typically devoted to amines: naming amines, making amines, reactions of amines etc.. This emphasizes the differences between functional groups. Our course has tried to focus on the similarities.

Now that we are beginning "organic synthesis" it is important that you review the functional groups we have encountered.

Critical Thinking Questions

12. Below are examples of molecules containing various functional groups.
   a) Match the name of the functional group with the appropriate molecule. Choose from: **Amine, Alcohol, Alkane, Alkene, Alkyl Halide, Alkyne, Carboxylic Acid, Ether, Nitrile (cyano), Nitro, and Thiol.**
   b) Next to the carboxylic acid, draw its conjugate base (a **carboxylate ion**).
   c) Next to the amine, draw its conjugate acid (an **ammonium ion**).

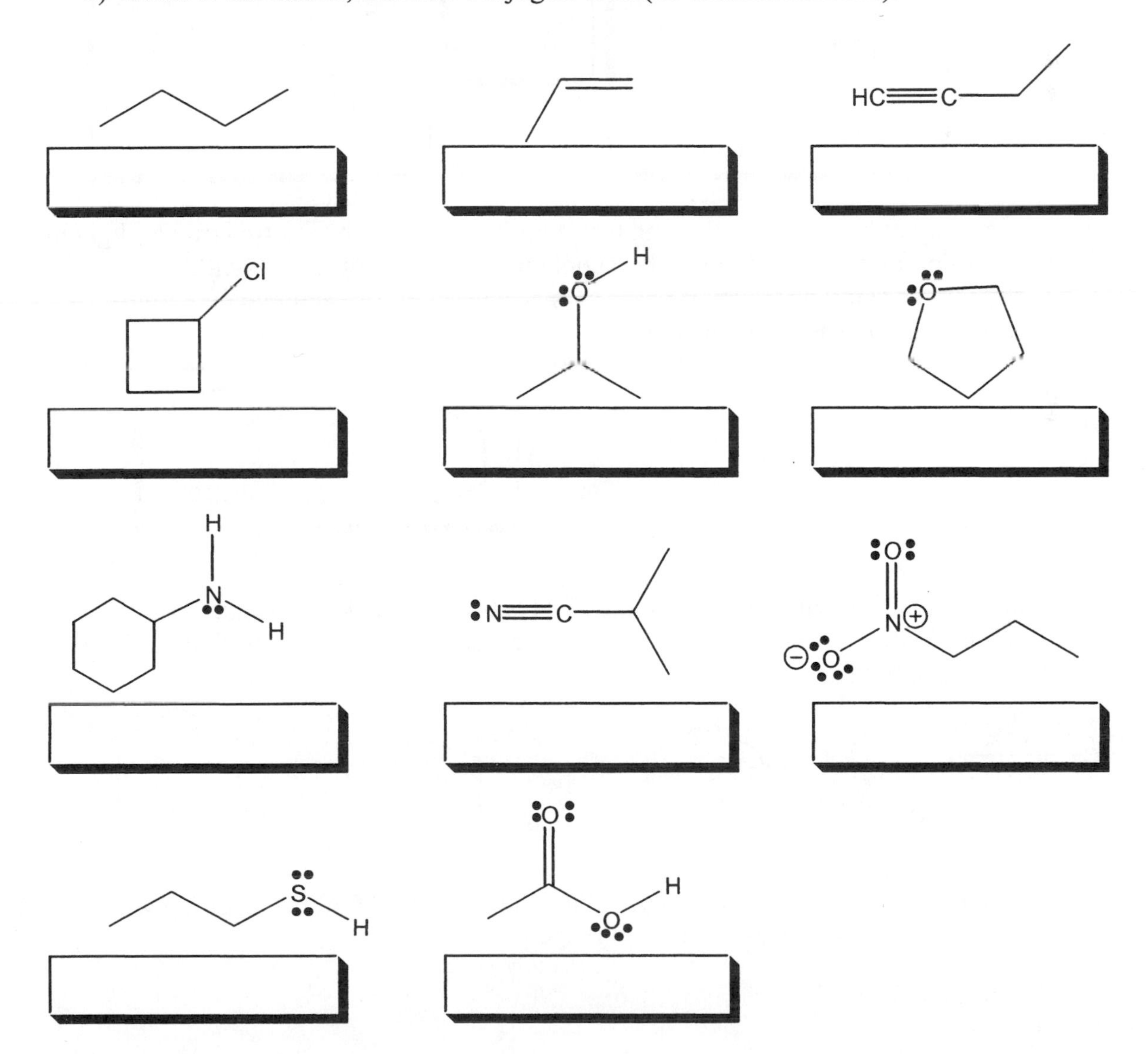

Exercises

1. The synthesis on the first page of this ChemActivity can be accomplished in four steps. Fill in empty boxes for reagents and products.

Notes:
By convention, reagents needed for a step in a synthesis are written **above the reaction arrow.**
As with all synthesis problems, you are <u>**not**</u> required to show the mechanism of each step!
Reagents for changing a nitro group into an amine are **Zn(Hg) & HCl**. We will learn this later.

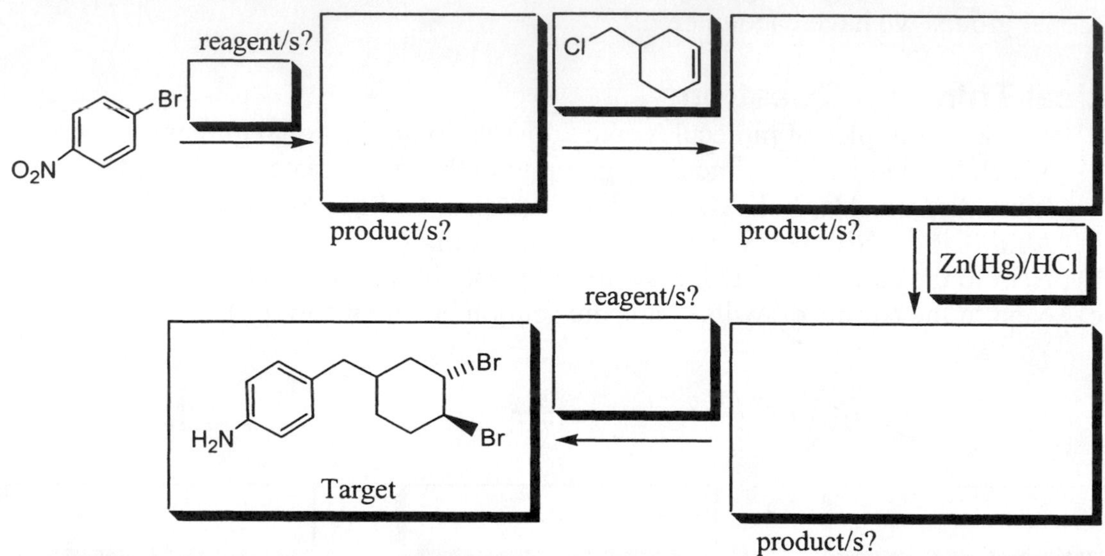

2. Design a synthesis of the following target molecule using the starting material given and... (You may use any non-carbon containing reagents in either synthesis.)
   a) any alkyne (hint: a reduction reaction is also necessary to achieve the target).
   b) any Grignard or lithium reagent.

Br

starting material

target

3. Which of the following are acceptable alkyl halides for making Grignard or lithium reagents? If they are not acceptable, show the side-reaction that would occur if each were treated with Li or Mg.

HC≡ ... Br HO ... Br H–N ... I Cl

Rule of Thumb: In general, an alkyl halide with any other functional group will not work for making a Grignard or lithium reagent.

4. Explain why the synthesis in Exercise 1 would not work if you started with 4-amino-1-bromobenzene instead of 4-nitro-1-bromobenzene. Draw the side reaction that completely consumes the carbon anion before it can be used in the desired second step.

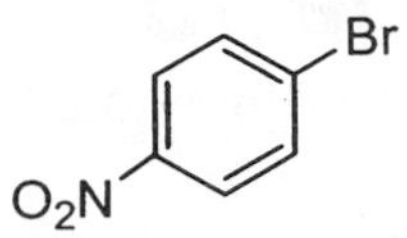

4-nitro-1-bromobenzene 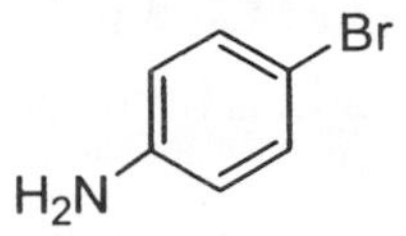 4-amino-1-bromobenzene

5. Consider the following acid-base reactions from CTQ 8:
   a) Use curved arrows to show each reaction.
   b) Draw the products of each reaction.
   c) Calculate ΔH_{rxn} for each reaction.
   d) Which reaction is downhill (favorable) and which reaction is uphill (unfavorable)?
   e) Explain why one of the methyl H's was not removed in each case.

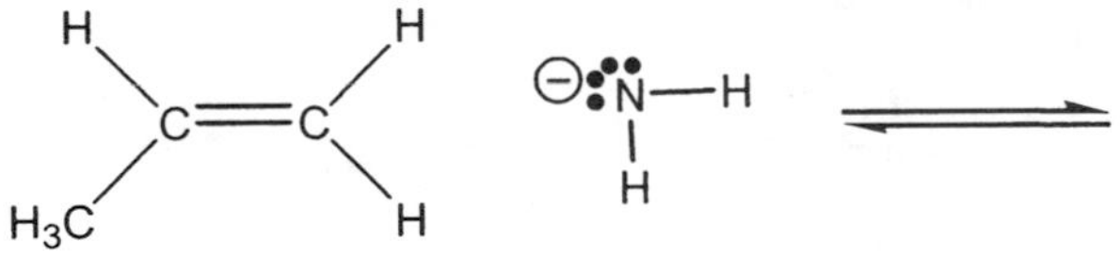

6. One method of preparing an alkyne is to heat a **geminal dihalide** in the presence of a base (such as hydroxide). ("**Geminal dihalide**" comes from Gemini meaning "twin" and refers to a carbon with two halides.) Construct a reasonable mechanism for the following reaction.

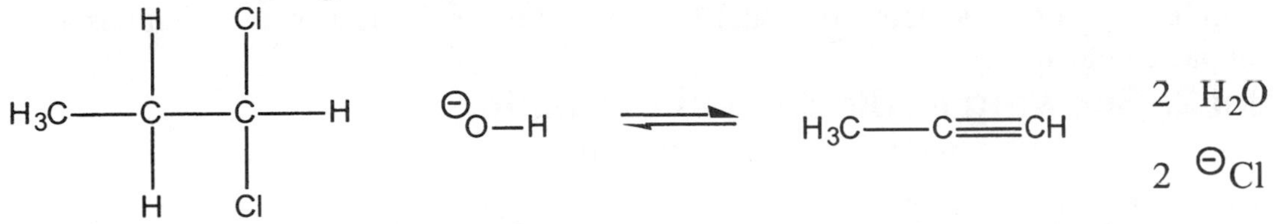

7. Circle and name each functional group in the following multi-functional group molecules.

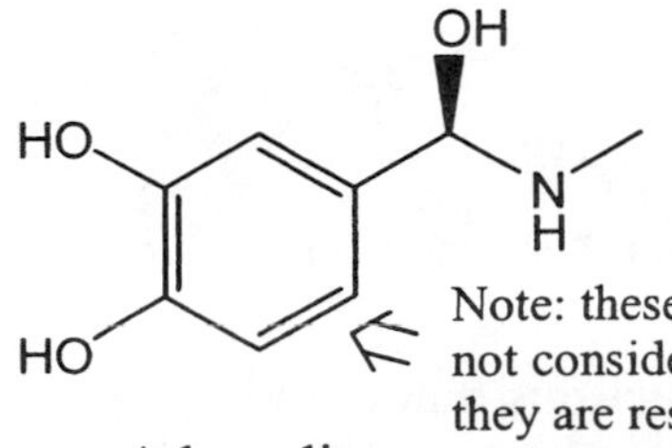

Note: these three π bonds are not considered alkenes because they are resonance stabilized-and so do not react like alkenes

Adrenaline

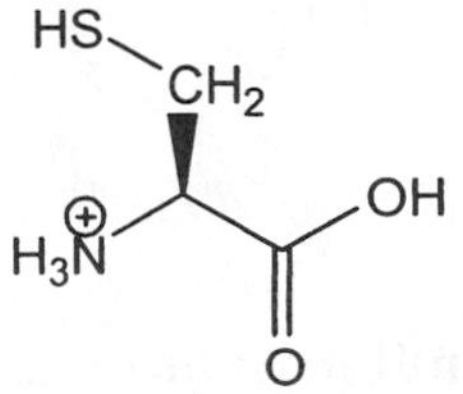

Cysteine (an amino acid)

8. Read the assigned pages in the text and do the assigned problems.

ChemActivity 20

Part A: Radical Halogenation Reactions

(How can we add a functional group to an un-functionalized alkane?)

Model 1: Polar and Non-polar Bond Breakage

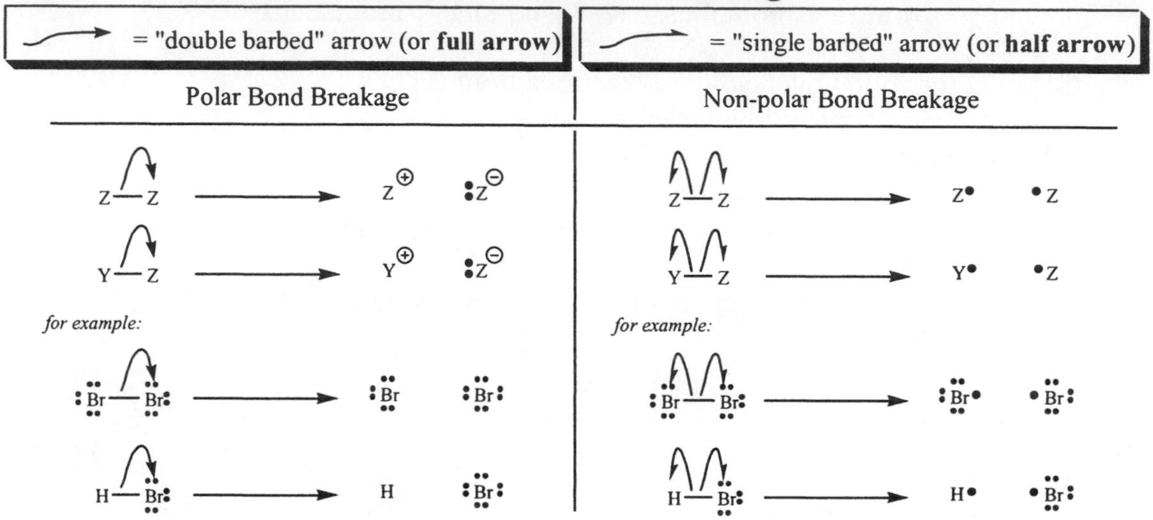

= "double barbed" arrow (or **full arrow**)

= "single barbed" arrow (or **half arrow**)

Polar Bond Breakage	Non-polar Bond Breakage
Z—Z ⟶ $Z^{\oplus}$:$Z^{\ominus}$	Z—Z ⟶ Z• •Z
Y—Z ⟶ $Y^{\oplus}$:$Z^{\ominus}$	Y—Z ⟶ Y• •Z
for example:	*for example:*
:Br—Br: ⟶ :Br :Br:	:Br—Br: ⟶ :Br• •Br:
H—Br: ⟶ H :Br:	H—Br: ⟶ H• •Br:

Critical Thinking Questions

1. Add formal charges, if necessary, to the *example* species in Model 1 (everything below the words "*for example.*")
2. A **full arrow** depicts the movement of a pair of electrons. What does a **half arrow** depict?

3. Circle any **radical species** in Model 1. (A **radical species** is a species that has an unpaired electron.)

Model 2: Showing Radical Bond Formation

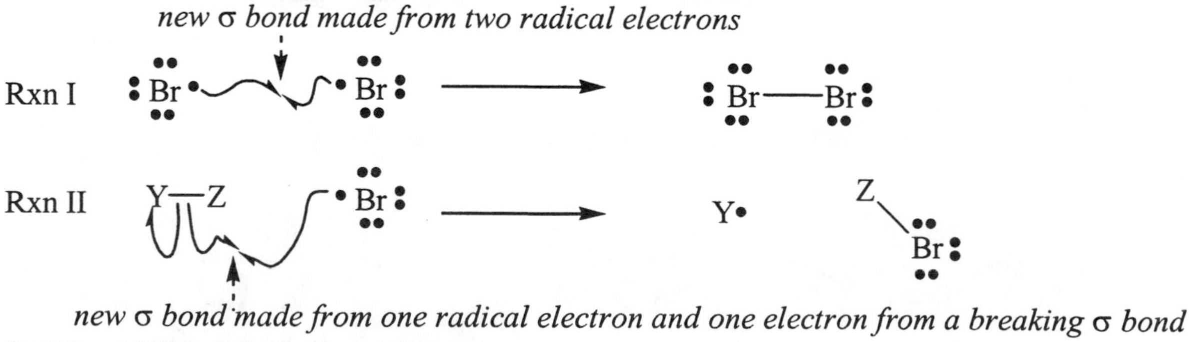

new σ bond made from two radical electrons

Rxn I :Br• •Br: ⟶ :Br—Br:

Rxn II Y—Z •Br: ⟶ Y• Z—Br:

new σ bond made from one radical electron and one electron from a breaking σ bond

Critical Thinking Questions

4. Rxn II in Model 2 is one of two sigma bond forming reactions that could have taken place with these two reactants. Use curved **half**-arrows to show the mechanism of a reaction that yields Y—Br and Z radical from these same reactants.

Model 3: Reactions of Bromine Radical with Various Alkanes

For each reaction below, the most likely products are shown.

Rxn III $Br\cdot$ + $H_3C{-}CH_3$ ⟶ Br—H + $\cdot CH_2{-}CH_3$

Rxn IV $Br\cdot$ + $H_3C{-}CH_2{-}CH_3$ ⟶ Br—H + $H_3C{-}\dot{C}H{-}CH_3$

Rxn V $Br\cdot$ + $H_3C{-}H_2C{-}CH(CH_3){-}CH_3$ ⟶ Br—H + $H_3C{-}H_2C{-}\dot{C}(CH_3){-}CH_3$

Critical Thinking Questions

5. Label the carbon radicals in Model 3 as methyl, primary, secondary, tertiary, allyl or benzyl.
   a) Which carbon radical in Model 3 is closest to having an octet? Explain your reasoning. (Recall that alkyl groups are electron donating.)

   b) The following is a list of radicals from highest to lowest potential energy. Is this list consistent with your conclusion in part a)? Explain.

highest V.E. •H

•CH_3

•1°

•2°

•3° or •allyl

• benzyl

• X (halogen) lowest V.E.

   c) Which reaction in Model 3 is most likely? Explain

6. Add curved half arrows to Rxn III in Model 3 so as to illustrate the mechanism of product formation.
   a) As in Model 2, draw the other sigma bond forming reaction that could have taken place in Rxn III, but did not. (Note: all 6 H's of ethane are equivalent.)

   b) Construct an explanation for why Rxn III in Model 3 happens instead of the reaction you drew in part a).

Model 4: Light Induced "Homolytic" Bond Cleavage

When a solution of Br_2 is exposed to a strong light source a very small percentage of the Br_2 molecules break apart to form two bromine radicals.

Br—Br Strong Light Source (hν) ⇌ Br• •Br

billions and billions	(ratio of reactants to products)	very few

"Strong Light Source" is usually abbreviated with the letters **hν**.

Information

No evidence for methyl or 1° carbocations has been found to date. However, the same is not true for methyl and 1° radicals: both are known. In fact there is evidence for all of the following radicals (listed below from highest to lowest potential energy).

highest energy •H •CH_3 •2° •3° •allyl •benzyl •X (halogen) lowest energy

Critical Thinking Questions

7. When a solution of Br_2 and methane is exposed to a strong light source one of the major products is CH_3Br.

CH_4 + Br—Br (excess), hν → → → CH_3Br + other products

Use curved arrows to show a reasonable **multi-step** mechanism for this reaction.

Model 5: Collision Statistics

A reaction between two radicals may seem very likely from an energy standpoint because in such a reaction two high P.E. species combine to form one low P.E. species.

for example:

$$H_3C\bullet \ \ \bullet Br \xrightarrow[\textbf{but statistically unlikely!!}]{\text{rxn is very very down hill}} H_3C\text{—}Br$$

However, radicals are **so** high in energy that they react with anything they "bump into." This means a radical has a short life-span and usually does not have time to "find" another radical.

Critical Thinking Questions

8. Picture a crowded stadium with 100,000 people. Five of these people are violently reactive and will start a fight with the first person they bump into-getting themselves thrown out of the stadium. What are the chances that two of the five "violently reactive people" get in a fight with one another? (Assume they enter at different gates.)

9. The first two steps of the mechanism from CTQ 7 are drawn for you below.

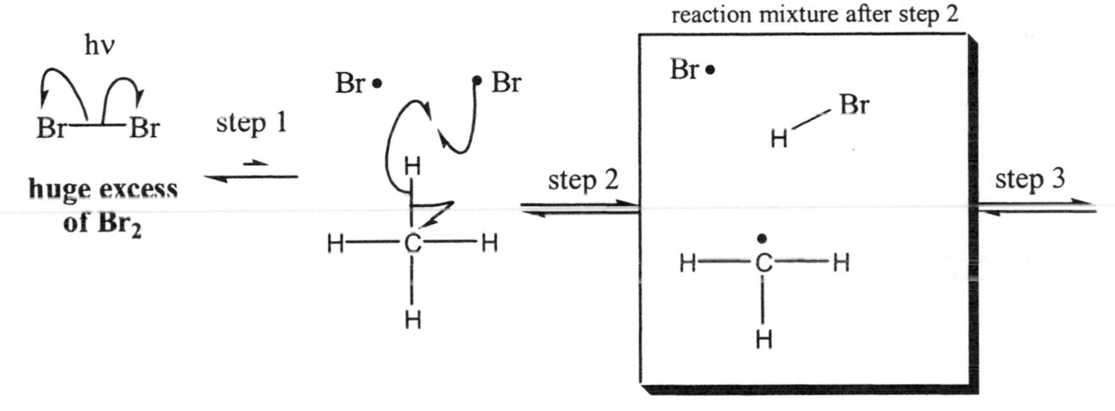

a) Based on the information in Model 5, circle any step in your mechanism on the previous page that has a low probability of occurring.

b) $\cdot CH_3$ is like the "fighter" in the stadium analogy. It is a very reactive radical, and it will react with the first thing it bumps into. What is the most prevalent species in the reaction mixture ("stadium")? Add this species to the box labeled "reaction mixture after step 2."

c) Show the mechanism of the statistically most likely reaction involving $\cdot CH_3$ and draw the products of step 3. **Notes**: This reaction leads to CH_3Br. **There are radical species left over at the end of step 3.**

Part B: Radical Chain Reactions

(What is the mechanism of a radical halogenation chain reaction?)

Model 6: The Three Parts of a Chain Reaction

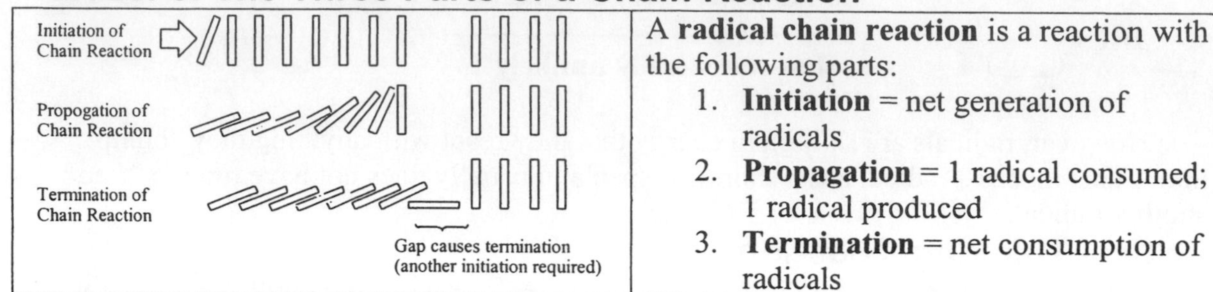

A **radical chain reaction** is a reaction with the following parts:

1. **Initiation** = net generation of radicals
2. **Propagation** = 1 radical consumed; 1 radical produced
3. **Termination** = net consumption of radicals

Critical Thinking Questions

10. The reaction in CTQ 7 (shown again below) is a radical chain reaction.

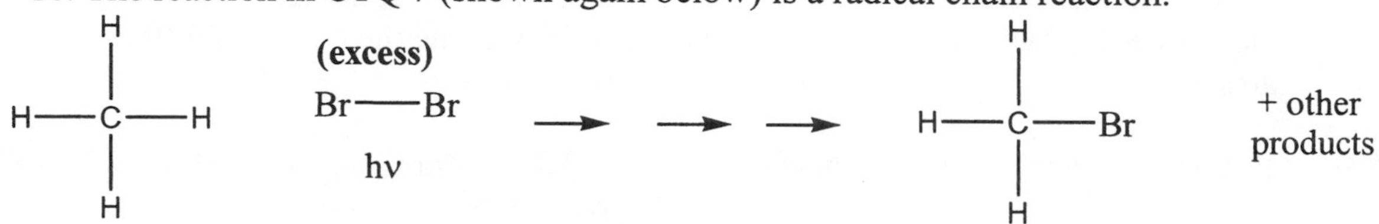

a) Use curved arrows to show the **initiation step.**

b) Use curved arrows to show the most likely **propagation steps** leading to the product, H_3C-Br.

c) A collision between two radicals is unlikely, but it does happen every once in a while. Eventually such reactions will stop the chain reaction. List at least two possible **chain termination steps** (they don't have to lead to H_3C-Br).

d) A variety of **less favorable** propagation reactions lead to side products. List one possible propagation step that leads to a product other than H_3C-Br.

11. Which of the following diagrams best tracks the products formed from a single initiation event (breaking apart of one Br_2 molecule) in a radical chain reaction.

Br_2 | •Br | •Br + Products | •Br + Products | etc. | Chain Termination

Br_2 → •Br | •Br + Products | •Br + P | etc. | Chain Termination; •Br | •Br + P | etc. | C.T.

Br_2 → •Br → •Br + P, •Br + P; •Br → •Br + P, •Br + P; etc. etc. etc. etc. etc. etc. etc. etc.; C.T. C.T. C.T. C.T. C.T. C.T. C.T. C.T.

Model 7: Which H Will Be Replaced by X?

- In a radical halogenation reaction a halogen (F, Cl, Br or I) replaces an H.
- In most cases there is more than one type of H to choose from.

For example the hydrocarbon below has two different types of H's:

$(CH_3)_2CHCH_3$ + X_2 $\xrightarrow{h\nu}$ $(CH_3)_3CX$ or $(CH_3)_2CHCH_2X$

X = F, Cl, Br or I

Three factors determine which H on a hydrocarbon will be replaced:

i. The type of H. (benzyl, allyl, 3°, 2°, 1° or methyl)
ii. The number of H's of a given type.
iii. The identity of X. (We will discuss this on the next page.)

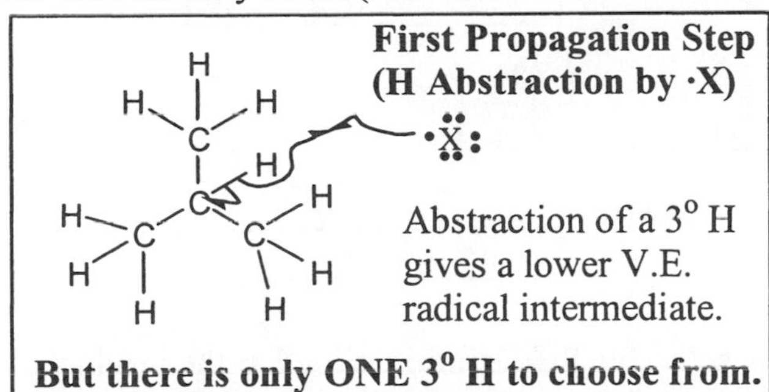

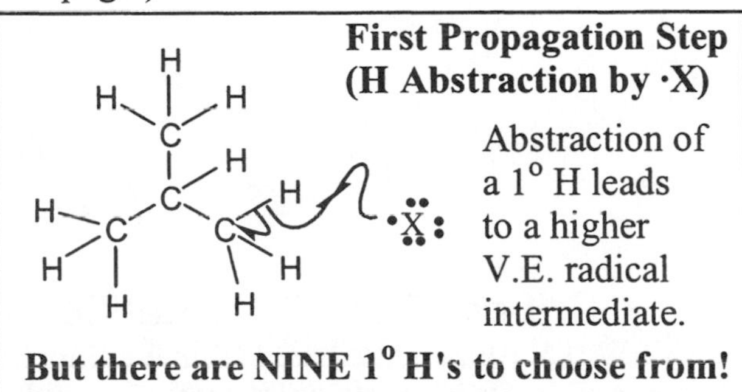

Critical Thinking Questions

12. Explain why, for the example above, you would expect a **1:9** ratio of 3° to 1° H substitution products if a 3° radical had about the same potential energy as a 1° radical.

13. But a 3° radical is LOWER in potential energy than a 1° radical. When X = Cl, a 3° radical intermediate is **5 times more favorable** than a 1° radical intermediate. Given this, explain why, for the example above, a **5:9** ratio of 3° to 1° H substitution products is expected. (Consider both the type of H and the number of identical H's.)

Model 8: Selectivity of the Photo-Halogenation Reaction

- Radical halogenation with F_2 is unselective, violent and dangerous limiting the usefulness of this reaction in organic synthesis.
- Radical halogenation with I_2 is so slow as to be useless in organic synthesis.
- **Radical halogenation with Br_2 is *just right!*** It is very useful in organic synthesis. The rate is manageable and reactions with Br_2 are very **selective**.

(A **selective reaction** gives close to 100% of a single product. An unselective reaction gives a mixture of products.)

Table 8: Relative Preference of Halogens for 1°, 2 °, 3 °, Allyl & Benzyl Sites

Halogen (X)	X sub. for H in a **1°** Position	X sub. for a H in a **2°** Position	X sub. for a H in a **3°** Position	X sub. for a H in **allyl** Position	X sub. for a H in **benzyl** Position
Fluorine (F)	1	1.2	1.4	1.5	1.6
Chlorine (Cl)	1	4	5	7	10
Bromine (Br)	1	100	2000	10,000	100,000

For Example:

Rxn VI: $(CH_3)_2CH{-}CH_3$ + F_2 $\xrightarrow{h\nu}$ $(CH_3)_3C{-}F$ + $(CH_3)_2CH{-}CH_2{-}F$

1.4	product ratio	9

Rxn VII: $(CH_3)_2CH{-}CH_3$ + Cl_2 $\xrightarrow{h\nu}$ $(CH_3)_3C{-}Cl$ + $(CH_3)_2CH{-}CH_2{-}Cl$

5	product ratio	9

Rxn VIII: $(CH_3)_2CH{-}CH_3$ + Br_2 $\xrightarrow{h\nu}$ $(CH_3)_3C{-}Br$ + $(CH_3)_2CH{-}CH_2{-}Br$

2000	product ratio	9

Critical Thinking Questions

14. Fluorination and Chlorination are **less selective** than Bromination. Do the product ratios for Rxn VI-VII support this statement? Explain.

15. Explain how the ratio 1.4 to 9 was arrived at in Rxn VI in Model 8.

16. Is your explanation consistent with the 2000:9 ratio in Rxn VIII?

17. Consider the following two step synthesis. Note that the identity of the halogen (X_2) is not specified.

X_2, hν → X– ; $^{\ominus}CN$ → NC–

a) A student chooses Cl_2 as the halogen. In the first step, he gets a mixture of three mono-chlorinated products. Use the information in Model 8 to calculate the ratio of these three products.

Cl Cl Cl

only this one gives the desired product upon S_N2 reaction with NC^-

b) If the mixture above is treated with cyanide ion (NC^-) a mixture of three different cyano-products is observed in a ratio of 5:8:6. Is this ratio consistent with your answer to part a)?

c) Explain why replacing Cl_2 with a different halogen (specify which one) would give a much better yield of the desired product.

d) Calculate the ratio of desired mono-CN product to other mono-CN products if this other halogen were used.

Exercises for Part A

1. Along with the methyl radical, there is a Br radical in the box in CTQ 9. What happens if Br radical reacts with the <u>most abundant species in the reaction mixture</u> (Br-Br)? Draw this reaction and explain why it not very interesting.

2. Construct an explanation for why the solvents on the left are suitable for radical halogenation reactions, but the solvents on the right are not. (Note that benzene is resonance stabilized, making it very unreactive.)

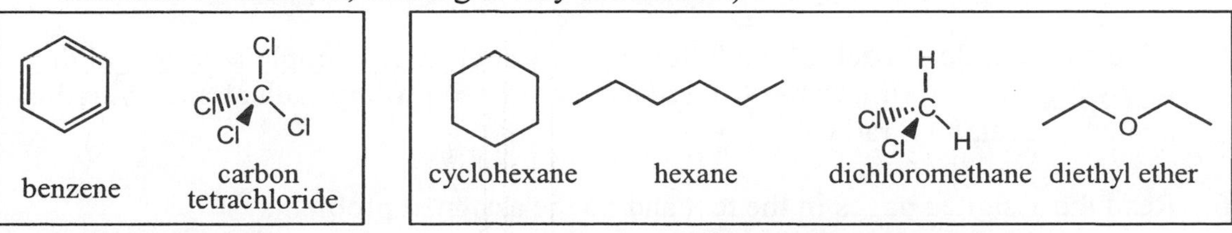

suitable solvents **NOT suitable solvents**

3. Consider the partial resonance representation of a benzyl radical species.

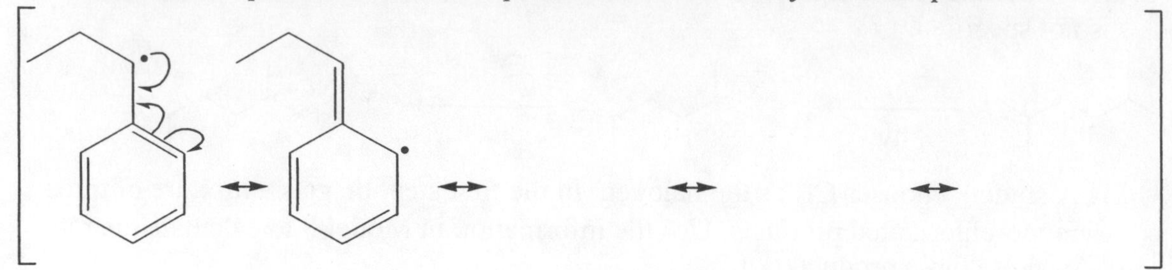

   a) Use curved half-arrows to generate the other three resonance structures.
   b) On the incomplete composite structure of this benzyl radical (shown below), mark each carbon that has partial radical character with a **delta radical**.

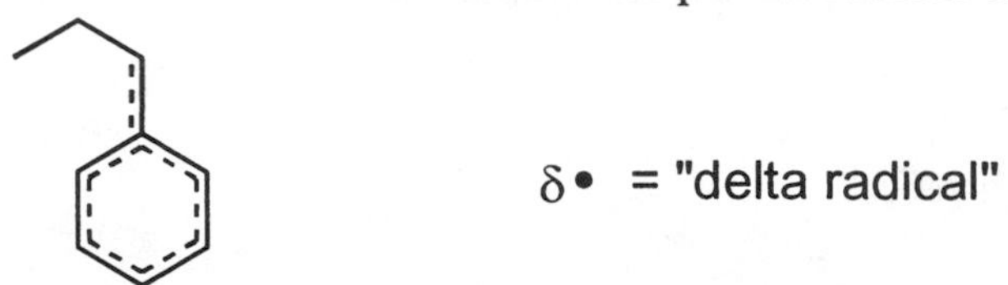

   c) Explain why a benzyl radical is much lower in potential energy than the radical drawn below. (Hint:are there important resonance structures for the species below that show the spreading out of the radical over several atoms?)

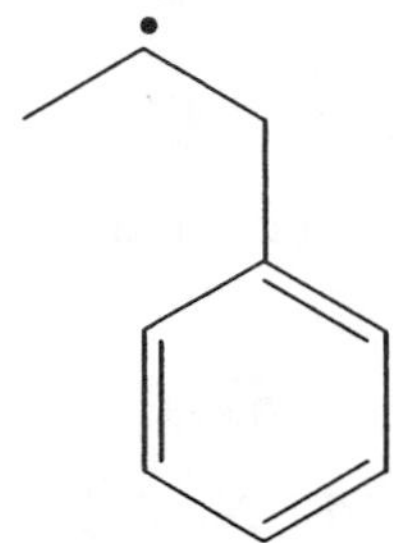

4. Based on the analogy to carbocations, circle the species below that is lower in potential energy.

   a) Construct an explanation for your choice. (Hint: use curved arrows to draw any other resonance structures for each of the radical species above.)
   b) Which one of these radicals is an allyl radical? Explain why.
   c) Draw an allyl carbocation and explain the similarity between an allyl radical and an allyl carbocation.

5. Give an example of each of the following and label each example with one of the following words: primary radical, secondary radical, tertiary radical, methyl radical, benzyl radical, allyl radical.

6. Read the assigned pages in the text and do the assigned problems.

Exercises for Part B

7. Consider the following reaction.

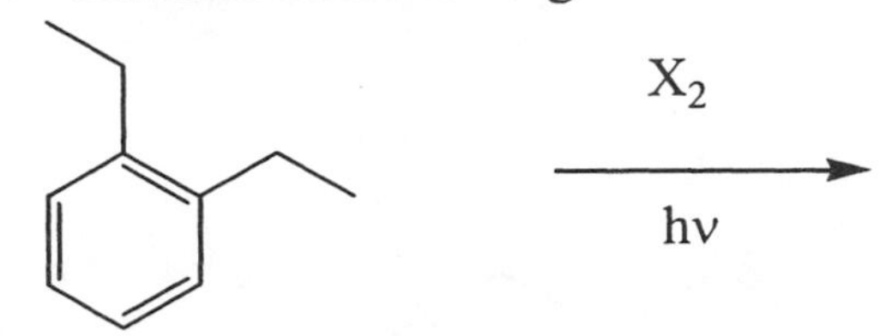

Note: only H's attached to sp^3 C's can be substituted for by X

a) There are two different H's in this molecule that can be replaced by X. Add them to the drawing above and label them H_a and H_b.
b) There are two different H's on this molecule that will not react with X radical. Label them H_c and H_d.
c) Specify whether H_a and H_b are primary, secondary, tertiary, allyl or benzyl.
d) Calculate the relative amounts of each of the two different products when the reaction is run with Cl_2.
e) Calculate the relative amounts of each of the two different products when the reaction is run with Br_2.
f) Which reaction is more selective? **Rxn in part d)** or **Rxn in part e)** [circle one].

8. Consider the following reaction:

Br_2 hν →

Product I　　Product II

product ratio

a) Use curved arrows to show the mechanism for formation of Product I (include the initiation step and at least one termination step).
b) Use curved arrows to show the mechanism for formation of Product II (include the initiation step and at least one termination step).
c) Using the data in Table 8, predict the ratio of Product I:Product II and put the appropriate numbers in the box labeled "product ratio."

9. A **selective reaction** gives mostly a single product. An **un-selective reaction** gives a mixture of various products. In general, which reaction below is most selective?
   a) radical fluorination of an alkane
   b) radical chlorination of an alkane
   c) radical bromination of an alkane

10. Use curved half arrows to show the most likely radical reaction including the most likely mono-halogenation product using…
    a) Cl_2 and light
    b) Br_2 and light.

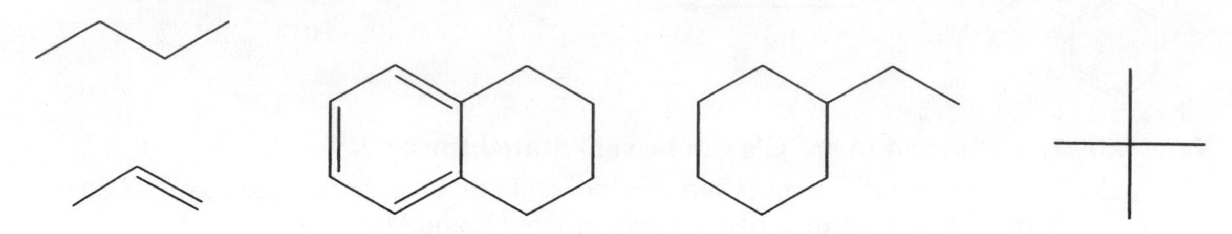

11. S-3-methylhexane undergoes radical bromination to form a racemic mixture of two products. Draw both products and explain why a racemic mixture is formed rather than pure S or pure R.

12. Design a synthesis of each of the following target molecules starting from cyclohexane. You may use any reagents in your proposed syntheses (carbon or non-carbon containing).

+ enantiomer

+ enantiomer

13. Read the assigned pages in the text and do the assigned problems.

ChemActivity 21

Anti-Markovnikov Addition of H-Br

(How does the placement of Br differ in radical vs. polar addition of HBr to an alkene?)

Model 1: Radical Reactions of Alkenes

Figure 1a: Radical Chain Reactions with Br_2 and H—Br

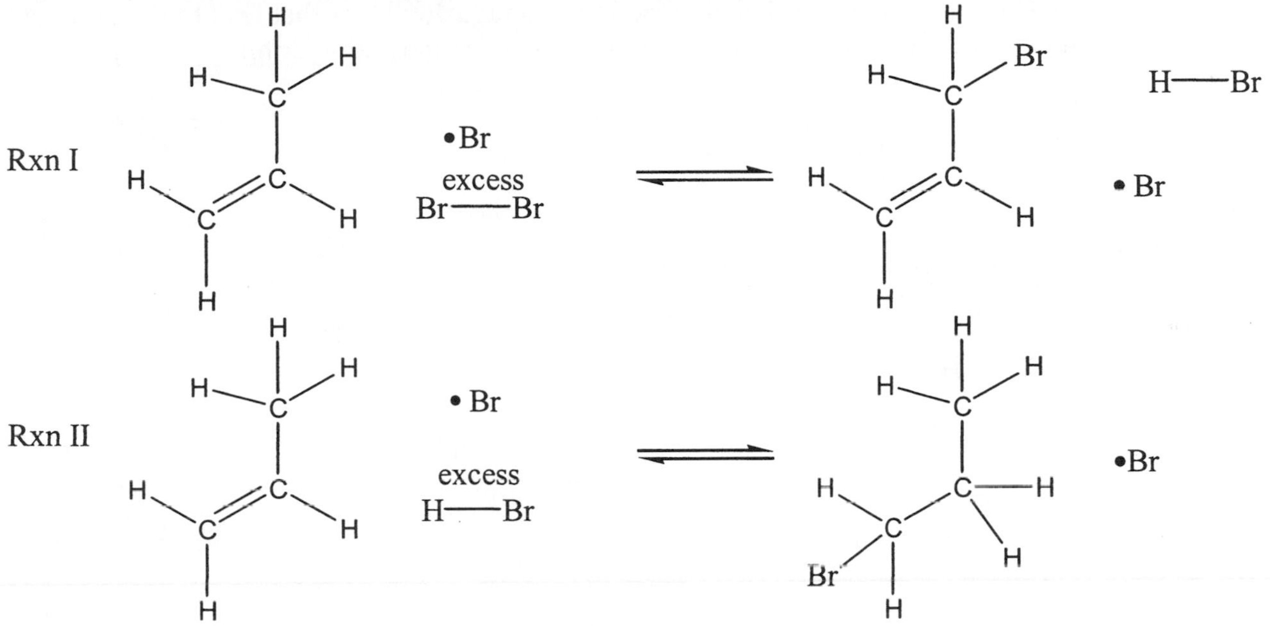

Rxn I

•Br

excess

Br—Br

• Br

H—Br

Rxn II

• Br

excess

H—Br

•Br

Critical Thinking Questions

1. Review: Use curved arrows to show **propagation** steps that lead to the products shown in Rxn I. Do not show any termination steps.

Figure 1b: Reaction of a π Bond with a Radical

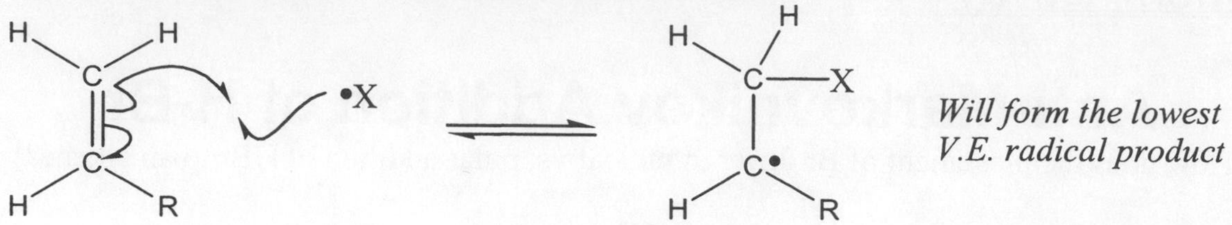

Will form the lowest V.E. radical product

2. Devise a series of radical chain reaction **propagation** steps that lead to the products shown in Rxn II (redrawn for you below). <u>Do not show any termination steps.</u>

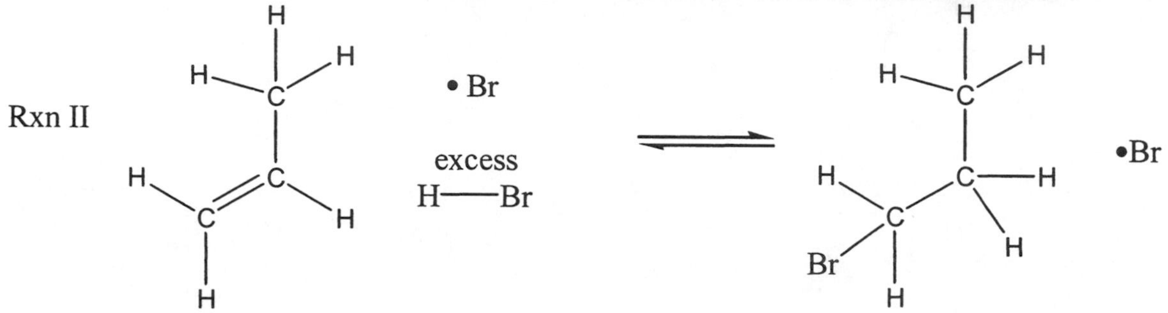

3. Explain why the following radical is <u>not</u> produced during Rxn II.

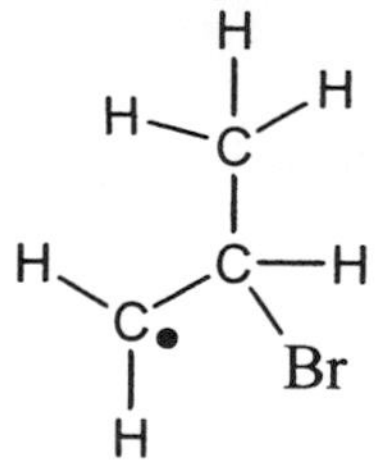

Model 2: Radical Chain Reaction Initiators

Previously we have seen that a strong light source can be used to initiate a radical chain reaction by breaking a Br—Br bond.

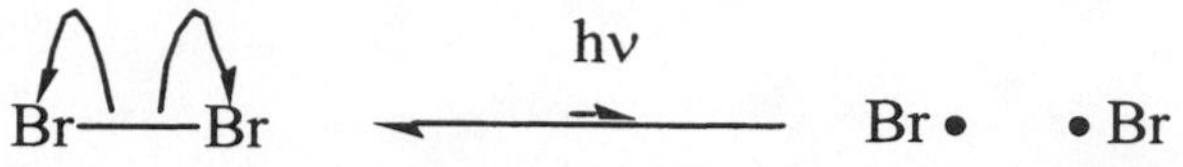

However, this method can only be used when there is a high concentration of Br_2 in solution. Rxn II (above) has a high concentration of H-Br instead of Br-Br.

In cases where Br–Br concentration is low, **peroxides** are used as **radical initiators** because the O-O bond is easy to break. Benzoyl peroxide is a common radical initiator.

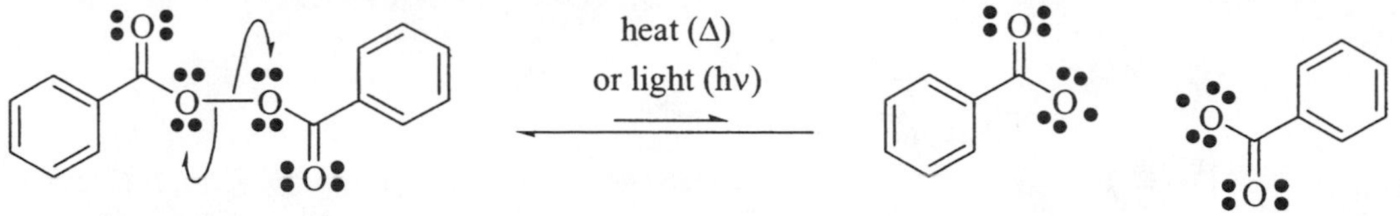

heat (Δ)
or light (hν)

benzoyl peroxide

Note: Δ is an abbreviation indicating the addition of heat to a reaction.

Bond dissociation energy (BDE) is the amount of energy required to break a bond in a non-polar fashion ("homolytic bond cleavage"), with one electron going to each of the two atoms involved in the bond. The BDE of the O–O bond of...

hydrogen peroxide (HO–OH) = 51 kcal/mole (kcal = a measure of energy)
benzoyl peroxide = 18 kcal/mole

Critical Thinking Questions

4. Which is easier to break, the O–O bond of hydrogen peroxide or the O–O bond of benzoyl peroxide? Cite data from Model 2 to support your answer.

5. Construct a reasonable mechanism for the following reaction.

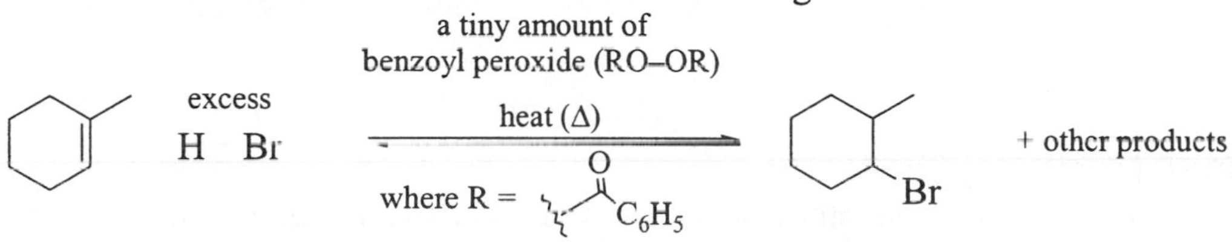

excess H Br — a tiny amount of benzoyl peroxide (RO–OR), heat (Δ) → Br + other products

where R = $C(=O)C_6H_5$

Show...

i. an initiation step.
ii. propagation steps that lead to the product shown.
iii. two possible termination steps.

Model 3: Radical and Polar Addition of HBr to an Alkene

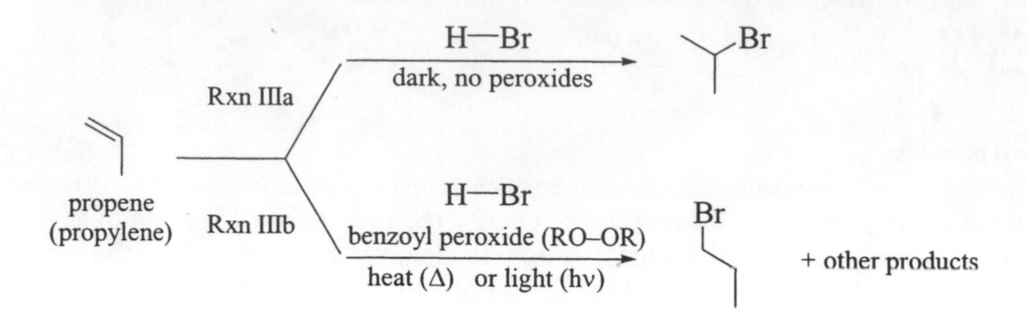

H—Br

dark, no peroxides

Rxn IVa

styrene

Rxn IVb

H—Br

benzoyl peroxide (RO–OR)

heat (Δ) or light (hν)

Critical Thinking Questions

6. Consider the regiochemistry of reactions IIIa and IIIb and label each product **Markovnikov** or **anti-Markovnikov**, as appropriate.
   a) Show the mechanism of Rxn IVa in the space below.

   b) Draw the major carbon containing product of Reactions IVa at the top of the page.

7. Summarize the pattern you see regarding which reagents/conditions produce Markovnikov products and which produce anti-Markovnikov products.

8. Without going through the mechanism, draw the product of Rxn IVb, above, and label it as **Mark**. or **anti-Mark**., as appropriate.

9. A student proposes the following **multi-step synthesis** of the **target molecule** starting from 2-methylpropene. (By convention, in a synthesis, reagents are placed above each arrow, and non-carbon containing products are not shown.)

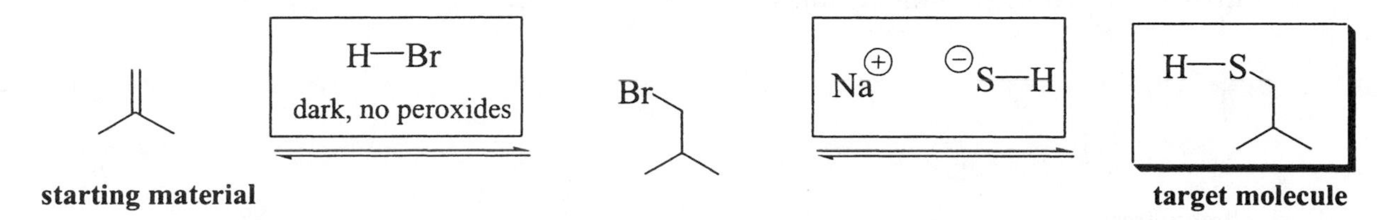

a) One step in this synthesis is wrong. Cross out the reagents for this step.
b) Correct the student by writing in reagents that will accomplish this step.

10. To each box below add reagents that will accomplish the transformation shown.

Br

Br

CN

11. Write the appropriate reaction name below each reaction arrow in CTQ 9: **Choose from:** S_N2, E2, radical bromination, anti-Markov. addition of HBr.

12. Add reagents to the box to complete the multi-step synthesis below.

R–C(=O)–O–O–H

NEW RXN

starting material

O⊖ OH

+ enantiomer

dilute H–Cl

OH OH

+ enantiomer

target molecule

Note: The first step is a new reaction found on reaction sheet R4. You do not need to know the mechanism of this new reaction.

Exercises

1. Design a synthesis of the following target molecules beginning with the starting material given. Write the reagents needed for each transformation above the reaction arrows for that step in the synthesis.

a)

starting material

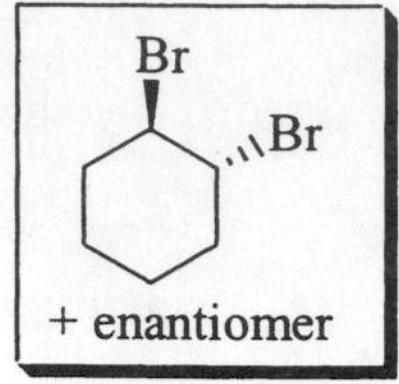

target molecule

b)

HO

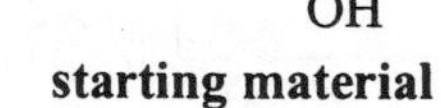

OH

starting material

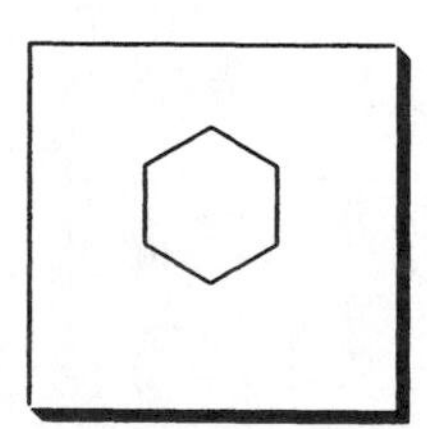

target molecule

c)

starting material

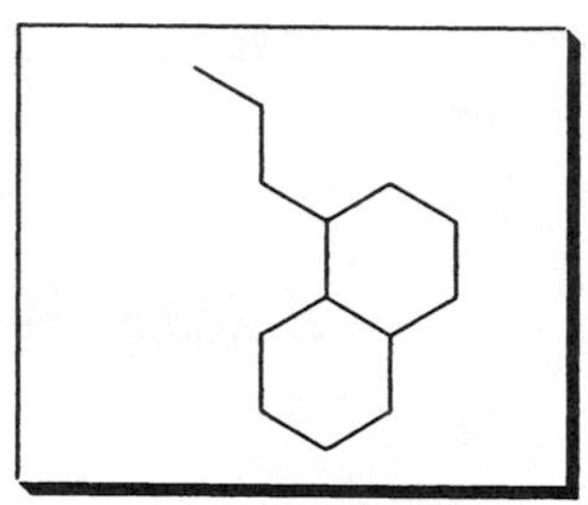

target molecule

2. Over each reaction arrow, write in reagents that would give the product shown?

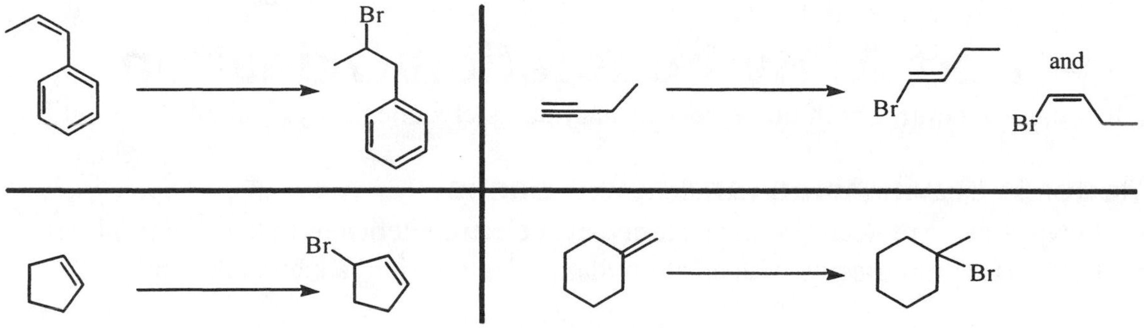

Br

and

Br

Br

Br

Br

3. Write the mechanism of each reaction in Exercise #2.

4. Oxygen—oxygen single bonds are relatively easy to break. Construct an explanation for why benzoyl peroxide has an especially small B.D.E. compared to other peroxides? (*see* Model 2)

5. Read the assigned pages in the text and do the assigned problems.

ChemActivity 22

Part A: Hydroboration/Oxidation

(How can we control both the regiochemistry and stereochemistry of alcohol formation?)

Model 1: Lewis Acids (and Lewis Bases)

- Lewis Acid = molecule with an atom that is electron deficient (a.k.a. electrophile).
- Lewis Base = molecule with an atom that is electron rich (a.k.a. nucleophile).

For a given molecule, the most **most Lewis acidic** atom = **most electron deficient atom.**

Table 1a: Selected Elements with (Electronegativities)

H (2.3)						
Li (0.9)	Be (1.6)	B (2.1)	C (2.5)	N (3.1)	O (3.6)	F (4.2)

Critical Thinking Questions

1. Formal charges can be misleading. For example, oxygen is NOT the most electron deficient atom on hydronium ion.

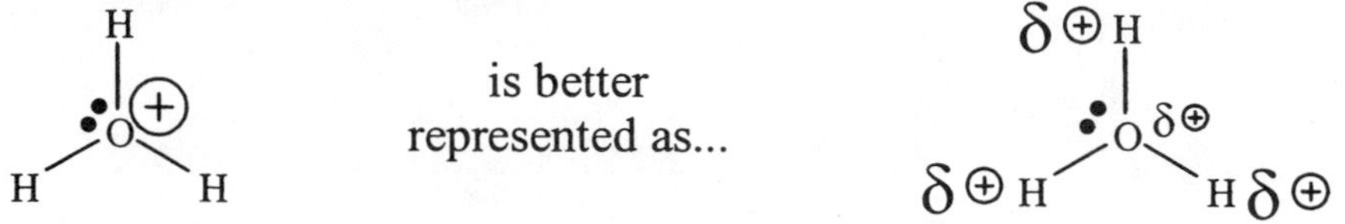

   a) What is the most Lewis acidic atom on hydronium ion?

   b) Are the electronegativities in Table 1 consistent with the fact that oxygen "hogs" the electron density in each O-H bond, making itself almost neutral? Explain.

   c) Why can't carbon do the same thing and make itself neutral?

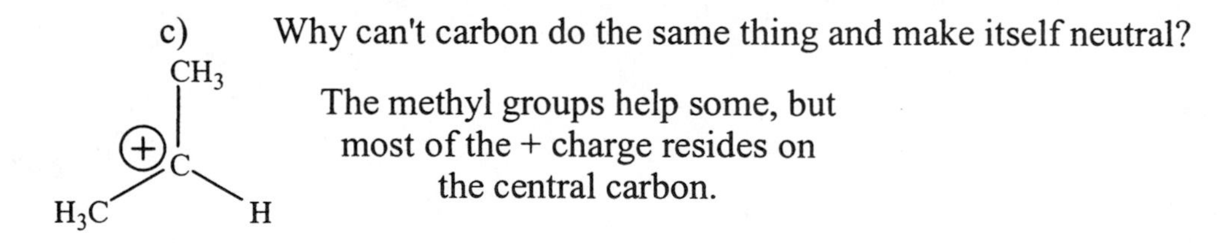

2. Draw a Lewis structure of borane (BH_3). **<u>Note that the boron atom does not have an octet</u>**.

   a) Which is more electronegative, B or H?
   b) Which atom of borane is most Lewis acidic? Explain your reasoning.

3. Draw a carbocation that is likely to form in each reaction below. Note: In each case, the π electrons react with the most Lewis acidic atom on the electrophile.

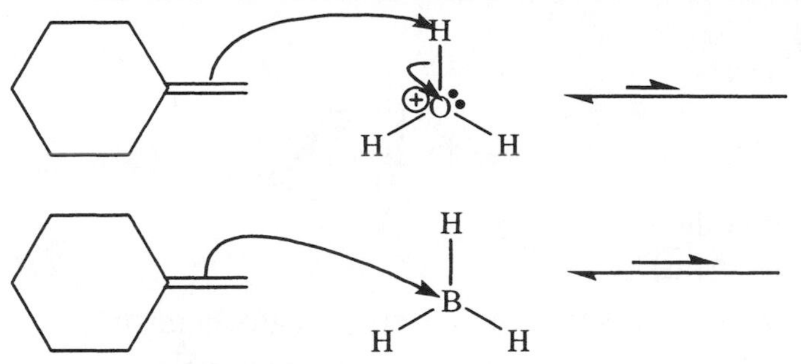

Model 2: Hydroboration (Addition of H & BH_2 to an Alkene)

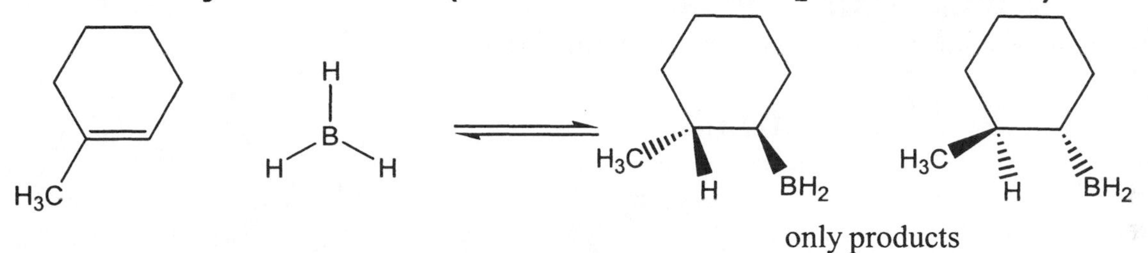

The following mechanism has been proposed to explain these products.

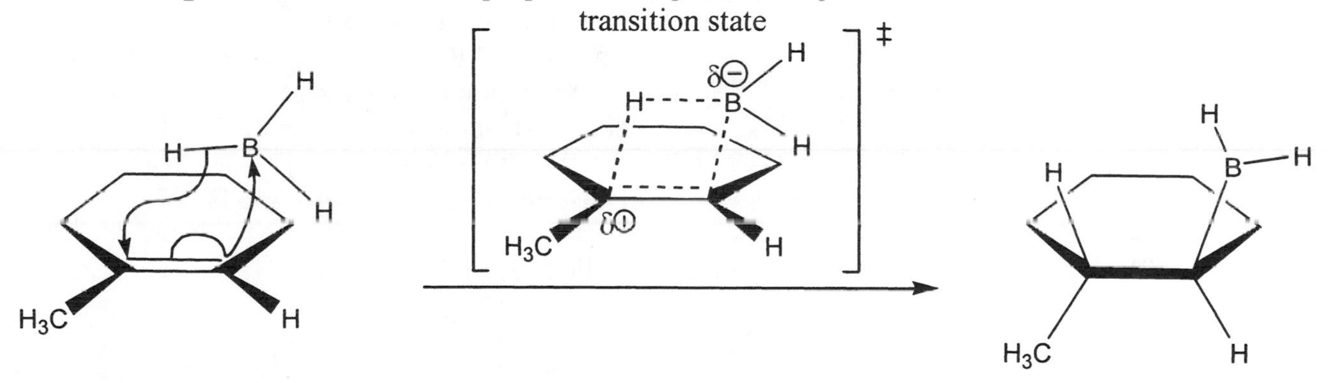

Critical Thinking Questions

4. Draw a product with the opposite regiochemistry from the ones shown in Model 2. (Regio-chemistry refers to which carbon the non-hydrogen atom (B) is bonded to.)

5. Stereochemistry: According to Model 2, does addition of BH_2 and H occur in a **syn** fashion (added to same side of ring) or an **anti** fashion (opposite sides of ring)?

6. Does the mechanism in Model 2 explain the regiochemsitry of the products?

7. Does the mechanism in Model 2 explain the stereochemistry of the products?

Model 3: Oxidation of BH_2 to OH

H_3C H BH_2 — H_2O_2 / NaOH / H_2O → H_3C H OH

hydrogen peroxide
(mild oxidizing agent)

- BH_2 can be easily oxidized to OH without affecting the stereo or regiochemistry.
- The mechanism of oxidation of BH_2 to OH is not covered in these materials.

Critical Thinking Questions

8. Consider the overall transformation from alkene to alcohol.

Rxn I H_3C — 1) BH_3 2) H_2O_2/NaOH/H_2O → H_3C H OH H_3C H OH

a) Label the products Markovnikov or anti-Markovnikov.
b) Circle one of the following summaries of Rxn I:
   i. Rxn I yields **Markov**. **syn** addition of OH and H.
   ii. Rxn I yields **anti-Markov**. **syn** addition of OH and H.
   iii. Rxn I yields **Markov. anti** addition of OH and H.
   iv. Rxn I yields **anti-Markov. anti** addition of OH and H.

9. Draw the mechanism and product/s of the following reaction.

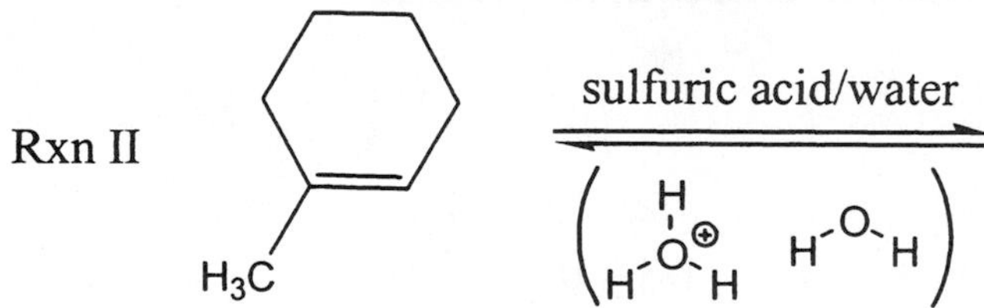

a) Label the products Markovnikov or anti-Markovnikov.
b) Circle one of the following summaries of the Rxn II:
   v. Rxn II yields **Markov**. addition of OH and H.
   vi. Rxn II yields **anti-Markov**. addition of OH and H.
c) Rxn II gives
   vii. **syn**. addition of OH and H.
   viii. **anti** addition of OH and H.
   ix. a 50/50 mixture of **syn** and **anti** addition of OH and H.

Part B: Other Oxidation Reactions

(Which functional groups can be oxidized?)

Model 4: Oxidative Bond Breakage

- C–H and O–H bonds (but not C-C bonds) can be easily broken with oxidizing agents such as PCC (*a mild reagent*) or $KMnO_4$ (*a stronger reagent*)

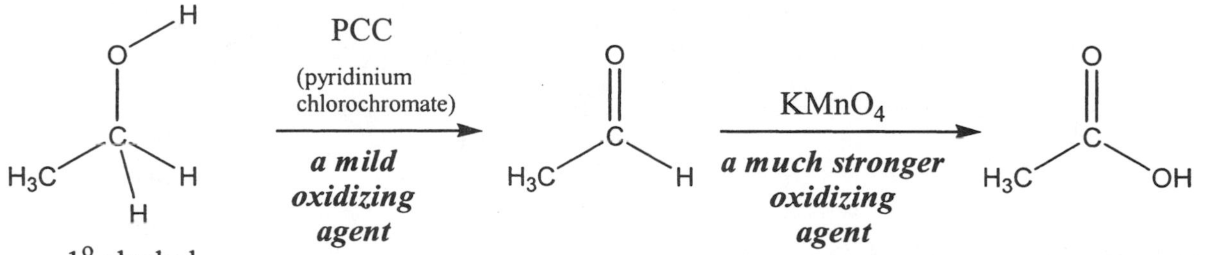

PCC (pyridinium chlorochromate) — ***a mild oxidizing agent*** — $KMnO_4$ — ***a much stronger oxidizing agent***

1° alcohol → aldehyde → carboxylic acid

Oxidation in organic chemistry = adding O's (or more bonds to O's) or taking away H's.
Reduction in organic chemistry = adding H's or taking away O's.

Critical Thinking Questions

10. Construct an explanation for why the alcohol in Model 4 is said to be "oxidized" in the first step even though no oxygen atoms were actually added to the molecule in this step.

11. Put an X through each bond that is broken during each oxidation reaction above.

12. According to Model 4, why can't a 2° alcohol be easily oxidized to a carboxylic acid?

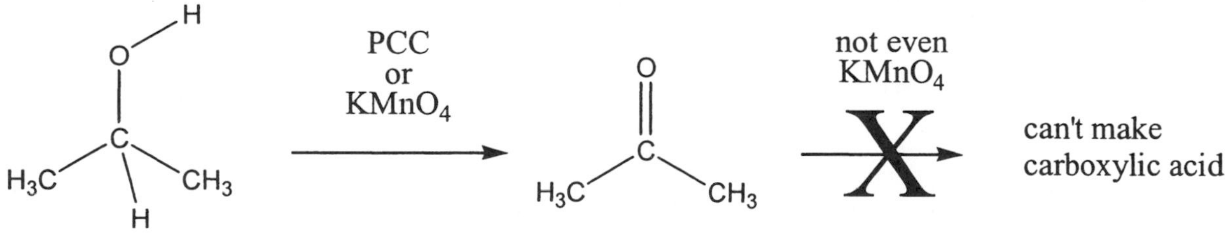

PCC or $KMnO_4$ — not even $KMnO_4$ — can't make carboxylic acid

2° alcohol → ketone

13. Consider the molecules below.
    a) Circle each molecule that can be oxidized without C-C bond breakage.
    b) Below each circled species, draw the oxidation product that would result.

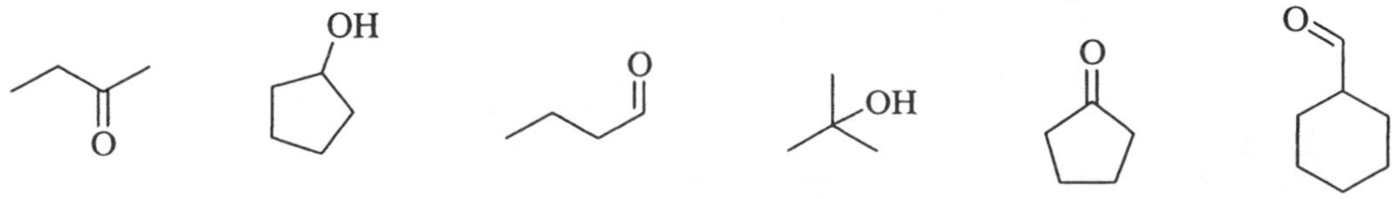

Model 5: Oxidative Cleavage of C-C π Bonds

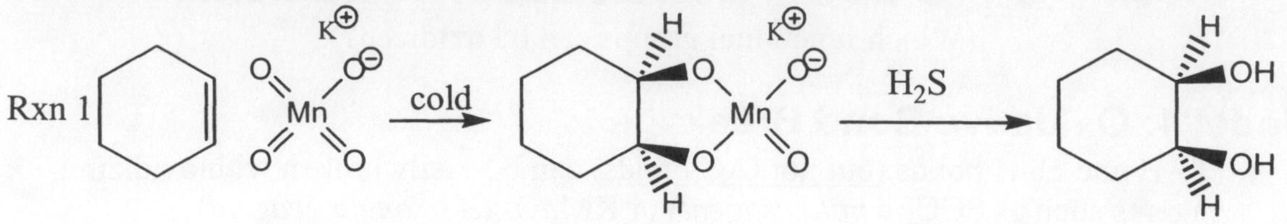

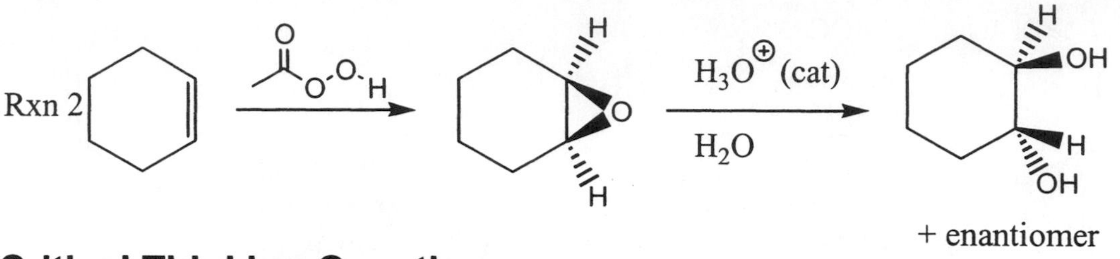

Rxn 1 cyclohexene + $K^{\oplus}$ $MnO_4^{\ominus}$ $\xrightarrow{\text{cold}}$ cyclic manganate ester $\xrightarrow{H_2S}$ cis-1,2-cyclohexanediol

Rxn 2 cyclohexene $\xrightarrow{\text{peroxyacetic acid}}$ epoxide $\xrightarrow[H_2O]{H_3O^{\oplus}\text{ (cat)}}$ trans-1,2-cyclohexanediol

+ enantiomer

Critical Thinking Questions

14. You do not need to know the mechanisms of Rxn 1 or the first step of Rxn 2. Draw the mechanism of the second step of Rxn 2 in Model 5.

15. Above each arrow, write reagent/s you would use to perform the transformation shown?

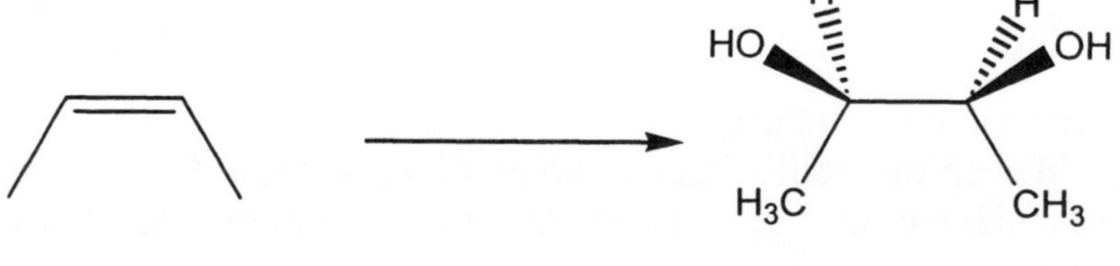

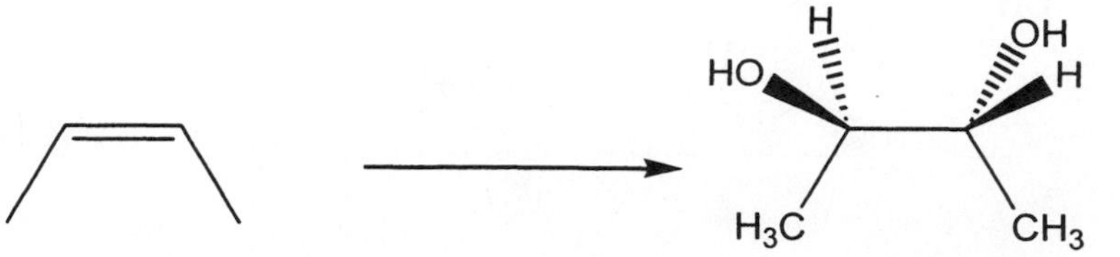

Model 6: Oxidative Cleavage of the Whole C=C Bond (σ & π)!!

$(H_3C)_2C{=}C(H)CH_3$ → 1) ozone (O_3) 2) Zn, acetic acid → $(H_3C)_2C{=}O$ + $O{=}C(H)CH_3$

$(H_3C)_2C{=}C(H)CH_3$ → 1) $KMnO_4$ (hot) 2) $H_3O^{\oplus}/H_2O$ → $(H_3C)_2C{=}O$ + $O{=}C(OH)CH_3$

Critical Thinking Questions

16. Put an X though any bonds in the starting material that are broken during each oxidation reaction above.
    a) Circle the most oxidized product in Model 6.
    b) Which is a stronger oxidizing agent: ozone or hot $KMnO_4$? Explain your reasoning.

17. Draw the most likely product/s that result from treatment of 1-methylcyclohexene with each reagent listed.

1-methylcyclohexene →
- 1) O_3 2) Zn, acetic acid →
- 1) $KMnO_4$ (hot) 2) $H_3O^{\oplus}/H_2O$ →
- $KMnO_4$ (cold) NaOH →
- 1) $CH_3C(=O)OOH$ 2) $H_3O^{\oplus}/H_2O$ →

Exercises for Part A

1. Draw the three products that DID NOT form in the hydroboration reaction in Model 3 (two have the wrong stereochemistry, but the right regiochemistry; one has the wrong regiochemistry.

2. Identify each product below as Markov. or Anti-Markov.; and syn or anti addition, as appropriate. Then place one of the following over each reaction arrow.
   - i. H_2O/H_2SO_4
   - ii. 1) BH_3 2) $H_2O_2/NaOH/H_2O$
   - iii. not possible to get this product with i or ii.

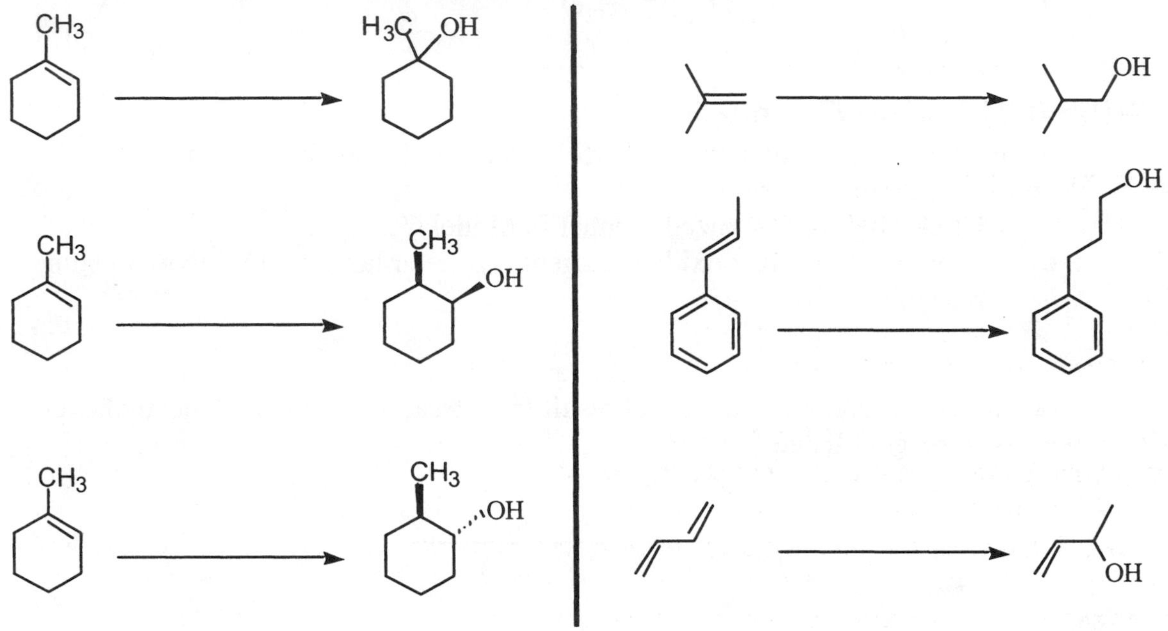

CH3 → H3C OH

CH3 → CH3 OH

CH3 → CH3 OH

→ OH

→ OH

→ OH

3. Draw all the products that will form in the following reaction. Use wedge and dash bonds to show the absolute stereochemistry of each chiral carbon in the products. (Do not draw any products twice!!)

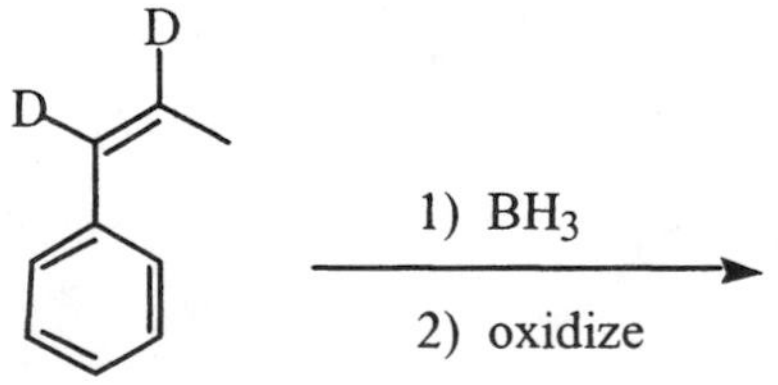

D D → 1) BH_3 2) oxidize

Assign R or S to each chiral carbon in your products.

4. Draw <u>all the products</u> that will form in the following reaction. Use wedge and dash bonds to show the absolute stereochemistry of each chiral carbon in the products. (Do not draw any products twice!!)

D D

$\xrightarrow[H_2O]{H_3O^{\oplus}}$

Assign R or S to each chiral carbon in your products.

5. Show a synthesis of each of the following target molecules from methylenecyclohexane. (Some syntheses may be only one step.)

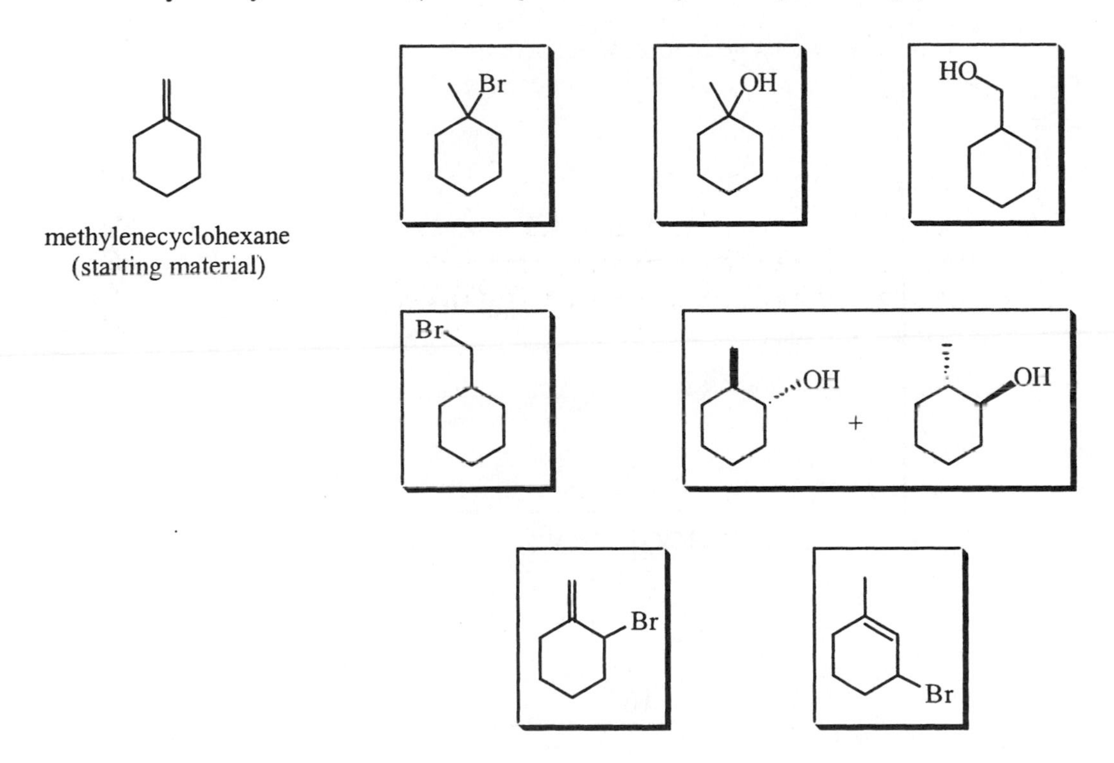

6. Read the assigned pages in your text and do the assigned problems.

Exercises for Part B

7. Draw the most likely product/s that result from treatment of 1-methylcyclohexene with each reagent listed.

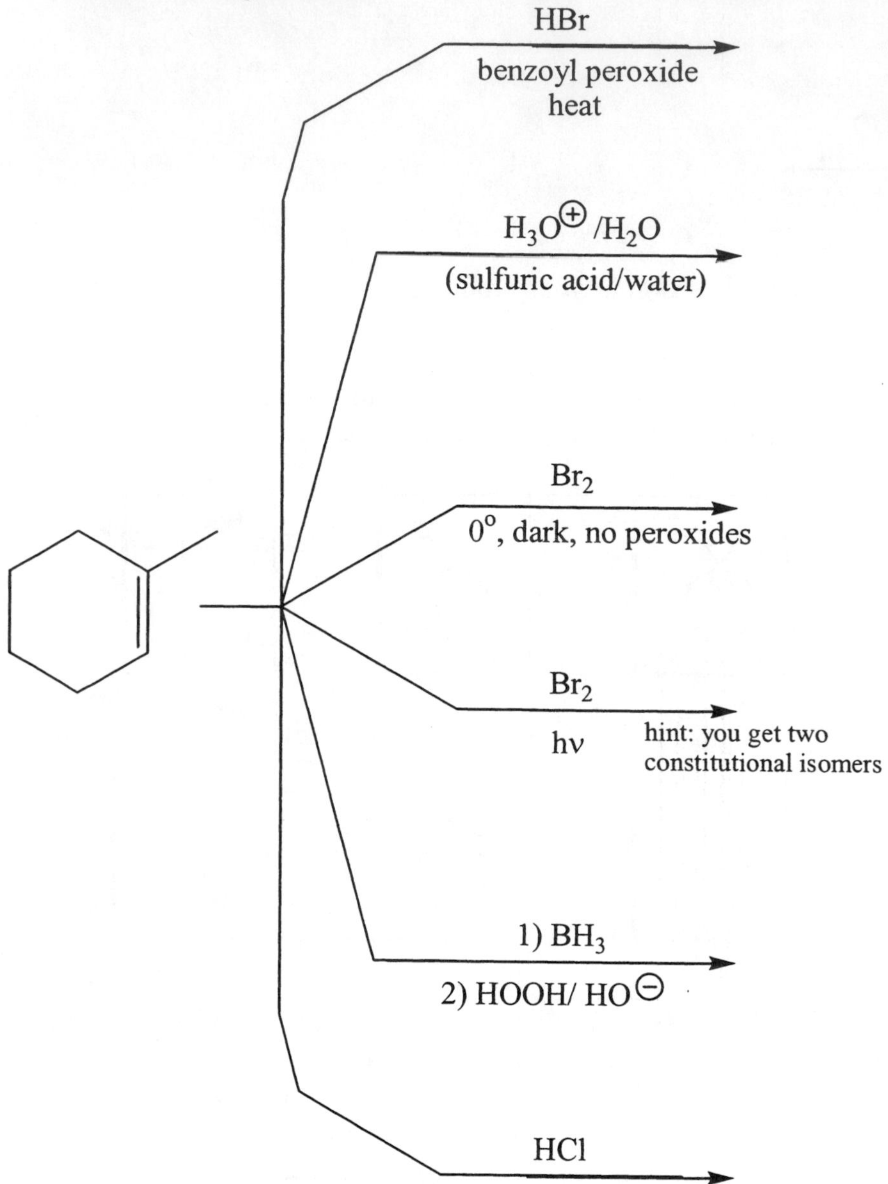

8. Draw the products of the following reaction.

1) $KMnO_4$ (hot)

2) $H_3O^{\oplus}$ /H_2O

9. Read the assigned pages in your text and do the assigned problems.

ChemActivity 23

Electrophilic Addition to Alkynes

(What is the mechanism by which an alkyne can be converted to a ketone?)

Model 1: Enol to Keto Tautomerization

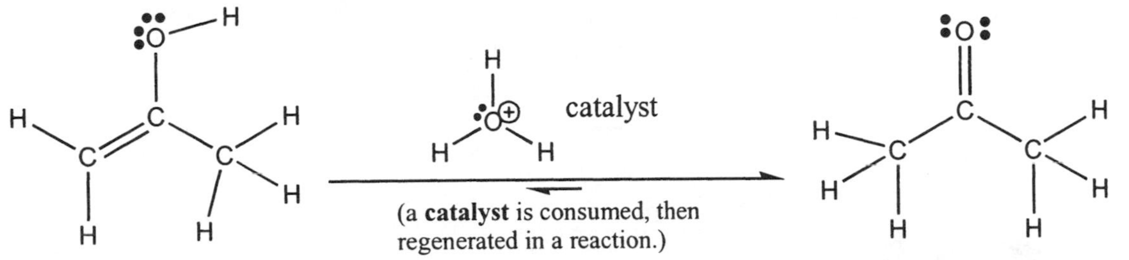

ene-ol
(also called "enol" form)

catalyst

(a **catalyst** is consumed, then regenerated in a reaction.)

ketone
(also called "keto" form)

Critical Thinking Questions

1. Construct a reasonable mechanism for the reaction shown in Model 1.

2. According to the information in Model 1, which is lower in potential energy: the keto form or the enol form of the molecule in Model 1?

3. Speculate about the etymology (origin) of the name "enol."

Information:

- Most enols are unstable and will **tautomerize** to the keto form, even without acid catalysis. In these cases the solvent carries the H from O to the C (or N).
- Unless you are specifically asked to do so, you need not show the mechanism of such tautomerizations. Simply write the word "tautomerize" over the reaction arrow...

enol form $\xrightarrow{\textbf{tautomerize}}$ keto form (this is an aldehyde)

4. Draw the keto form of each of the following enols.

enol form **keto form**

OH **tautomerize**

OH **tautomerize**

H O H C N H **tautomerize**

5. Construct an explanation for why the unusual enol below (left) is favored over its keto form. That is, the enol to keto tautomerization is uphill instead of downhill.

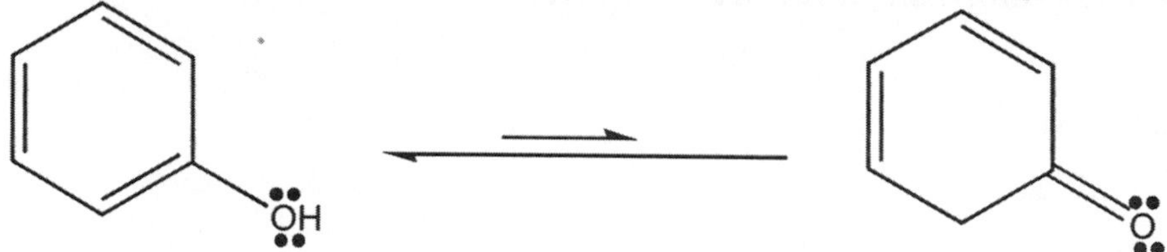

Model 2: Triple Bond Revisited

- A triple bond consists of a σ bond and two **orthogonal** π bonds (orthogonal = perpendicular and non-overlapping).
- Either or both π bonds in an alkyne (triple bond) can undergo electrophilic addition just like a π bond in an alkene (double bond).

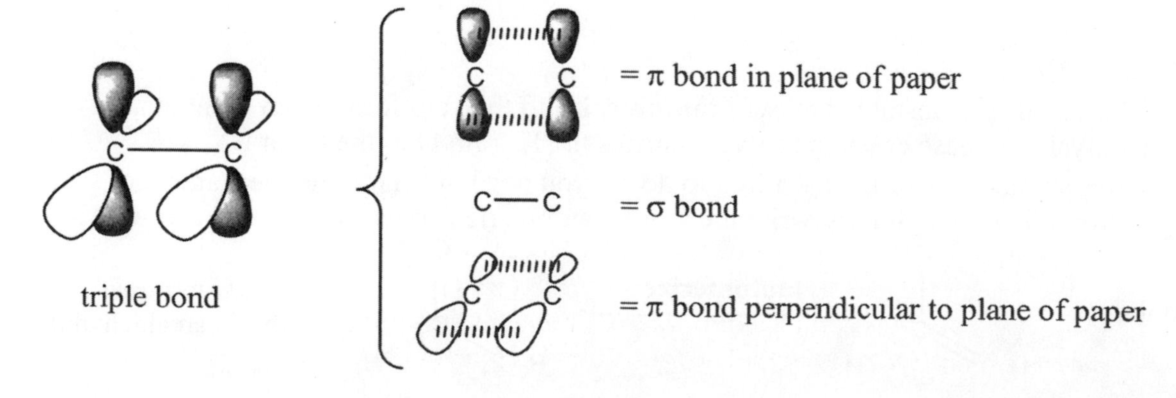

Critical Thinking Questions

6. Construct a reasonable mechanism for the following electrophilic addition reaction. (Do not show the mechanism of the tautomerization-just write "tautomerize.")

$H_3C—C{\equiv}C—CH_3$ + H_2O $\xrightleftharpoons{H_3O^{\oplus}\text{ (cat.)}}$ $H_3C—C(=O)—CH_2—CH_3$

2-butyne

2-butanone

7. Draw the mechanism and circle the more likely product of the following electrophilic addition reaction.

$H_3C—C{\equiv}C—H$ + H_2O $\xrightleftharpoons{H_3O^{\oplus}\text{ (cat.)}}$ $H_3C—C(=O)—CH_3$ or $H_3C—CH_2—C(=O)—H$

propyne

a) Explain why the circled product forms preferentially.

b) Is the enol leading to the circled product Markovnikov or anti-Markovnikov?

c) Draw the uncircled product at the end of the reaction below, then, in the box, draw the enol that would give rise to this product.

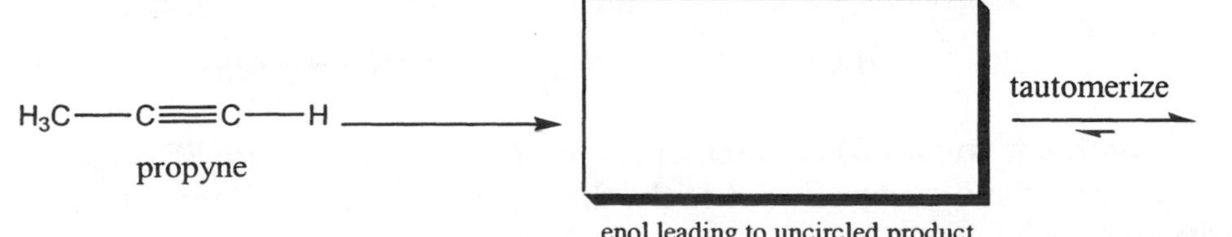

uncircled product

d) Is the enol in the box Markovnikov or anti-Markovnikov?

e) Write appropriate reagents over the reaction arrow in part c).

Exercises

1. When propyne is treated with H–Cl, two different products are observed (A & B).

- Product A is an intermediate product on the way to Product B.
- The molecular formula of each product is given below.
- Product B is only observed when excess HCl is added.

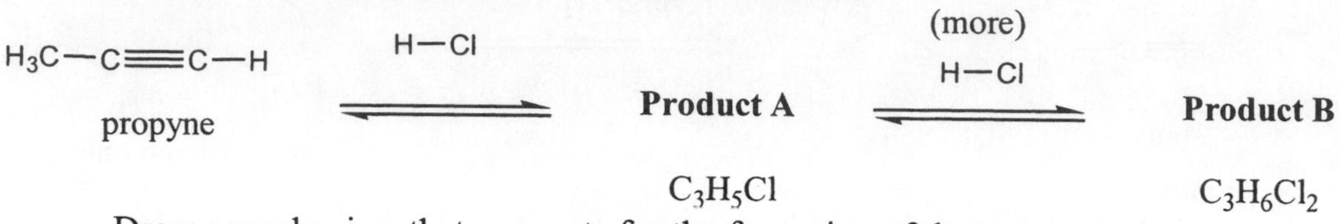

$$H_3C{-}C{\equiv}C{-}H \text{ (propyne)} \xrightleftharpoons{H{-}Cl} \textbf{Product A } (C_3H_5Cl) \xrightleftharpoons[\text{(more)}]{H{-}Cl} \textbf{Product B } (C_3H_6Cl_2)$$

Draw a mechanism that accounts for the formation of these two products.

2. One method of preparing an alkyne is to heat a **geminal dihalide** in the presence of a base (such as hydroxide). ("**Geminal dihalide**" comes from Gemini meaning "twin" and refers to a carbon with two halides.) Construct a reasonable mechanism for the following reaction.

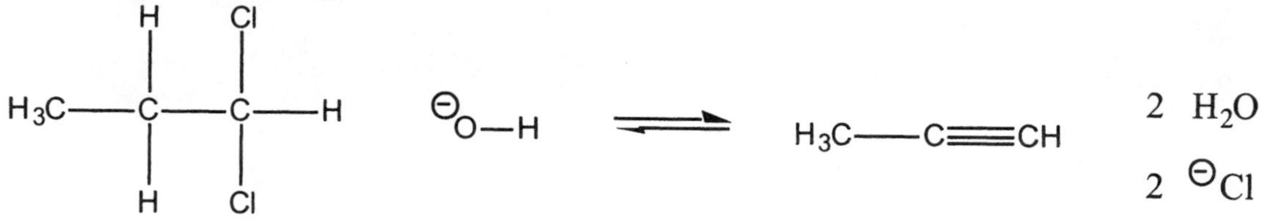

$$H_3C{-}CH_2{-}CHCl_2 \quad {}^{\ominus}O{-}H \rightleftharpoons H_3C{-}C{\equiv}CH \quad 2\ H_2O \quad 2\ {}^{\ominus}Cl$$

3. Over each reaction arrow, write the reagent/s that could be used to produce the products shown? (In each case, label the products shown Markov. or anti-Markov.?)

a)

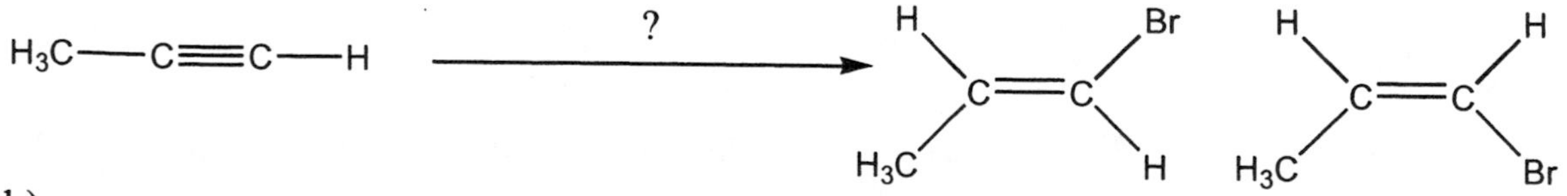

$$H_3C{-}C{\equiv}C{-}H \xrightarrow{?} (H)(H_3C)C{=}C(Br)(H) \quad (H)(H_3C)C{=}C(H)(Br)$$

b)

$$H_3C{-}H_2C{-}C{\equiv}C{-}D \xrightarrow{?} (Br)(H_3C{-}CH_2)C{=}C(H)(D) \quad (Br)(H_3C{-}CH_2)C{=}C(D)(H)$$

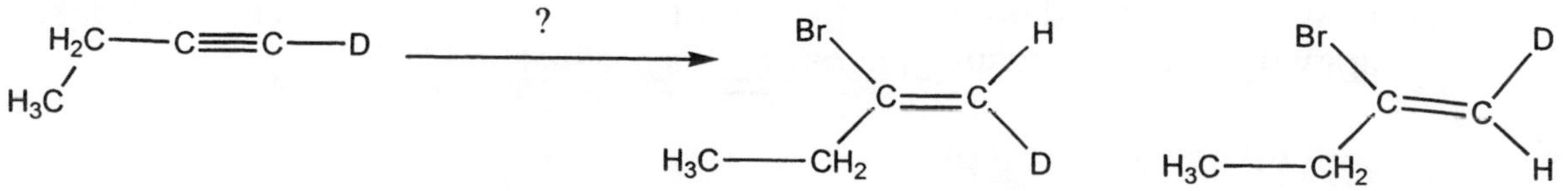

c) The two products in part b) are niether *cis* nor *trans*. Read "**Rule 5**" on *p.* 441 of your text and assign E or Z to the these products. (Recall that D takes priority over H.)

4. The nucleophile produced by deprotonation of a terminal alkyne is very similar to a Grignard reagent. Strong nucleophiles such as Grignard reagents are known to react preferentially with the C of a C=O (carbonyl group).

a) Construct an explanation for why the C of the carbonyl group (C=O) is electrophilic (has a partial + charge to attract the nucleophile).

b) Use curved arrows to show reaction of the nucleophile (the deprotonated alkyne anion) with the carbonyl, and draw the resulting products.

c) Use curved arrows to show the reaction that takes place when the products of this reaction are neutralized with dilute HCl. (Note: dilute HCl cannot cause electrophilic addition to the π bonds of the triple bond.)

d) Why must the dilute HCl be added in a second step, after the carbon nucleophile has been allowed to react? In other words, what would happen if you added the dilute HCl at the same time as the nucleophile?

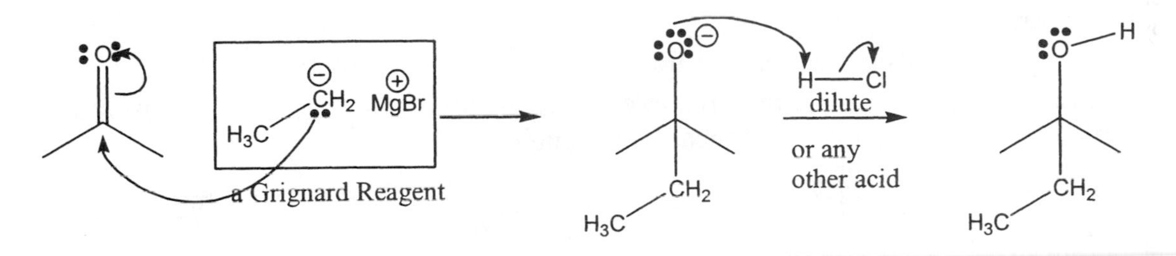

counterion

:O:

1) Li⊕ ⊖:C≡C—CH3

2) HCl dilute

5. Read the assigned pages in the text and do the assigned problems.

6. Do Nomenclature Worksheet II.

ChemActivity 24

Part A: Conjugated Double Bonds

(How can we explain the special stability of conjugated double bonds?)

Review:

Recall that at room temperature a single bond can freely rotate, but a double bond cannot rotate at room temperature. See the examples below:

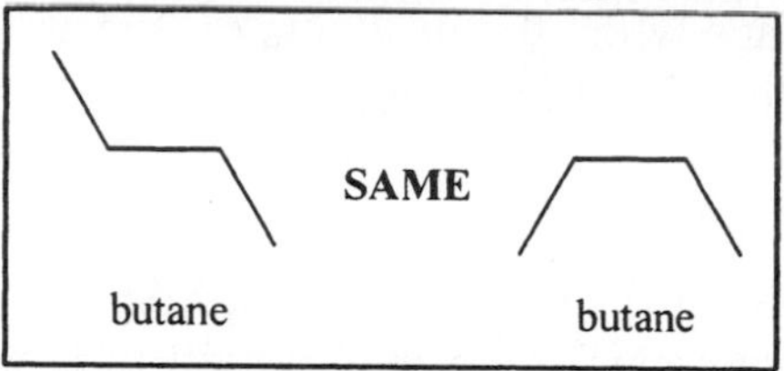

rapidly interconvert at room temperature

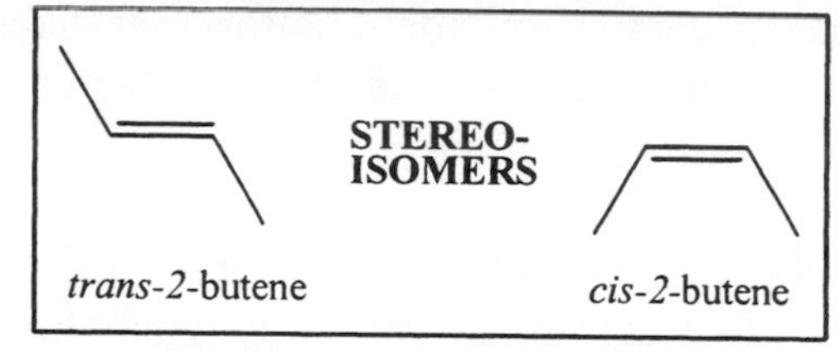

cannot interconvert at room temperature

Model 1: A Strange Single Bond

Observations:

- 1,3-butadiene exists in two distinct forms (*trans*-like and *cis*-like).
- 3.9 kcal/mole of energy are required to convert one to the other. (For comparison: it takes 30 kcal/mole to rotate a typical C=C double bond.)

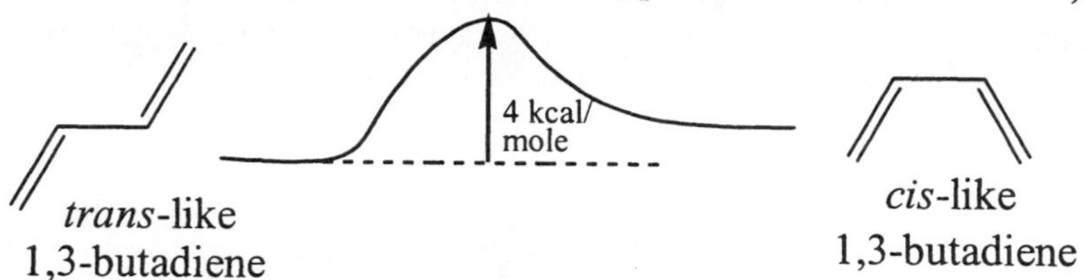

For comparison: rotation of C=C

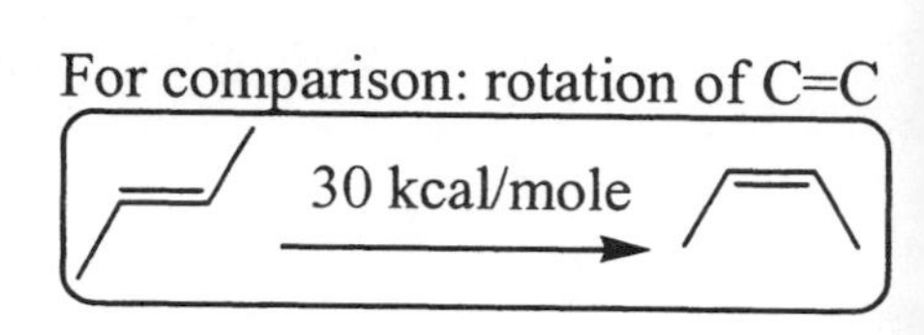

Critical Thinking Questions

1. Explain why the observations in Model 1 are consistent with a hypothesis that the C-C single bond of 1,3-butadiene has some double bond character?

2. Are the following bond length data consistent with the hypothesis that the C-C single bond of 1,3-butadiene has some double bond character? Explain.

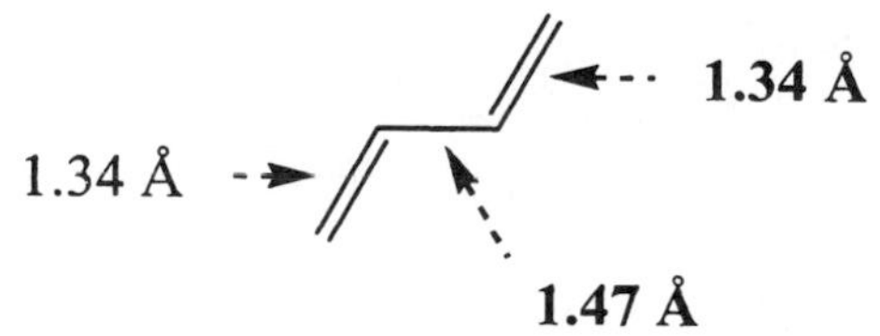

Information:

typical C-C single bond = 1.54 Å
typical C=C double bond = 1.31 Å

Note: there are 10 billion angstroms (Å) in 1 meter (10^{10} Å = 1 meter)

Model 2: Conjugated Double Bonds

When two double bonds are next to one another the two central p orbitals involved in the π bonds can overlap to form a partial double bond, as indicated by the arrows in the diagram below. This is called **conjugation**. To be conjugated together, all the ***p* orbitals must be lined up in the same plane!**

Figure 2a: Three Representations of 1,3-butadiene

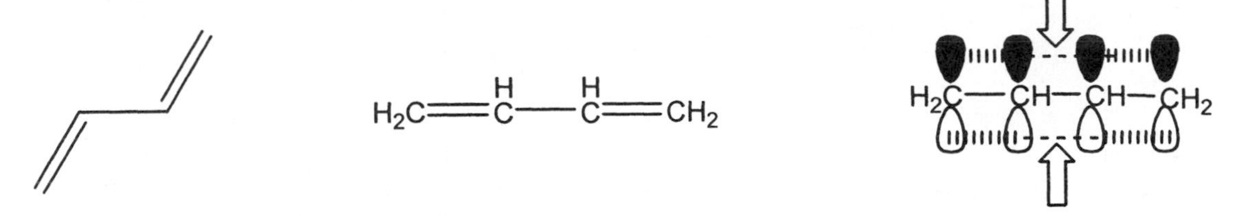

Critical Thinking Questions

3. Each thick dotted line (''''''''') on the farthest right representation of 1,3-butadiene represents approximately....[choose the best answer below].
   A. one double bond
   B. one π bond
   C. half of one π bond
   D. 6% of one π bond

4. According to Model 2, each thin dotted line in Figure 2a represents approximately... (Choose the best answer from the choices in the previous question.)

Review:

Bonding interactions lower the potential energy of a molecule. To break a bond you (the experimenter) must add energy to a molecule. This raises its potential energy. Forming a bond is the opposite. When a bond forms, energy is released (usually as heat).

making bonds = lowers potential energy
breaking bonds = raises potential energy

5. Consider the following di-enes:

   a) Explain why the diene on the right has more bonding interactions than the one on the left. Hint: consider any partial bonds.

   b) Predict which one will be lower in potential energy and explain your reasoning.

Figure 2b: Heats of Hydrogenation for Various Alkenes

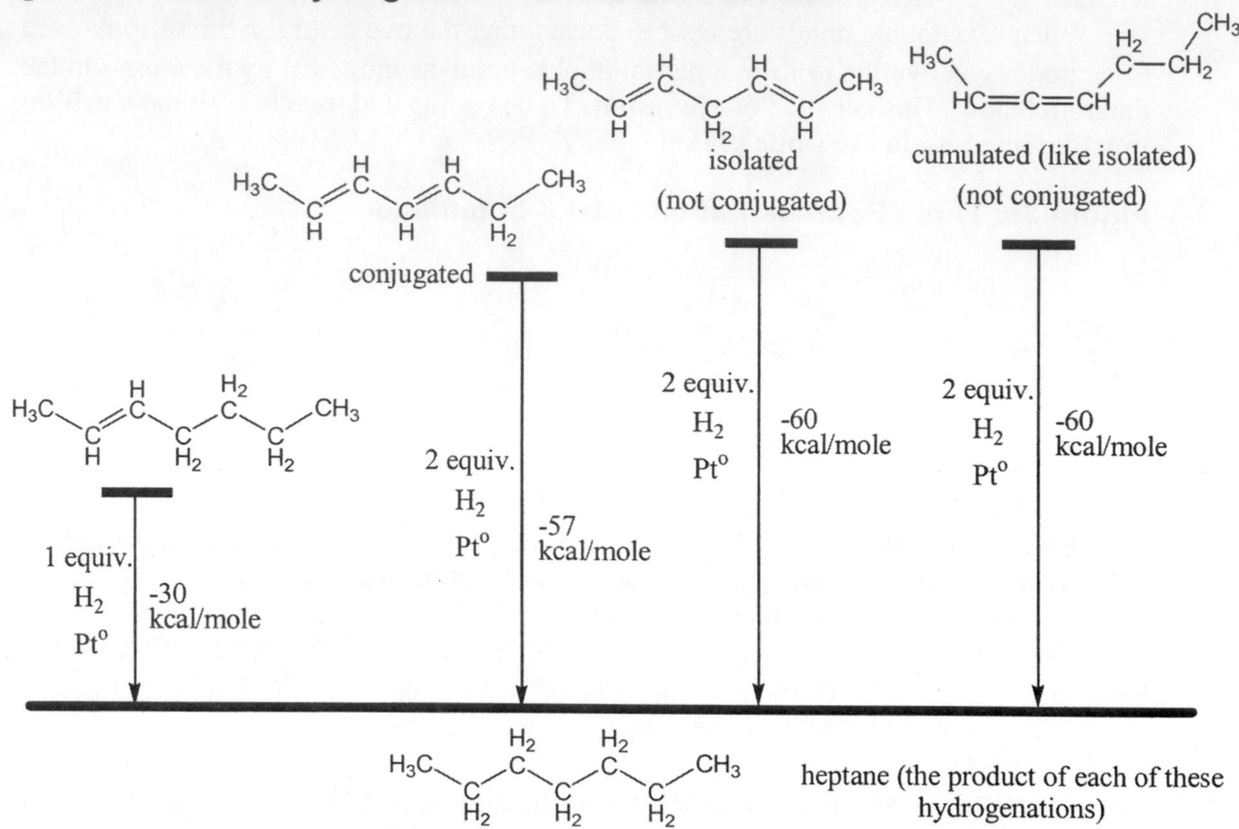

6. Consider the hydrogenation reactions in Figure 2b.
   a) T or F: When one molecule of H–H is added to a double bond (in the presence of Pt) one π bond is broken and two new C-H bonds are formed.
   b) T or F: A hydrogenation reaction is down-hill in terms of potential energy. That is, more energy is released from bonds forming than is consumed to break bonds.
   c) T or F: The P.E. of two isolated π bonds = 2 x P.E. of one π bond.
   d) According to the reactions above, which is higher in potential energy 2,4-heptadiene or 2,5-heptadiene?
   e) Is your answer in part d) consistent with your answer to CTQ 5b?

7. The following are representations of a molecule with a pair of **cumulated** double bonds.

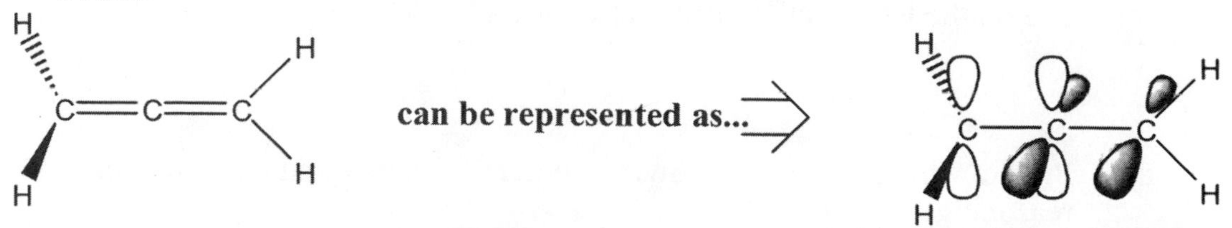

Explain why the two π bonds in this molecule are isolated from one another. (They cannot overlap as with a pair of conjugated double bonds.)

Part B: Molecular Orbital Theory

(What can MO theory tell us that resonance and hybridization theories cannot?)

Information: History of Theories Describing Bonding

Lewis theory was developed in the late 1800's to explain observations that were possible then. Since that time, experiments have yielded deeper insight into the nature of molecules—properties and behavior that **Lewis theory cannot explain**. *For example: How do we explain the existence of the following apparent 1° carbocation?*

Answer: We invented resonance theory.

How do we explain the non-planar geometry of allene?

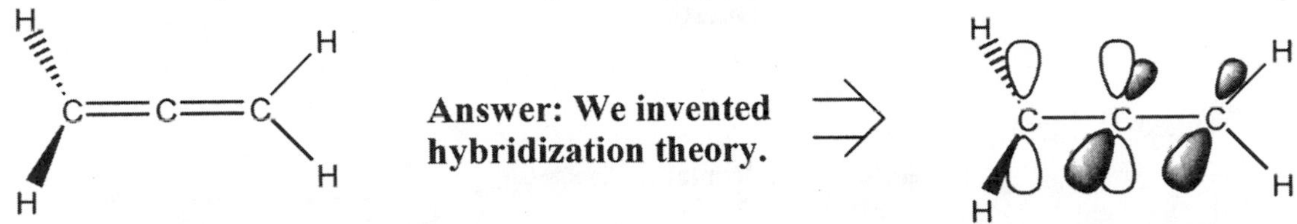

In this ChemActivity you will learn about **molecular orbital theory**. Just as hybridization and resonance did not replace Lewis theory, MO (molecular orbital) theory does not replace the other methods of prediction you have learned in this course. We will see that MO Theory has specific uses.

Model 3: Molecular Orbital Theory (MO Theory)

- MO theory makes a big deal of the wave nature of electrons.
- A wave has two possible phases: up and down.
- For orbitals, we can represent one phase as black and the other phase as white.

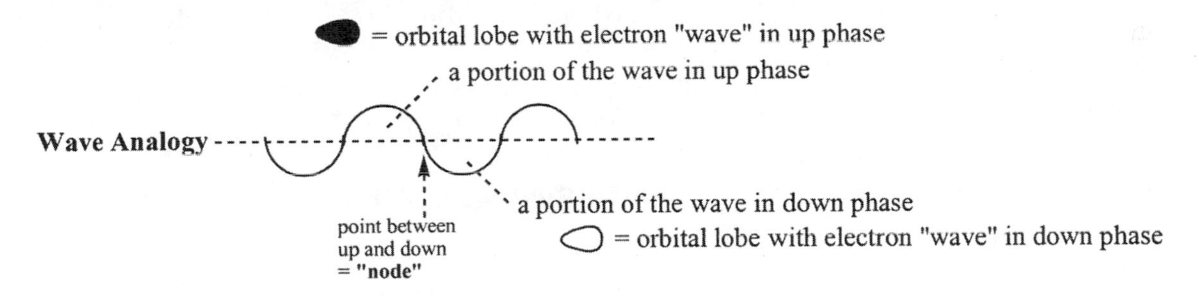

- If orbital lobes have **same phase = bonding** (attraction).
- If orbital lobes have **opposite phase = anti-bonding** (repulsion).

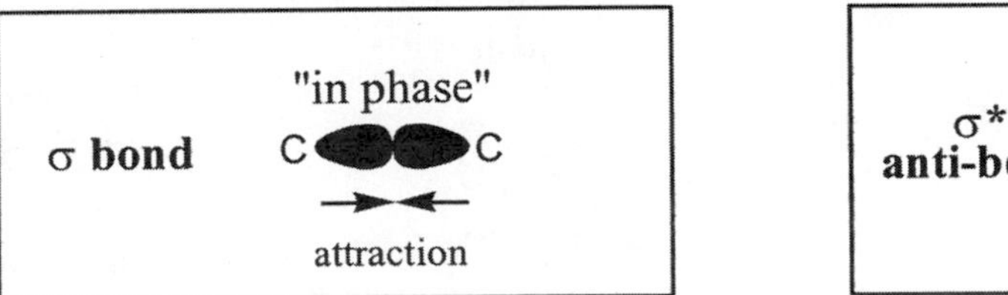

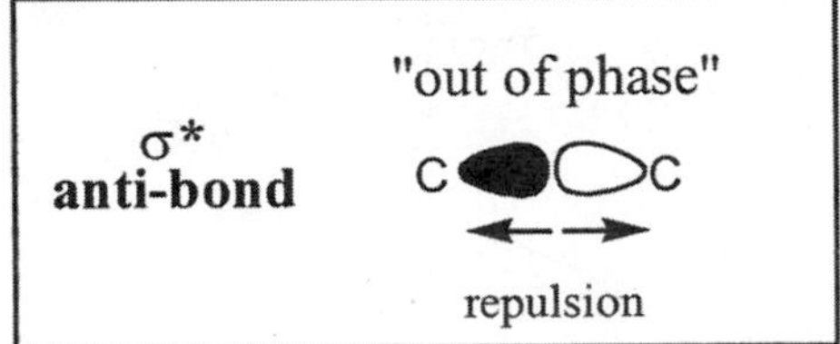

- One σ* anti-bond (if present) exactly cancels the attractive forces of one σ bond.

Critical Thinking Questions

8. Label one of the overlaps below with "π **bond**" and the other with "π* **anti-bond**."

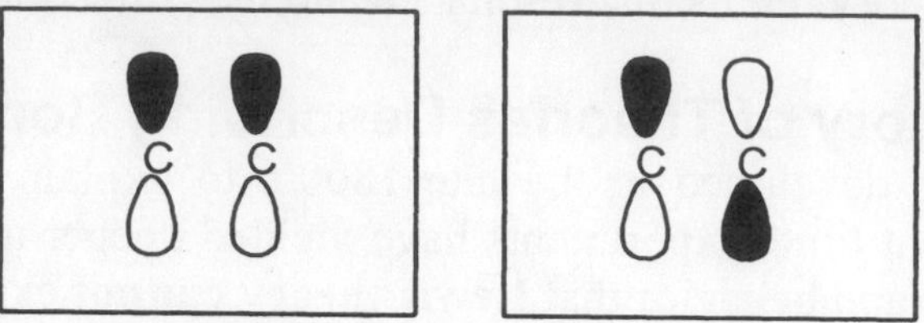

Figure 3a: MO (Molecular Orbital) Diagram of a Carbon-Carbon σ bond

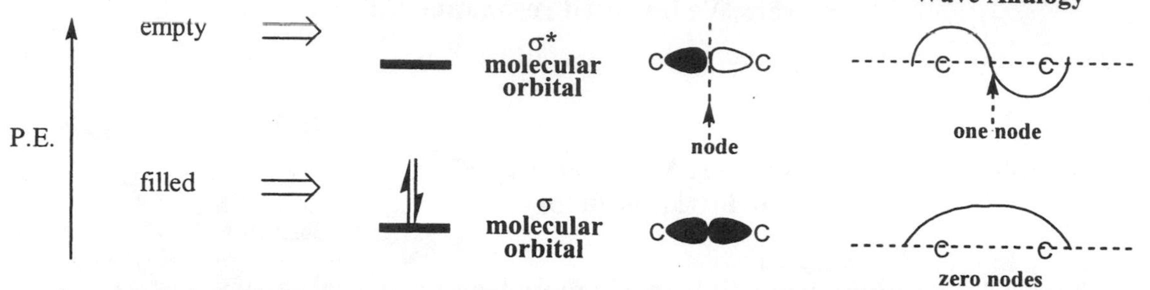

- Any orbital (including a "molecular orbital") can hold **zero, one or two electrons.**
- A "node" is a line of zero electron density between an up-phase (black) orbital and a down-phase (white) orbital.
- A σ bond consists of a filled σ bonding MO
- If the σ* anti-bonding MO were filled it *would* exactly cancel out the attraction of the bonding MO. In this case the σ* anti-bonding MO is empty so it has no effect at all.

9. What is the maximum number of electrons that can fit in a molecular orbital (MO)?

10. Which is higher in potential energy: a σ* **anti-bonding MO** or a σ **bonding MO**?

11. According to MO theory, would it strengthen or weaken the C-C attraction if we added one electron to the MO diagram shown in Figure 1a? Explain.

Figure 3b: Incomplete MO Diagram of π System of Ethene

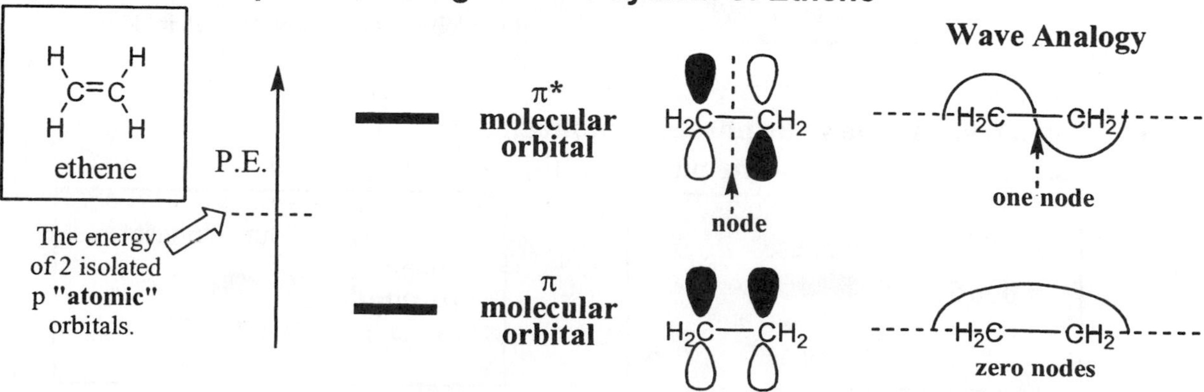

12. How many electrons are there in the π system (*p* orbitals) of ethene?
13. Add these electrons to the MO diagram of ethene, beginning with the lowest potential energy molecular orbital. Does the π* orbital have any effect in this case?

Figure 3c: Incomplete MO Diagram of π System of Allyl Carbocation

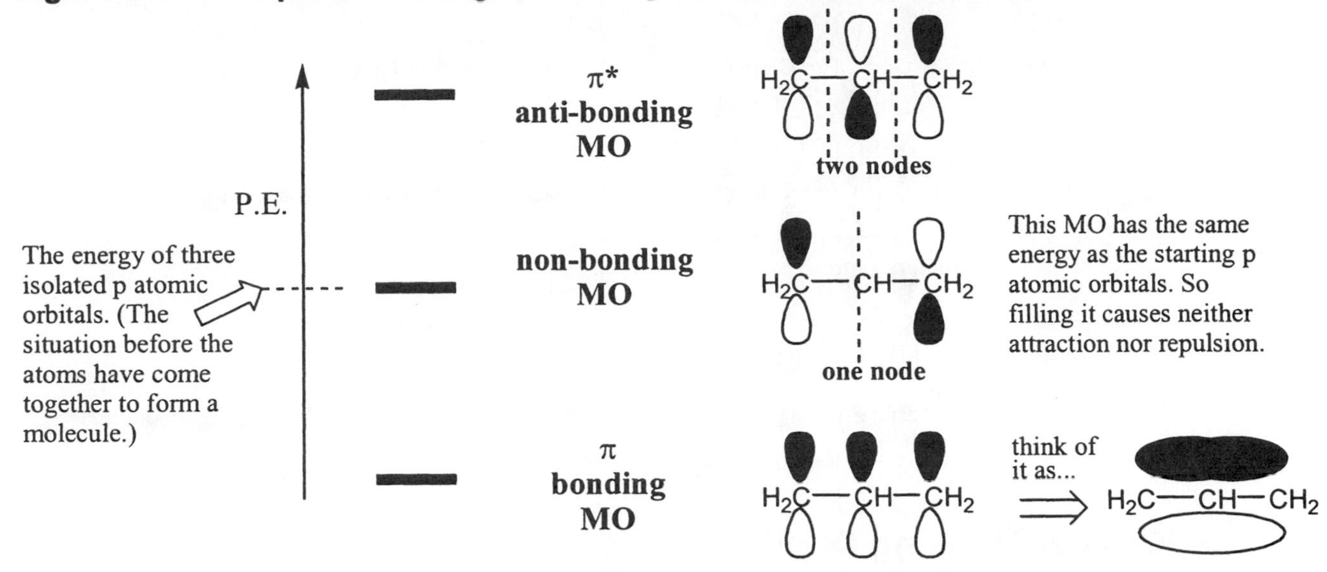

14. Add two electrons to the MO diagram above to make it into a description of the pi system of the allyl carbocation shown below.

[$H_2C{=}CH{-}\overset{+}{C}H_2$ ↔ $H_2\overset{+}{C}{-}CH{=}CH_2$]

15. Add one more electron (to give a total of three) to the MO diagram above. This MO diagram now describes a new species with three pi electrons called allyl radical. Replace the word "Carbocation" in the title at the top of the page with "Radical."

resonance description of **allyl radical** [$H_2C{=}CH{-}\dot{C}H_2$ ↔ $H_2\dot{C}{-}CH{=}CH_2$]

a) According to your new MO diagram of allyl radical at the top of the page, which C or C's have "radical character?" (Note: The unpaired electron is spread evenly through all lobes of the molecular orbital it resides in.)

b) Is your prediction in part a) consistent with a prediction based on the resonance description of allyl radical?

c) According to MO theory, do you expect the carbon-carbon bonds of allyl radical to be...
   A. Stronger than the C-C bonds of allyl carbocation
   B. Weaker than the C-C bonds of allyl carbocation
   C. The same strength as the C-C bonds of allyl carbocation.

d) Draw a resonance description of the species that results if you add a fourth electron to the π system shown in Figure 3c.

Model 4: Molecular Orbital Description of 1,3-Butadiene

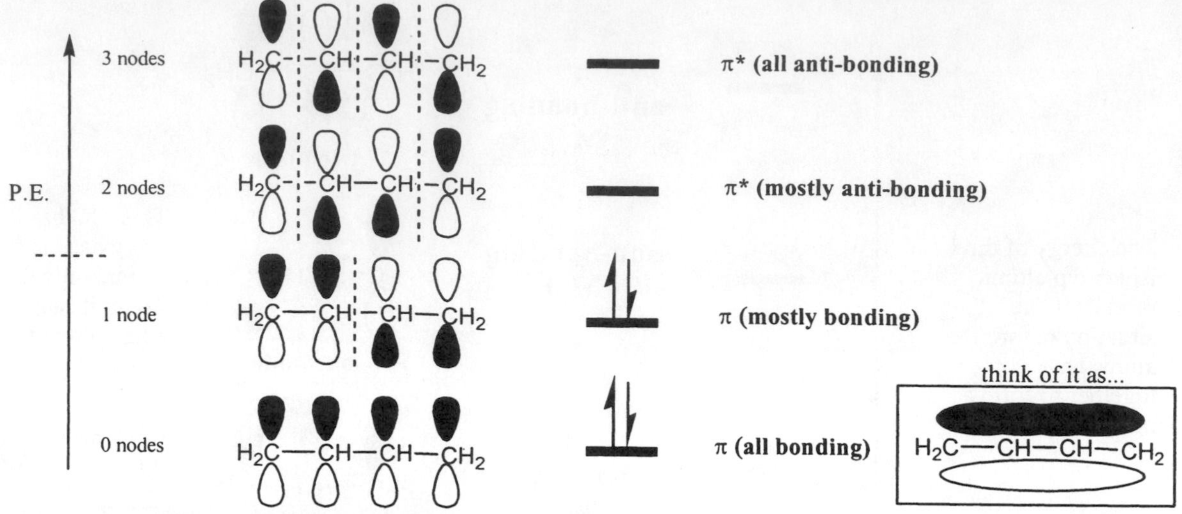

Critical Thinking Questions

16. According to previous Models in this ChemActivity, what does the dotted line half-way up the potential energy scale above represent?

17. Use the MO description in Model 4 to explain why the C_2-C_3 bond of butadiene is shorter than a typical single bond, but longer than a typical double bond.

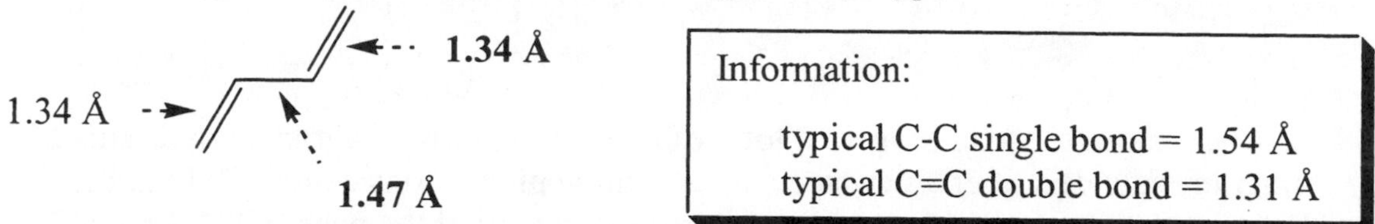

Model 5: Excited States of 1,3-Butatdiene

- Molecular orbital theory's greatest advantage over other theories is that it can be used to make predictions about excited states of molecules.
- If you shine light of the correct energy/frequency on a sample of 1,3-butadiene, you can excite one electron from the **HOMO (highest occupied molecular orbital)** up to the **LUMO**.

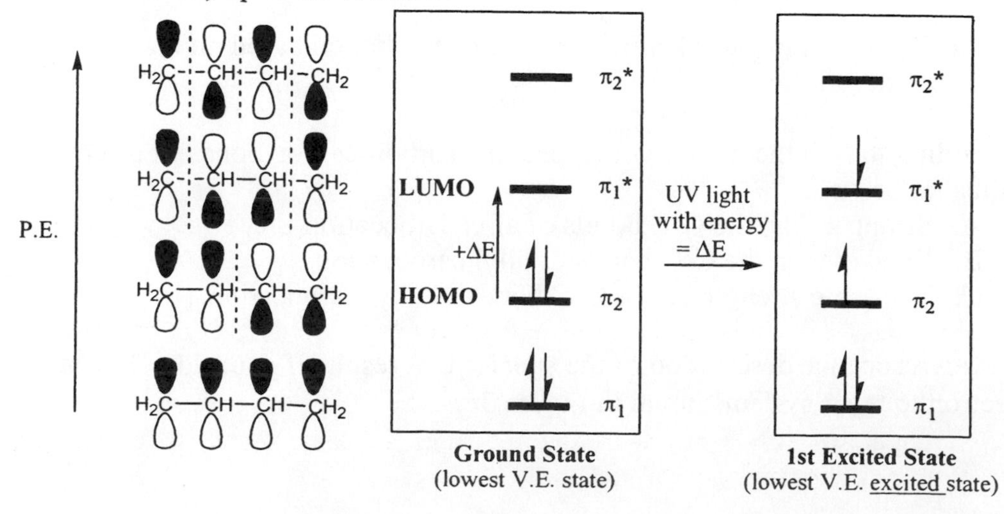

Model 6: UV-Visible Spectroscopy

- Light in the ultraviolet and visible range of the spectrum often has the correct energy to excite an electron from a bonding orbital to an anti-bonding orbital.

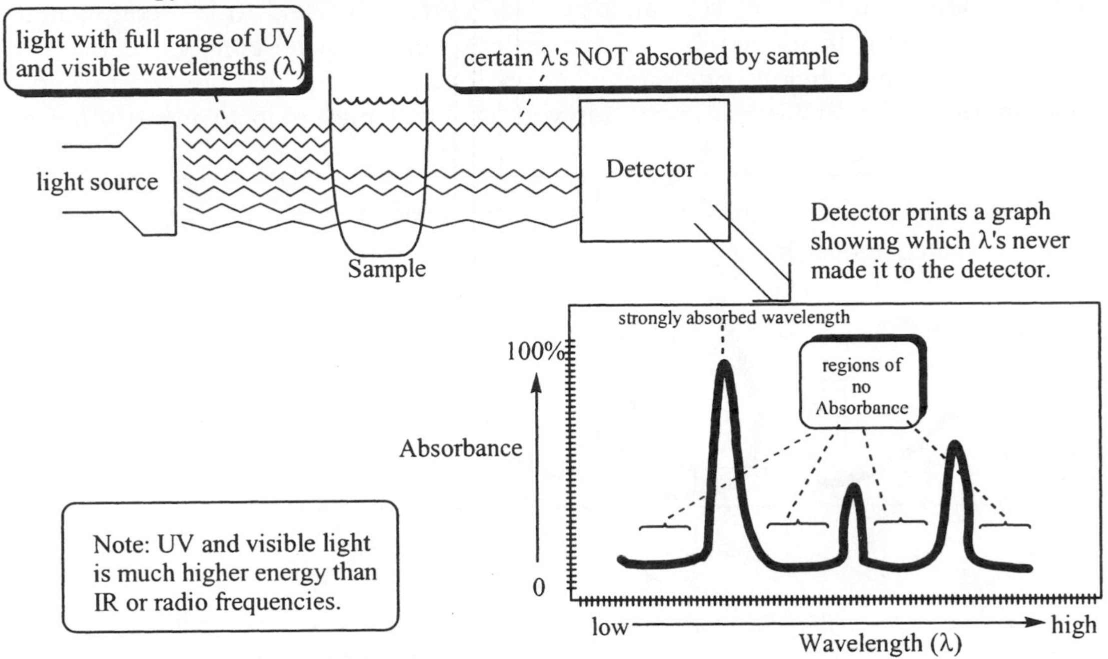

Critical Thinking Questions

18. Consider the MO diagrams of the ground state and 1st excited state of 1,3-butadiene, below. The distances **x** and **y** represent two different energy/frequencies of light. The first excited state occurs when the sample absorbs some light of frequency x.
    a) Which bonds do you expect this excitation to weaken and/or strengthen?
    b) Fill in electrons to complete the MO diagram of the second excited state, which occurs when the sample absorbs some light of frequency y.

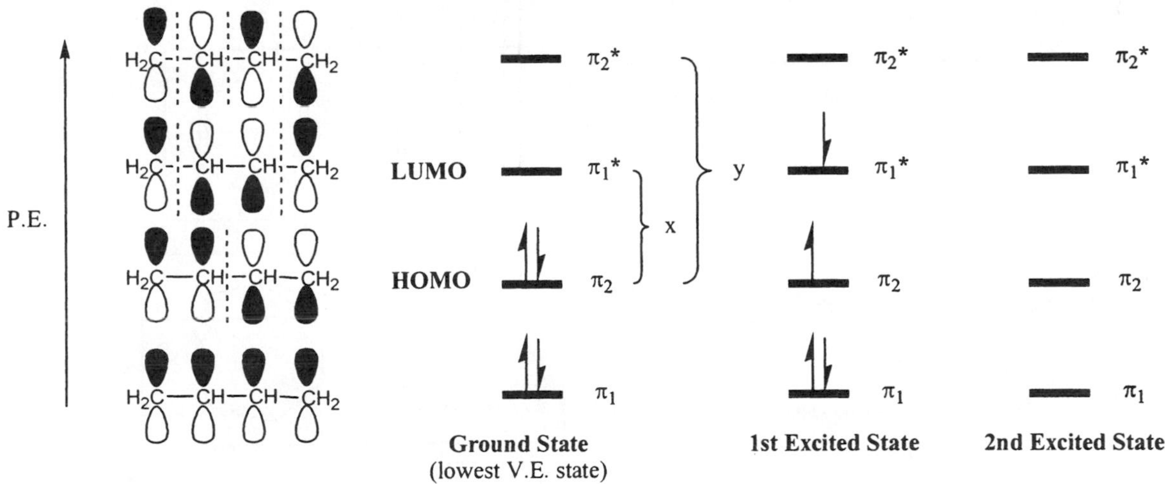

Notes:

- From looking at the UV-vis spectrum of a molecule you can determine the energy difference between its HOMO and its LUMO.
- This can tell you something about the structure of the molecule.
- This type of spectroscopy is not nearly as important as NMR and IR spectroscopy, and we will use it sparingly in this course.

Exercises for Part A:

1. Explain why 1,3-butadiene is planar in its lowest potential energy conformation.

2. Double bonds that are not separated by a sp^3 carbon are said to be "conjugated." There is no limit to the number of double bonds that can be conjugated together. For each molecule below, circle together any double bonds that are conjugated to one another. The first one is done for you. (Hint: it may help to identify the sp^3 carbons.)

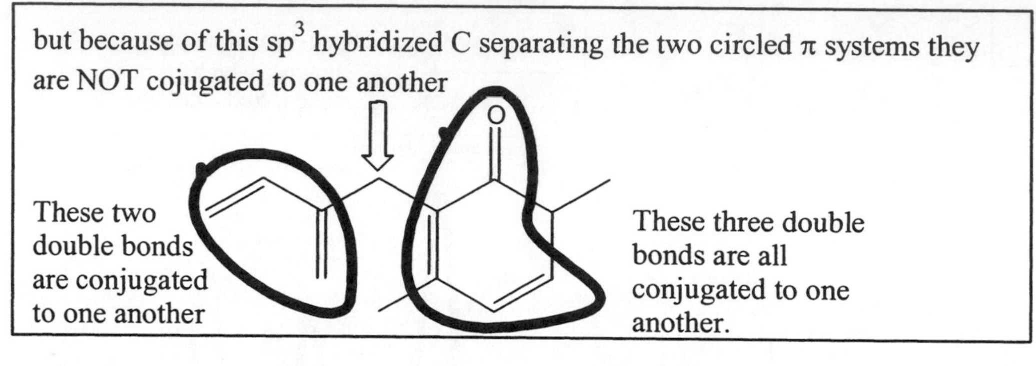

3. For each box, predict which of the two species will be lower in potential energy.
4. For each box, predict which of the two bonds indicated with an arrow will have a shorter bond length.

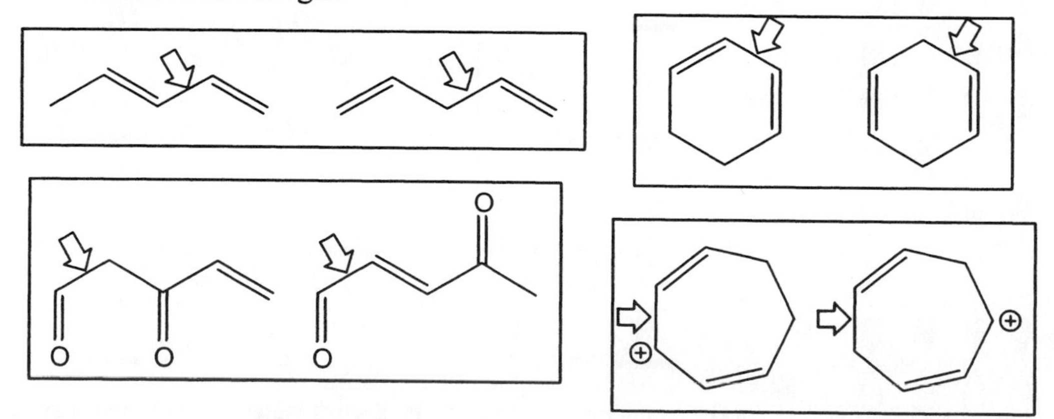

5. Draw an orbital representation of 1,3,5-hexatriene. Indicate each pi bond as a pair of bold dotted lines and each partial pi bond as a light dotted line.
6. Exactly two of the pi bonds in 1,2,3-butatriene (shown below) are conjugated to one another. Indicate which two and explain why the third pi bond is not conjugated to the other two.

$$H_2C{=}C{=}C{=}CH_2$$

7. Using the heats of hydrogenation in Figure 2b circle the most likely heat released (in kcal/mole) during catalytic hydrogenation of each tri-ene below. Assume complete hydrogenation so that the product is nonane in each case.

A. 93 kcal/mole
B. 90 kcal/mole
C. 87 kcal/mole
D. less than 84 kcal/mole

A. 93 kcal/mole
B. 90 kcal/mole
C. 87 kcal/mole
D. less than 84 kcal/mole

8. Read the assigned pages in your text and do the assigned problems.

Exercises for Part B

9. Speculate what the acronym LUMO stands for.

10. Predict whether the excitation shown in Model 5 will lengthen or shorten each of the following bonds of 1,3-butadiene...
   a) the C_1-C_2 bond.
   b) the C_2-C_3 bond.
   Explain your reasoning.

11. Add electrons to Figure 3c (redrawn below) so that it represents the MO diagram of allyl anion. Usc thc MO diagram to predict which C's will have the most electron density (and therefore hold the excess negative charge). Is your prediction consistent with a prediction based on a resonance description of ally anion?

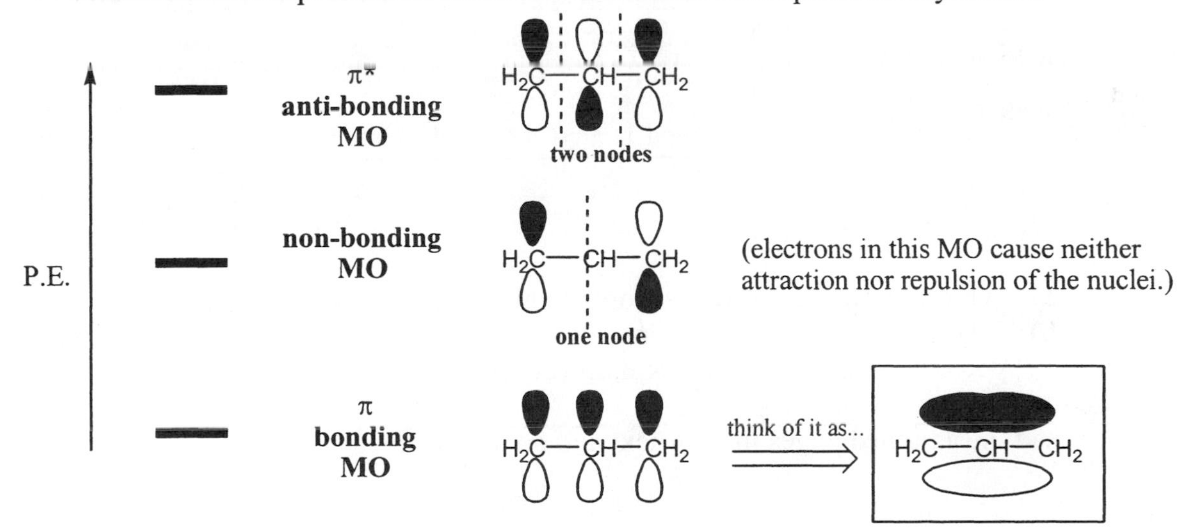

12. Draw an arrow on the ground state MO diagram below showing the magnitude of ΔE_2, the energy required to excite an electron in the HOMO of 1,3-butadiene to the highest anti-bonding orbital. Then complete the MO diagram of this 2nd excited state.

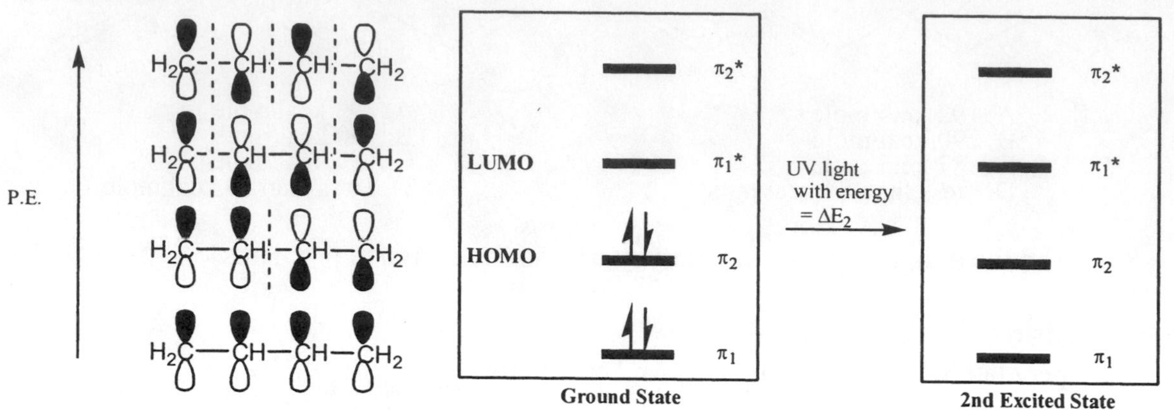

13. Sunburn is caused, principally, by two wavelengths of UV light called UVA and UVB. The energy of each of these is represented by an arrow, below.

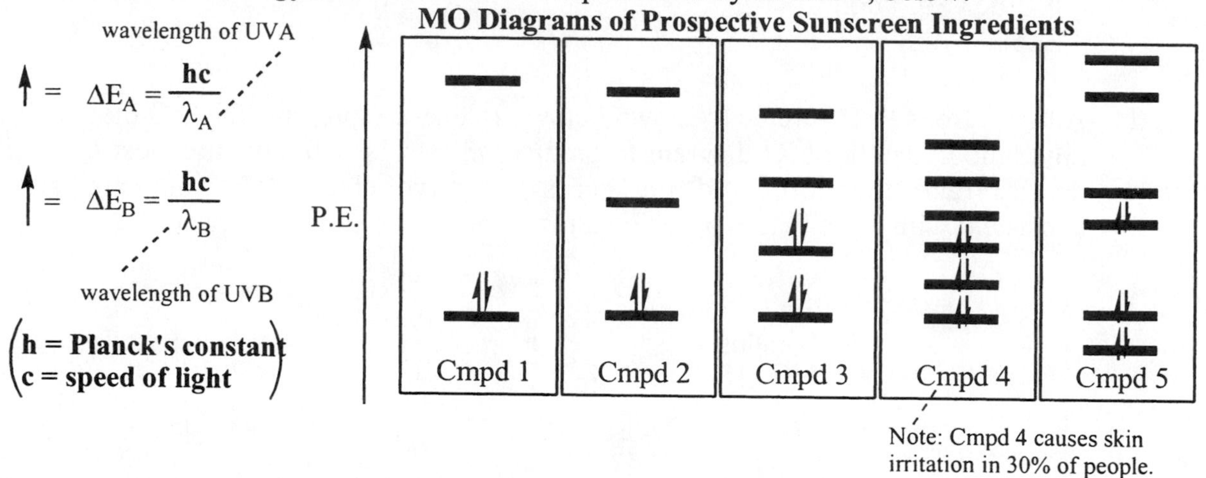

a) Which Compound above absorbs light of the shortest wavelength (λ)?
b) If you were developing a sunscreen, which compound or compound/s would you put in your product? Explain your reasoning.

14. Read the assigned pages in your text and do the assigned problems.

Supplement: Rules for Drawing MO Diagrams of π Systems

- # *p* atomic orbitals in π system = # of molecular orbitals in MO diagram. (*For example: For 1,3-butadiene, start with 4 p orbitals and get 4 MO's.)*
- MO's have zero, one, two, three...etc. nodes (going from lowest to highest P.E.)
- Placement of nodes must be symmetrical. (*e.g. If 1 node, it must be in the middle.*)
- Fill MO's with electrons starting from lowest P.E. MO.

Example:

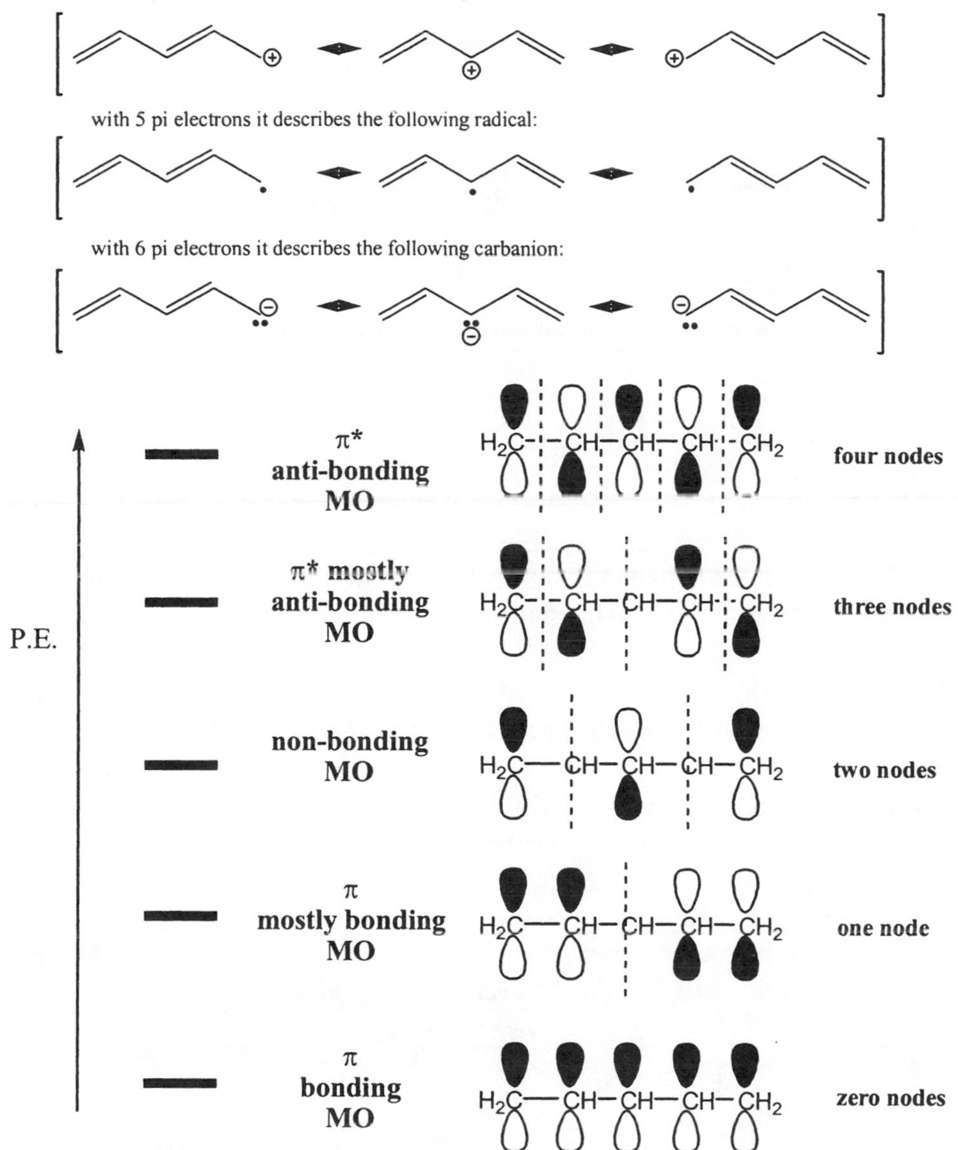

Practice Questions

Use the rules for drawing MO diagrams to construct an MO diagram for 1,3,5-hexatriene.

ChemActivity 25

Part A: Aromaticity

(What structures are likely to exhibit the special stability known as aromaticity?)

Model 1: Amazing Benzene!

Figure 1: Heats of Hydrogenation

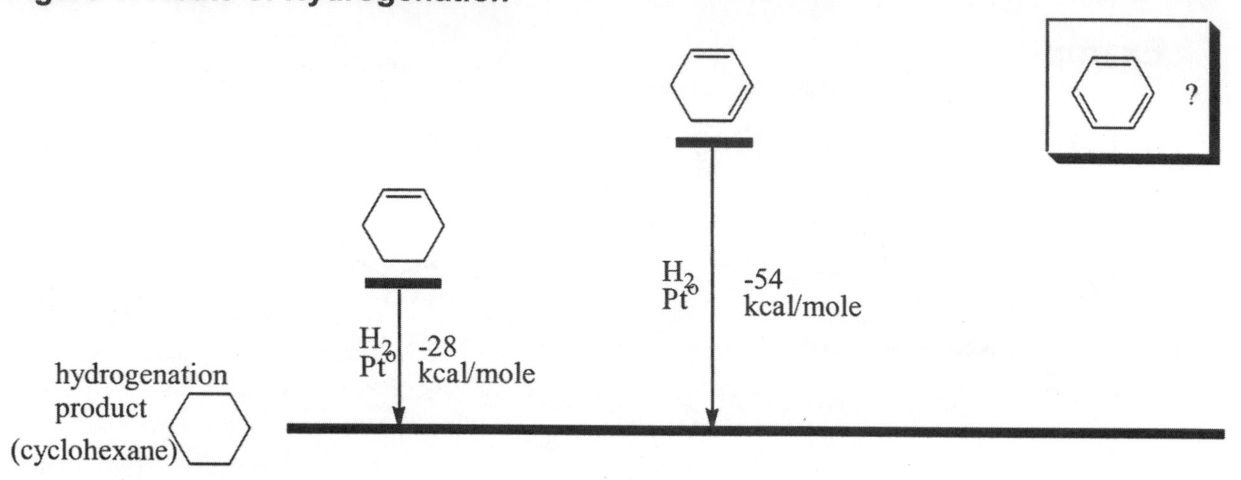

Critical Thinking Questions

1. Based on the data in Figure 1, estimate the heat of hydrogenation for benzene (heat released when benzene is reduced to cyclohexane).

2. The heat of hydrogenation for benzene is -49kcal/mole. Add a solid line to Figure 1 indicating this value for benzene. Is benzene lower in P.E. than you expected?

Model 2: Aromaticity

- Benzene and other similar molecules exhibit an **almost magical stability** (lower than expected potential energy).
- Before it was understood, this "magical stability" was named **aromaticity** because many molecules in this class have a strong **aroma** (smell).
- Resonance alone cannot explain this amazing stability.

The Racetrack Explanation: *The electrons of benzene "think" they are in a huge box (the larger = lower P.E..) because they can race endlessly around a track of conjugated p orbitals. This racetrack of p orbitals is called a* ***cyclic conjugated pi system****.*

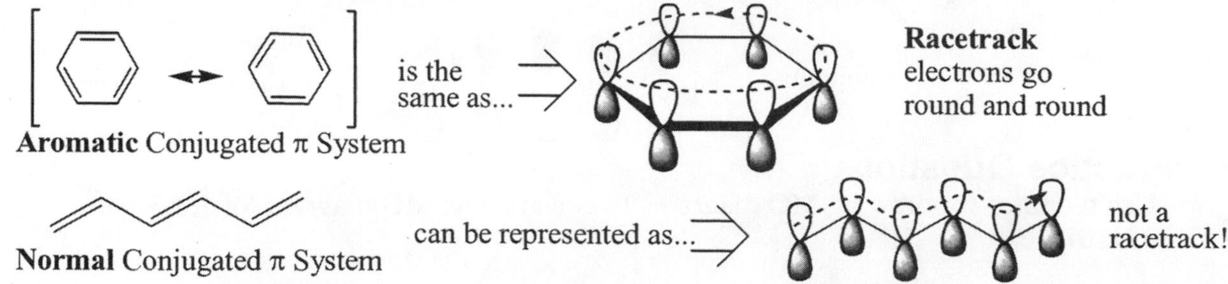

Figure 2: Not all cyclic conjugated pi systems are aromatic!

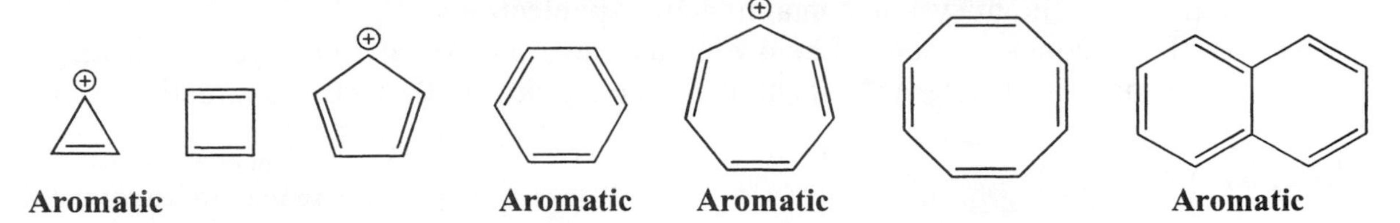

Critical Thinking Questions

3. Confirm that there is a *p* orbital (empty or half-filled) at each carbon in Figure 2.
   a) Label each cyclic conjugated pi system in Figure 2 with the total # **of pi electrons** in the "racetrack." (These are the electrons in *p* orbitals.)
   b) Certain "magic numbers" of pi electrons give rise to aromaticity. According to the data in Figure 2, circle the "magic numbers" in the list below.

   2 4 6 8 10 12 14 16 18 20

   c) Of 12-20, which do you expect to be "magic" numbers that allow aromaticity.

4. In 1938 German chemist Erich Hückel noticed this series and wrote what has come to be called "**Hückel's Rule**," which states: "**a continuous, cyclic, conjugated pi system exhibits aromaticity (special stability) if it contains 4n + 2 pi electrons** (where n can = 0, 1, 2, 3…etc.)."
   a) Do the numbers you circled in part b), above, agree with Hückel's Rule?
   b) Only one of the following molecules is aromatic. Label it **aromatic**.

5. A molecule is considered aromatic if it contains an aromatic pi system as part of it.
   i) **Cross out** the three molecules below that are **not aromatic**.
   ii) Trace each aromatic pi system ("racetrack") in the remaining (aromatic) molecules. The first two are done for you. (Note that the second example has two **aromatic rings**. Even though these two aromatic rings are conjugated together, they are considered separate "racetracks.")

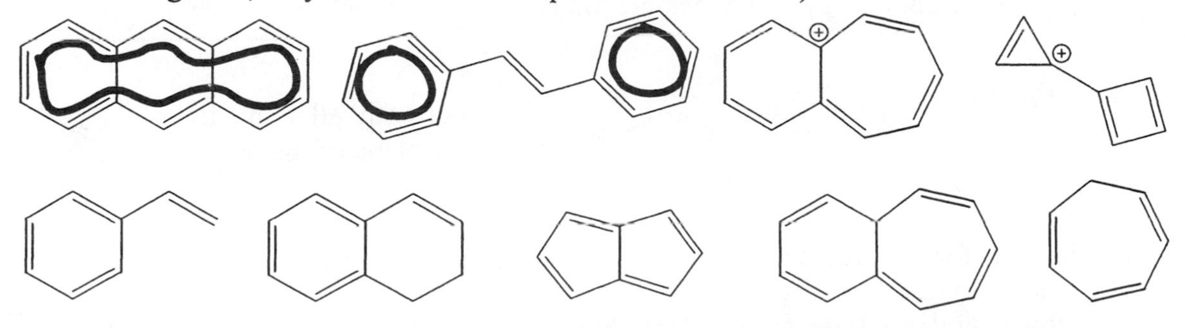

6. A molecule is considered aromatic if you can draw a resonance structure with a continuous cyclic pi system containing 4n+2 pi electrons.
   a) The resonance structure below has a 6-electron aromatic pi system. Draw a resonance structure of this molecule containing a 10-electron aromatic system.

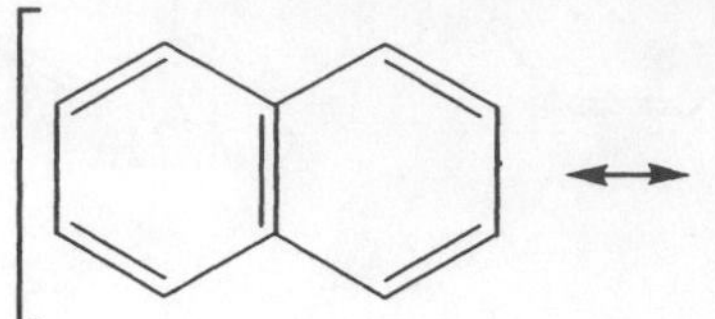

Note: there are a total of three resonance structures in this set. Draw the third one at home for extra practice.

   b) The resonance structure below **does not contain an aromatic pi system.** Draw a resonance structure of this molecule that is aromatic (contains an aromatic pi system).

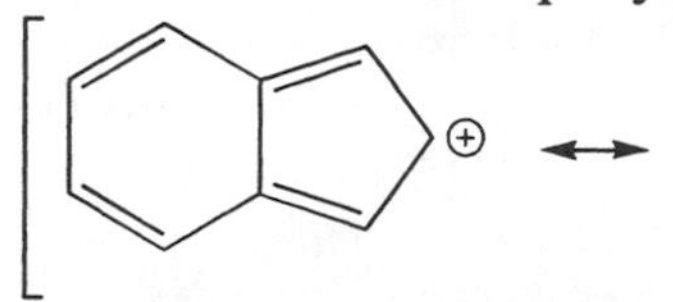

Note: there are a total of 11 resonance structures in this set. Only four of them contain an aromatic pi system. Draw them at home for practice.

   c) How many pi electrons are involved in the aromatic resonance structure you drew in part b?
   d) Which of the two resonance structures you drew in each set is more important? Explain your reasoning.

7. Consider the following orbital representations of benzene.

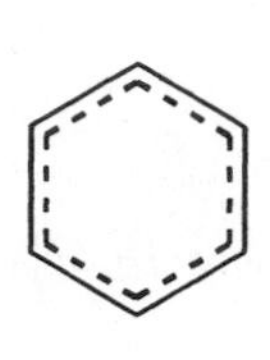

benzene

=

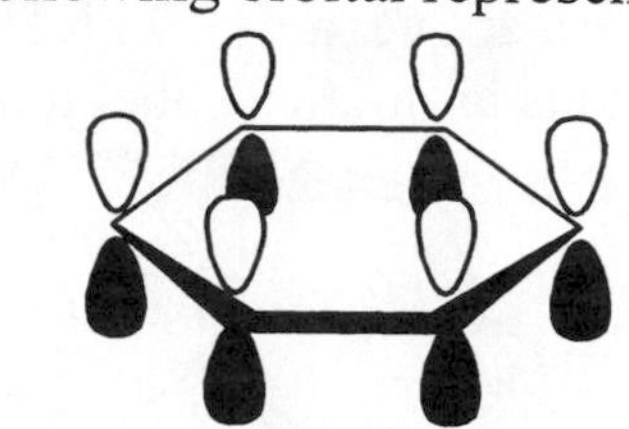

planar

or

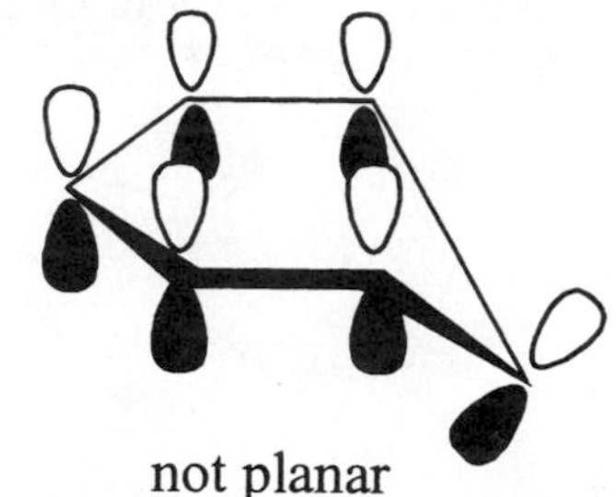

not planar

   a) Which pi system (planar or not planar) has better *p* orbital racetrack?
   b) Is your conclusion consistent with the fact that benzene is always planar?
   c) Explain why all conjugated pi systems (not just aromatic ones) are lower in potential energy when they are planar.

For Example:

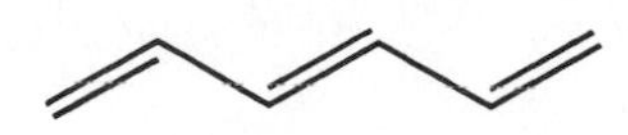

For both: all atoms lie in the plane of the paper.

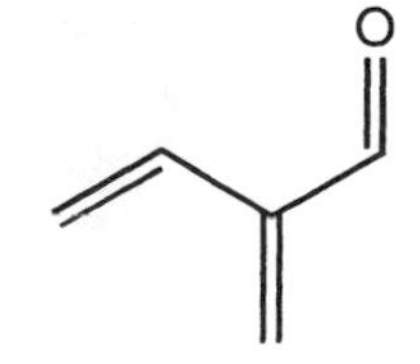

8. Review: for a molecule to be aromatic it must have a pi system…
   that is **cyclic**
   that is **conjugated and uninterrupted** (no sp^3 hybridized carbons in the ring)
   that has **4n+2 pi electrons** (2 or 6 or 10 or 14 etc.)
   whose overall **geometry** is ____________ ***(fill in the blank)***.

Part B: MO Explanation for Aromaticity

(How can we explain anti-aromaticity using molecular orbital theory?)

Model 3: Molecular Orbital Explanation for Hückel's Rule

- Rule of thumb: the MO diagram of a cyclic conjugated pi system always takes the shape of the ring involved, with one vertex pointing down. For example...

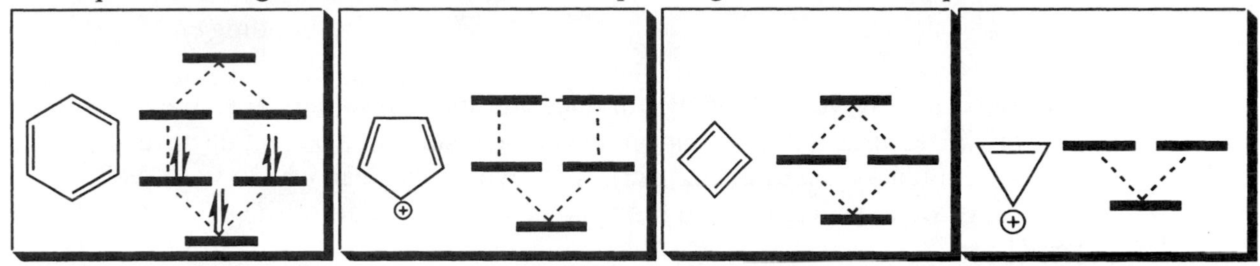

MO Diagram of Benzene (completed)

MO Diagrams of Other Cyclic Conjugated Pi Systems (electrons not filled in yet)

- Molecular orbitals on the same MO diagram with the same energy are called **degenerate orbitals**.
- Degenerate orbitals must all be half-filled before any one is filled. For example...

Valence Electron Orbital Occupation for a Oxygen Atom

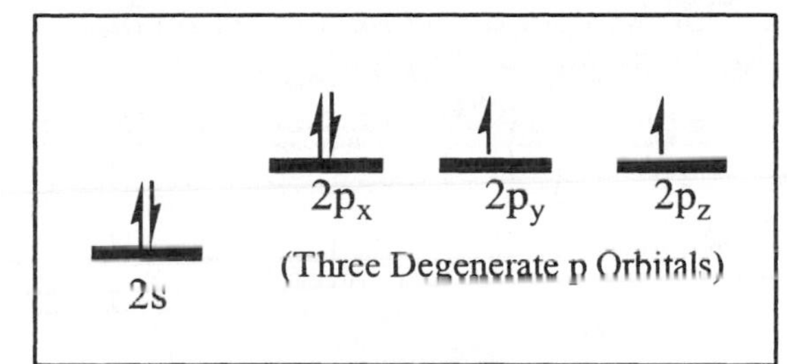

CORRECT

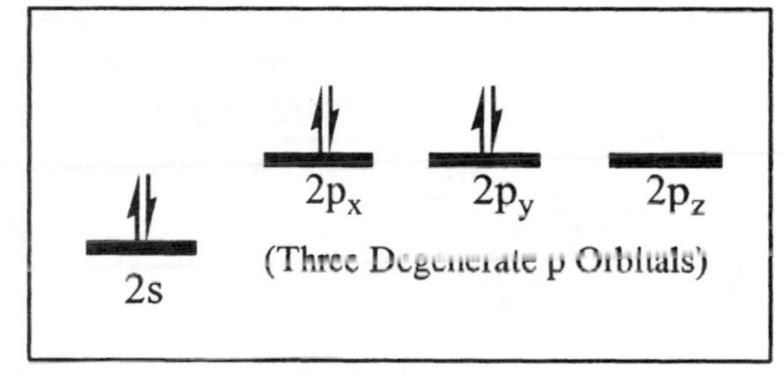

NOT CORRECT

Critical Thinking Questions

9. How many pairs of degenerate molecular orbitals are there in the MO diagram of benzene? **Label them.**

10. Add the appropriate number of electrons to the three other MO diagrams in Model 3.

Information

The situation of an unpaired electron in an orbital is unfavorable. We will say that a species with an unpaired electron has "**radical character**" to remind us that, like other radicals, it will be very reactive.

a) According to the information above, label two compounds at the top of the page with the words "**has radical character-very reactive.**"

b) Employing Hückel's Rule, label aromatic compounds at the top of the page with the words "**aromatic-very stable.**"

11. The anion below, left, has two pi bonds, which contain four pi electrons. **Place four electrons in the MO diagram for this species**, shown below, right.

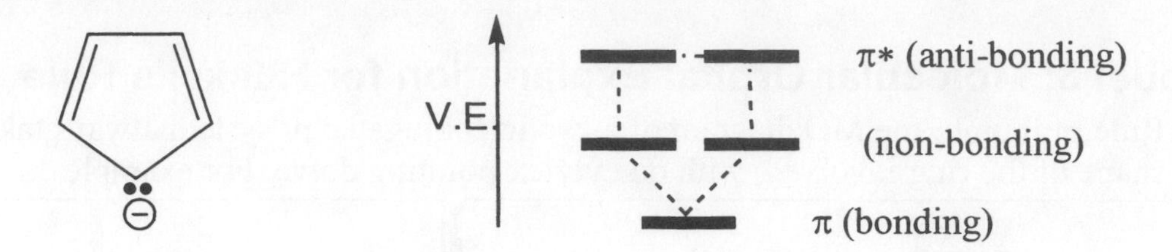

a) According to your MO diagram does this anion have radical character?
b) NOW, assume the lone pair on this anion resides in a *p* orbital (instead of a sp^3 orbital as you might expect). **Add two more electrons to your MO diagram** to reflect that the lone pair electrons are part of the pi system.

Information

A lone pair will "choose" to reside in a *p* orbital <u>if doing so makes the molecule aromatic</u>.

c) Do you expect the lone pair on this anion to reside in a *p* orbital as we assumed in part b)? Explain your reasoning.

d) Is this cyclic anion aromatic?

12. For each column, check one or the other box. You will get the same answer no matter which cyclic MO diagram you use to figure out your answer. (MO diagrams for 6, 5, 4 and 3 member rings are shown on the previous page; 7,8,9 & 10 are shown below.)

# Of Electrons in the π System	2	4	6	8	10	12
Has Radical Character According to MO						
Aromatic (according to Hückel's Rule)						

MO Diagrams for 7, 8, 9 and 10 Member Rings

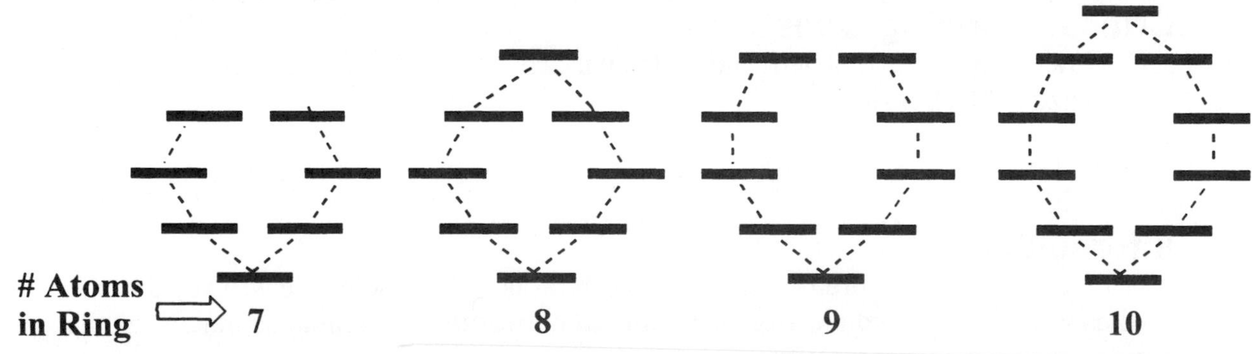

13. The first half of Hückels' Rule states that a cyclic conjugated pi system with 4n+2 electrons will be very stable. **The second half states that a cyclic conjugated pi system with 4n electrons (4,8,12, etc.) will be very <u>unstable</u>**. Explain the second part of Hückel's Rule.

Model 4: Anti-Aromatic Molecules

Cyclobutadiene can be isolated and studied at temperatures less than -200°C. At temperatures above -200°C it immediately reacts with itself to form other more stable products.

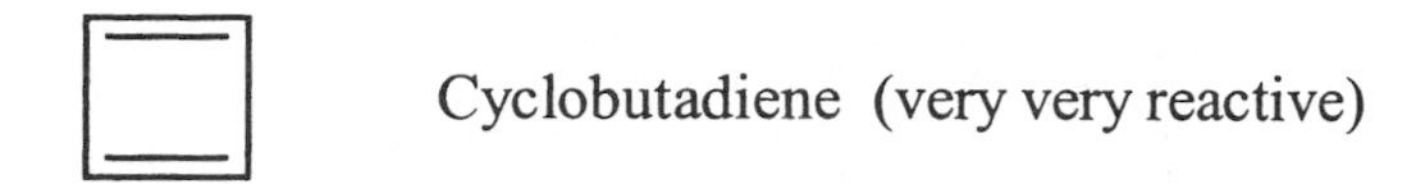
Cyclobutadiene (very very reactive)

- If a pi system obeys all the rules for aromaticity, but has 4n electrons (where n = 1,2,3...etc.) then it is called **anti-aromatic**.
- Anti-aromatic molecules display an almost magical **<u>instability</u>** (due to radical character). Think of it as the evil twin of aromaticity.
- A molecule will "do" anything it can to avoid being anti-aromatic.

Critical Thinking Questions

14. List four things about cyclobutadiene that makes it anti-aromatic. (Hint: see CTQ 8.)

15. Cyclooctatetraene is quite stable at room temperature. However, close examination of the structure reveals that **it is not planar**.

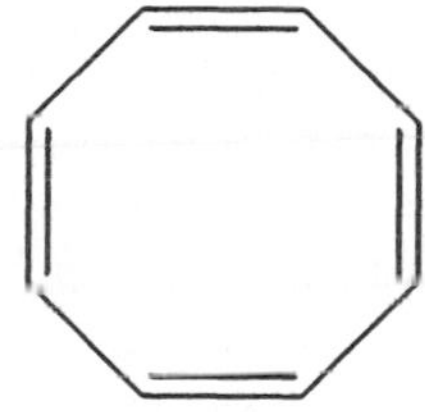
Cyclooctatetraene

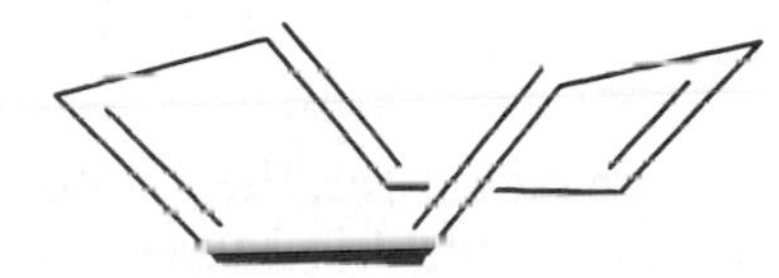
Cyclooctatetraene in its Preferred Conformation

a) Speculate about why this molecule does not "want" to be planar.

b) Cyclooctatetraene is considered **non-aromatic.** This means it is neither aromatic nor anti-aromatic. Explain.

c) Give an example of another non-aromatic molecule. (Hint: most molecules fall into this category.)

Note: It turns out that cyclobutadiene is just about the only anti-aromatic molecule. This is because most larger molecules can flex and twist to avoid being planar.

Exercises for Part A

1. Indicate the largest aromatic ring in each of the following aromatic molecules.

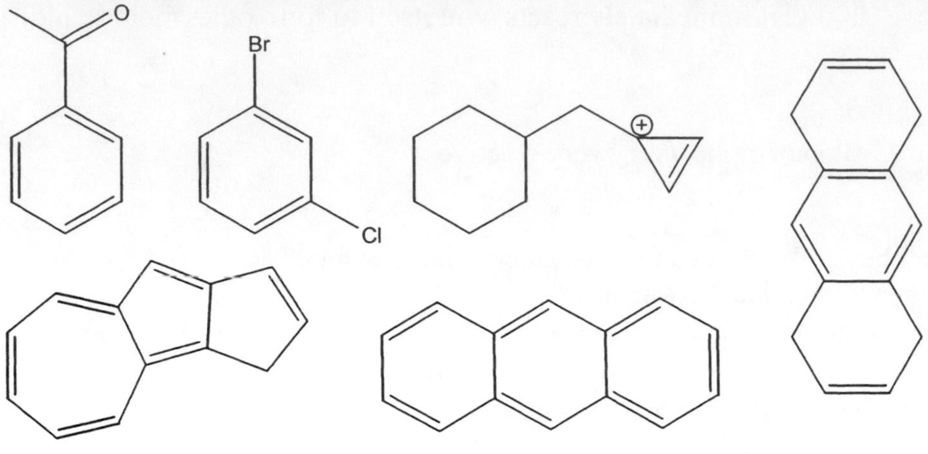

O

Br

Cl

2. Draw the enol form of the following ketone and construct an explanation for why the enol form is favored in this special case. (Note that normally the keto form is lower in potential energy and therefore favored.)

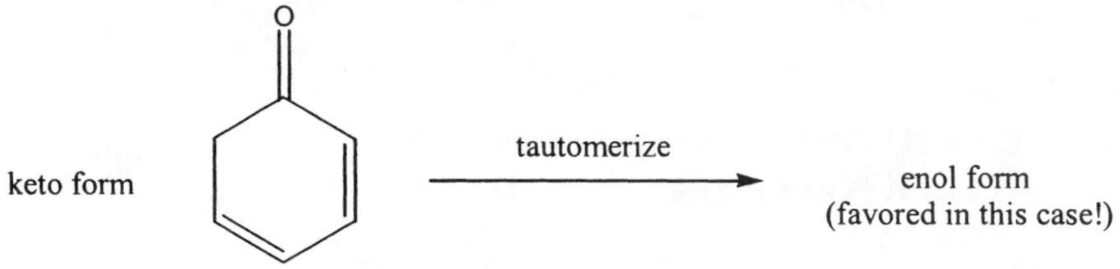

O

keto form — tautomerize → enol form (favored in this case!)

3. For a set of resonance structures, the members of the set that are aromatic are "lower energy" arrangements of the electrons. These aromatic resonance structures are therefore considered "more important" than their non-aromatic counterparts. Each of the following belongs to a set containing aromatic resonance structures. For each, draw at least one of these more important resonance structures.

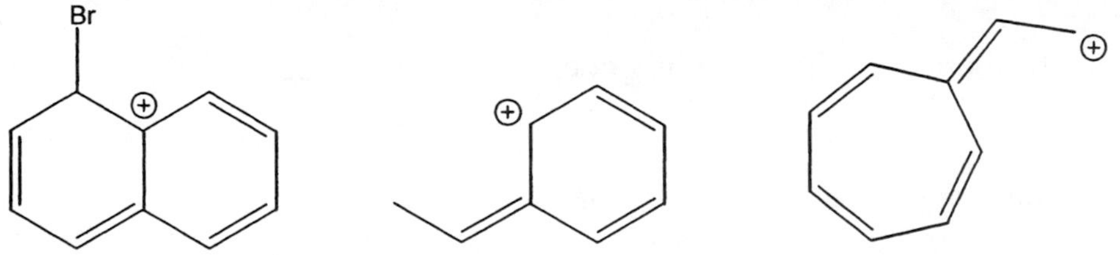

Br

4. Explain why each of the following species is NOT aromatic.

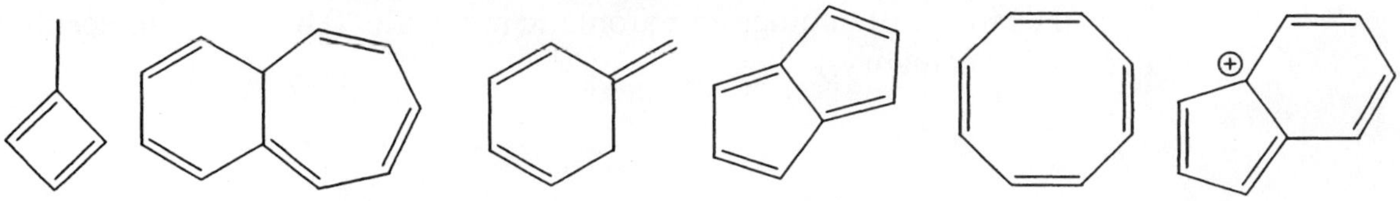

5. The following ketone actually exists as a zwitterion. (A zwitterion is a compound with both a + and - charge.) Construct an explanation for why the compound below prefers the zwitterion state over the neutral state.

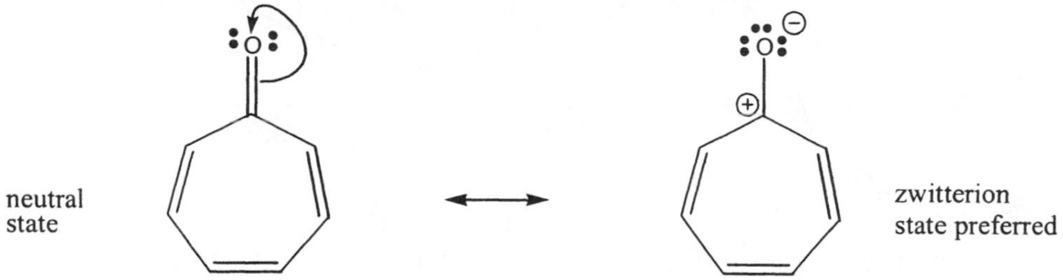

6. Read the assigned pages in the text and do the assigned problems.

Exercises for Part B

7. The molecule on the left has a lone pair which resides in a *p* orbital despite the fact that in a *p* orbital the lone pair resides closer (90° vs. 109.5°) to N-H and N-C sigma bonding electrons.

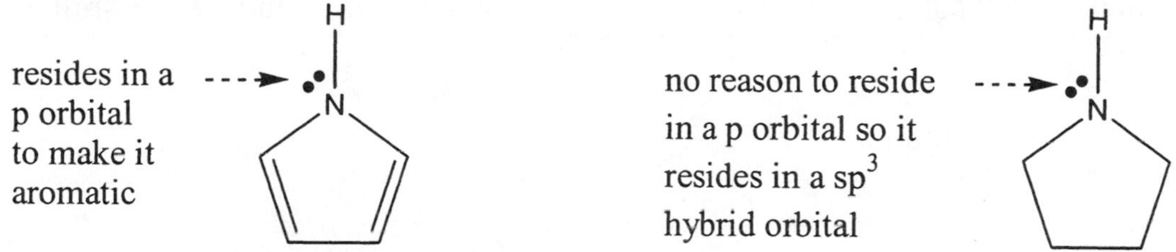

   a) Construct an explanation for why this lone pair resides in a *p* orbital.
   b) Each of the following molecules is aromatic. Label each lone pair with the name of the orbital it prefers to reside in.

8. Consider the two conformations of the 10-member ring shown below.

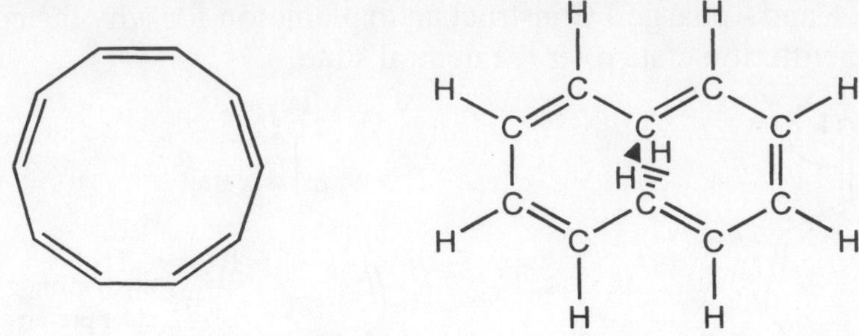

a) Construct an explanation for why the conformation on the left is very unfavorable. (Hint: recall that bond angles of sp^2 hybridized carbons prefer to be very close to 120°.)

b) In the conformation on the right, the two H's drawn with wedge and dash bonds cause the carbon backbone to flex out of planarity. Do you expect this molecule to be aromatic, anti-aromatic or non-aromatic? Explain.

9. Construct an explanation for why the pKa on the right is much lower than the pKa on the left. That is, explain why the reaction on the right is much less up-hill.

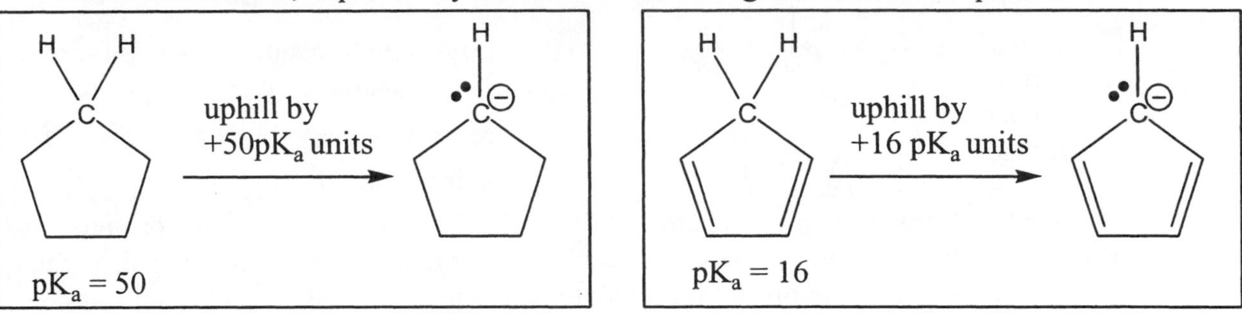

10. Explain why carbocation B is formed preferentially when water leaves. Then show the mechanism by which B is formed from A.

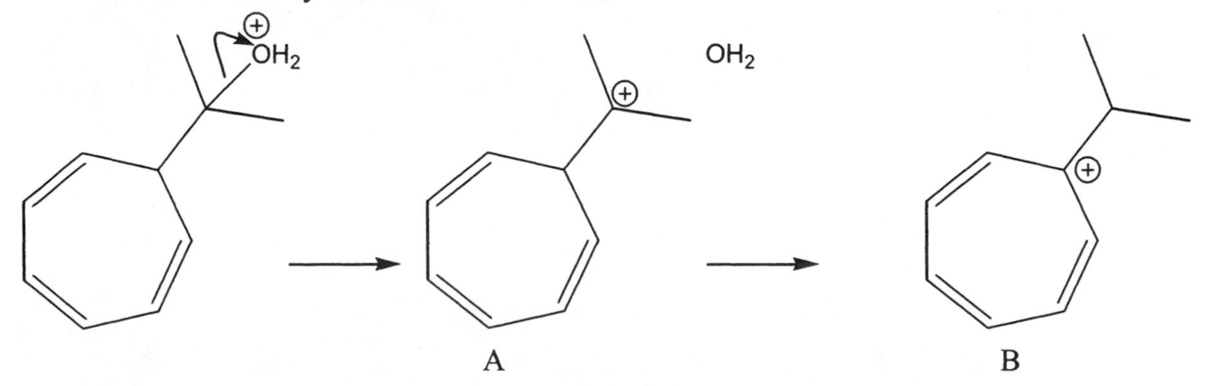

11. Read the assigned pages in the text and do the assigned problems.

ChemActivity 26

NMR of Aromatic Molecules

(Can the NMR spectrum of a molecule tell you if it is aromatic?)

Model 1: Chemical Shift (ppm) of Allyl and Vinyl H's

H ← - - allyl H (one away from C=C)

trans-3-hexene

H ← - - vinyl H (attached to C=C)

Figure 1a: (Idealized) Proton NMR of *trans*-3-Hexene

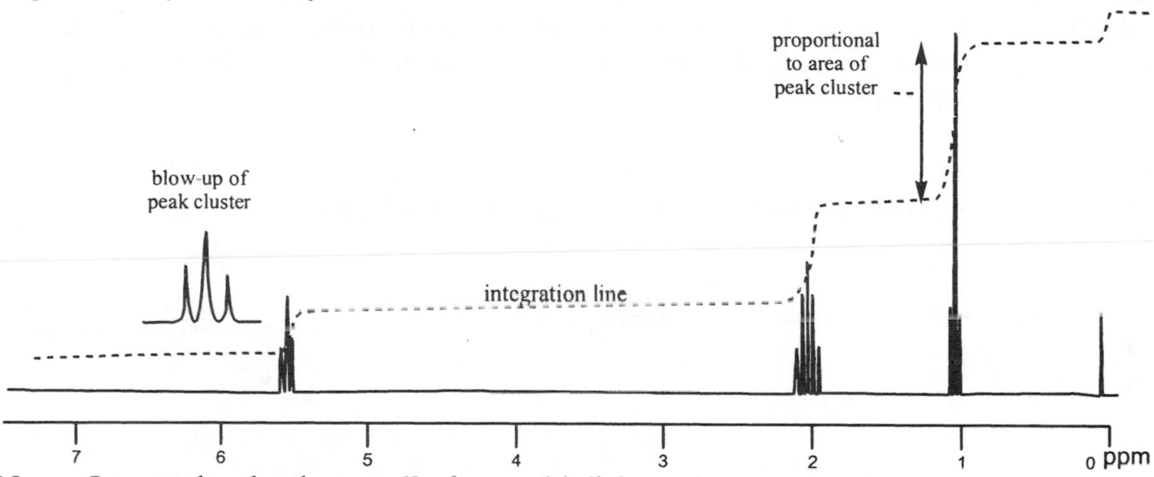

Note: It can be hard to tell the multiplicity of a peak without magnifying it. Magnifications of peaks sometimes appear above and to the left or right of a peak.

Critical Thinking Questions

1. Label each H on *trans*-3-hexene (at the top of the page) as vinyl, allyl or neither.
   a) Assign each H in *trans*-3-hexene a letter a, b, c…etc. Label equivalent H's with the same letter.
   b) Match each letter to a peak cluster in the proton NMR spectrum.
   c) Are your assignments consistent with the fact that vinyl H's show up on proton NMR spectra between 4 and 6 ppm. (Almost no other signals appear in this range, so a peak in this range is a good indicator of a double bond.)

2. Draw the structure of benzene, and predict the number of peak clusters (and each one's multiplicity and integration) in the proton NMR spectrum of benzene.

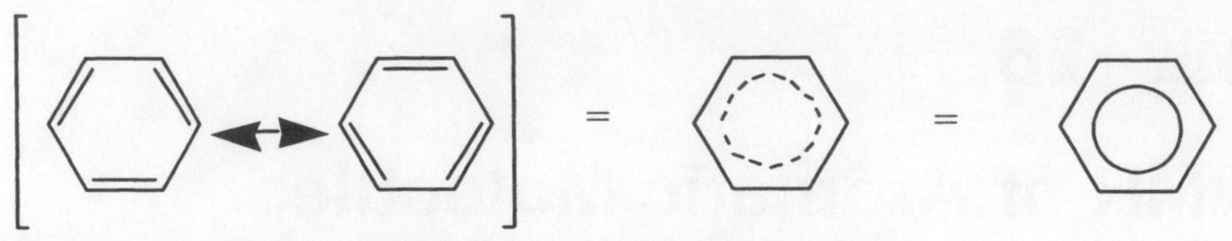

Figure 1b: Proton NMR Spectrum of Benzene

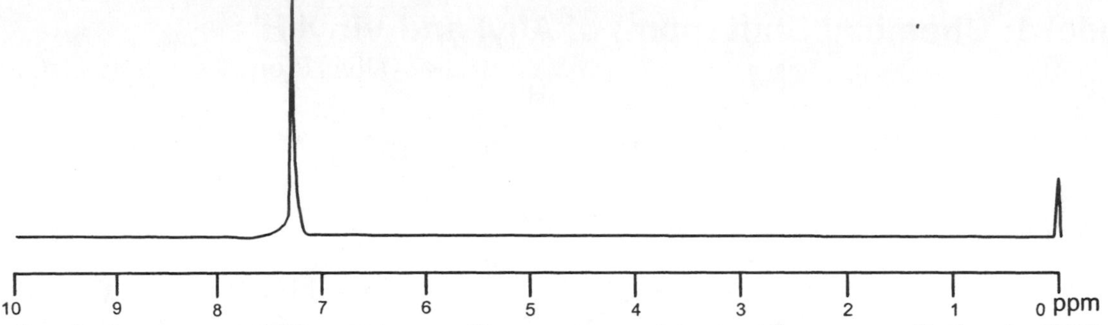

3. Is the proton NMR spectrum of benzene consistent with your prediction in CTQ 2?

4. What is the chemical shift (in ppm) of a H atom on benzene? (Such H atoms are often called **aromatic hydrogens** because they are directly attached to the ring of an aromatic pi system.)

Model 2: Nomenclature of Benzene Derivatives

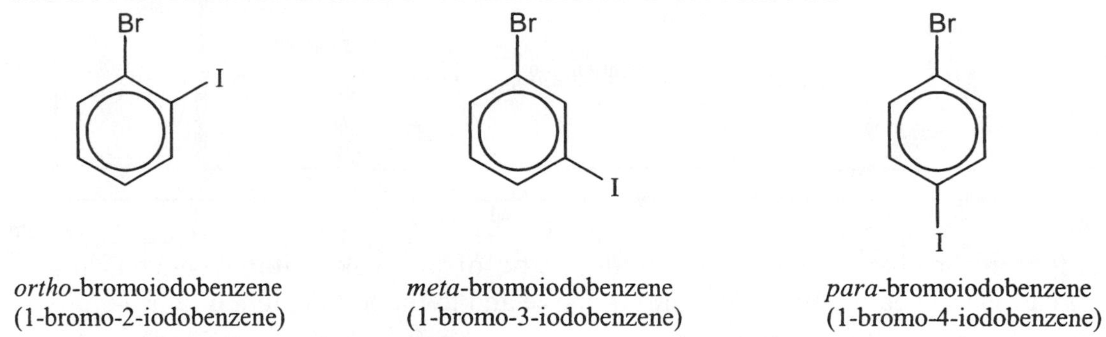

ortho-bromoiodobenzene (1-bromo-2-iodobenzene) | *meta*-bromoiodobenzene (1-bromo-3-iodobenzene) | *para*-bromoiodobenzene (1-bromo-4-iodobenzene)

Critical Thinking Questions

5. Draw in the H's on the structures above, and label them a, b, c…etc. to indicate equivalent H's. (Hint: look for symmetry. One of them is symmetrical, and two H's on either side of a symmetrical molecule are equivalent.)

6. The **multiplicity** and **integration** of ^{1}H NMR spectra can be described using a special shorthand.

For example: **[d (2H); d (2H)]** or **[s (1H); d (1H), t (1H), d (1H)]**

a) Identify which isomer from Model 2 is described by each shorthand example above.
b) Write a shorthand description of the *ortho* isomer from Model 2.

7. The proton NMR spectrum of which bromoiodobenzene isomer is shown below?

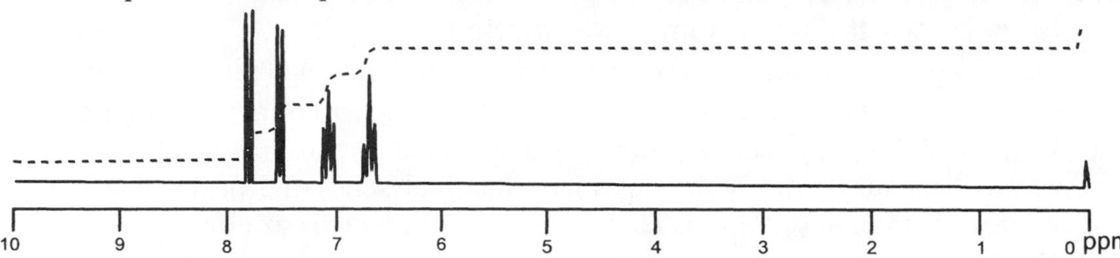

8. The concentration of a molecule in the sample tube determines the size of the NMR peaks. A sample solution such as the one above is usually about 0.1 molar in bromoiodobenzene. This means there are about 500 solvent molecules for each sample molecule. Construct an explanation for why CCl_4 is a much better solvent for NMR than CH_2Cl_2?

Model 3: The "Aromatic Region" of a Proton NMR Spectrum

Aromatic H's (and almost nothing else) show up on a ^{1}H NMR spectrum between **6 and 9 ppm**. This is called the **aromatic region**. The spectra in Model 2 are relatively easy to interpret because the peak clusters do not overlap, this is not always the case. In fact, several types of aromatic H's may get mushed together and show up as one broad singlet. An example of a messy aromatic region is shown below.

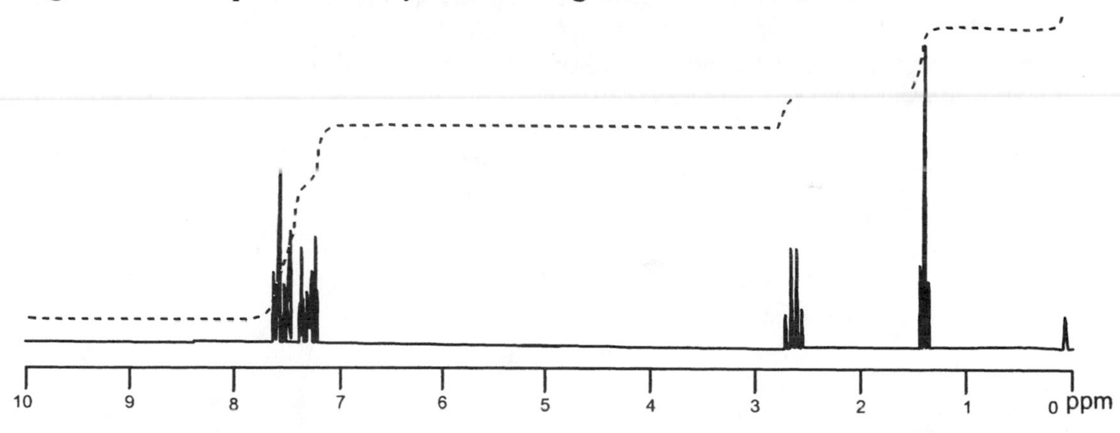

Critical Thinking Questions

9. Even though the aromatic region is very crowded, some information can still be gleaned from the aromatic region of a spectrum such as the one above.
   a) What is the ratio of aromatic to non-aromatic hydrogens (hint: look at the integration)?

   b) Which structure below best fits the proton NMR spectrum in Model 3?

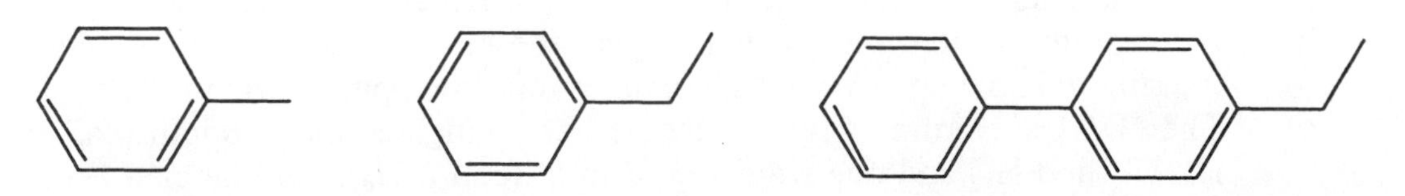

Model 4: Ring Current Effect
(An Explanation for the High ppm of Aromatic H's)

- NMR, simply put, measures the magnetic field near each unique H in a molecule.
- When a charged particle (such as an electron) moves it generates a magnetic field.
- The electrons in the pi system of benzene whizz around on their racetrack, generating a powerful local magnetic field. This magnetic field sends the peak cluster of an aromatic H to high ppm (6-9ppm). This is called the **ring current effect**.

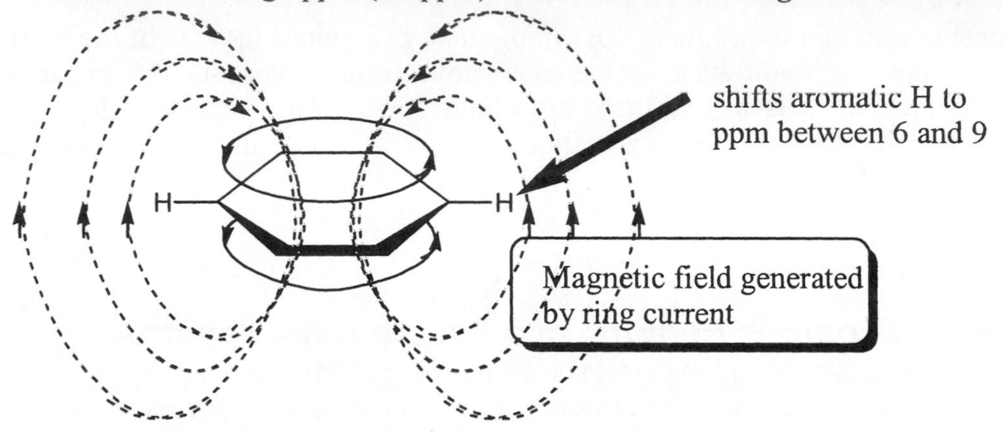

Critical Thinking Questions

10. A similar but larger ring current is generated by the 18 electron aromatic pi system shown below. This type of aromatic molecule is called a **porphyrin**. Similar aromatic molecules are at the active site of many biological molecules including hemoglobin, cyctochrome c and chlorophyll.

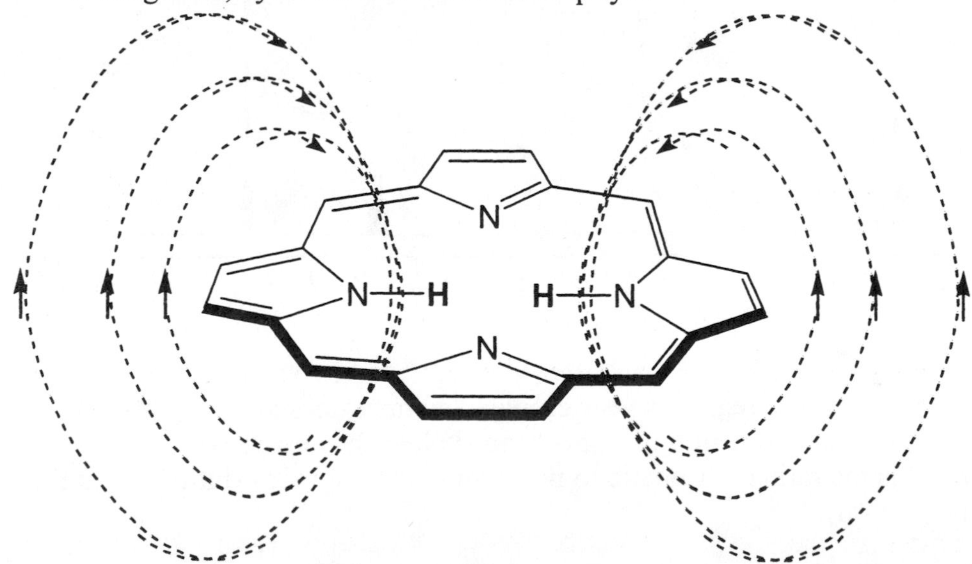

a) Trace the 18 electron aromatic system for the resonance structure above. (Be careful! There are 22 pi e^- so four do not participate in the aromatic system.)

b) By analogy to benzene, do you expect the highlighted H's at the center of the ring to show up at very high ppm or very low ppm? Explain.

c) The NMR spectrum of this molecule has 3 singlets above 8ppm (call them H_a, H_b, and H_c) and one singlet at -3 ppm (call it H_d). Add labeled H's (a-d) to the molecule above and predict the integration of H's a-d.

Exercises

1. Complete Nomenclature Worksheet II, Part A. Review Nomenclature Worksheet I. Also review ChemActivity 16 on NMR.
2. How many peak clusters would you expect in the proton NMR spectrum of…
   a) chlorobenzene?
   b) *para*-dichlorobenzene?
   c) *meta*-chlorotoluene?
3. How many peaks would you expect in the decoupled ^{13}C NMR spectrum of each molecule listed above?
4. Draw the structure that goes with each of the following names. (See Model 2 for definitions of *ortho*, *meta* and *para*.)
   a) *meta*-dichlorobenzene (also called 1,3-dichlorobenzene)
   b) *ortho*-methoxytoluene
   c) What is another name for the structure you drew for b) that has numbers and ends in "benzene."
   d) *para*-nitrobenzoic acid (a nitro group is a -NO_2 group)
   e) *ortho*-chloroaniline
   f) *para*-methyltoluene
   g) 3-chloro-4-hydroxybenzaldehyde
5. Label the circled H's on morphine as vinyl, allyl, benzyl, aromatic or none of these.

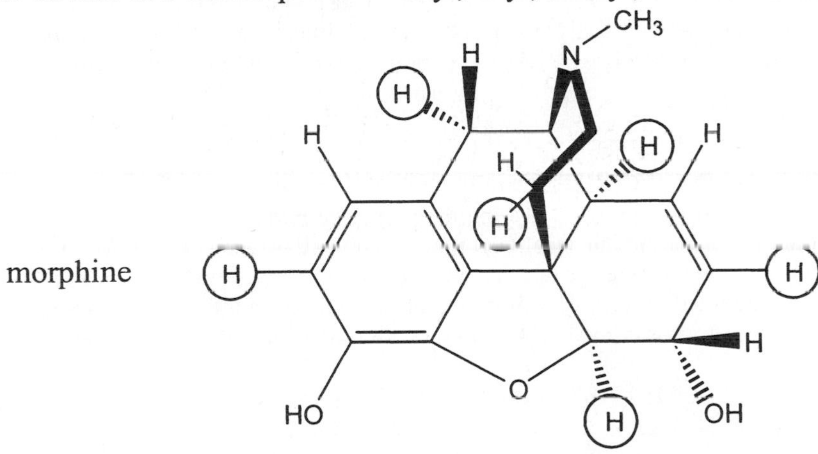

6. Rank the circled H's on morphine from lowest to highest expected proton NMR chemical shift (ppm value) using the following information.

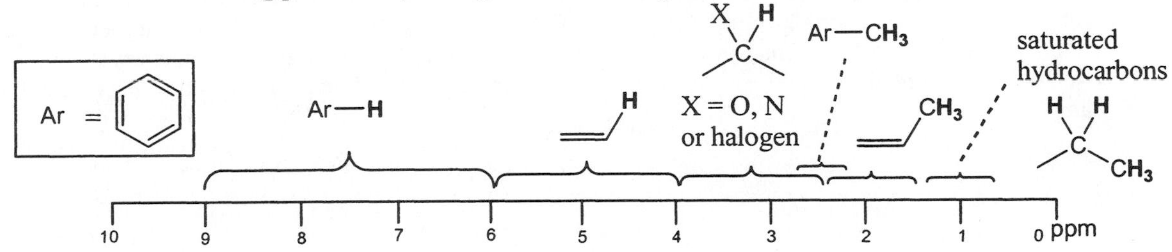

7. Read the assigned pages in your text and do the assigned problems.

8. Read the summary of NMR spectroscopy from CA 16 on the following page and Review the reading assignment from CA 16.

^{1}H NMR

- **ppm** or **chemical shift** (*given by location along the x axis*) tells you the amount of electron density around an H. The closer the H is to an electronegative element, the more "deshielded" it is and therefore the higher the ppm number of its peak cluster (farther left on the spectrum). Multiple bonds also cause the signal of nearby H's to be shifted to the left. Look in the index of your text for "chemical shifts" to find a list of common functional groups with their chemical shifts. Each chemically distinct H should have a unique chemical shift, though in practice different peak clusters sometimes overlap just by coincidence. This can make spectrum interpretation very difficult. Proton NMR peaks usually appear in the 0-10 ppm range.
- **Integration** or **peak area** *(given by integration line)* tells you the relative size of each peak and therefore the relative number of equivalent H's represented by each peak. Note that integration only gives you a ratio of peak areas, making it impossible to tell the difference between a 1H to 3H ratio and a 2H to 6H ratio.
- **Multiplicity** is the number of peaks in a peak cluster (also called **splitting** or **proton-proton coupling**). It tells you the number of nonequivalent neighbor H's within three bonds. For example, a doublet (two peaks) tells you there is exactly 1 non-equivalent H within three bonds of the H responsible for this signal. Equivalent H's don't split each other.
- **NMR Equivalent Hydrogens** = hydrogens that can be associated via an internal mirror plane or are the same distance away (through bonds AND through space) from every other atom or feature in the molecule. This means that a molecule and its enantiomer will produce the exact same NMR spectra. (Enantiomers are not distinguishable by NMR!!)

Important Notes: peaks at 0, 1.54 and 7.26 ppm on a proton NMR spectrum are common impurities.

- The 0ppm peak is TMS, which is added to the sample as an internal reference.
- The peak at 1.54ppm is water. (Water from the air contaminates most NMR spectra.)
- The peak at 7.26ppm is $CHCl_3$ which is a common contaminant in the NMR solvent $CDCl_3$. (Note that $CDCl_3$ has no H's!!)

^{13}C NMR

- **ppm** or **chemical shift** in ^{13}C NMR is similar to that of proton NMR. If a C is close to an electronegative element or involved in a multiple bond, or both, you will find the corresponding peak at higher ppm (farther left on the spectrum). Each chemically distinct C should have a unique chemical shift, though in practice different peak clusters sometimes overlap just by coincidence. This can make interpretation of coupled spectra very difficult. This is why decoupled spectra are usually taken. C-13 NMR peaks usually appear in the 0-100 ppm range, though some peaks are observed as high as 200ppm.
- **Peak area** has LITTLE MEANING in C-13 NMR. Peak area is a function of many things, one being whether the C is 1°, 2°, 3° or 4°. You will not be asked to interpret the meaning of peak area in C-13 NMR.
- **Multiplicity** in a proton coupled ^{13}C NMR spectrum tells you the number of H's attached to a given carbon. For example, a C that produces a doublet (two peaks) must have exactly 1 H attached to it. Very often chemists record "proton decoupled" ^{13}C NMR spectra. Such spectra have a singlet for each chemically unique carbon. This is useful for complex molecules for which the peak clusters would overlap. A decoupled ^{13}C NMR spectrum tells you the number of different carbon atoms present in the sample.

ChemActivity 27

Polymerization Reactions

(What mechanisms can explain the formation of polymers?)

Model 1: Polymers

Figure 1a: Possible Mechanism for ®Styrofoam (Polystyrene) Formation

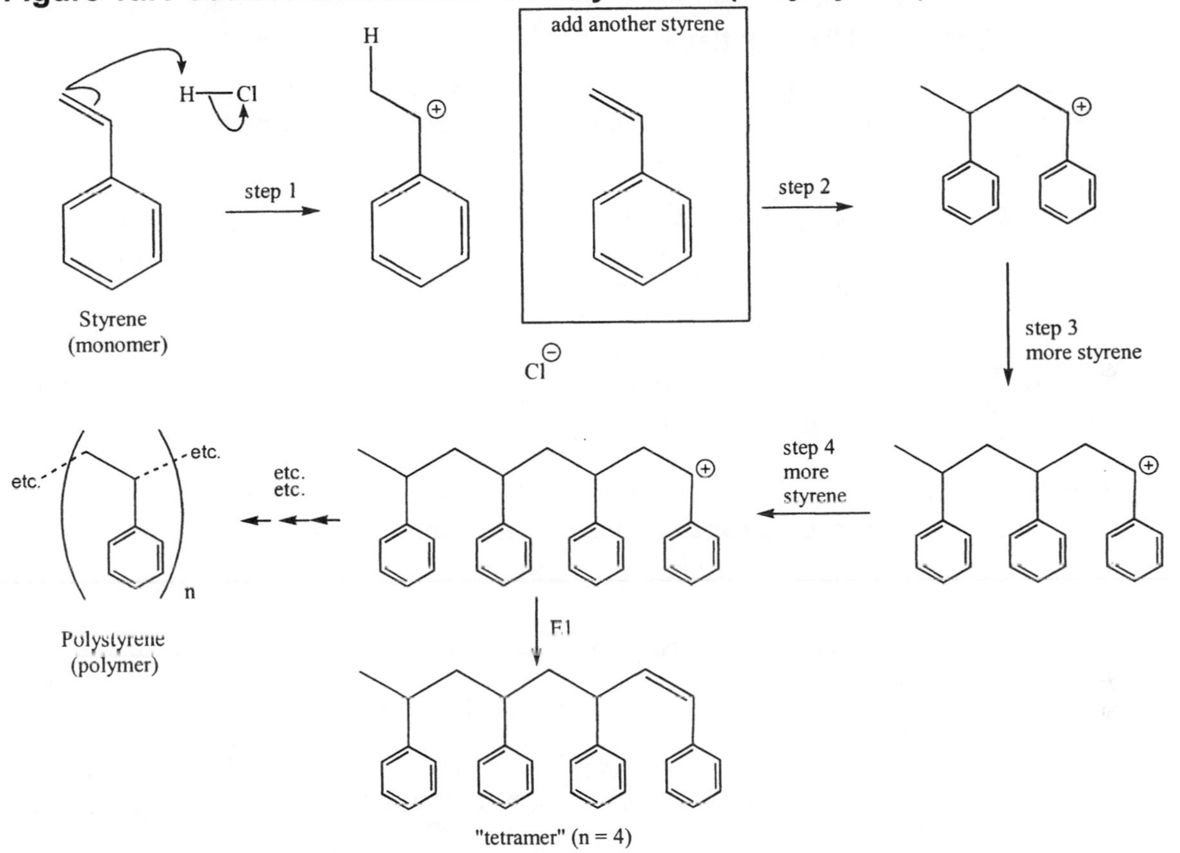

Styrene
(monomer)

step 1

add another styrene

step 2

step 3
more styrene

step 4
more
styrene

etc.
etc.

E1

Polystyrene
(polymer)

"tetramer" (n = 4)

The polar mechanism shown above is reasonable, but most polymerizations including the industrial synthesis of polystyrene are radical reactions. Instead of carbocations, the reaction precedes via radicals. For a review of radical reactions, see CA's 20-21.

Critical Thinking Questions

1. Use curved arrows to illustrate electron movement in steps 2-4 in Figure 1a.
   a) Assume Cl^- is a strong enough base to perform the E1 reaction in Figure 1a, and show curved arrows illustrating the reaction giving a tetramer of styrene.
   b) Is HCl a **Lewis acid catalyst** or a **Lewis base catalyst** [circle one]?
   c) Circle each monomer subunit (the C's that used to be an individual styrene molecule) in the styrene tetramer.
   d) A typical polystyrene molecule has a molecular weight between 50,000 and 100,000 amu (or g/mole). Based on this information, what is a typical range for n (the number of styrene monomer subunits in a polystyrene chain)?

2. Draw a tetramer of vinyl chloride. Do not show a mechanism for its formation. What would you name a polymer of vinyl chloride?

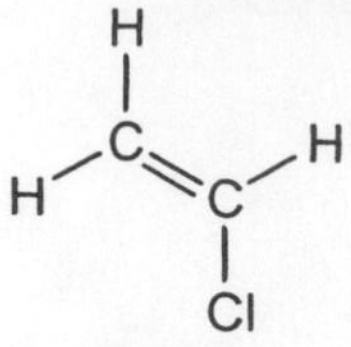

H
C
H C H
Cl

vinyl chloride

Model 2: Steroid Biosynthesis

- Under laboratory conditions it is statistically impossible that a bunch of monomers would line up and form a polymer in one concerted step as in the picture below.

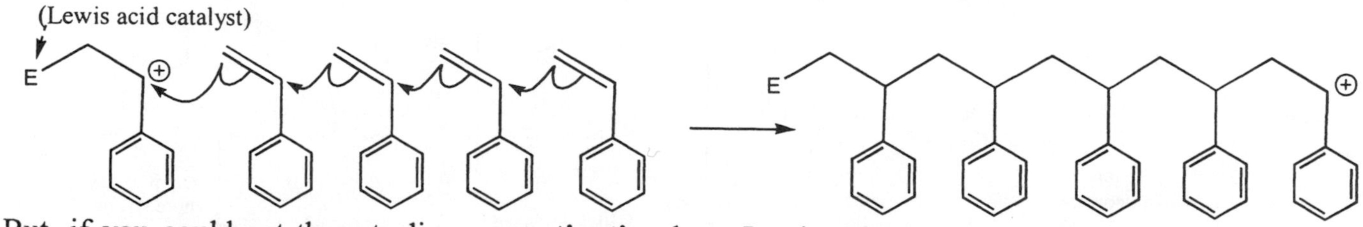

(Lewis acid catalyst)

E ⊕ → E ⊕

But, if you could get them to line up, **activation** by a Lewis acid catalyst would cause the polystyrene molecule to zip up like a zipper!

- The molecule below right is a **steroid**. Minor modification of lanosterol gives **cholesterol, testosterone, progesterone, estrogen** and other steroids.
- Lanosterol is made in your body in four steps from squalene.

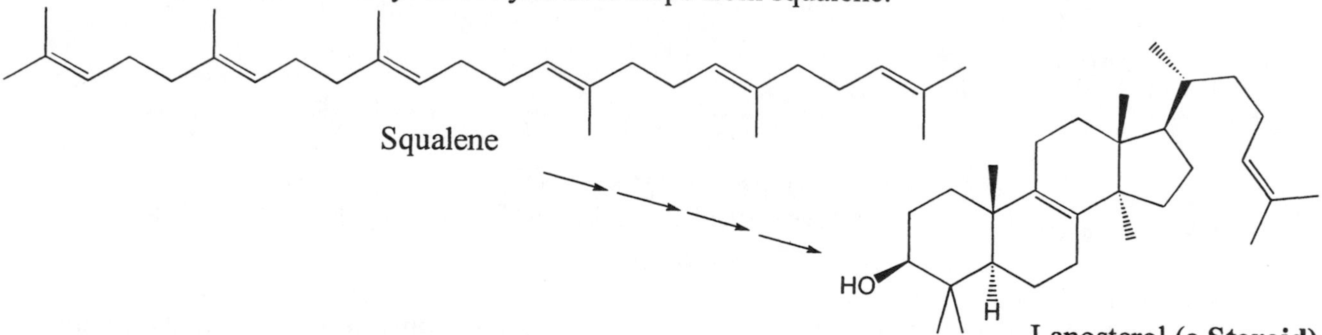

Squalene

HO H

Lanosterol (a **Steroid**)

The first step is selective epoxidation of just one of the double bonds. (Enzymes can direct a reaction to a specific location without affecting other parts of the molecule.)

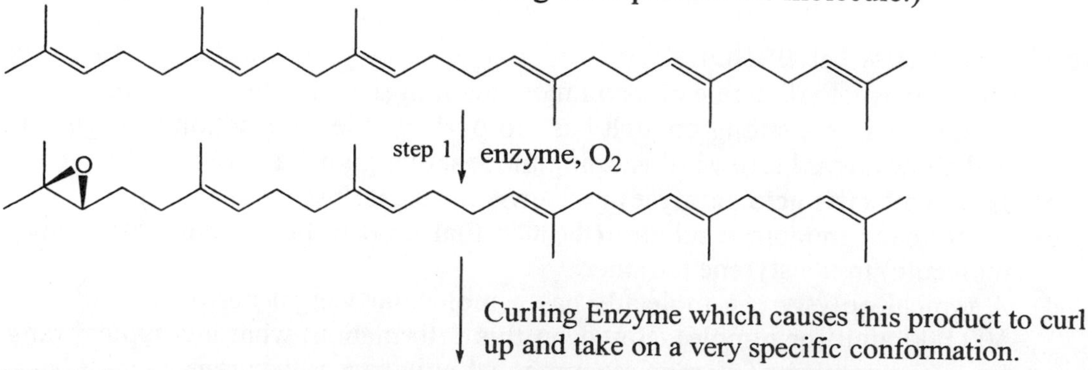

step 1 enzyme, O_2

O

Curling Enzyme which causes this product to curl up and take on a very specific conformation.

The product of this step is shown on the **top left** of the next page.

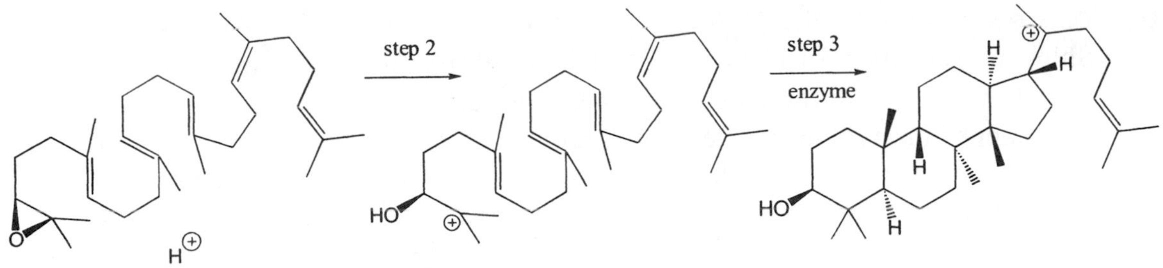

step 2

step 3

enzyme

Critical Thinking Questions

3. Confirm that the straightened out epoxide product of step 1 is the same as the curled up species above, left?
   a) Use curved arrows to illustrate steps 2 & 3.
   b) The enzyme that directs step 3 ensures that the reaction is **stereospecific**. Speculate what the term **stereospecific** means.
   (Hint: only the stereoisomer shown is formed, though there are 2^n possible stereoisomers of this product, where n = number of chiral centers.)

4. The final step of this biosynthesis is an E1 accompanied by a chain reaction of **stereospecific** hydride shifts and alkyl shifts (shown below).

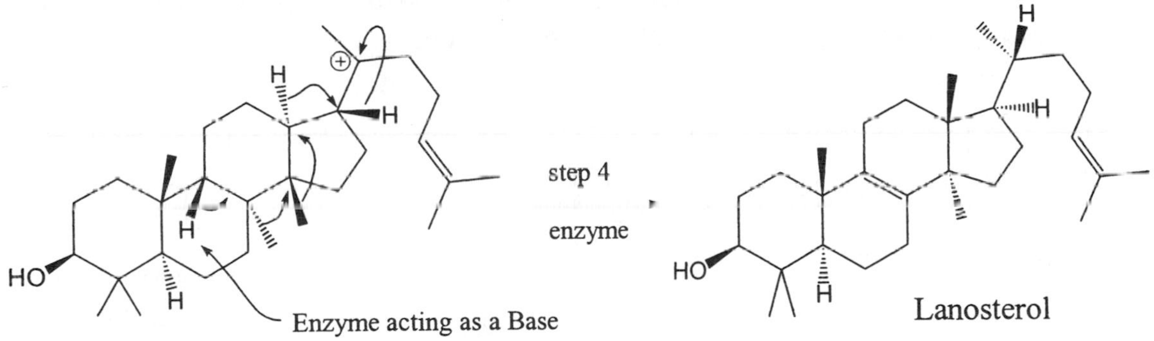

step 4

enzyme

Enzyme acting as a Base

Lanosterol

Enzymes which mediate this and other reactions are able to **control the stereochemistry** of each step. Nature developed this amazing chemistry over billions of years of evolution. We do not yet fully understand how enzymes do this—it is one of the most active areas of research in chemistry today.

Currently, we can do some stereospecific chemistry in the laboratory, but these reactions are very difficult and very expensive (you will learn about some of these). If we were to try and sythesize lanosterol using conventional (non-stereospecific) chemistry we would get a mixture of **R** and **S** at each chiral center.

a) Mark each chiral center in lanosterol with an *.
b) How many chiral centers are there?
c) How many different steroisomers of lanosterol could exist? Use the formula in CTQ 3b.

(Note: Only the stereoisomer shown above and labeled lanosterol has biological activity. Any one of the other 128 stereoisomers would not work in your body.)

Exercises

1. Show a mechanism for formation of a tri-mer of the following monomers.

- A C-F bond is very unreactive. A polymer of tetrafluoroethene is called teflon.
- A polymer of propylene is called "polypropylene." This is a fabric used for expedition long-underwear. Its structure is essentially a long chain of propane subunits.

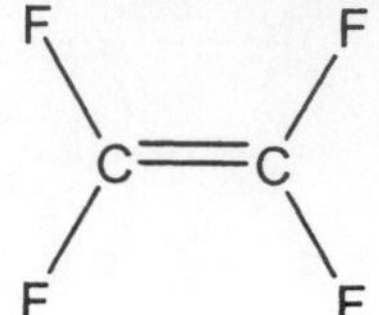

tetrafluoroethene

propene (common name = propylene)

a) When something is "sticky" this indicates the formation of surface-to-surface covalent bonds (or hydrogen bonds) being formed. Why is Teflon non-stick?

b) What is a common use for the gas, propane?

c) Why is it important to stay away from the fire when wearing polypropylene underwear?

2. A polymer can be represented by drawing a small segment such as a di-mer tri-mer or tetra-mer, or you can show just the repeating part of the chain as is shown below for Styrofoam. Plexiglas is a polymer of the monomer shown below. Draw a tri-mer and the shorthand representation of the repeating unit of Plexiglas.

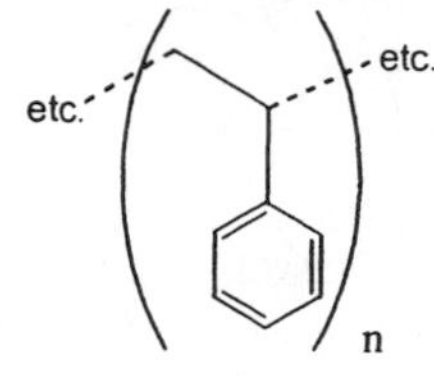

Polystyrene (polymer)

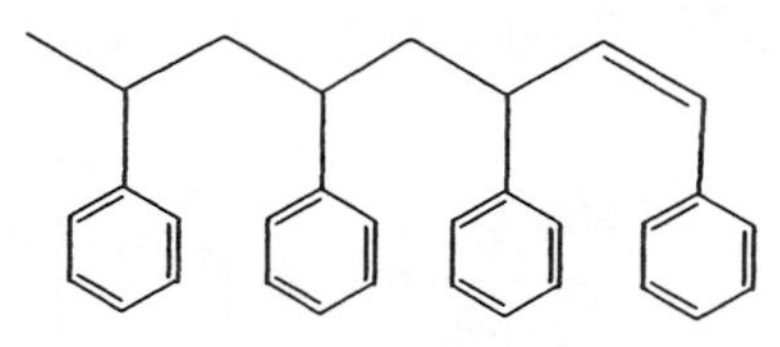

"tetramer" (n = 4) of Polystyrene

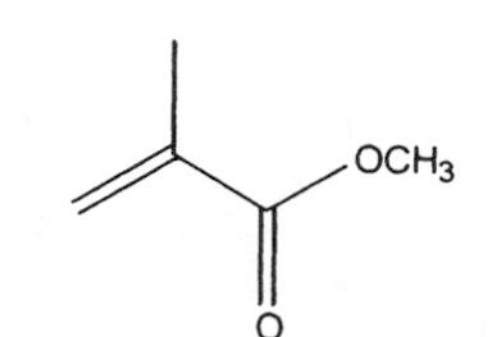

monomer from which Plexiglas is made

3. Read the assigned pages in your text and do the assigned problems.

4. Review the reactions on Reaction Sheets R1-R5.

ChemActivity 28

Diels-Alder Cylcloaddition Reactions

(What mechanisms can explain the Diels-Alder products?)

Model 1: The Diels-Alder Reaction

heat

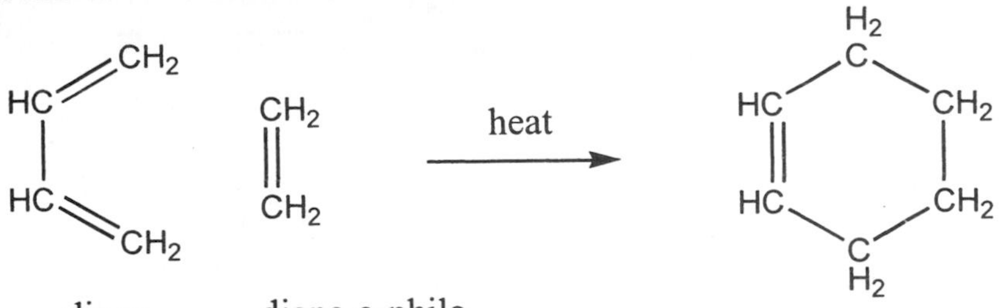

diene diene-o-phile

Critical Thinking Questions

1. Use curved arrows to account for bonds formed and broken in the Diels-Alder reaction above. (This is a one-step reaction.)
2. Draw a likely transition state for the Diels-Alder reaction (---- is a partial bond).

   a) How many electrons are either forming or breaking a pi bond in this t.s.?
   b) Explain why the Diels-Alder transition state is said to have aromatic character.

Model 2: Stereospecificity of the Diels-Alder Reaction

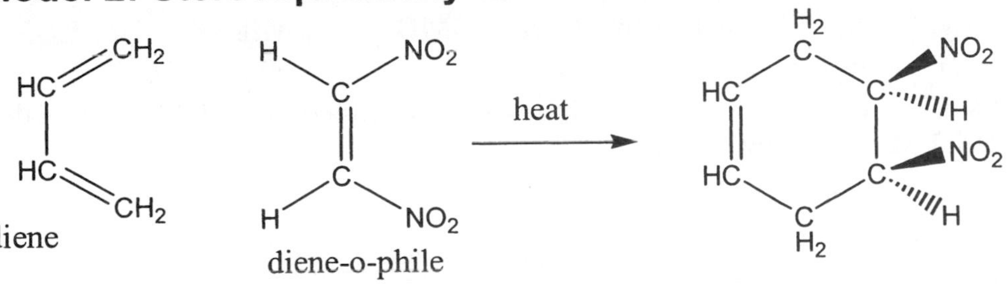

heat

diene diene-o-phile

heat

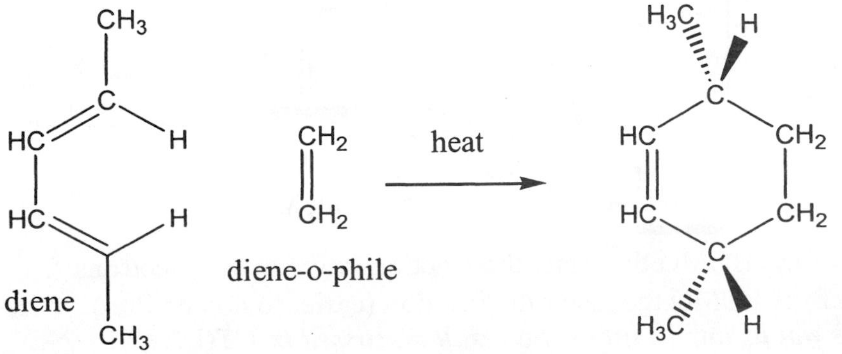

diene diene-o-phile

Critical Thinking Questions

3. Does a *cis* dienophile generate a *trans* or *cis* product?

4. Draw a diene and dienophile that will give rise to each of the following Diels-Alder products.

? —heat→ (cyclohexene with NO_2, H, H, NO_2 substituents)

? —heat→ (cyclohexene with H, CH_3 and H_3C, H substituents)

Model 5: Orbital Description of Bonding in Diels-Alder Reaction

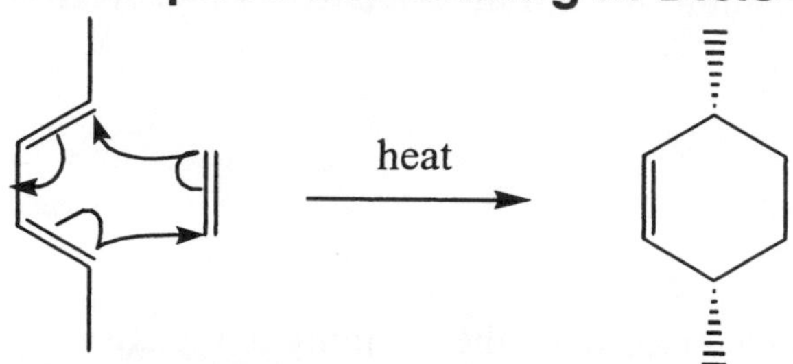

Our simple curved arrow accounting of bonds formed and broken does not explain the stereochemistry of the Diels-Alder reaction.

To go further we must examine the molecular orbitals involved.

- In MO theory, a bond is formed when a **filled HOMO mixes with an empty LUMO**.
- This forms a new filled bonding MO and a new empty anti-bonding MO.
- In the Diels-Alder reaction, a principle mixing is between the **diene HOMO** and the **dienophile LUMO** (as shown below).

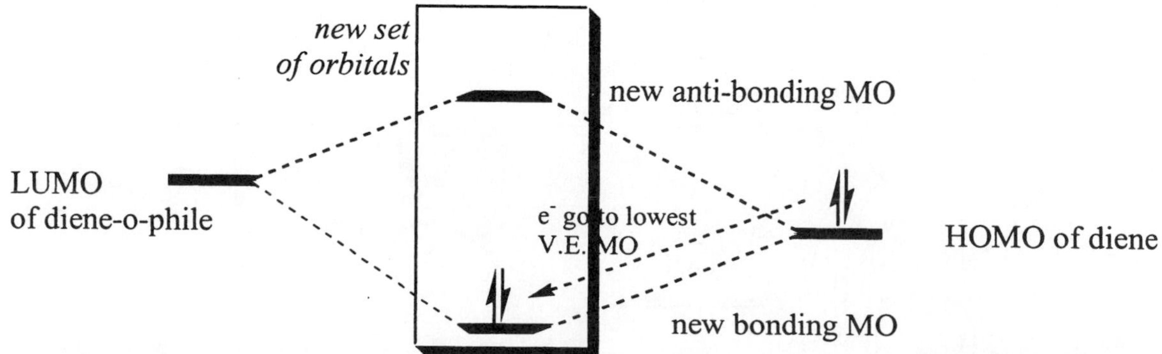

- The higher the P.E. of the HOMO the faster the reaction (more active electrons!).
- The lower the P.E. of the LUMO the faster the reaction (easier to donate into).

This MO explanation is not as important as the result discussed in CTQ 7.

Critical Thinking Questions

5. Why can't a filled HOMO mix with a filled HOMO to form a new bond? (Hint: complete the MO diagram by showing electron occupation of the new orbitals.)

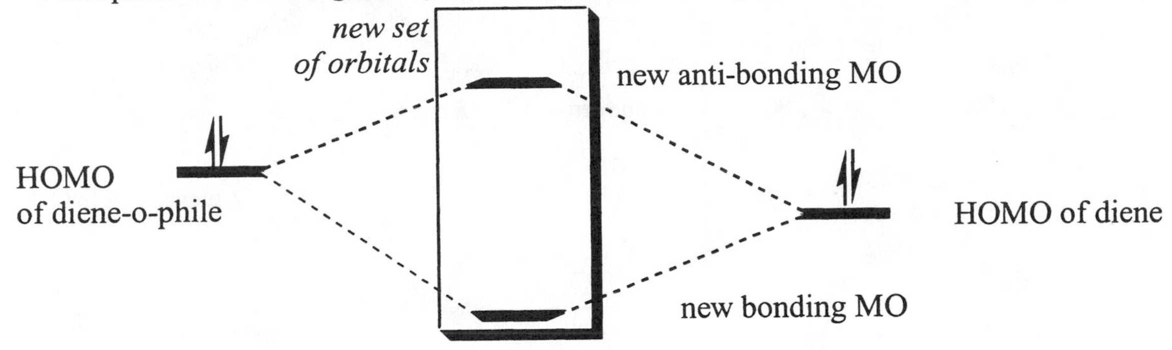

6. Why can't two empty LUMO's mix to make a new bond? (Again consider the number of electrons involved.)

Information

- An **electron withdrawing group** such as nitro (NO_2) **lowers the potential energy** of all the molecular orbitals of the molecule it is attached to.

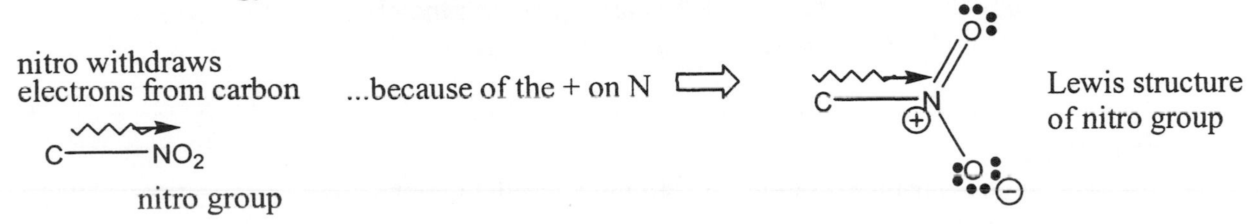

- An **electron donating group** such as methyl ($-CH_3$) does the opposite and **raises the P.E.** of all the molecular orbitals of the molecule it is attached to.

7. To activate (speed up) the Diels-Alder reaction it is preferable to… **[circle two]**
   A. have electron withdrawing groups on the diene.
   B. have electron withdrawing groups on the dienophile.
   C. have electron donating groups on the diene.
   D. have electron donating groups on the dienophile.

8. Which of the following diene/dienophile pairs will undergo the fastest Diels-Alder reaction? Explain your reasoning.

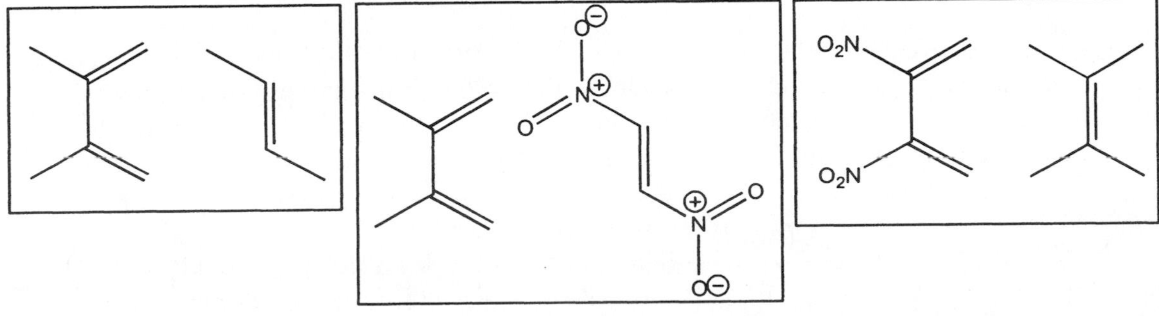

9. Consider the following picture of orbital overlap between the diene HOMO and the dienophile LUMO.

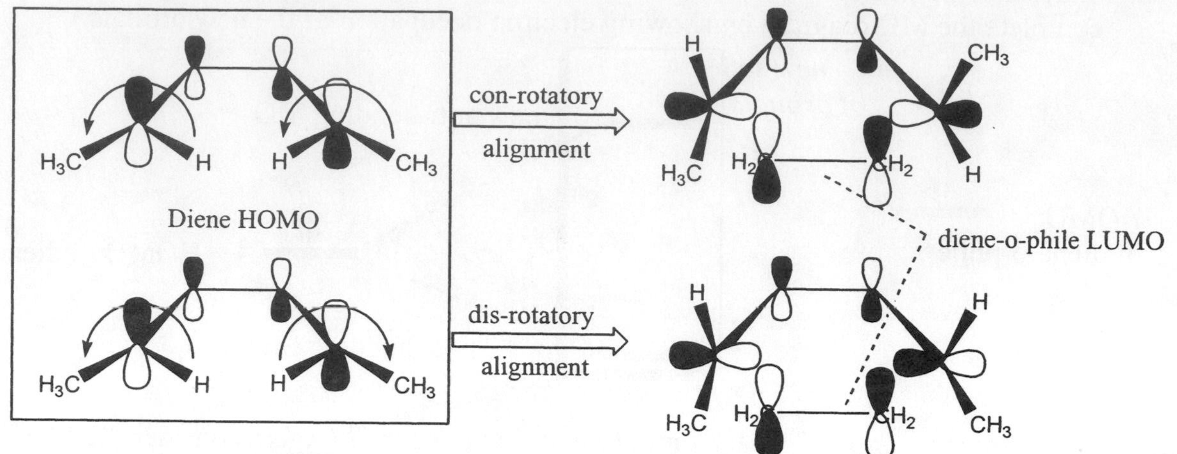

a) What do the terms **conrotatory** and **disrotatory** refer to?

b) Which type of rotation yields more bonding (in-phase) orbital overlap between the diene HOMO and the dienophile LUMO?
conrotatory <u>or</u> **disrotatory** [circle one]

c) Is your conclusion in part b) consistent with the observed stereospecificity of the Diels-Alder reaction (summarized in Model 2)?

Exercises

1. Based on your interpretation of Model 2, predict which of the two products will form in the reaction below. Circle the correct product.

+ CN CN Δ (heat) → CN CN or CN CN

2. Draw a diene/dienophile reactant pair that will result in the other product.

3. A diene in its s-*trans* configuration cannot react with a dienophile. Construct an explanation for why only the s-cis configuration can react with a dieneophile.

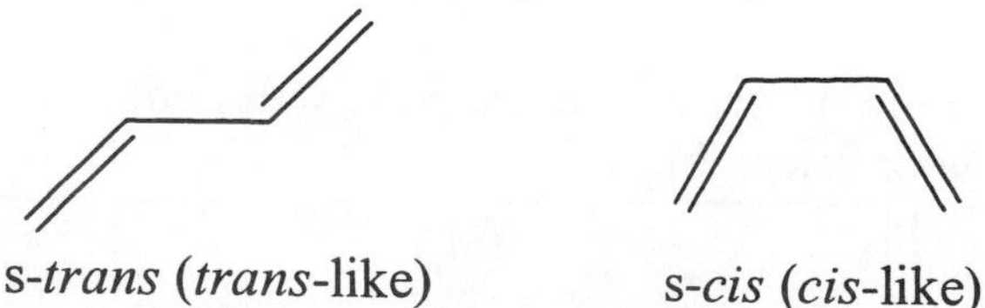

s-*trans* (*trans*-like) s-*cis* (*cis*-like)

4. The statement in Exercise 3 is TRUE, yet upon heating a sample of s-*trans*-1,3-butadiene in the presence of cyanoethene, Diels-Alder cycloaddition products are observed in high yield. Explain this observation.

+ CN Δ (heat) → CN + CN

5. Experiment has shown that the dienophiles in Box I (below) undergo fast Diels-Alder cyclization with a suitable diene. The dienophiles in Box II undergo VERY fast Diels-Alder cyclization with a suitable diene.

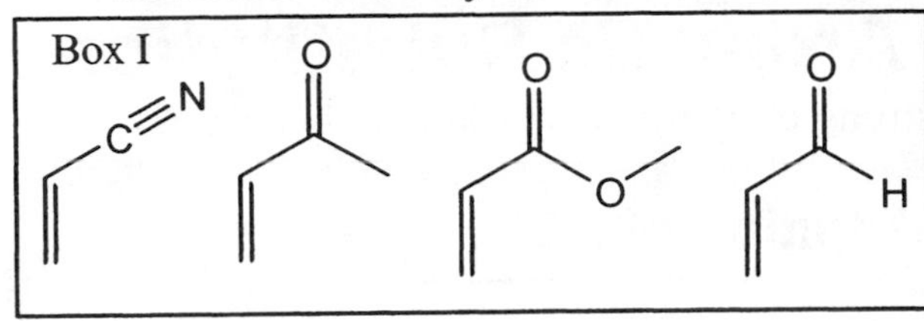

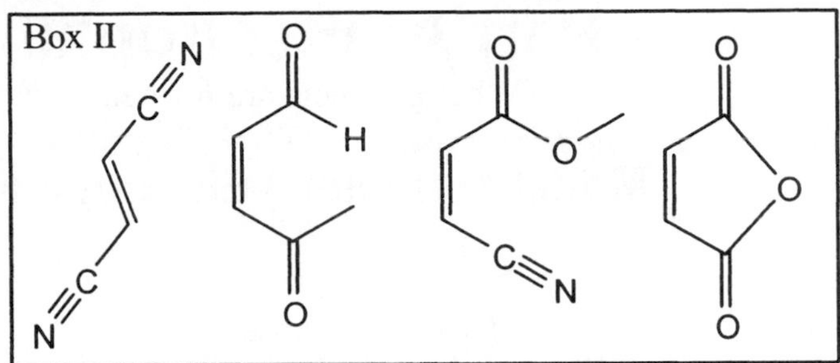

a) Explain why each vinyl carbon above has a partial positive charge.

b) Based on the examples above, a Diels-Alder reaction is sped up by **electron withdrawing groups** or **electron donating groups** on the dieneophile [circle one]?

6. Reaction (I) is very slow, reaction (II) is very fast, and reaction (III) yields no cycloaddition products at all. Explain these observations.

(I) + CN / CN — Δ (heat) — VERY SLOW → CN / CN

(II) + CN / CN — Δ (heat) — VERY FAST → CN / H / CN / H "exo product"; H / CN / H / CN "endo product"

(III) + CN / CN — Δ (heat) — NO REACTION

7. Read the assigned pages in your text and do the assigned problems.

ChemActivity 29

Part A: Electrophilic Aromatic Substitution

(What products are formed when a strong electrophile is added to benzene?)

Model 1: (review) Electrophilic Addition of HCl

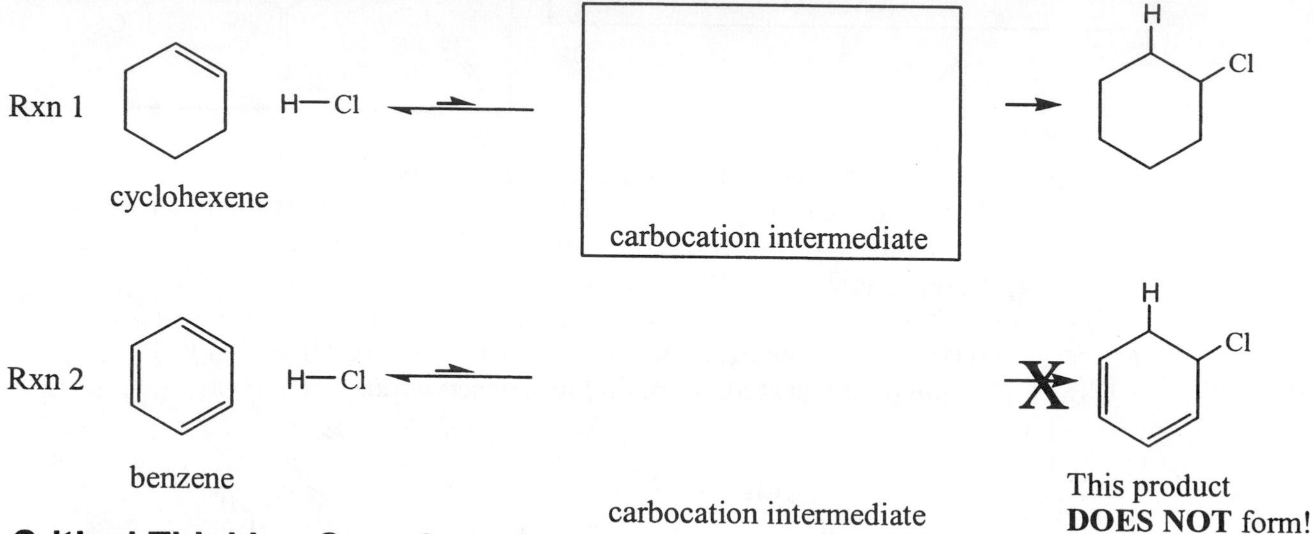

Critical Thinking Questions

1. For Rxn 1 (above) draw curved arrows showing the mechanism of electrophilic addition of HCl. Include an appropriate carbocation intermediate in the box above.

Figure 1: Reaction Diagrams for Electrophilic Addition of HCl

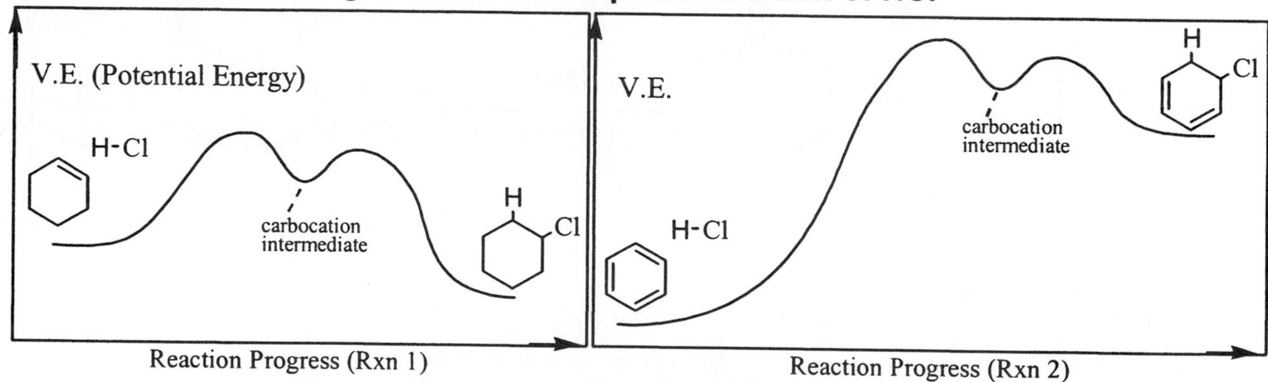

2. Rxn 1 is slightly down-hill in terms of energy. Rxn 2 is very up-hill in terms of energy (*see* Figure 1). **Construct an explanation for the large difference in energy between the reactants and the product in Rxn 2.**

3. Draw the carbocation that would form in Rxn 2. Explain why this carbocation goes back to the starting material (H–Cl and benzene) instead of forming the product.

Model 2: Electrophilic Aromatic Substitution

- Recall that deuterium (D) has nearly identical reactivity as hydrogen (H).
- In the following reaction D–Cl reacts the same way as would H–Cl.
- When benzene is treated with D-Cl, an H is replaced with a D. (With excess D-Cl this continues until all the H's have been replaced and the product is C_6D_6.

Rxn 3 C_6H_6 + D—Cl ⇌ ⟶ C_6H_5D + H—Cl

Critical Thinking Questions

4. Use curved arrows to show a reasonable mechanism for the reaction in Model 2. (Hint: the first step is formation of a carbocation intermediate, just as in Rxn 2.)

5. The energy diagram for Rxn 2 (using DCl in place of HCl) is shown below using a *dotted line.* On this same set of axes, draw a **solid line** to show an energy diagram for Rxn 3 in Model 2. (Note: Rxns 2 and 3 have the same carbocation intermediate.)

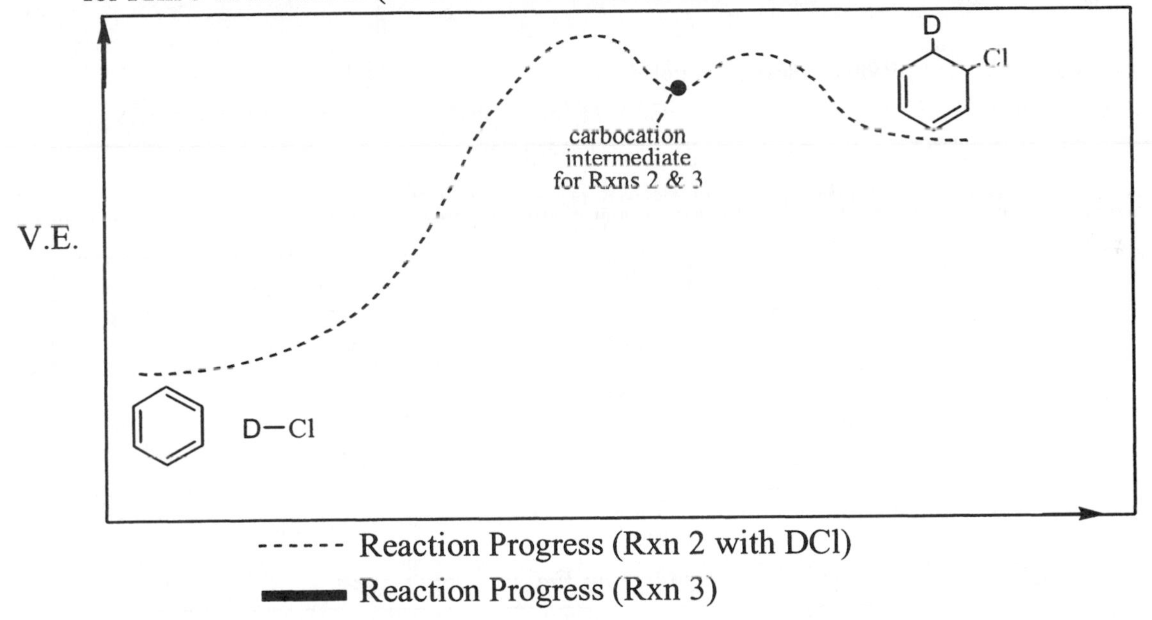

6. Construct an explanation for why Rxn 3 is much more likely to occur than Rxn 2. (Note: Rxn 2 does not occur under normal circumstances.)

7. Draw a generalized mechanism for EAS (Electrophilic Aromatic Substitution) showing the substitution of an electrophile (E^+) for one of the H's of benzene in the presence of a mild base (B). (Note: in Rxn 3, $E^+ = D^+$)

H E⊕ :B ⇌ E H—B⊕

Information: Not All Electrophiles are Created Equal

- A very strong electrophile (E^+) is required to break up an aromatic ring.
- The H of a strong acid will work since it exists essentially as H^+.

Know the following strong acids.

δ+ H—Cl δ–	δ+ H—Br δ–	δ+ H—I δ–	δ+ H—O(δ–)—S(=O)(=O)—OH	δ+ H—O(δ–)—N⊕(=O)—O⊖
hydrochloric acid	hydrobromic acid	hydroiodic acid	sulfuric acid	nitric acid

- Suitable electrophiles (E^+) for EAS (Electrophilic Aromatic Substitution) include an H^+ or D^+ donated by a strong acid, or the electrophiles (E^+) listed below.

Table 2: Other Suitable Electrophiles (E^+) for EAS

Reactant	E^+ =	Reagents	Product/s
benzene or other aromatic ring	O=N⊕=O	H_2SO_4 and HNO_3	benzene–NO_2
	⊖O—S⊕(=O)=O	anhydrous sulfuric acid	benzene–SO_3H
	Br⊕ or Cl⊕ *	Br_2 and $FeBr_3$ or Cl_2 and $FeCl_3$	benzene–Br or benzene–Cl
	R⊕ *	R-X, AlX_3 (X = Cl or Br)	benzene–R; R = alkyl group
	⊕C(=O)—R *	X—C(=O)—R AlX_3 (X = Cl or Br)	benzene–C(=O)—R; R = alkyl group

*The last three rows of electrophiles are generated only in the presence of a Lewis acid catalyst such as FeX_3 or AlX_3. A supplementary activity on Lewis acid catalysts is assigned for homework.

Part B: Substituent Effects

(How do substituents on an aromatic ring direct the placement of E^+ and rate of EAS?)

Model 3: Electrophilic Aromatic Substitution with Toluene

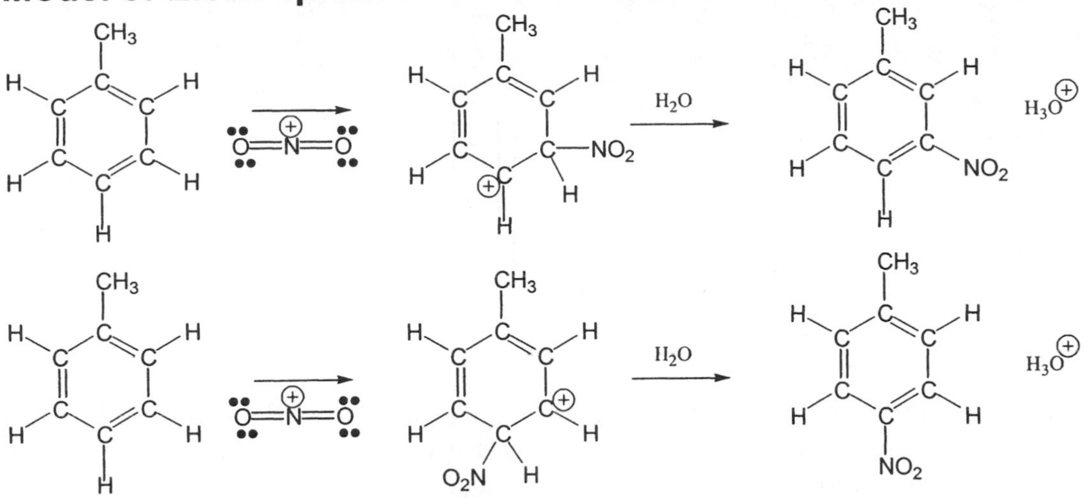

Critical Thinking Questions

8. Add curved arrows showing the movement of electrons in the two reactions above.

   a) Draw any missing important resonance structures for <u>each</u> of the two carbocation intermediates. (One from each set is re-drawn for you below.)

meta

para

   b) In the *meta* set, none of the three resonance structures stand out as being most important, but in the *para* set one resonance structure stands out as being **most important**. Explain this statement and circle the most important res. structure.

   c) Of the two products at the top of the page, one forms and the other does not. **Circle the product that forms and explain your reasoning.**

9. Draw the other two resonance structures of the intermediate that forms if the nitro group adds to the *ortho* position (as shown in the mechanism below).

a) The carbocation above is very close in potential energy to the intermediate in the *para* set on the previous page. In what way are these two sets of resonance structures similar?

b) Of the *ortho* intermediate and the *para* intermediate, one is slightly lower in potential energy due to steric effects. Which one do you expect to be lower in potential energy? ***ortho* intermediate** or ***para* intermediate** [circle one].

c) The following energy diagram shows pathways to the ***ortho, meta,*** and ***para*** products. Label each pathway with the correct name.

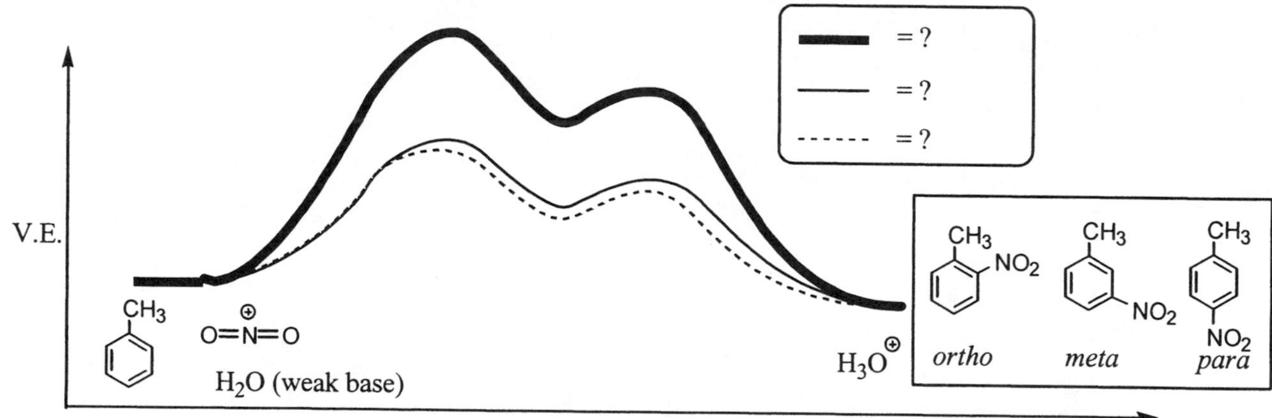

d) When toluene is mixed with nitric acid and sulfuric acid (giving $^{+}NO_2$) two of the three products in the box above are formed. Circle these two and cross out the other one.

e) A methyl group, or any other alkyl group (R) is said to be an "**ortho & para director**." Explain why the R group "directs" the nitro to the *o* or *p* positions.

Model 4: Steric vs. Electronic Effects in EAS

- For the reaction below, chemists say that the ***meta* product is NOT formed because of unfavorable electronic effects**. (The electron arrangement in the *meta* intermediate is not as favorable as in the *ortho* and *para* intermediates.)
- Generally, electronic effects are much more powerful than steric effects, but steric effects can have an impact.
- With large E^+ like nitro ($^+NO_2$), the *para* product is favored over *ortho* product.

R, O=N$^+$=O, H_2O ⇌ O_2N, R, R, NO_2, R, NO_2

R = $-CH_3$	*ortho* (63 %)	*meta* (~3 %)	*para* (34 %)
R = —CH(CH_3)CH_3	*ortho* (12 %)	*meta* (~3 %)	*para* (85 %)

Critical Thinking Questions

10. Explain the very strong preference for the *para* product when R = *iso*propyl (as compared to when R = methyl).

11. "Without steric effects, you would expect the *ortho* product to be twice as abundant as the *para* product." Explain this statement (*see* the Hint below).

R, O_2N, H, H, H, H — *ortho* (66 %)
R, H, H, H, NO_2, H — *meta* (~0 %)
R, H, H, H, H, NO_2 — *para* (33 %)

Expected Ratio if there were no Steric Effects

R, H, NO_2, H, H, H

R = alkyl group

Hint: What is the difference between this product and the *ortho* structure in the box at left?

12. An alkyl group (R) is slightly electron donating. This means it donates electron density into an attached aromatic ring. Based on this information, which of the following EAS reactions do you expect to be faster? Explain your reasoning.

R = alkyl group

R —(H_2SO_4/HNO_3 (gives $^+NO_2$))→ O_2N—R + R, NO_2

—(H_2SO_4/HNO_3 (gives $^+NO_2$))→ NO_2

Hint: Which is better for the EAS reaction, to have a ring that more is electron rich or more electron poor?

Part C: Resonance Donating and Withdrawing Groups

(How do π donating and withdrawing groups affect rate and placement in EAS?)

Model 5: Second Order Resonance Structures

Second order resonance structures are like "**assistant resonance structures**." They are not as important as regular (first order) resonance structures, but they can tell us some information about the arrangement of electrons in a complicated molecule.

Nitrobenzene

Second Order Resonance Structures of Nitrobenzene

Critical Thinking Questions

13. Explain why the three **second order resonance structures** shown in Model 5 are not full fledged resonance structures. That is, why are they "not as important" as a regular, first order resonance structure such as the one at the far left, above?

14. This set of second order resonance structures tells us that the nitro group withdraws electron density into its p orbitals very strongly from certain carbons on the aromatic ring. Complete the composite drawing of nitrobenzene below by placing a δ+ on appropriate carbons in the ring.

Composite Drawing of Nitrobenzene
(showing a combination of all first and second order resonance structures)

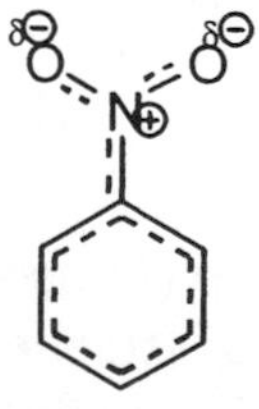

15. Add curved arrows to nitrobenzene at the top of the page showing how you would change it into one of the second order resonance structures. Then use curved arrows to generate each of the subsequent second order resonance structures.

16. In the first step of an electrophilic aromatic substitution (EAS) reaction **the aromatic ring is acting as a nucleophile** and reacting with the electrophile (E^+).

Z Z Z
$E^{\oplus}$ Base
Step One Step Two
E H E

a) Which is more likely to react with an electrophile, **a carbon that is electron rich**, or **a carbon that is electron poor and holds a δ+ charge** [circle one]?
b) Assume that Z = nitro group and add δ+ to appropriate ring carbons on the structure of the starting material above. (Hint: *see* Model 5.)
c) Construct an explanation for why, when "Z" is an **electron withdrawing group** such as nitro, **the *meta* product is the major product**.

$E^{\oplus}$
Base
NO_2
E
major product

NO_2 E NO_2 E
minor products

17. Any molecule with a lone pair next to the aromatic ring will have second order resonance structures. The first in a set of three for aniline is shown below.
a) Add curved arrows to aniline showing how you would change it into the second order resonance structure shown. Then use curved arrows to generate the two missing second order resonance structures.

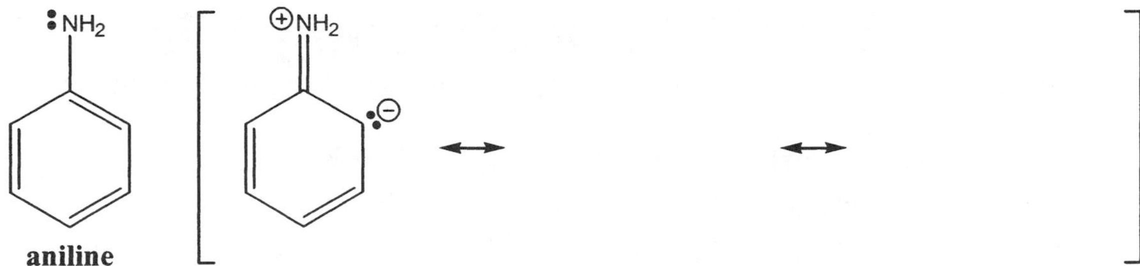

Second Order Resonance Structures for Aniline

b) In terms of electrons, does a NH_2 group **donate into** or **withdraw from** [circle one] the aromatic ring?
c) Complete the composite drawing of aniline below by placing a δ– on appropriate carbons in the ring.

Composite Drawing of Aniline
(showing a combination of all first and second order resonance structures)

H H
N δ⊕

18. Recall that, in the first step of an EAS reaction **the aromatic ring is acting as a nucleophile** and reacting with the + charged electrophile (E^+).

Z E⊕ Step One Z ⁑Base H E Step Two Z E

a) Assume that Z = an amino group (NH_2) and add δ– to appropriate ring carbons on thc structure of the starting material above.

b) Construct an explanation for why, when Z = a **strong electron donating group** such as NH_2, the major products are ***ortho*** and ***para***.

E⊕ ⁑Base NH_2 E NH_2 E

major products

minor product

19. To say a group on the ring is a "*meta* director" means it directs an E^+ (in the next EAS reaction) to add the to the *meta* position on the ring. Complete the following:

a) A strong electron withdrawing group is...
an *ortho/para* director <u>or</u> **a *meta* director** [circle one].

b) A strong electron donating group (with a lone pair to donate into the ring) is...
an *ortho/para* director <u>or</u> **a *meta* director** [circle one].

Part D: Summary of Directing Effects

(Why are halogens "deactivators" but *ortho/para* directors in an EAS reaction?)

Model 6: Inductive Effects vs. Resonance Effects

- An inductive effect is the donation or withdrawal of electron density through sigma bonds. (As shown in the diagram below, left.)
- A resonance effect is the donation or withdrawal of electron density as demonstrated by 1st or 2nd order resonance structures (as shown below, right).

Example of an Inductive Effect

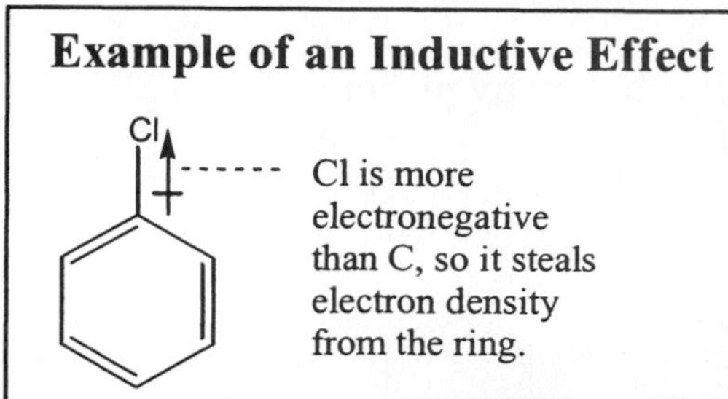

Example of a Resonance Effect

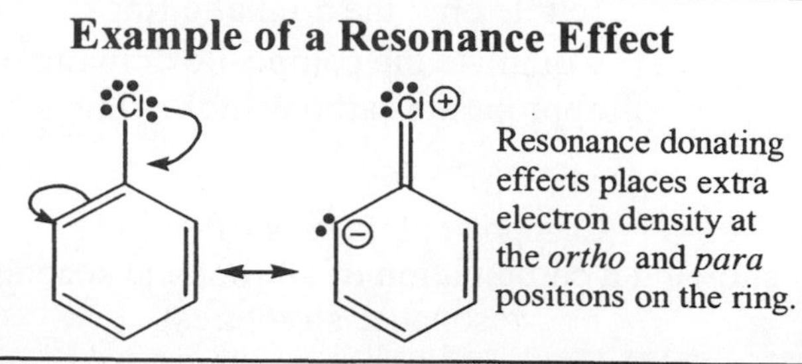

Note: A halogen is an inductive withdrawing group <u>and</u> a resonance donating group.

Table 6: Directing Effects of Various R Groups in EAS Reactions

R–benzene + $E^{\oplus}$ / :Base ⇌ *ortho* + *para* + *meta*

Identity of R	Inductive Effects	Resonance Effects	Product Regio-chemistry	Relative Reaction Rate
Amines $-NH_2$, $-NHR$, $-NR_2$	weak e^- **withdraw**	strong e^- **donation**	*ortho & para*	very very fast
Phenols $-OH$, $-OR$	moderate e^- **withdraw**	strong e^- **donation**	*ortho & para*	very fast
Alkyl groups $-CH_3$, $-CH_2CH_3$ etc.	weak e^- **donation***	weak e^- **donation***	*ortho & para*	moderatel y fast
Benzene $-H$	—	—	—	1
Halogens $-I$, $-Br$, $-Cl$, $-F$	strong e^- **withdraw**	weak e^- **donation**	*ortho & para*	slow
Acyl groups $-C^{\delta+}(=O^{\delta-})R'$	strong e^- **withdraw**	strong e^- **withdraw**	*meta*	very slow
Sulfate group $-S^{\delta+}(=O^{\delta-})_2-OH^{\delta-}$	strong e^- **withdraw**	strong e^- **withdraw**	*meta*	very slow
Nitro group $-N^{\oplus}(=O)-O^{\ominus}$	strong e^- **withdraw**	strong e^- **withdraw**	*meta*	very slow
Cyanide group $-C^{\delta+}\equiv N^{\delta-}$	strong e^- **withdraw**	strong e^- **withdraw**	*meta*	very slow
Ammonium $-NR_3^{\oplus}$	strong e^- **withdraw**	strong e^- **withdraw**	*meta*	very slow

**Electron donation by an alkyl group can be considered an inductive or a pseudo-resonance effect (see hyperconjugation). Hyperconjugation cannot be demonstrated with 2nd order resonance structures.*

Critical Thinking Questions

20. Summarize how Column 3 (Resonance Effects) is related to Column 4 (Product Regiochemistry)?

21. Summarize how Column 5 (Reaction Rate Relative to Benzene) is related to the overall level of electron withdrawal or donation? (consider both inductive and resonance effects)

22. According to Model 6, are halogens *o/p* directors or *m* directors?

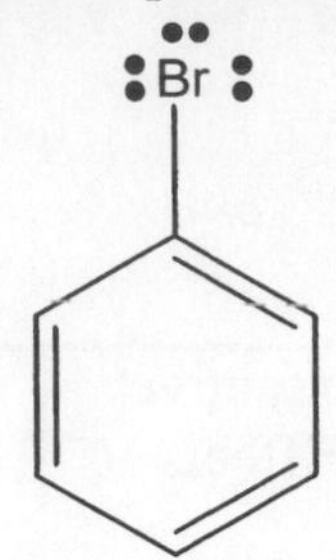

a) Halogens are powerful inductive withdrawing groups. This means that a halogen removes electron density from all carbons of the ring. Put a δ+ next to each ring carbon in the structure of bromobenzene drawn above.

b) But, a halogen has a lone pair to donate into the ring. This means they are resonance donating groups. On the same drawing of bromobenzene, put a small δ– next to the three carbons that receive electron donation from Br.

c) These two effects (inductive withdrawal and resonance donation) compete with one another, but the **inductive withdrawing effects win**. Is this consistent with the reaction rate for bromobenzene (on Table 6) relative to plain benzene? Explain.

d) If you were an electophile (E^+), which carbons on bromobenzene would you be most attracted to? Explain your reasoning.

e) Is your answer in part d) consistent with the data in Table 6?

Information:

activating group = group that makes the rate of EAS faster than with benzene
deactivating group = group that makes the rate of EAS slower than with benzene

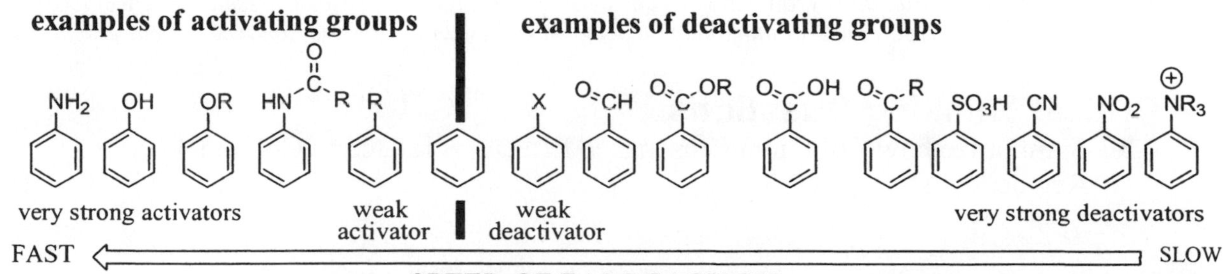

Model 7: EAS Reactions with Di-Substituted Rings

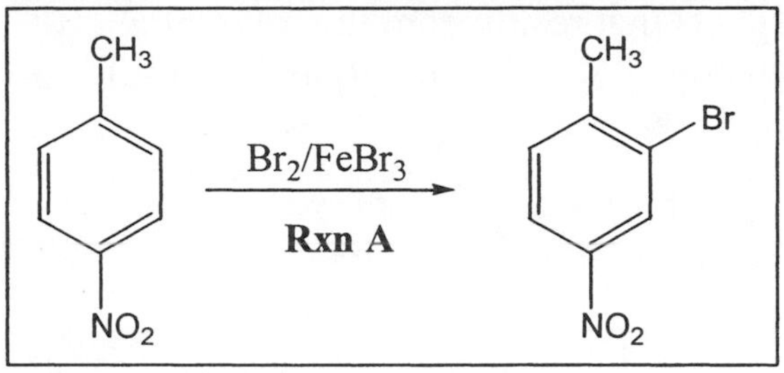
CH3
Br2/FeBr3
Rxn A
NO2
CH3
Br
NO2

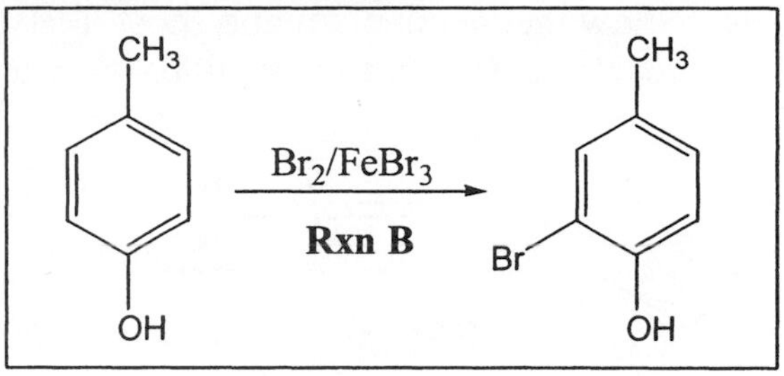
CH3
Br2/FeBr3
Rxn B
OH
CH3
Br
OH

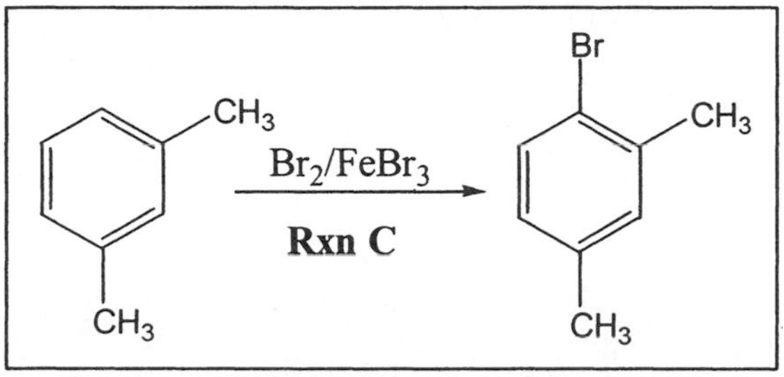
CH3
CH3
Br2/FeBr3
Rxn C
Br
CH3
CH3

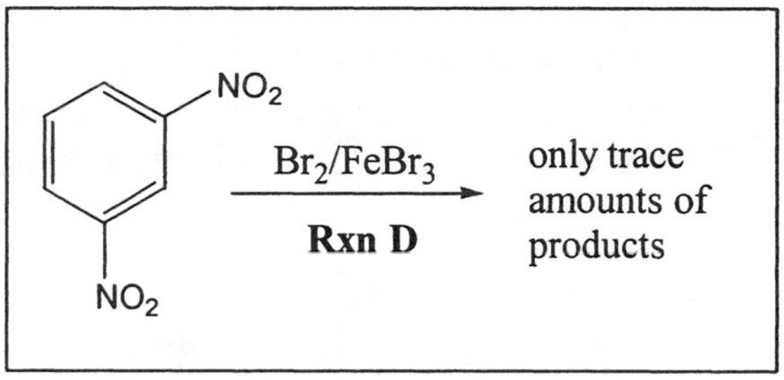
NO2
NO2
Br2/FeBr3
Rxn D
only trace
amounts of
products

Critical Thinking Questions

23. Consider the starting material in **Rxn A**, in Model 7.
    a) Mark the position/s on the ring where the CH_3 group would direct an E^+.
    b) Mark the position/s on the ring where the NO_2 group would direct an E^+.
    c) Are these two directing effects **opposed to one another** or **in agreement** [circle one]?

24. Consider the starting material in **Rxn B**, in Model 7.
    a) Mark the position/s on the ring where the CH_3 group would direct an E^+.
    b) Mark the position/s on the ring where the OH group would direct an E^+.
    c) Are these two directing effects **opposed to one another** or **in agreement** [circle one]?
    d) Label each group on the ring (OH and CH_3) as a strong activator, weak activator, weak deactivator or strong deactivator.
    e) Construct an explanation for why the product shown is the major product.

25. Consider the starting material in **Rxn C**, in Model 7.
    a) Mark the position/s on the ring where each CH_3 group would direct an E^+.
    b) Construct an explanation for why the product below is formed in very small amounts compared to the product shown above.

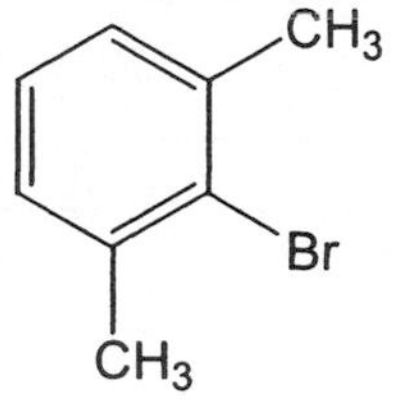
CH3
Br
CH3

Minor Product

26. Which reaction/s in Model 7 demonstrate the general rule that Friedel-Crafts reactions are extremely slow with a deactivated aromatic ring.

Exercises for Part A

1. Show the mechanism and most likely products that result from the following reactants. (Note: two weak bases, water and bisulfate ion are also in solution.)

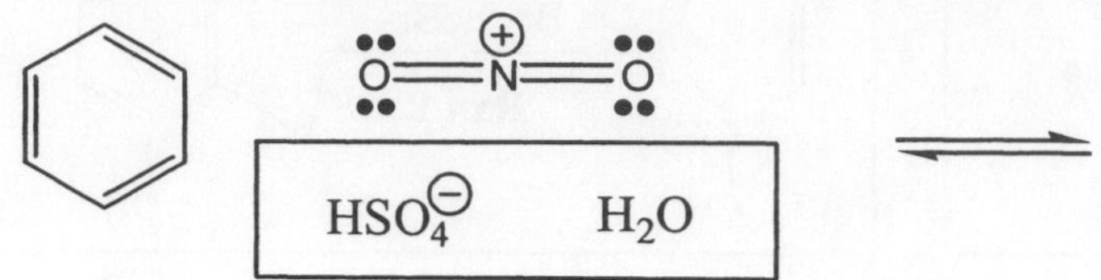

2. Sulfuric acid with absolutely no water in it is called **fuming sulfuric acid** and contains small amounts of the powerful electrophile SO_3 (one resonance structure is shown below). Construct a mechanism for the following reaction. Hint: the final step is an intramolecular H atom transfter.

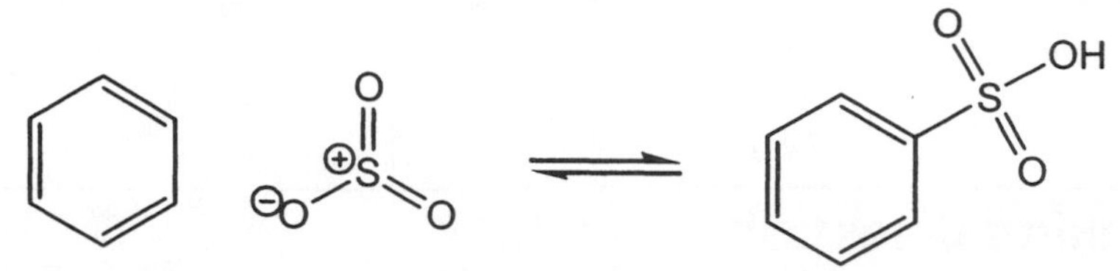

3. Draw all possible resonance structures for the carbocation intermediate in Model 2.

4. $^{+}NO_2$ is formed when nitric acid (HNO_3) and sulfuric acid (H_2SO_4) are mixed. Draw the Lewis structure of each and construct a mechanism that explains formation of $^{+}NO_2$. Hint: water and HSO_4^- are the other products formed in this reaction.

5. When toluene is treated with sulfuric and nitric acids under special conditions, three nitro (NO_2) groups are substituted for hydrogens (at the 2, 4 and 6 positions on the ring). The product is a highly explosive substance commonly known by a three letter name. Draw the structure and write the common name and the chemical name for this explosive substance.

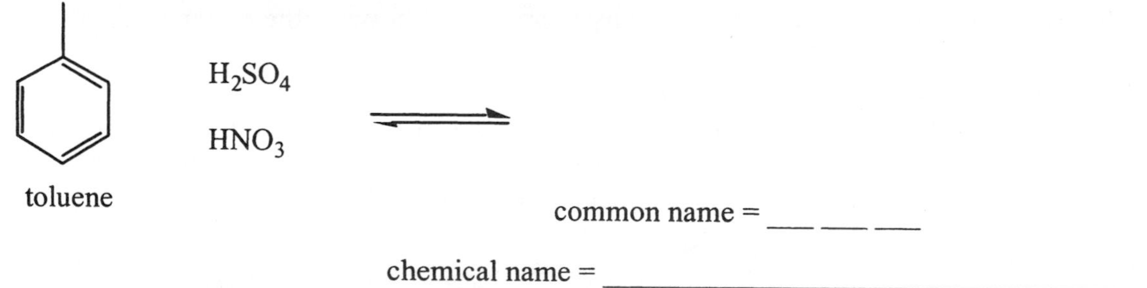

common name = ___ ___ ___

chemical name = ______________________________

6. Construct a reasonable mechanism for the following reaction called a Friedel-Crafts alkylation.

Note: when R-X and AlX_3 are mixed, you can assume the result is $R^{\oplus}$ and $^{\ominus}AlX_4$

assume this to be...

$AlCl_3$ (cat.)

Cl–CH_2CH_3

$^{\oplus}CH_2CH_3$

$^{\ominus}AlCl_4$

CH_2CH_3

H–Cl

7. Draw the mechanism (use curved arrows) and most likely product/s that would result from the following EAS reaction called a Friedel-Crafts acylation.

assume these species are present

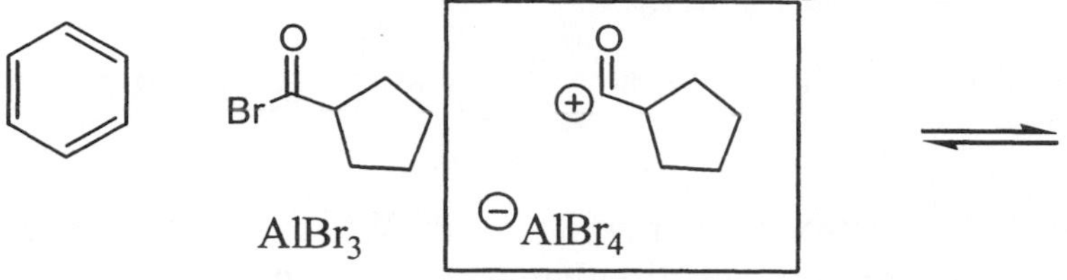

Br
O
$AlBr_3$
O
⊕
⊖$AlBr_4$

8. Draw a mechanism to explain the formation of each of the two Friedel-Crafts products. Hint: think of (and draw) the $R^{\delta+}$ group in the R–X–$AlCl_3$ complex as a carbocation (R^+), then think about possible carbocation rearrangements.

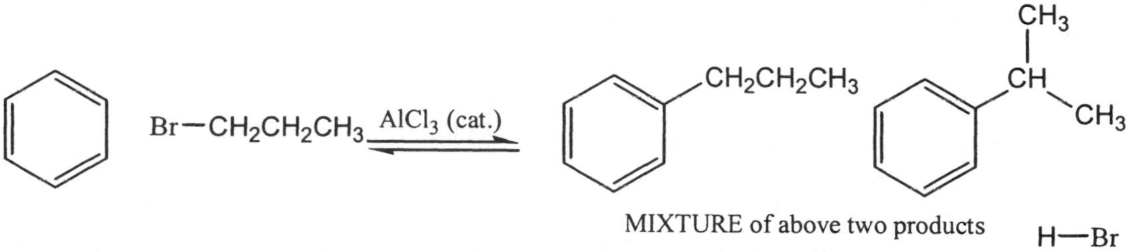

Br—$CH_2CH_2CH_3$ $AlCl_3$ (cat.)
$CH_2CH_2CH_3$
CH_3
CH
CH_3
MIXTURE of above two products
H—Br

9. Give an example (not appearing in this ChemActivity) of...
   a) an alkyl halide (R–X) that will likely undergo rearrangement during a Friedel-Crafts alkylation.
   b) an alkyl halide (R–X) that will NOT undergo rearrangement during a Friedel-Crafts alkylation.

10. Shown below are two ways of making the same target product starting from benzene. Synthetic pathway b gives a higher % yield of the desired product. Explain why.

a
Br
$AlCl_3$ (cat.)

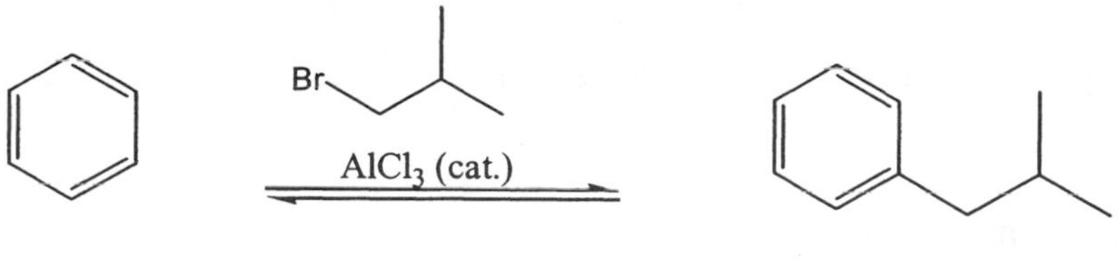

b
Br
O
$AlCl_3$
O
HCl heated in Zn/Hg amalgum
(you are NOT responsible for this mechanism)

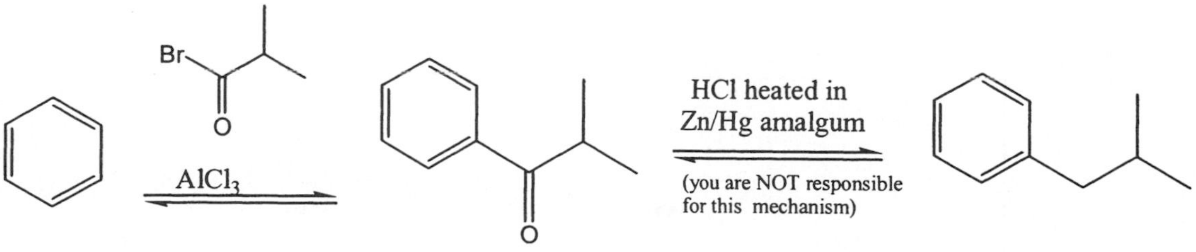

11. Read the assigned pages in your text and do the assigned problems.

12. Complete the mini-activity on Lewis acid catalysts found at the end of this ChemActivity.

Exercises for Part B

13. Of the choices in brackets, circle the word or phrase that makes the sentence true.
    a) The pi system of the ring acts as [**a nucleophile** *or* **an electrophile**] in an EAS reaction.
    b) The [**more** *or* **less**] electron rich the pi system of the aromatic ring, the faster the rate of EAS.

14. A nitro group is a very powerful electron withdrawing group. Which do you expect will undergo EAS reaction faster: **benzene** or **nitrobenzene** [circle one], and explain your reasoning.

15. Construct an explanation for the following finding: Even with the best electrophile (E^+), di-nitro benzene undergoes EAS extremely slowly. So slowly that not even trace amounts of product are observed.

O_2N … NO_2 — $E^{\oplus}$:Base, very very very slow → NO PRODUCTS OBSERVED

16. When a flask containing a mixture of 1 mole of nitrobenzene and 1 mole of toluene is treated with one mole of D-Cl, deuterium is incorporated into the toluene ring, but not the nitrobenzene ring. Explain. That is, why does the toluene "hog" all the D-Cl, while the nitrobenzene does not get any?

This notation says that D is on the ring, but does not specify which position on the ring.

1 mole toluene (CH_3) + 1 mole nitrobenzene (NO_2) — 1 mole D—Cl → D-toluene (CH_3): observed; D-nitrobenzene (NO_2): not observed

17. Read the assigned pages in your text and do the assigned problems.

Exercises for Part C

18. Draw three 2nd order resonance structures for phenol.

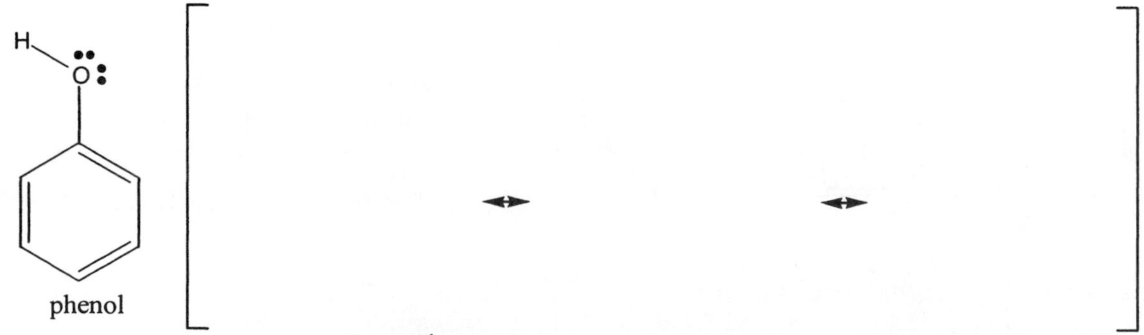

    a) Explain why the 2nd order resonance contributors are less important than 1st order resonance structures, and contribute only a small amount to our overall understanding of phenol.

b) Explain the following statement: The 2nd order resonance structures help explain why R groups with a lone pair (such as –OH) activate the *ortho* and *para* positions toward electrophilic aromatic substitution (EAS).

19. Draw three 2nd order resonance structures for benzoic acid.

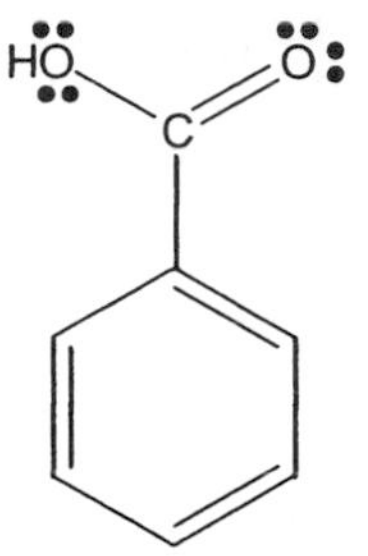

HO C O

benzoic acid

↔ ↔

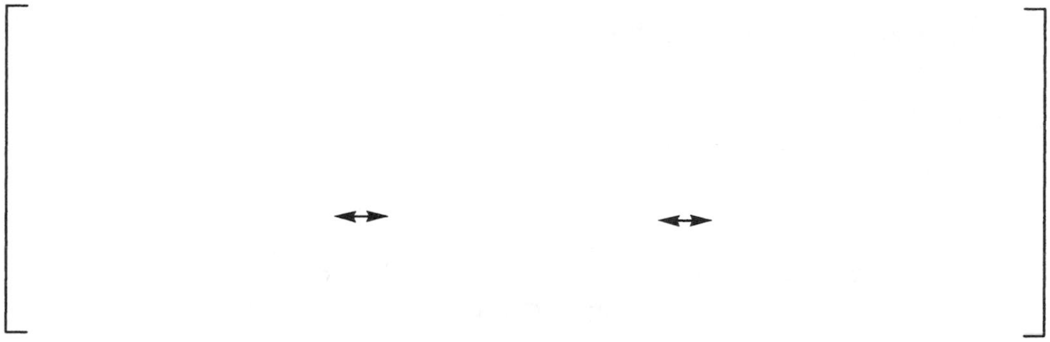

a) Based on these 2nd order resonance structures, do you expect the carboxylic acid group (COOH) to be a resonance donating group or a resonance withdrawing group?
b) Explain the following statement: The 2nd order resonance structures in part c) help explain why a carboxylic acid group **deactivates** the *ortho* and *para* positions toward electrophilic aromatic substitution (EAS).

Information

So far we have argued that 2nd order resonance structures can be used to predict the regiochemistry of an EAS reaction. The following questions ask you to consider the potential energies of intermediates on the EAS reaction pathways. As with all reactions with a small change in energy between reactant and product, it is the height of the activation barrier that determines which product will form. The potential energy of the intermediate is an excellent approximation (according to the Hammond Postulate) of the height of the activation barrier. Simply put...
The pathway with the most favorable carbocation intermediate will likely dominate.

20. Consider electrophilic aromatic substitution (EAS) performed on nitrobenzene.

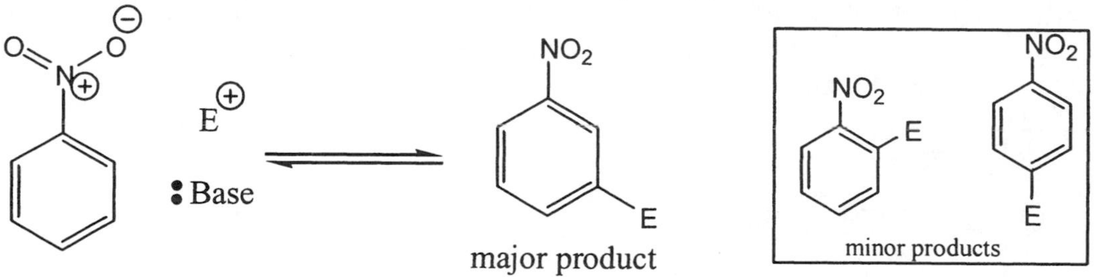

O N⊕ O⊖

$E^{\oplus}$

:Base

NO_2 E

major product

NO_2 E NO_2 E

minor products

a) How does the placement of E^+ on the nitrobenzene ring differ from an EAS reaction starting with toluene or aminobenzene (aniline)?

b) Draw the intermediate on the reaction pathway to the major (*meta*) product. (Be sure to include **all important resonance structures**.)

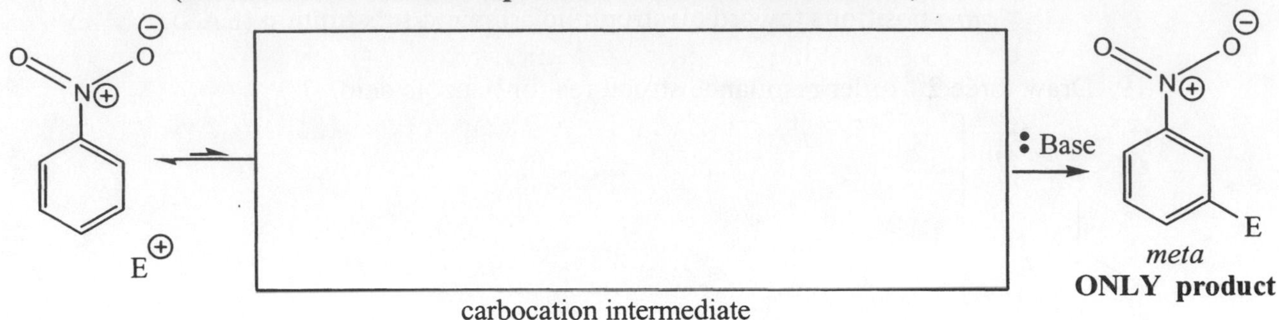

c) Draw the intermediate on the reaction pathway to the *para* product. (Be sure to include **all important resonance structures**.)

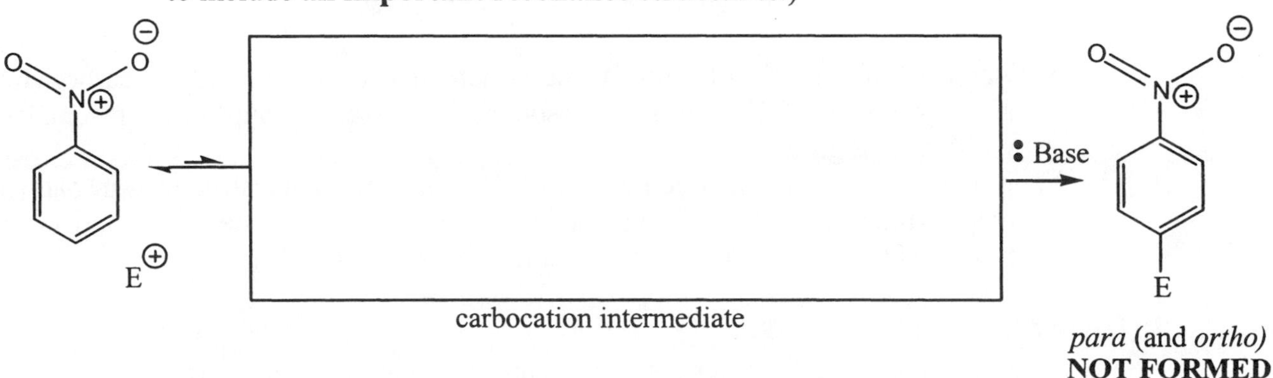

d) Any resonance structure in which two + charges are next to each other is very unfavorable. Circle all unfavorable resonance structures above.

e) Construct an explanation for why the intermediate on the reaction pathway to the *meta* product is lowest in potential energy.

f) Draw an energy diagram showing all three pathways (*ortho, meta* and *para).*

g) Construct an explanation for why the *meta* product is strongly favored in this reaction over the *para* (and *ortho*) product.

21. Aniline gives only *ortho* and *para* products in an EAS reaction.

a) Draw the intermediate on the reaction pathway to the *para* product. Be sure to draw all **FOUR** important resonance structures.

$\ddot{N}H_2$ $\quad$ $E^{\oplus}$ $\quad$ aniline $\quad$: Base $\quad$ $\ddot{N}H_2$ $\quad$ E

b) Draw the intermediate on the pathway to the *meta* product.

c) Construct an explanation for why the intermediate on the pathway to the *meta* product is higher in potential energy than the intermediate on the pathway to the *para* product.

d) Draw an energy diagram showing all three pathways (*ortho, meta* and *para).*

22. Aniline reacts with a given electrophile 100 times faster than toluene and 1000 faster than benzene.
    a) Explain why an EAS intermediate for aniline such as the one in part a) above is lower in potential energy than the intermediate you drew for *para*-substitution of toluene in Model 3.
    b) Construct an explanation for why aniline undergoes EAS much faster than toluene.

23. Read the assigned pages in your text and do the assigned problems.

Exercises for Part D

24. Consider the following reactions:

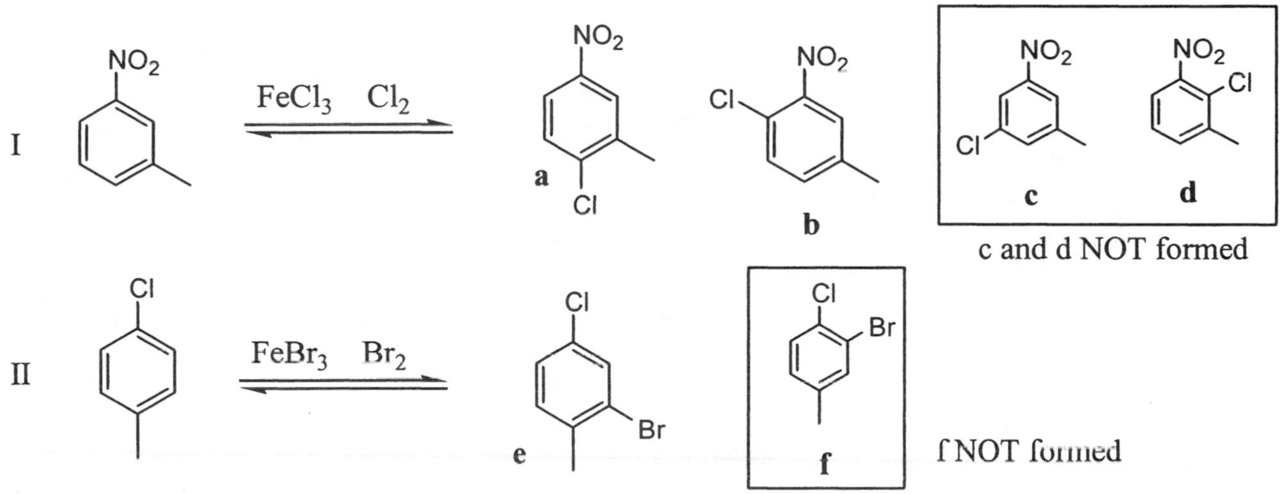

I $\xrightarrow{FeCl_3 \quad Cl_2}$ a, b (c, d)

c and d NOT formed

II $\xrightarrow{FeBr_3 \quad Br_2}$ e (f)

f NOT formed

a) Construct an explanation for why Product **c** is not formed in Rxn I.
b) Construct an explanation for why Product **d** is not formed in Rxn I.
c) Construct an explanation for why Product **f** is not formed in Rxn II.

25. In the reaction below, two major products are observed.
    a) Draw them and construct an explanation for why they are formed instead of the other two possibilities.
    b) Which of the two major products do you expect to dominate and why?

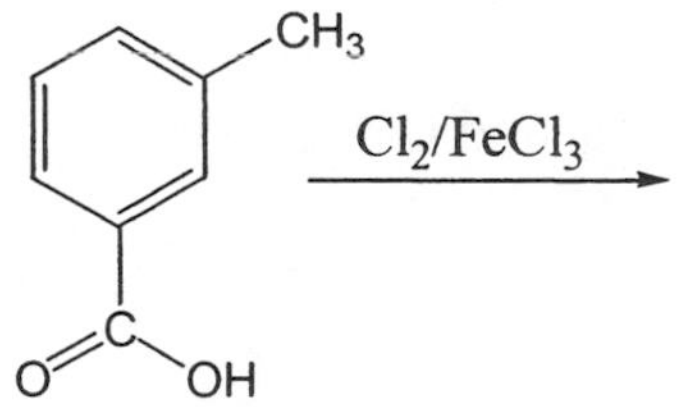

$\xrightarrow{Cl_2/FeCl_3}$

26. Mark each of the following statements True or False based on your current understanding. **(If false,** cite an example of a substituent for which it is false.)
    a) T or F: A strong activator overpowers the directing effects of a weak activator.
    b) T or F: A weak activator overpowers the directing effects of a deactivator.
    c) T or F: All activators are *o/p* directors.
    d) T or F: All deactivators are *m* directors.

27. Circle each of the following that help explain why EAS with fluorobenzene is slower than EAS with phenol (hydroxybenzene)?
    I. F is more electronegative than O, generating stronger inductive effects.
    II. F holds it lone pairs very tightly, making it a weaker pi donator than O.
    III. F has more lone pairs than O.
    IV. F is smaller in size than O.

28. Consider the following reactions:

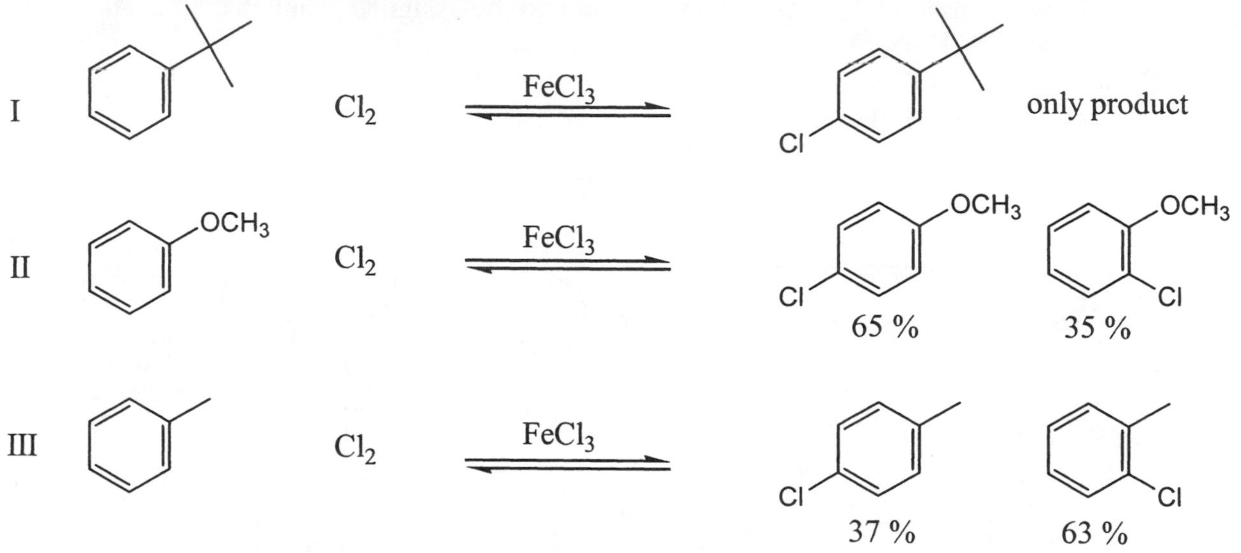

I   (tert-butylbenzene) Cl_2 $\xrightarrow{FeCl_3}$ (para-chloro product)   only product

II   (anisole, OCH_3) Cl_2 $\xrightarrow{FeCl_3}$ (para-Cl, OCH_3) 65 %   (ortho-Cl, OCH_3) 35 %

III   (toluene) Cl_2 $\xrightarrow{FeCl_3}$ (para-Cl) 37 %   (ortho-Cl) 63 %

a) Construct an explanation for why only the *para* product is observed in reaction I.
b) Fact: Neglecting steric effects, the expected ratio of *para:ortho* is 1:2 for reactions I, II and III. Construct an explanation for this fact.
c) Why is Reaction III closest to the expected 1:2 ratio?

29. According to the note at the end of Part D of this ChemActivity, which of the following starting materials will yield detectable products in a Friedel-Crafts alkylation or acylation.

benzene **bromobenzene** **phenol** **nitrobenzene**
***ortho*-chloroaniline** **2-bromo-6-chlorophenol**

Note: for the last two compounds the ring is net activated since NH_2 and OH are strong enough activators to overpower even two weak deactivators.

30. Read the assigned pages in your text and do the assigned problems.

Mini-Activity on Lewis Acid Catalysis

(What catalyst is needed to generate F^+ Cl^+ Br^+ I^+ and C^+ electrophiles?)

Information: Lewis Acid Catalysts

- Any molecule or atom that wants more electrons (lacks electrons) is a **Lewis acid**.
- Any molecule or atom that wants to share its electrons (excess e^-) is a **Lewis base**.
- A **catalyst** helps increase the reaction rate, but is not consumed in the reaction.

Figure A: Common Lewis Acids

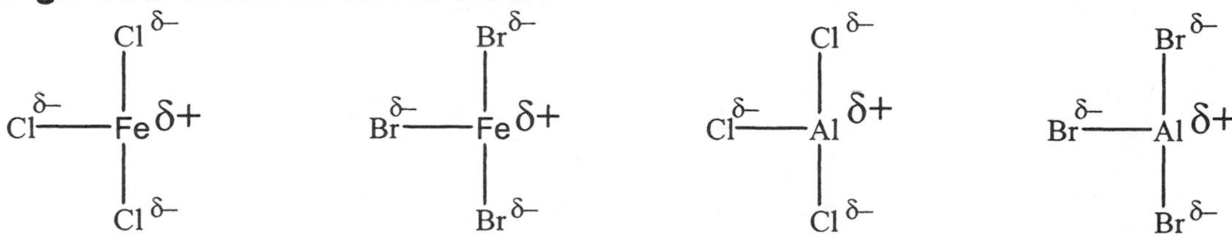

iron chloride iron bromide aluminum chloride aluminum bromide

In each molecule above, the electronegative halogens steal electron density from the central metal, leaving the Fe or Al with a lack of electrons and almost a full + charge.

Critical Thinking Questions

31. Circle the most Lewis acidic atom in each molecule in Figure A.

Information

- The electron cloud around bromine is normally symmetrical (below, left).
- A passing molecule with a dipole moment will polarize this cloud (below, middle)
- An even stronger effect is generated when a Lewis acid catalyst such as $FeBr_3$ binds to one of the Br atoms of Br_2 (below, right)

Outline of Undisturbed Electron Cloud of Br_2

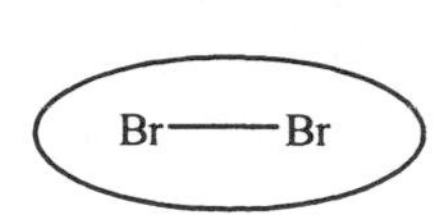

Outline of Electron Cloud of Br_2 near transient + charge disturbance

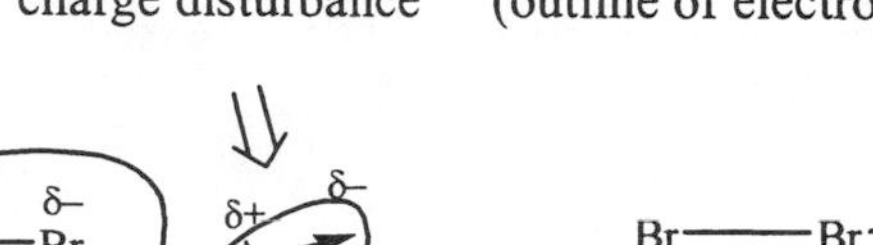

passing molecule with a small dipole

Br_2 with partial bond to $FeBr_3$ (outline of electron cloud not shown)

Br—Br- - - -Fe(Br)(Br)—Br

32. Consider the picture above, right, showing Br_2 bound to $FeBr_3$.
    a) Add to this drawing a depiction of the shape of the electron cloud of Br_2 when it is bound to $FeBr_3$ (above right). Don't include the electron cloud of $FeBr_3$ in your drawing.

    b) Add δ+ and δ– where appropriate to the Br_2 portion of this complex (above right).

33. Consider the reactions below:

I. 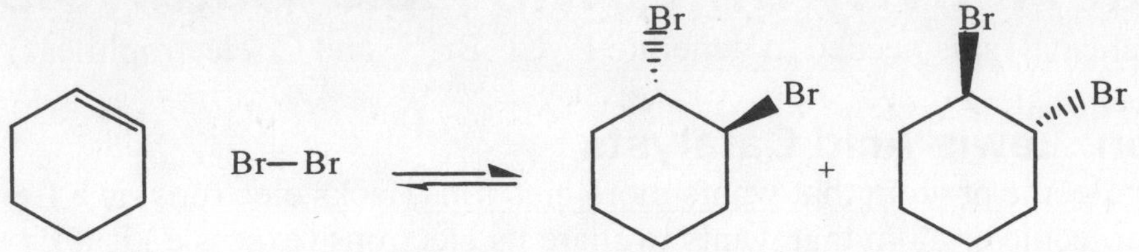 cyclohexene + Br—Br ⇌ Br, Br (trans-1,2-dibromocyclohexane) + Br, Br

II.  benzene + Br—Br ⇌ **NO REACTION**

III. 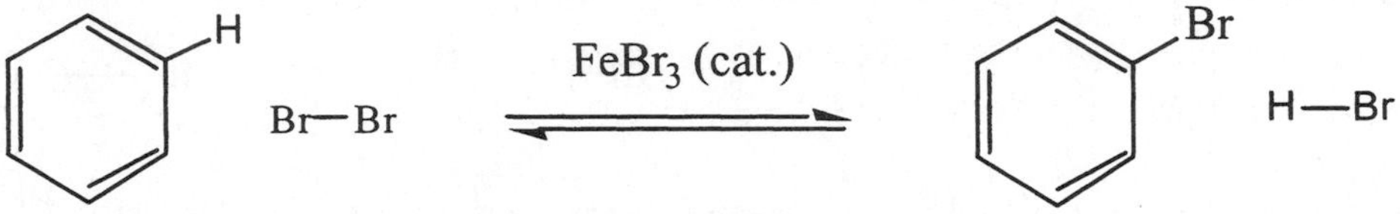 benzene (H) + Br—Br ⇌ (over arrow: $FeBr_3$ (cat.)) bromobenzene (Br) + H—Br

a) Why does Br_2 react with cyclohexene but not with benzene? (*see* Rxns I & II, above)

b) Construct a reasonable mechanism for Reaction III that shows the role of the catalyst. For simplicity, in your mechanism show…

Br----Br---$FeBr_3$	as	⊕Br and Br—⊖$FeBr_3$

Note: in the last step $FeBr_4^-$ acts as a base, generating $FeBr_4H$, which decomposes into HBr and $FeBr_3$, regenerating the Lewis acid catalyst (as shown below).

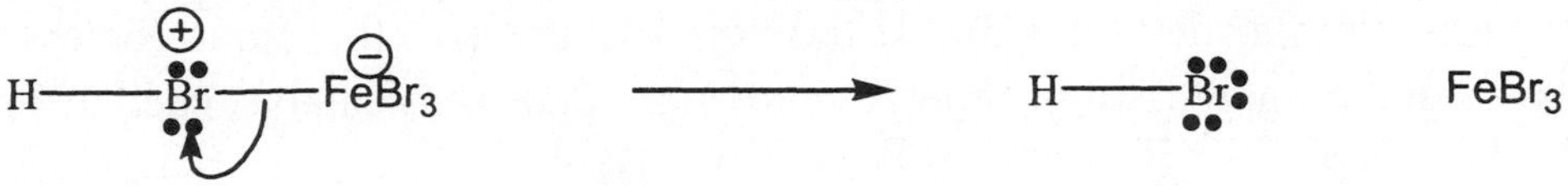

H—⊕Br—⊖$FeBr_3$ ⟶ H—Br $FeBr_3$

Information: Friedel-Crafts Alkylation and Acylation

- One of the most difficult and important objectives in organic synthesis is the formation of new carbon-carbon sigma bonds.
- Electrophilic aromatic substitution (EAS) is one of the few ways to do this.
- As with the EAS bromination in Model 4, a Lewis acid catalyst (usually $AlCl_3$) is required to make a sufficiently strong electrophile (*see* $H_3C^{\delta+}$ below left).
- For simplicity, you can think of this methyl electrophile as a methyl carbocation (*see* below right), though 1° and methyl carbocations <u>do not</u> exist.

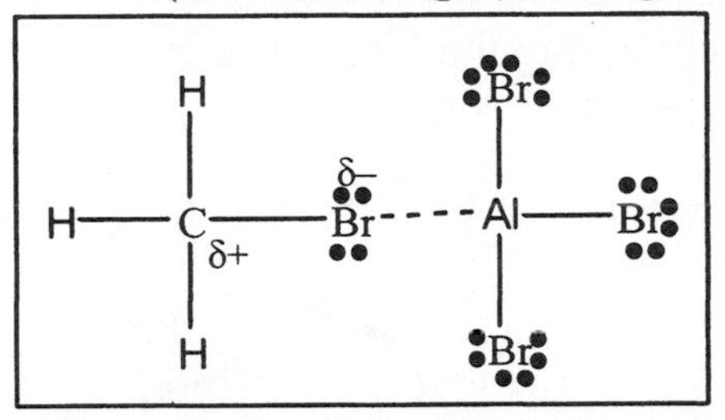

≈

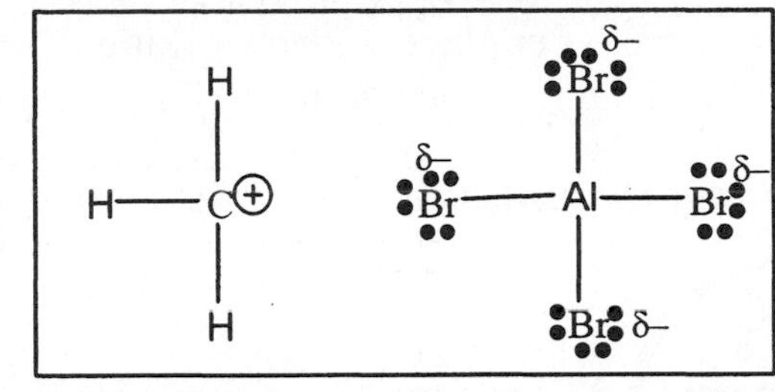

Note: Al should have a formal - charge but this representation is more accurate.

This technique for making carbon-carbon sigma bonds was discovered accidentally by Charles Friedel (1832-1899) and James Crafts (1839-1917). The two were carrying out reactions involving $AlCl_3$ using benzene as a solvent (which they erroneously thought be totally inert). Friedel-Crafts Reactions work with two different types of carbon groups, below designated as R.

$X_3Al^{-}\cdots X^{\delta-}$—$R^{\delta+}$

X = Cl or Br

R = alkyl or acyl group

R = alkyl group	R = acyl group
–CH_3 methyl; –CH_2CH_3 ethyl; –CH(CH_3)CH_3 *isopropyl*; etc.	–C(=O)–CH_3 ethanoyl (acetyl); –C(=O)–CH_2CH_3 propanoyl; etc.

34. What reagents would you use to carry out the following reactions?

ChemActivity 30

Synthesis Workshop I

(What is retrosynthesis? How can I change an *o, p* director into a *m* director?)

Model 1: Starting From Both Ends (Synthesis & Retrosynthesis)

- Two teams of engineers, one working from each end, dug the tunnel under the English Channel (the Chunnel). They met in the middle.

Figure 1: The Chunnel

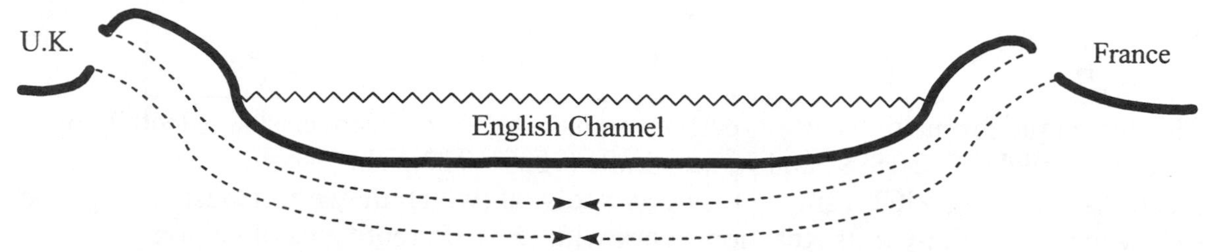

- Use the same strategy when designing an organic synthesis. Start from the starting material and work forward **AND think backward from the target.**
- Organic chemists call this "thinking backwards" **retrosynthesis**.
- The target molecule below can be made in two steps from benzene. One possible synthetic pathway is shown below.
- Forward steps are shown with regular reaction arrows, retrosynthetic steps are shown with "retro" arrows.

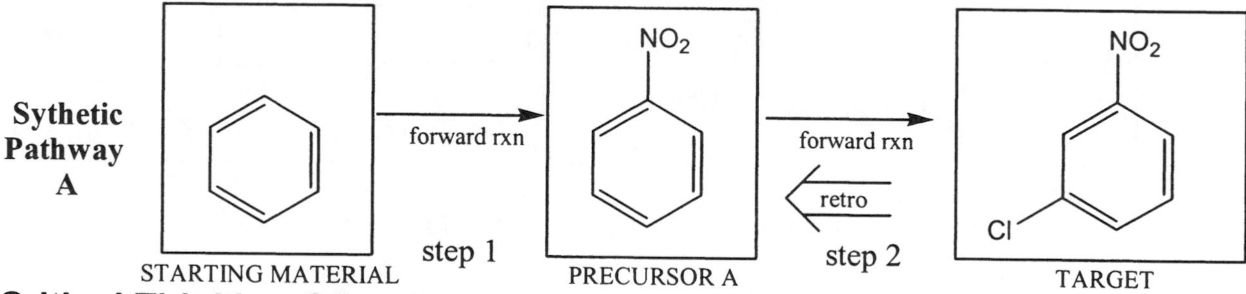

Critical Thinking Questions

1. Over the reaction arrow in Step 1, write the reagents necessary to accomplish this forward step.

2. This synthesis is so simple that we could just keep going forward, but for practice, lets go backwards from the TARGET. **Retrosynthesis is like pulling apart a model of the TARGET**, and it is very useful with long syntheses.
   a) Put an X through the bond you would break if you were pulling apart a model of the TARGET to get back to PRECURSOR A.
   b) Over the reaction arrow in Step 2, write the reagents necessary to accomplish this forward step.

3. Normally, in between products are not supplied, so during retrosynthesis you have to choose which group to pull off. In this case there were two groups to choose from. Draw the precursor that would result if you pulled off the nitro group.

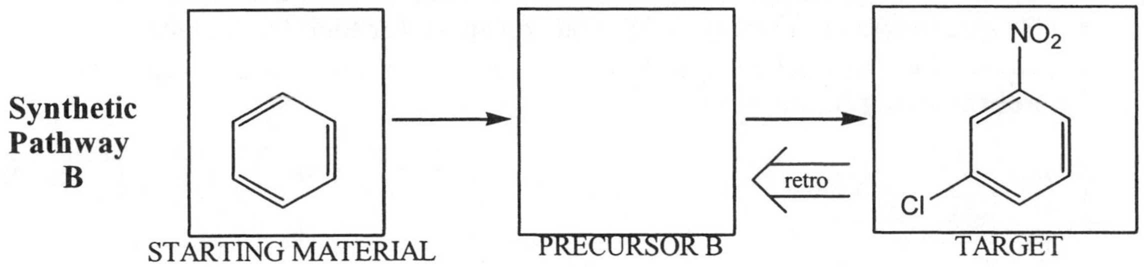

4. Explain why Pathway A, on the previous page, is much better than Pathway B. Hint: consider the reagents (if any) that would accomplish the forward steps in B.

5. A student proposes the following step as part of a synthesis. Explain why it will not work.

NH_2 $\xrightarrow{Cl_2/FeCl_3}$ NH_2, Cl

Model 2: Changing an *o,p* to a *m* Director, and Vice Versa

NO_2 — Zn(Hg), HCl (reduction) or H_2, Pd metal catalyst (reduction) → NH_2

NH_2 — CF_3COOOH (oxidation) → NO_2

O, R — Zn(Hg), HCl (reduction) or H_2, Pd metal catalyst (reduction) → H, H, R

H, H, R — CrO_3, H_2SO_4, H_2O (oxidation) → O, R

Critical Thinking Questions

6. Label the group on each of the four benzene derivatives in Model 2 as "*o, p* director" or "*m* director," as appropriate.

7. Design an efficient synthetic pathway to replace the student's faulty step in CTQ 5.
   - This will require three synthetic steps.
   - Use a combination of forward synthesis and retrosynthesis.
   - Be sure to write reagents over the reaction arrow for each forward step.
   - **Hint: The first backward step involves changing one of the attached groups, not removing it.**

NH_2 → → ⇐ retro → NH_2, Cl

starting material — target

Model 3: Protecting Groups

A sulfone group can be easily added or removed. Do not confuse a reverse reaction with retrosynthesis, which is only a thought experiment. This reverse reaction is real.

SO_3H

SO_3, H_2SO_4 (forward) ; H_3O^{+} H_2O (reverse)

Critical Thinking Questions

8. If you treat the starting material below with H_2SO_4/HNO_3, you get much more *para* product than *ortho* product. Explain why *para* is favored so much over *ortho* in this case.

H_2SO_4/HNO_3 → NO_2 *para* 95% ; NO_2 *ortho* $<5\%$

9. Propose a synthesis that **gives mostly *ortho* product**.
   - Be sure to include reagents over the forward reaction arrows.
   - **Hint: the last step involves removing a sulfone protecting group.** This means the last step, backwards, would involve sticking a sulfone group on the target.

Model 4: Other Useful Oxidation and Reduction Reactions

R = alkyl group

$\xrightarrow{KMnO_4}$

benzoic acid
(carboxylic acid)

Doesn't work if no benzyl H. For example...

$H_3C-C(CH_3)-CH_3$ $\xrightarrow{KMnO_4}$ No Rxn

Y = any group

H_2/Pt at 2000psi
or
H_2/Rh (Rhodium metal)

Critical Thinking Questions

10. Which reaction in Model 4 is an oxidation and which reaction is a reduction.

11. For the reduction reaction, are the conditions very mild or very harsh. Explain your reasoning.

12. If it can, the group Y will also react with H_2/Rh.
    a) Give an example of a benzene ring with a group, Y, in which Y will also react with H_2/Rh metal.
    b) Give an example of a benzene ring with a group, Y, in which Y will NOT react with H_2/Rh metal.

The reactions in this activity are summarized on reaction sheet R6.

Exercises

1. Propose a solution to the following synthetic problem. You may use retrosynthesis if you like. Be sure to include reagents over the forward reaction arrows. Hint: see CTQ 7.

starting material

Cl

target

2. Propose a solution to the following synthetic problem that **does not produce large amounts of isomers of the target**. You may use retrosynthesis if you like. Be sure to include reagents over the forward reaction arrows. (Hint: this synthesis will take at least 6 steps!)

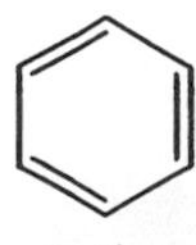

starting material

NO_2 NH_2

target

3. Propose a solution to the following synthetic problem that **does not produce large amounts of isomers of the target**. You may use retrosynthesis if you like. Be sure to include reagents over the forward reaction arrows.

OCH_3

starting material

OCH_3 H_2N NH_2 Br

target

4. Propose a solution to the following synthetic problem that **does not produce large amounts of isomers of the target**. You may use retrosynthesis if you like. Be sure to include reagents over the forward reaction arrows.

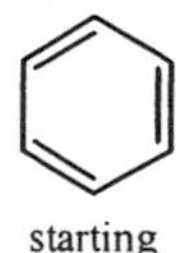

starting material

NO_2 Cl Cl Br

target

5. Read the assigned pages in your text and do the assigned problems.

6. Make up a synthesis problem using reactions from Reaction Sheet R6. If it is a good synthesis, email it to your instructor (write a description of it in words) and there is a chance it will show up on the next quiz.

ChemActivity 31

Part A: Nucleophilic Addition to C=O

(What is the mechanism by which a aldehyde or ketone can be converted to an alcohol?)

Review

nucleophile = "nucleus (+) lover" **nucleophilic** = "nucleus (+) loving"
electrophile = "electron (–) lover" **electrophilic** = "electron (–) loving"

Model 1: Nucleophilic Addition with a Very Strong Nucleophile

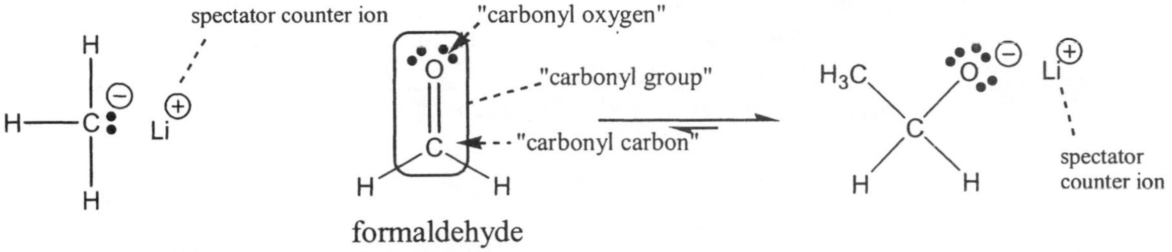

spectator counter ion
"carbonyl oxygen"
"carbonyl group"
"carbonyl carbon"
formaldehyde
spectator counter ion

Critical Thinking Questions

1. Draw a δ+ and a δ- on appropriate atoms of formaldehyde to show the polarity of the C=O bond.

2. Use curved arrows to show a mechanism for the reaction in Model 1.

3. Draw the alcohol product that results when an aldehyde or ketone is
   i) treated with a Grignard or Lithium reagent (strong nucleophile)
   ii) then neutralized with dilute acid

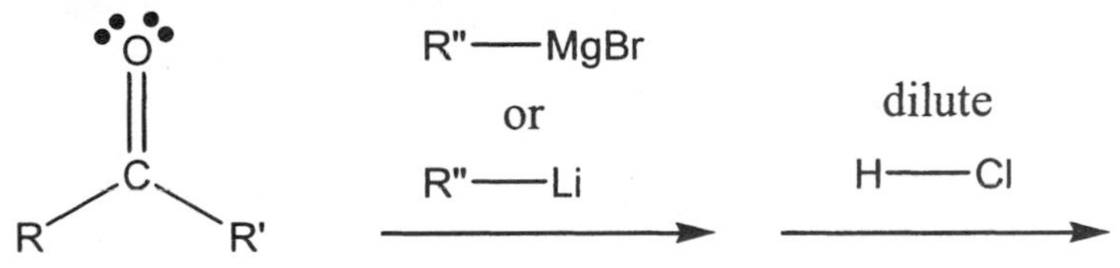

R"—MgBr or R"—Li → dilute H—Cl →

(aldehyde when R or R' = H)

4. According to the following factors, which do you expect to be more susceptible to reaction with a nucleophile: the carbonyl carbon of an aldehyde or a ketone?
   i. Based on sterics, which is more reactive: **aldehyde C** <u>or</u> **ketone C**?
   ii. Based on electronics, which is more reactive: **aldehyde C** <u>or</u> **ketone C**? (Hint: recall that alkyl groups are electron donating.)

example of an aldehyde (propanal) $H_3CH_2C-C(=O)-H$

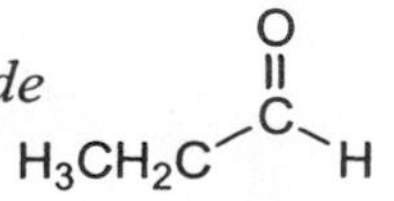

example of an ketone (2-butanone) $H_3C-C(=O)-CH_2CH_3$

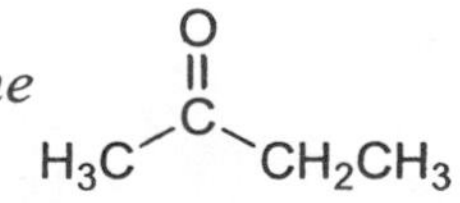

Model 2: Nucleophilic Addition with a Weak Nucleophile

H_3C H aldehyde H H H (catalyst) H H H_3C H H hydrate

Under normal conditions, most aldeydes and ketones are stable in water. However, when a small amount of acid is added as a catalyst, a di-alcohol product called a **hydrate** forms.

Critical Thinking Questions

5. Use curved arrows to construct a mechanism for the reaction in Model 2 such that **no intermediate has a negative charge on H, C or O**.
   Hint: the acid catalyst is involved in the first step.

6. The first intermediate in the reaction above has two resonance forms. Draw the other resonance structure of this intermediate.

Model 3A: Singles Dance *circa* 1930

The singles dance where my grandparents met provides a model for understanding how a system at equilibrium responds to stresses like adding more reactants, or taking away products. There were about 300 singles at the Synagogue that evening: roughly 200 men and 100 women. (Gramps beat the odds!) Because of teen shyness and limited space on the dance floor, by eight o'clock only 50 male-female dance couples had formed.

Starting Conditions

200 single men 100 single women

After the system has come to equilibrium

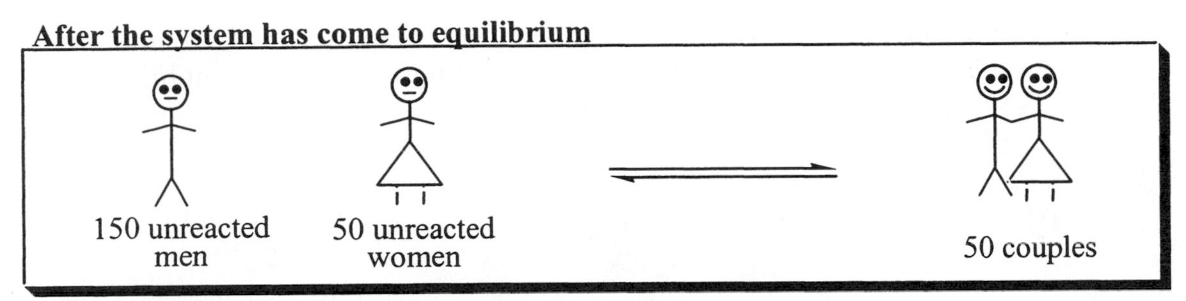

Critical Thinking Questions

7. At 9:00pm, a busload of 100 more single men arrived, and soon after there were 70 male-female couples crammed onto the dance floor. Construct an explanation for this new equilibrium position.

8. How many male-female couples do you expect on the dance floor if 1000 single men showed up (instead of 100 single men)?

9. Did the addition of men cause the reaction (shown with arrows in Model 3) to go forward (from left to right) or go backwards (right to left).

Model 3B: Le Chatelier's Principle

French chemist Henri Louis Le Chatelier (1850-1936) noticed that **a system in equilibrium responds to stress by trying to move back toward equilibrium**.

Critical Thinking Questions

10. Applying Le Chatelier's principle to the singles dance in Model 3A, what would you expect to happen if (instead of any men arriving) a bunch of couples left at 9pm—creating space on the dance floor.

11. Imagine a hypothetical chemical reaction in equilibrium:

$$A + B \rightleftharpoons C + D$$

For each stress, state whether the reaction will go **forward** or **backward** or **neither**.

a. add more A.
b. add more D
c. take away some D
d. take away some B
e. add more A and more B
f. add more A and take away a similar amount of D
g. add more C and add a similar amount of B

12. Consider an energy diagram for the overall reaction from Model 2 (shown below). Because the reaction is neither up-hill nor down-hill in energy, at equilibrium, you would expect a 1:1 ratio of aldehyde to hydrate.

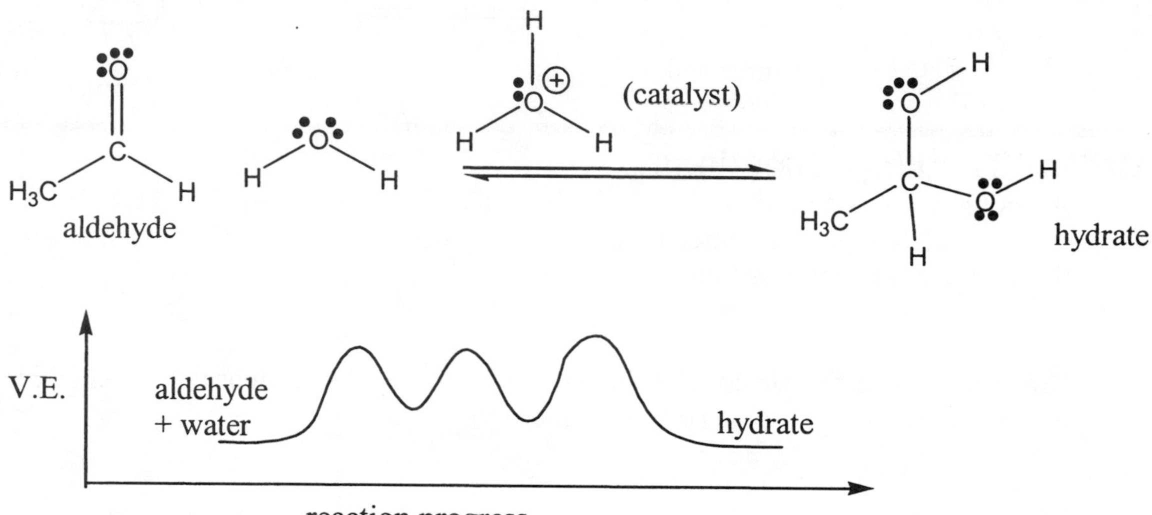

According to Le Chatelier's Principle, we can control the final equilibrium ratio of aldehyde to hydrate by…

adding water or **removing water**.

Which of these actions will shift the equilibrium toward more aldehyde and which will shift the equilibrium toward more hydrate? Explain.

Part B: Nucleophilic Addition-Elimination

(What is the mechanism of imine formation?)

Model 4: Nucleophilic Addition-Elimination

Grignard Reagent MgBr−CH_3 ≅ $^{\oplus}$MgBr $^{\ominus}$:CH_3

Step 1

Step 2

spectator counter ion

acid chloride

Critical Thinking Questions

13. Use curved arrows to show the mechanism of Step 2 of the reaction in Model 4.

14. Do you expect this reaction to be more favorable from **Left to Right (forward)** or from **Right to Left (reverse)** [circle one]? Hint: consider the leaving group that is liberated in each direction.

15. The reaction below is similar to the reaction in Model 4, except that Step 2 does not occur.

Step 1

Step 2

THESE PRODUCTS DO NOT FORM!

Construct an explanation for why Step 2 is very unfavorable in this case? Hint: is H^- a good leaving group?

16. Is the following overall transformation up-hill or down-hill in terms of potential energy? (Hint: consider in which direction does the better leaving group get eliminated?)

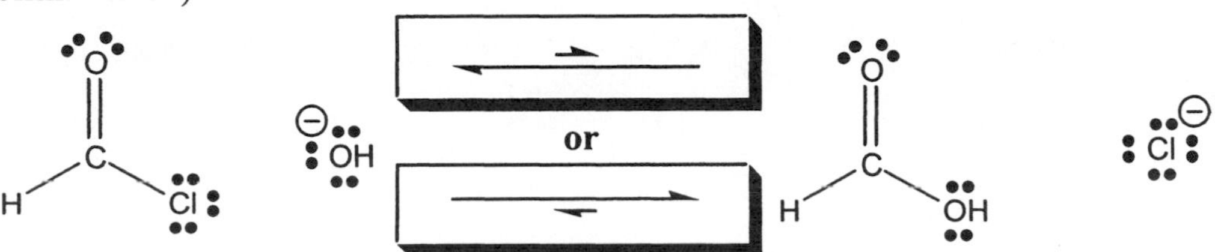

Explain your reasoning.

Model 5: Water as a Leaving Group in Addition-Elimination

- Recall that an OH group can be protonated with acid to make it into an excellent leaving group = OH_2 (water!).

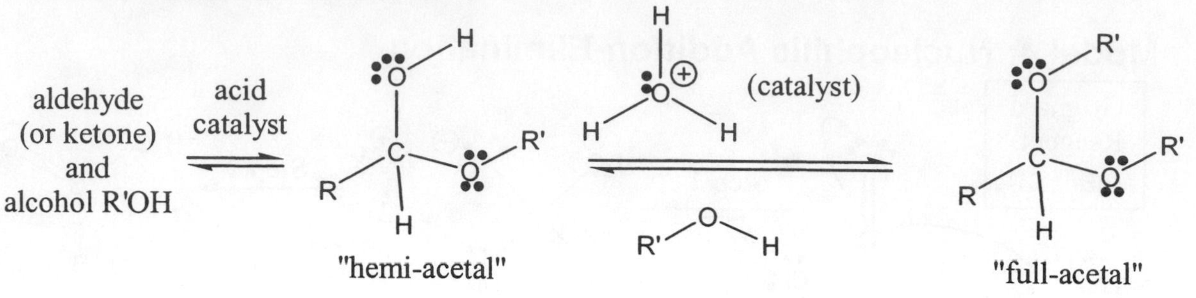

aldehyde (or ketone) and alcohol R'OH ⇌ (acid catalyst) "hemi-acetal" ⇌ (H_3O^+ (catalyst), R'OH) "full-acetal"

Critical Thinking Questions

17. A hemi-acetal can be converted to an acetal (also called a "full-acetal") in the presence of excess alcohol and acid catalyst. Show the mechanism of this reaction. Hint: acid is used in the first step to change the OH group into a good leaving group.

18. Use curved arrows to construct a mechanism for the following reaction such that **<u>no intermediate has a negative charge on H, C, O or N</u>**. (Hints: this mechanism requires at least five steps. One of these steps is the elimination of water as a leaving group. All reagents needed are shown.)

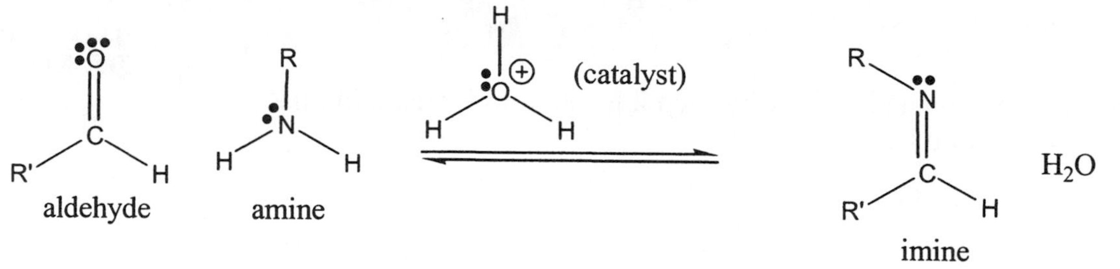

aldehyde + amine ⇌ (H_3O^+ (catalyst)) imine + H_2O

19. If you had an imine and you wanted to change it into an aldehyde what reagents could you add to the imine and in what amounts?

Exercises for Part A:

1. Construct a mechanism for the following reaction.

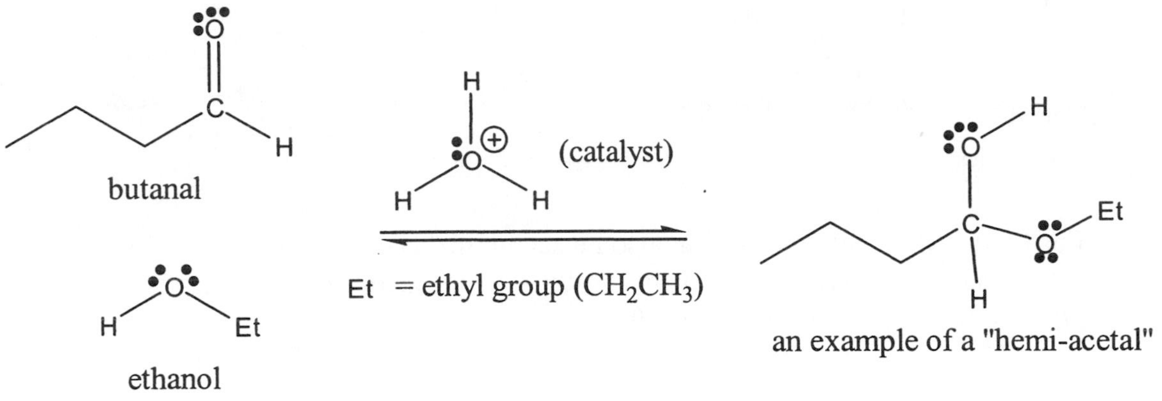

butanal

ethanol

(catalyst)

Et = ethyl group (CH_2CH_3)

an example of a "hemi-acetal"

What could you do to push the reaction toward a high yield of hemi-acetal?

2. The reagent Sodium Borohydride ($NaBH_4$) reacts exactly the way you would expect H^- to react (H^- is a very strong nucleophile). Given this, predict the product that results when butanal is treated with $NaBH_4$ followed by dilute HCl.

3. What reagents could you use to convert this alcohol back to butanal. Hint: this reaction is NOT reversible. Review your reaction sheet labeled oxidation of alchols.

4. Most strong nucleophiles will react with any aldehyde (and most ketones). Strong necleophiles we have discussed in this class include:

Grignard and Lithium reagents (essentially alkane anions, R_3C^-)
The conjugate base of a terminal alkyne (RCC^-)
Hydride ion, H^- (produced by $LiAlH_4$ or $NaBH_4$)
Hydroxide ion, HO^- (NaOH or KOH)
Cyanide ion, NC^- (NaCN or KCN)
Thiolate ion, RS^- (e.g. $NaSCH_3$)

Draw the product of a reaction between cyclohexanone and a representative from each category of nucleophiles above.

5. Construct mechanisms for the following reaction.

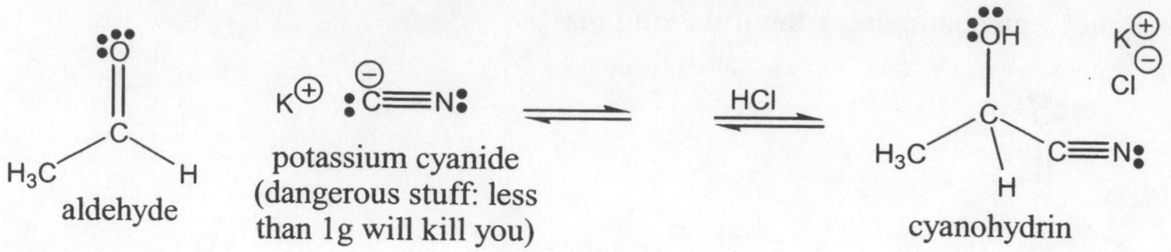

H_3C–CHO (aldehyde) + $K^{\oplus}$ $^{\ominus}$:C≡N: (potassium cyanide (dangerous stuff: less than 1g will kill you)) ⇌ (HCl) ⇌ H_3C–CH(OH)–C≡N: (cyanohydrin) + $K^{\oplus}$ $Cl^{\ominus}$

aldehyde

potassium cyanide
(dangerous stuff: less
than 1g will kill you)

HCl

cyanohydrin

6. Formation of a cyanohydrin can also be accomplished using HCN (cyanide gas). Construct a reasonable mechanism for this reaction. Hint: the first step is protonation of the carbonyl oxygen.

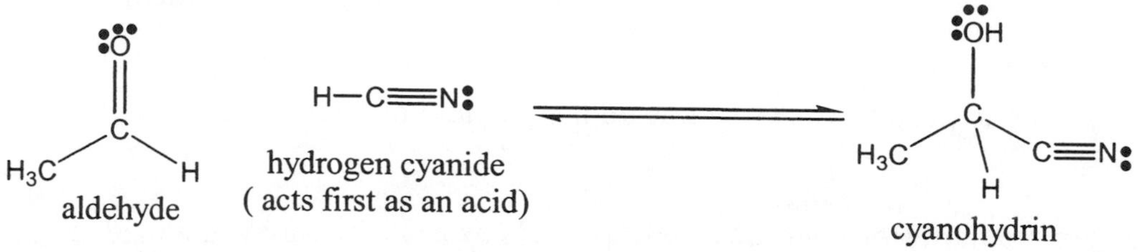

H_3C–CHO (aldehyde) + H–C≡N: ⇌ H_3C–CH(OH)–C≡N: (cyanohydrin)

aldehyde

hydrogen cyanide
(acts first as an acid)

cyanohydrin

7. Show the mechanism by which the following hemi-acctal can be converted into an aldehyde, and draw the resulting aldehyde.

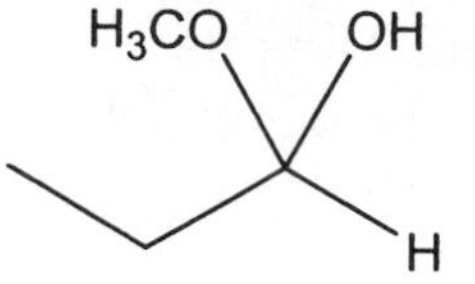

H_3CO OH

H

8. You are expected to know the following about the NMR and IR of aldehydes and ketones.

   i. In H^1 NMR, an aldehyde H shows up near 10ppm (to the far left).
   ii. In C^{13} NMR a carbonyl C shows up as a small peak above 200ppm.
   iii. Aldehydes and ketones show a C=O stretch absorption band on an IR spectrum between 1660-1770cm^{-1}. The exact location of the band is determined by conjugation as follows:
      a. C=O not conjugated, then band at high end ~1730cm^{-1}.
      b. C=O conjugated to one C=C, then band near 1705cm^{-1}.
      c. C=O conjugated to benzene ring, then band at low end ~1685cm^{-1}.

9. Read the assigned pages in your text and do the assigned problems.

10. Complete Nomenclature Worksheet II, Part B.

Exercises for Part B

11. Most nucleophiles will react with an acid chloride, even a relatively weak nucleophile such as water (see part a, below). **Each of these acid-chloride reactions takes place in at least TWO STEPS!**

   a) Use curved arrows to show the mechanism of the following reaction.

acid chloride + HOH ⇌ carboxylic acid + Cl–H

   b) Use curved arrows to construct a mechanism for the following reaction.

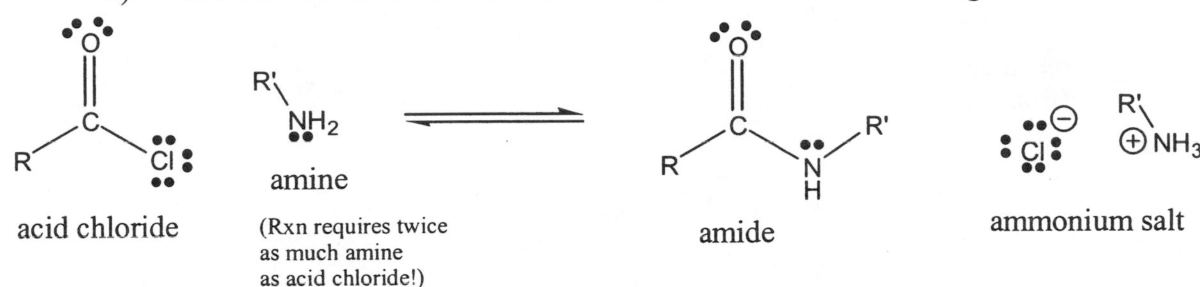

   c) Construct an explanation for why you need twice as much amine as acid chloride in the reaction in part b. (Hint: an amine is a good base.)

   d) Draw the most likely products of the following reaction.

acid chloride + alcohol (R'–OH) ⇌

12. Design a mechanism for each of the following reactions

a) acetone (an aldehyde or ketone) + CH_3OH (methanol) $\xrightleftharpoons{\text{HCl (cat.)}}$ hemi-acetal (unstable)

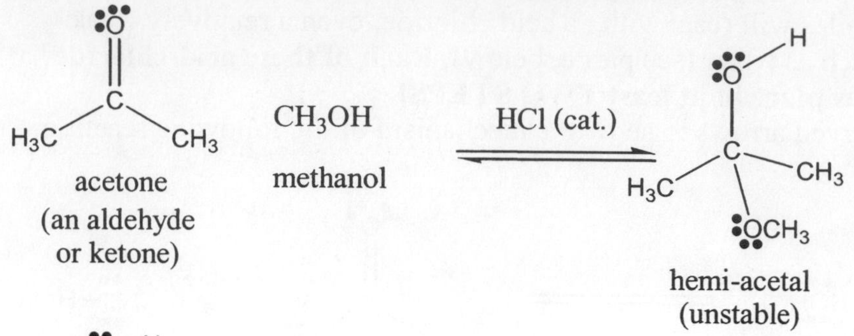

b) hemi-acetal (from part a) + excess CH_3OH (methanol) $\xrightleftharpoons{\text{HCl (cat.)}}$ acetal (stable) + H_2O

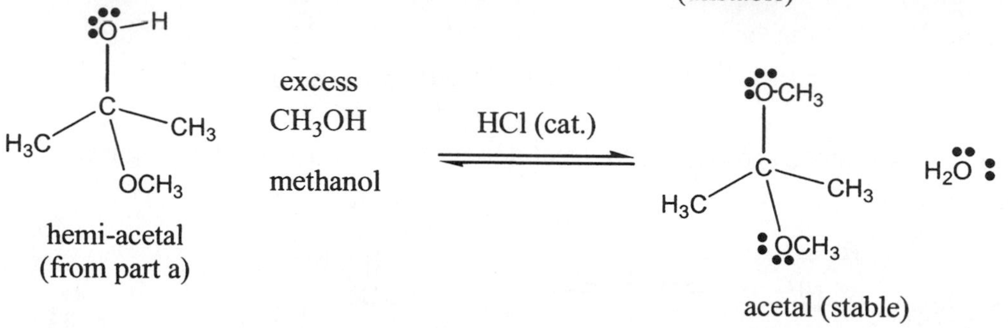

c) (an aldehyde or ketone with an alcohol 3 or 4 carbons away) $\xrightleftharpoons{\text{HCl (cat.)}}$ cyclic hemi-acetal (stable)

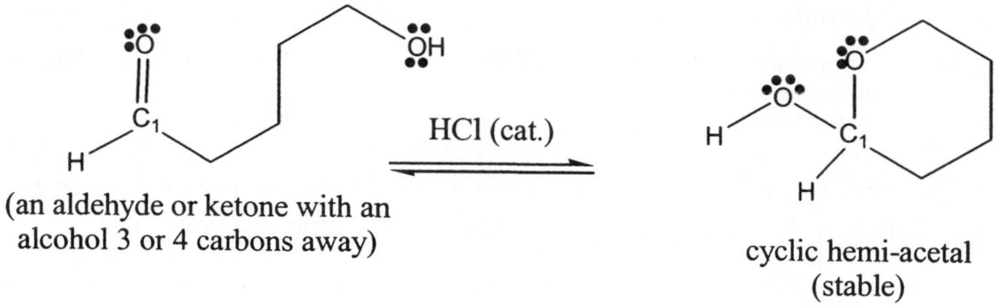

d) H_3C–C(=O)–CH_3 + HO–CH_2CH_2–OH $\xrightleftharpoons{\text{HCl (cat.)}}$ acetal (stable) + H_2O

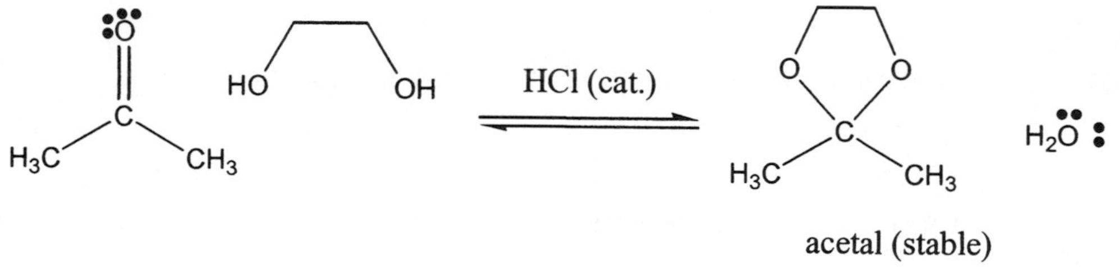

13. Read the assigned pages in your text and do the assigned problems.

ChemActivity 32

Protecting Groups

(What is a protecting group and how can an acetal be used as a protecting group?)

Model 1: Mono-Dentate ("One-Toothed") Attachment

- Imagine black and white balls bouncing around in a box with no gravity or air.
- Each black ball can bind to exactly one white ball (if the two collide).
- They will stay bound together for an average of 10-60 seconds, then separate.

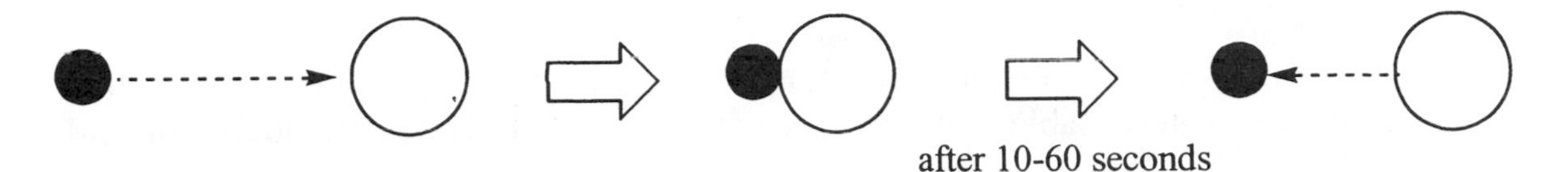

Critical Thinking Questions

1. If we tether two black balls together with a short cord, once one is bound to the white ball, the other has a very high probability of binding soon after. Explain.

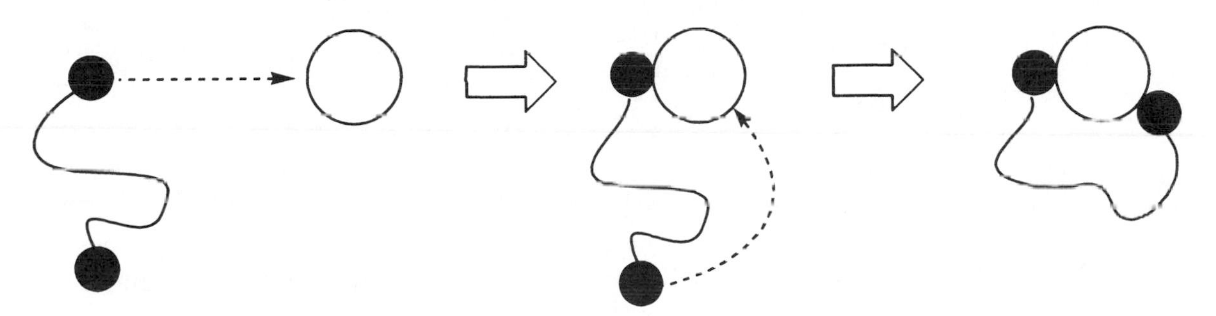

2. The system on the left is much more stable than the system on the right. Explain. (Hint: think about what happens in each case when one of the black balls "lets go" from the white ball.)

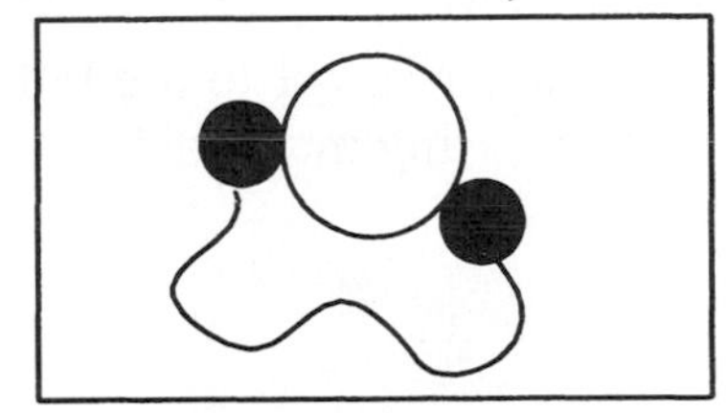

more stable
(3 balls tend to stay in the same vicinity)

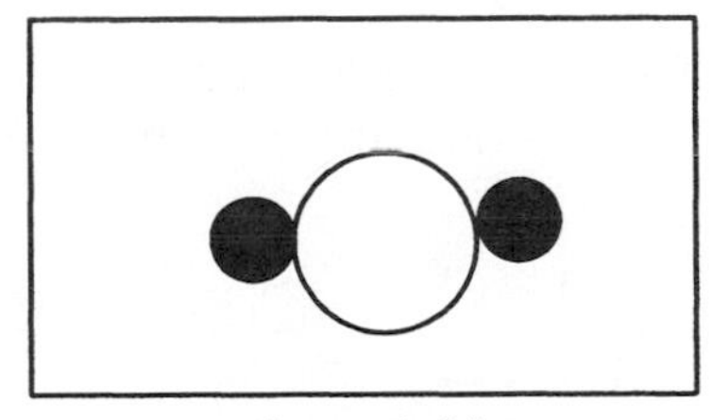

less stable
(3 balls will soon go their separate ways)

3. According to Model 1, each solo black ball is "mono-dentate," whereas the tethered system is "bi-dentate." Explain these terms. (Think about "biting" the white ball.)

Model 2: Cyclic Acetals

O

R R HO OH

H—Cl
(catalyst)
Rxn 1

O O H H
O
R R

R = alkyl group or H
(aldehyde or ketone)

Rxn 2 most any nucleophile

NO REACTION

- Acetals are stable in the presence of most bases and nucleophiles (without acid).
- Acetals are NOT stable in the presence of even a catalytic (small) amount of acid.

Critical Thinking Questions

4. Explain why a cyclic acetal (above) **forms more readily** and is **more stable** than a normal, non-cyclic acetal (shown below).

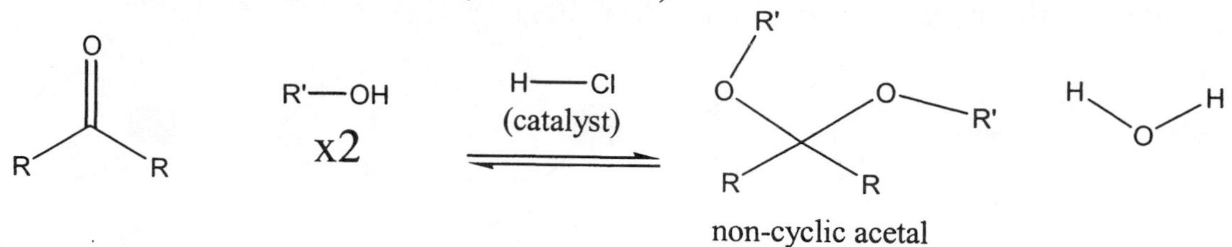

5. Assume Rxn 1 in Model 2 is neither up-hill nor down-hill in terms of potential energy.
   a) If you were given an aldehyde or ketone, how could you drive Rxn 1 to the right, producing a high yield of the cyclic acetal product?

   b) If you were given a cyclic acetal, how could you drive Rxn 1 to the left, producing a high yield of the aldehyde or ketone starting material?

6. According to the information in Model 2, which is more susceptible to reaction with a good nucleophile: an **aldehyde/ketone** or a **cyclic acetal** [circle one]?

Model 3: Reactivities of Aldehydes and Ketones

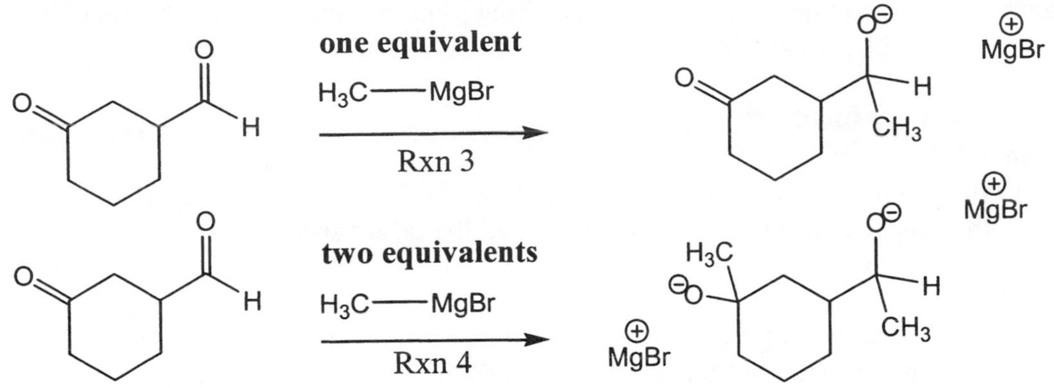
one equivalent
H_3C—MgBr
Rxn 3
two equivalents
H_3C—MgBr
Rxn 4
MgBr
CH_3
H
H_3C

Critical Thinking Questions

7. Based on the reactions in Model 3, which is <u>**more reactive**</u> toward nucleophilic addition: an **aldehyde carbonyl** <u>or</u> a **ketone carbonyl** [circle one]?

8. Is your answer above consistent with both steric and electronic arguments?

9. Will a diol such as the one in Model 2 react to form a cyclic acetal more readily with an **aldehyde** <u>or</u> a **ketone** [circle one]?

10. Explain why the following simple synthesis of the target shown will not work.

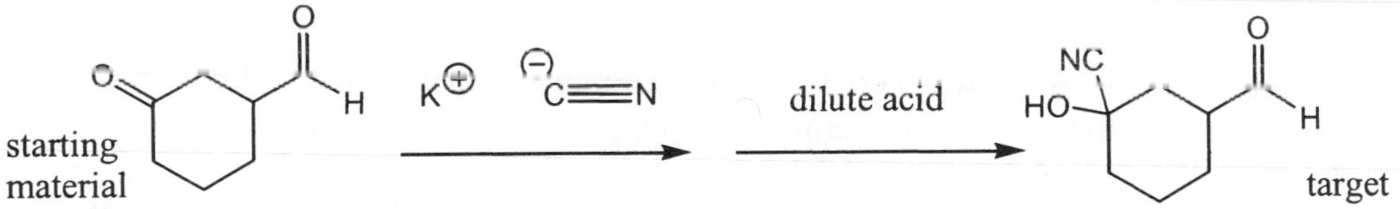
starting material
K
C≡N
dilute acid
NC
HO
H
target

11. Shown below is a successful synthesis of this same target molecule. Explain what is accomplished by each step in the synthesis. (In particular, what is the purpose of introducing the diol in the first step? Hint: see CTQ 6.)

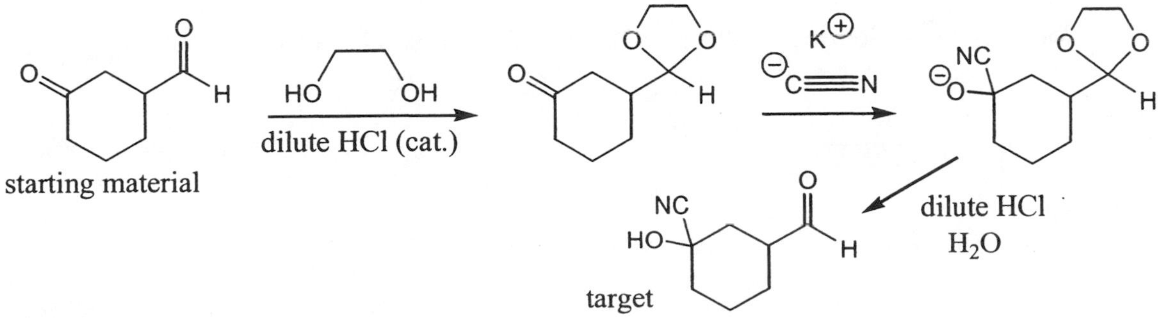
starting material
HO
OH
dilute HCl (cat.)
K
C≡N
NC
dilute HCl
H_2O
target

Model 4: Protecting Groups

A protecting group is a **removable group that prevents an unwanted reaction** in one part of the molecule while desired chemistry is taking place on another part. At the end of the desired reaction, the protecting group is removed.

Critical Thinking Questions

12. Consider the following synthesis.
    a) Explain why a shorter synthesis consisting of treatment of the starting material with sulfuric acid/nitric acid will not yield the target shown.

    b) Is there a protecting group being used in this synthesis? If so, what is its identity and what reaction is it preventing?

SO_3 → SO_3H ; H_2SO_4, HNO_3 → NO_2, SO_3H ; 1) $H_3O^{\oplus}$ 2) $HO^{\ominus}$ → NO_2

target

Exercises

1. Explain why three black balls (see Model 1) tethered together would form a complex that is much more permanent than even the two-ball tethered system. What would you call such a three-ball system?

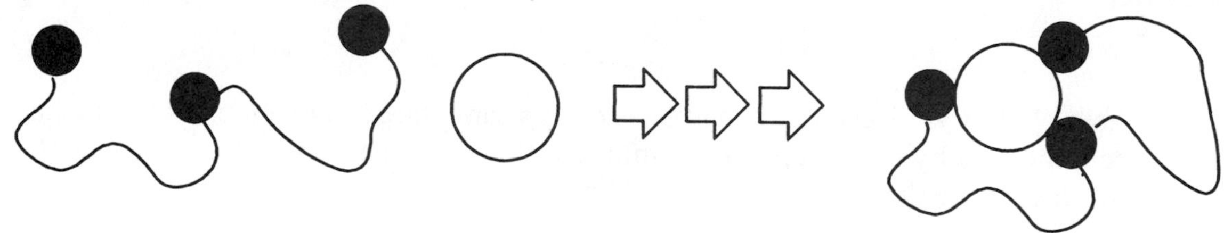

2. Consider the following reaction, called the Williamson Ether synthesis:

R–O–H (alcohol) + $K^{\oplus}$ $^{\ominus}$O–H ⇌ R–$O^{\ominus}$ + H_2O ; + $R'CH_2X$ (alkyl halide, X = Cl, Br or I) ⇌ (S_N2) R–O–CH_2R' (ether) + $X^{\ominus}$

a) Use curved arrows to show the mechanism of this reaction.

b) Using the starting material given below, design a Williamson ether synthesis of the target that does not yield any isomeric ethers.
Hint: Use acetone or formaldehyde to temporarily protect the two OH groups you do not want to react as a cyclic acetal.

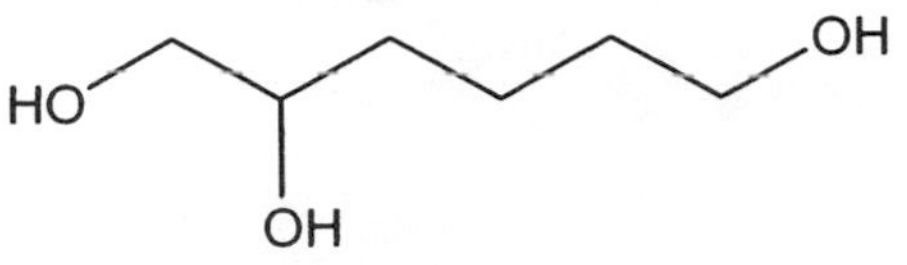

starting material 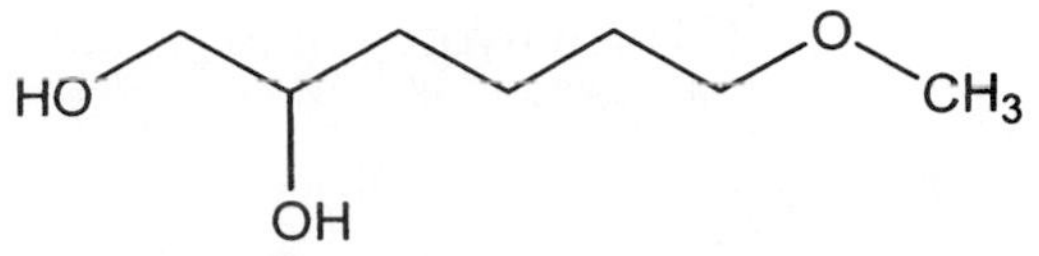 target

3. Design a synthesis of each of the following targets. Only one of them requires the use of a protecting group.

Starting Material $\Longrightarrow$ Target

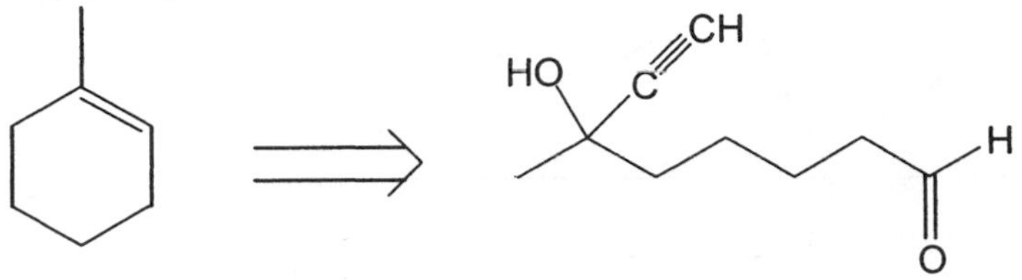

Starting Material $\Longrightarrow$ Target

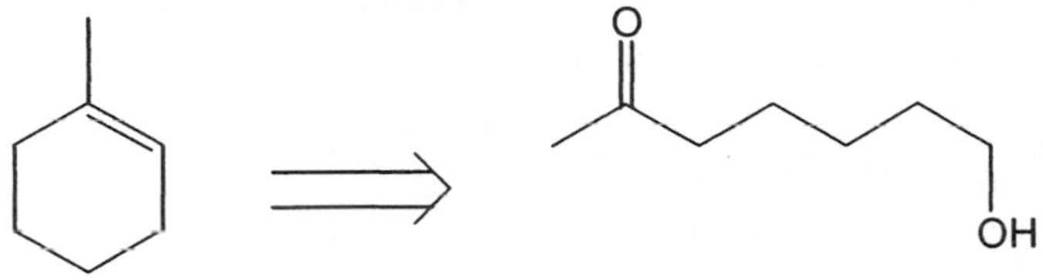

Starting Material $\Longrightarrow$ Target

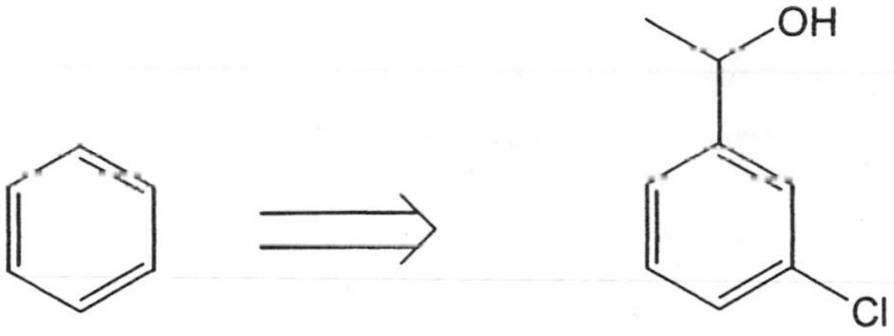

Starting Material $\Longrightarrow$ Target

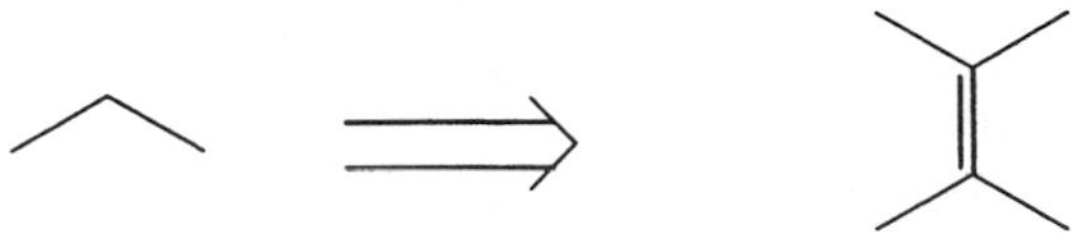

Starting Material $\Longrightarrow$ Target

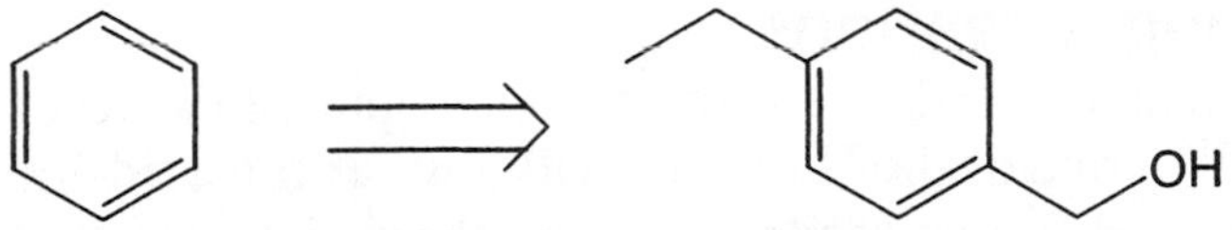

4. Read the assigned pages in your text and do the assigned problems.

ChemActivity 33

Part A: Acidity of the α-Hydrogen

(Which hydrogen of a ketone or aldehyde is most acidic?)

Review

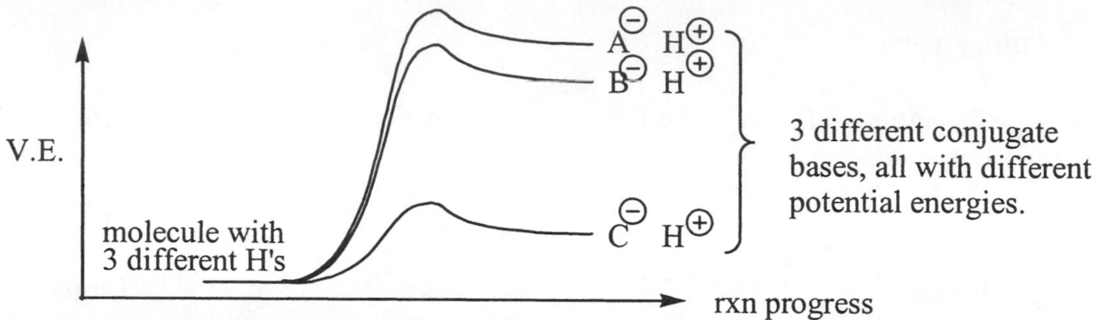

Critical Thinking Questions

1. The energy diagram above describes a molecule with three different possible conjugate bases. Which one is most likely to form (A, B or C)? Explain.

Model 1: Acidity of Propanal

$$CH_3CH_2CHO \xrightarrow{\ :\ddot{N}H_2^{\ominus}\ \ Li^{\oplus}\ }$$

propanal

> Recall the hierarchy of effects:
> 1) Formal Charge Effect (-, neutral, or +?)
> 2) Electronegativity Effect (C, N, O, halogen?)
> 3) Resonance Effect (resonance stabilization?)
> 4) Inductive Effect (inductive stabilization?)

Draw all three conjugate bases of propanal

Critical Thinking Questions

2. In the reaction above, H_2N^- can act as a nucleophile (and react with C=O) or as a base (and pull of one of the H's). **We will look at the acid-base reaction first**:
   a) How many different kinds of H's are there in propanal? Label them A, B...etc.
   b) Draw all the different conjugate bases that could result from an **acid-base reaction** between H_2N^- and propanal. (Do not draw the Nuc. Add'n product.)
   c) At least one of the conjugate bases of propanal has an important resonance structure. DRAW any missing important resonance structures.
   d) Circle the lowest potential energy conjugate base. Explain your choice.
   e) Circle the H on propanal that takes the least energy to remove. This is the "**most acidic**" H.

3. H_2N^- is a strong base, but also an excellent nucleophile. Show the mechanism of the nucleophilic addition reaction which will surely compete with the acid-base reaction in Model 1.

$^{\ominus}NH_2$ $Li^{\oplus}$

propanal

Model 2: Lithium Di-isopropyl Amine

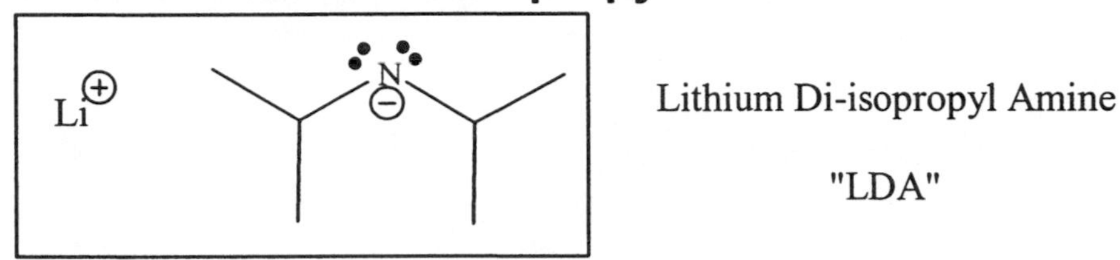

Lithium Di-isopropyl Amine

"LDA"

Critical Thinking Questions

4. Do you expect LDA to be a strong base, strong nucleophile, both or neither? Explain your reasoning.

5. Show the mechanism of the most likely reaction between butanal and LDA.

6. Carbon atoms in a carbonyl compound are labeled according to how far they are from the carbonyl carbon. For some reason we use the Greek alphabet (α, β, δ, γ...) as shown in pentanal, below.

   a) An H attached to an α-carbon is called an α-hydrogen. Circle and label each α-hydrogen in propanal.
   b) Essentially, a base can remove only one kind of H on a carbonyl compound. Which type of H is acidic and can be removed? $\mathbf{H_\alpha}$ $\mathbf{H_\beta}$ $\mathbf{H_\delta}$ $\mathbf{H_\gamma}$

Model 3: Alkylation and Halogenation α to a Carbonyl

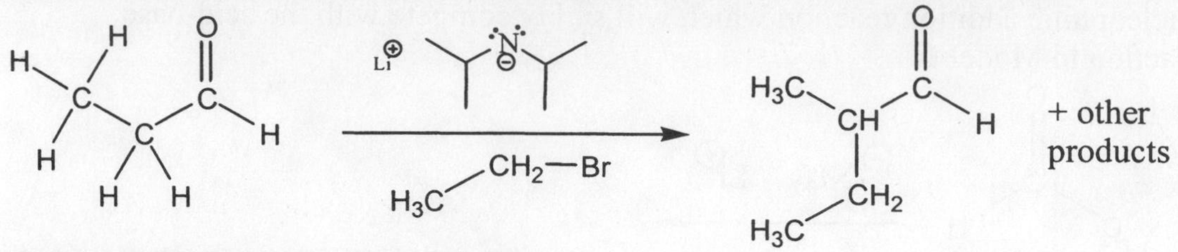

$Li^{\oplus}$ (lithium diisopropylamide); H_3C-CH_2-Br

+ other products

Critical Thinking Questions

7. Propose a mechanism for the reaction in Model 3.

8. Propose a mechanism for the following reaction.

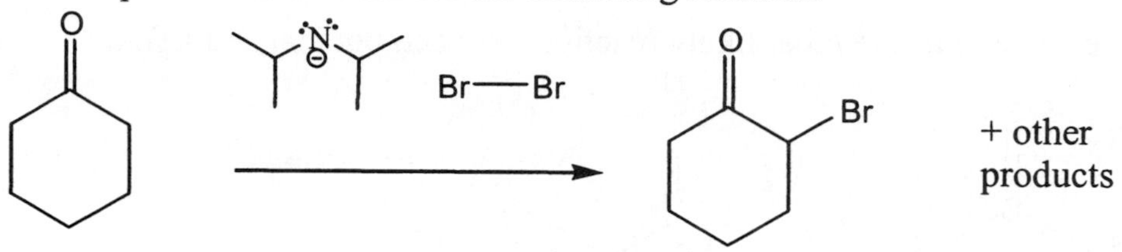

Br—Br

+ other products

Part B: Enol vs. Enolate

(What is the mechanism of acid-catalyzed nucleophilic substitution α to a carbonyl?)

Model 4: Enol vs Enolate

In Part A we learned that an aldehyde or ketone, when treated with a base such as LDA, forms a resonance stabilized α-C anion. Chemists call such anions **enolate ions**.

LDA

aldehyde or ketone with α-H's

Enolate Ion

Critical Thinking Questions

9. Draw the structure that results if you add an H^+ to the resonance structure shown above, **right**.

   a) Recall that an **enol** is a functional group with an **ene** attached to an **ol**. Is the structure you drew above considered an **enol**?

   b) T or F: An enolate ion is the conjugate base of an enol.

   c) T or F: An enolate ion is the conjugate base of an aldehyde/ketone.

Model 5: Keto-Enol Tautomerization

- Keto and Enol forms are in equilibrium with each other.
- Acid or base catalysis speeds inter-conversion between keto and enol.
- The keto form usually makes up >99% of the mixture but there is **always enough enol in solution such that the enol can undergo reactions.**

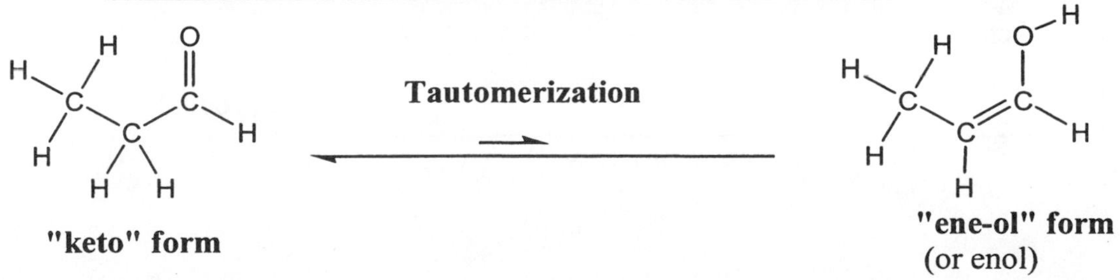

"keto" form

"ene-ol" form
(or enol)

Critical Thinking Questions

10. Use curved arrows to construct a reasonable mechanism for both steps of the following reaction. Note that the first step is an acid catalyzed tautomerization.

H_3O^+ (cat.) — tautomerization — Br—Br — H—Br

11. The following is an energy diagram for the reaction above.
    a) Before addition of Br_2, which do you expect to be the dominant form of the aldehyde: **keto** or **enol**? (Is this consistent with Model 5?)

    b) Explain, using Le Chatelier's Principle, how this reaction can yield mostly product even though the enol to keto ratio is very small.

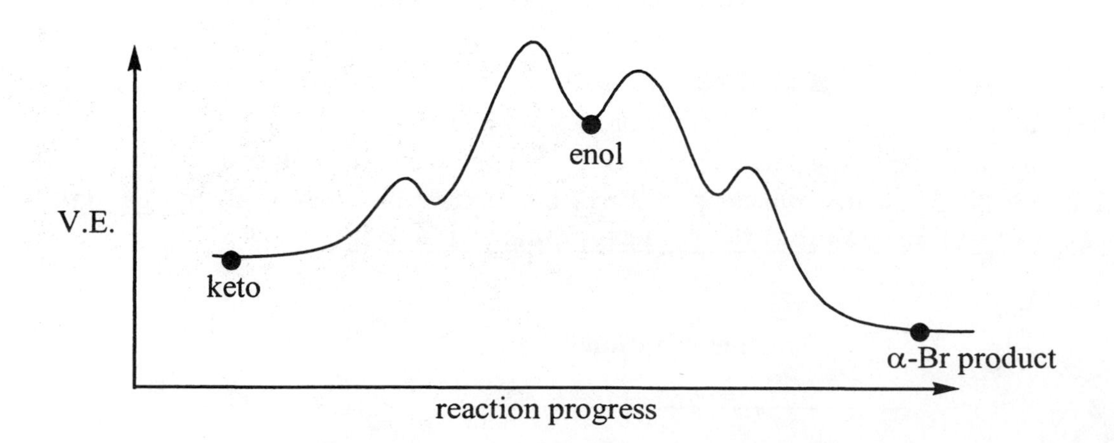

Model 6: Problems with Carboxylic Acids

- Students tend to forget that a carboxylic acid has a very acidic H (thus the name!).
- Functional groups that have a carbonyl, but are not an aldehyde or ketone are called **carboxylic acid derivatives** (examples shown below).
- The three types of CA derivatives that we will focus on (**esters, amides, and acid chlorides**) are shown below.

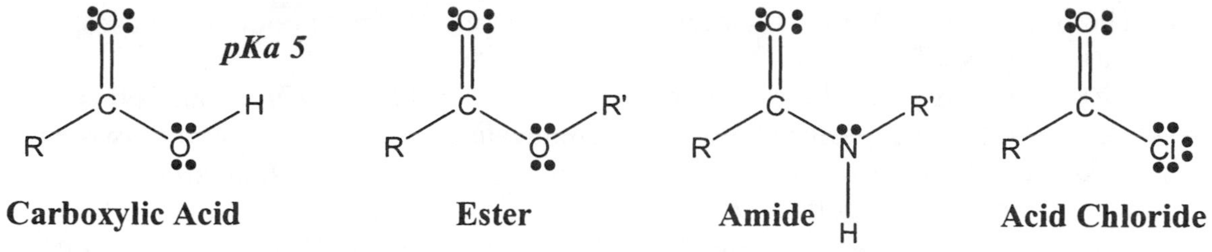

pKa 5

Carboxylic Acid **Ester** **Amide** **Acid Chloride**

Critical Thinking Questions

12. Draw the conjugate base of the carboxylic acid shown in Model 6. Be sure to include both important (1st order) resonance structures.

13. Shown below is the conjugate base of an aldehyde (enolate ion). Explain why it is much easier to remove an H from a carboxylic acid than from an aldehyde.

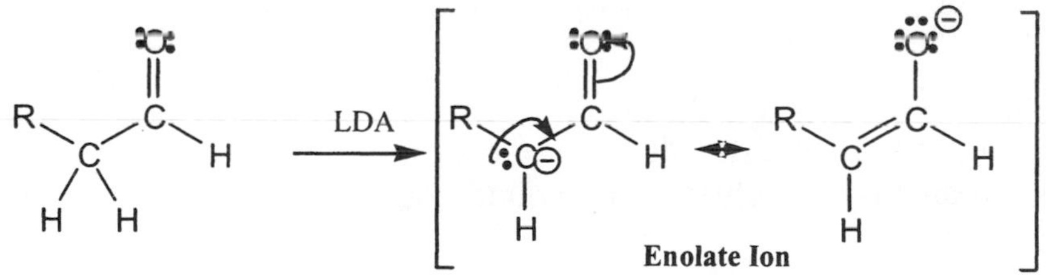

LDA

Enolate Ion

14. Construct an explanation for why Rxn A does NOT work, but Rxn B does work.

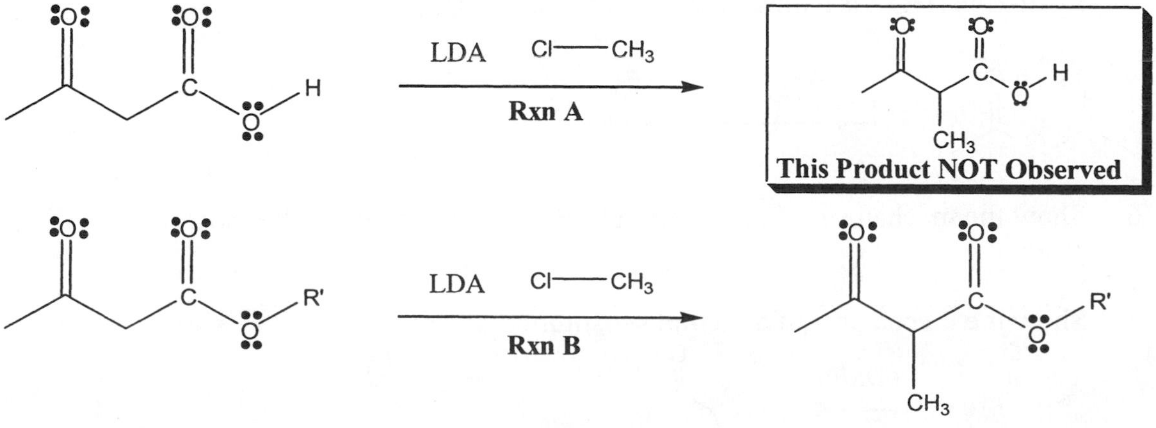

LDA Cl—CH_3

Rxn A

This Product NOT Observed

LDA Cl—CH_3

Rxn B

Exercises for Part A

1. Explain why the examples below are called **β-keto** aldehydes or ketones.

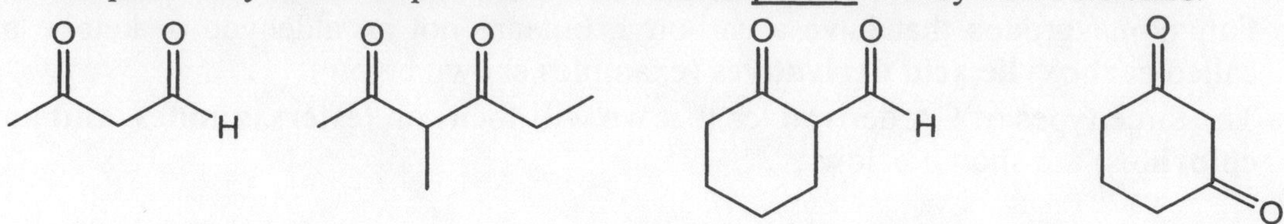

2. The β-keto aldehyde on the far left, above, has three different conjugate bases.
   a) Draw all three.
   b) One conjugate base is not resonance stabilized at all. Which one is this?
   c) Of the two resonance stabilized conjugate bases, one has a total of two resonance structures, and the other has a total of three. Draw all resonance structures of both.
   d) Draw in the most acidic H (and label it "most acidic H") on the far left β-keto aldehyde in Exercise 1.

3. Draw in the most acidic H on the other β-keto compounds in Exercise 1.

4. Consider the following pKa's:

pKa 9 pKa 17

   a) Which compound is more acidic?
   b) Circle the H on each structure that the pKa refers to.
   c) Construct an explanation for why the compound you chose in part a) is more acidic.

5. Are your answers above consistent with the fact that α-alkylation of a β-keto compound can be accomplished using the reagents shown below? Explain. (Note: hydroxide (HO^-) is a much weaker base than LDA.)

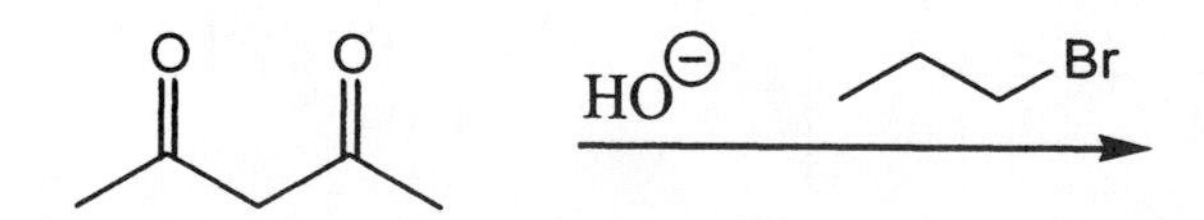

6. Show the mechanism of the reaction in Exercise 5, and draw the most likely product.

7. Show the mechanism of a second substitution of Br:

LDA/Br_2

8. Which reaction is faster? Rxn 1 or Rxn 2. Explain your reasoning. (Hint: In each reaction the rate limiting step is removal of the H by the base LDA--so consider which H is more acidic.)

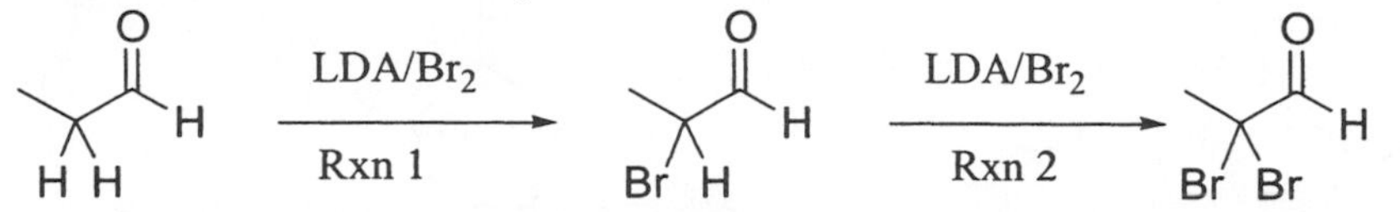

LDA/Br_2 Rxn 1   LDA/Br_2 Rxn 2

H H   H   Br H   H   Br Br   H

9. Is your answer above consistent with the fact that α-halogenations with LDA are very difficult to control, and often result in all the α-H's being replaced with halogen atoms? Explain.

Note: In Part B we will learn a method of α-halogenation that is easily controlled.

10. Read the assigned pages in your text and do the assigned problems.

Exercises for Part B

11. Show the mechanism of the following reaction:

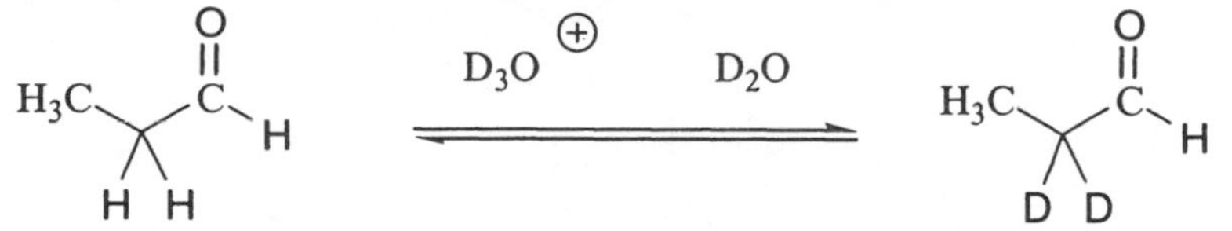

H_3C H H H $D_3O^{\oplus}$ D_2O H_3C D D H

12. A solution of pure **R** chiral aldehyde (shown below) was accidentally exposed to a small amount of acid. The next day, the solution was found to contain a racemic mixture (1:1 mix of **R** and **S**). Explain how this could have occurred.

H CH_3 $H_3O^{\oplus}$ H CH_3 H_3C H

pure R stereoisomer    racemic mixture (1:1)

13. Draw the keto form of phenol.
    a) Explain why phenol is more stable in its enol form than it keto form. (Note that phenol is one of the very few aldehydes or ketones for which the enol form is favored.)
    b) Draw another keto-enol pair that you expect will break the general rule and favor the enol over the keto form.

14. Use curved arrows to show a mechanism for the tautomerization in Model 5, first assuming acid catalysis (H_3O^+), then assuming base catalysis (HO^-).

15. Both enols and enolates can act as nucleophiles. Which one is a better nucleophile? Explain.

16. Design a reasonable synthesis of each of the following target molecules from the starting material given.

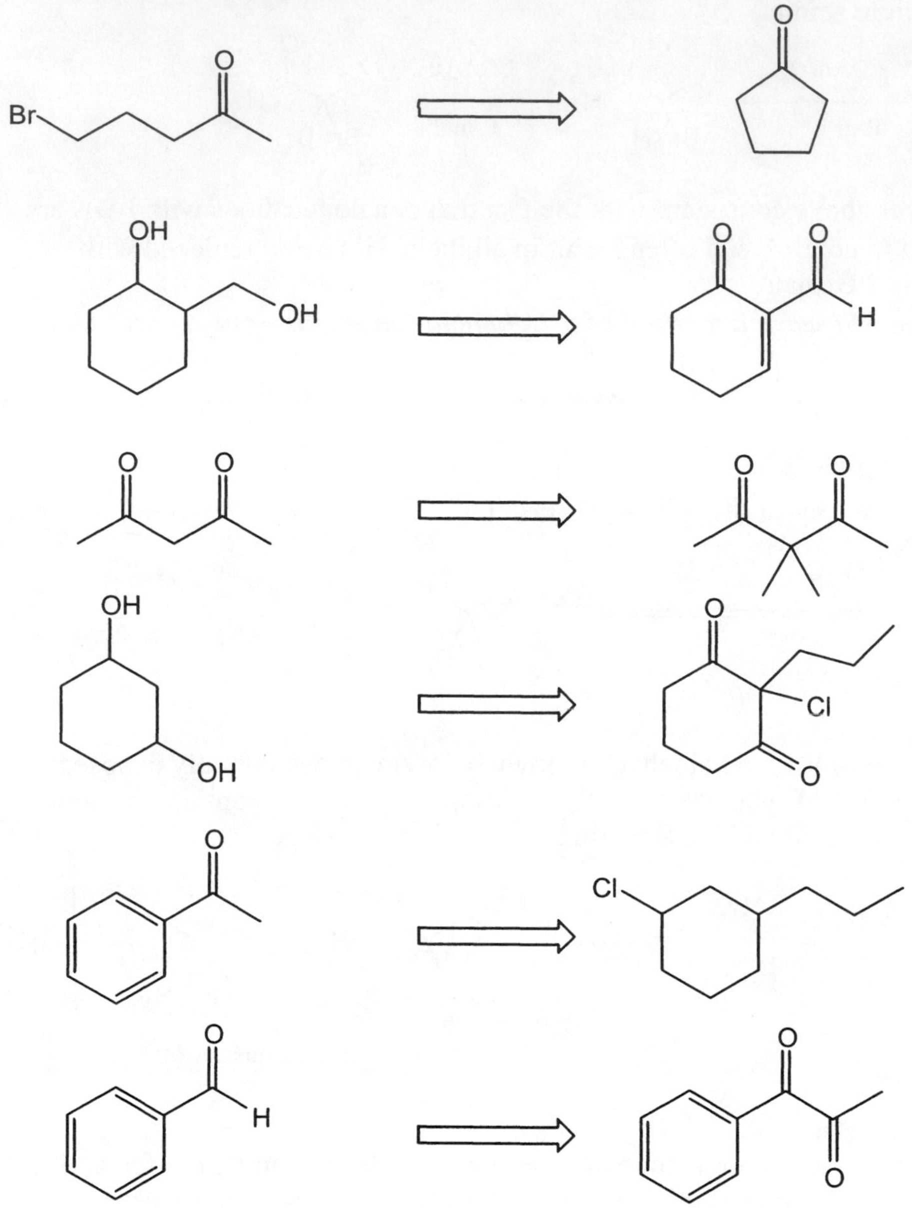

Br ⟹ O

OH OH ⟹ O O H

O O ⟹ O O

OH OH ⟹ O Cl O

O ⟹ Cl

O H ⟹ O O

17. Read the assigned pages in your text and do the assigned problems.

ChemActivity 34

Part A: Aldol Reaction

(What is the mechanism of the aldol reaction?)

Model 1: Aldol Reaction

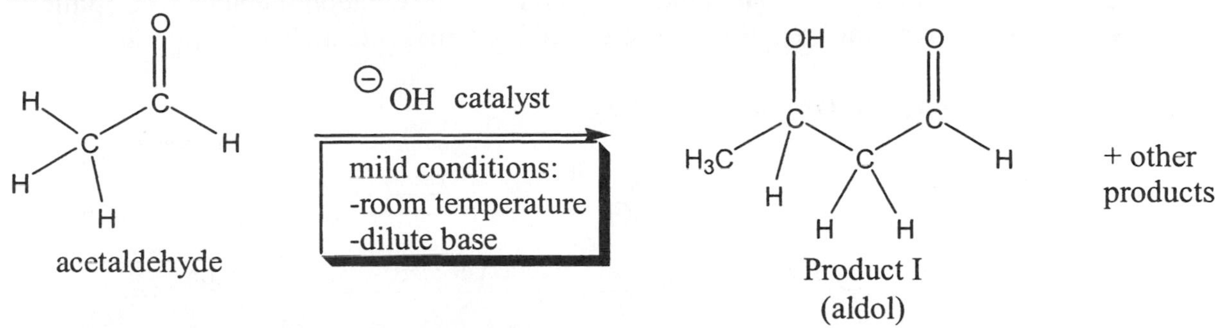

Critical Thinking Questions

1. Consider the reaction in Model 1.
   a) Note that Product I contains two different functional groups. Construct an explanation for the name "**ald-ol**."

   b) How many carbons are there in Product I?
   c) Estimate the number of acetaldehyde molecules required to make one molecule of Product I?
   d) Label the "**new carbon-carbon bond**" in Product I and circle each acetaldehyde subunit.
   e) Devise a reasonable mechanism for the reaction in Model 1. Be sure to include important resonance structures of any intermediate products.

Model 2: Analysis of an Aldol Product Using Retrosynthesis

- Most C-C bond formations are the result of an nucleophilic carbon ($C^{\delta-}$) joining with an electrophilic carbon ($C^{\delta+}$).
- Retrosynthetic analysis can be used to "pull apart" the product (on paper) into two pieces: the **electrophilic piece** and the **nucleophilic piece**.

Step 1: Find the new carbon-carbon bond and draw a line through it.
Step 2: Of the two C's in this bond, decide which was the $C^{\delta-}$ and which was the $C^{\delta+}$.
Step 3: Draw real molecules that could have served as the nucleophile and electrophile.
Step 4: Draw in a forward reaction arrow and add any reagents that are required.

Example I: Analysis of a Grignard Reaction

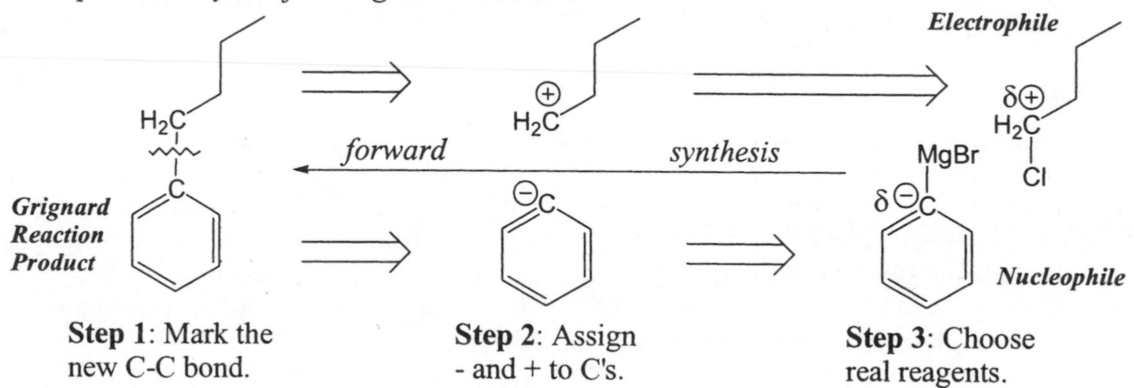

Step 4: In this case no extra reagents are needed to make the Nuc^- react with the $Elec^+$.

Example II: Analysis of an Aldol Reaction

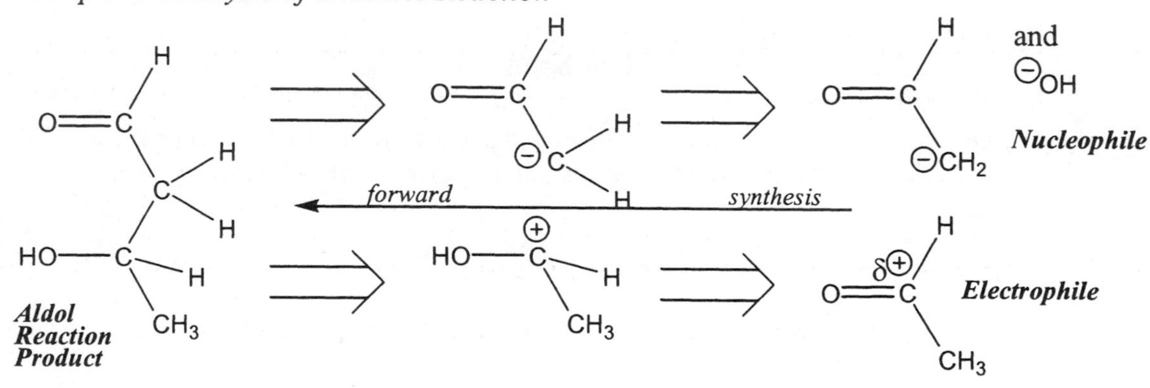

Step 4: The nucleophile is prepared by mixing acetaldehyde and a small amount of HO^-.

Critical Thinking Questions

2. According to Example 1, name one example of a functional group that makes a C nucleophilic and one example of a functional group that make a C electrophilic.
3. For the aldol reaction above, complete **Step 1** by marking, on the product, the new carbon-carbon bond.
4. On an aldehyde, which C is electrophilic and which C is nucleophilic (with addition of hydroxide)?

5. Follow each step in Model 2 (retrosynthesis) and determine the reagents necessary to make each of the following aldol products.

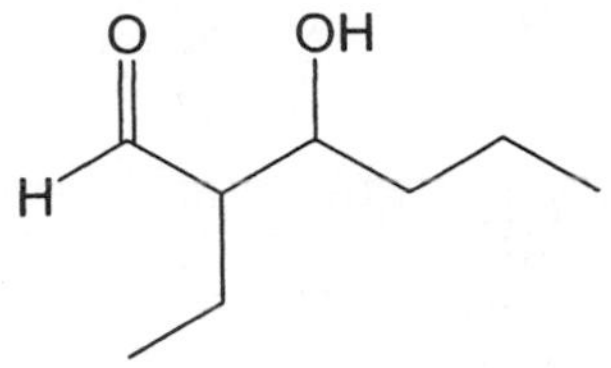

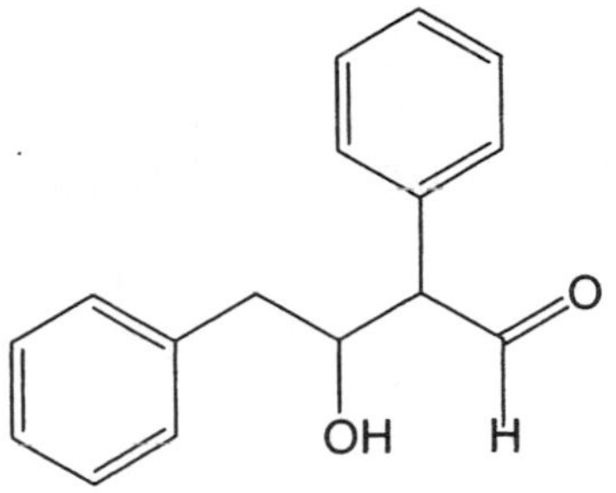

Caution: The next one is tricky!

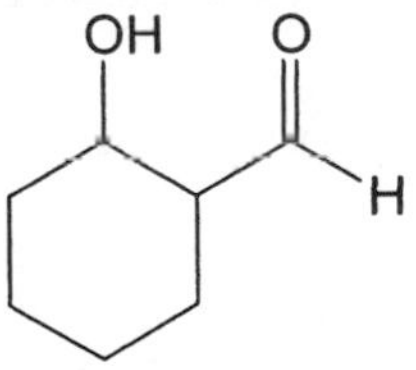

6. Explain why each of the following aldehydes **cannot** undergo an aldol reaction (even in the presence of hydroxide).

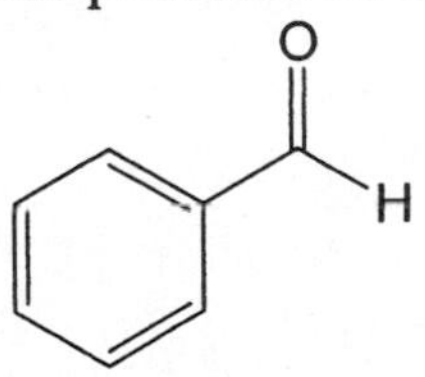

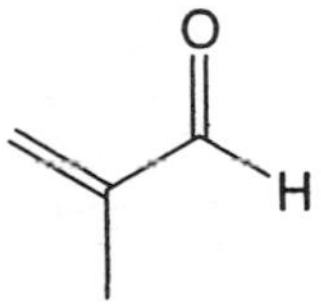

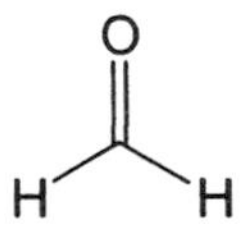

7. Complete the sentence: To undergo an aldol reaction, an aldehyde must have… (Hint: it is something that the aldehydes in CTQ 6 do not have.)

8. Give an example (not appearing in this activity) of an aldehyde that…
   a) Cannot undergo an aldol reaction.
   b) Can undergo an aldol reaction.

Part B: Mixed Aldol Reactions

(How can steric factors impact the yield of an aldol reaction?)

Review from Part A: In an Aldol Reaction…

- one molecule of aldehyde reacts with base and serves as nucleophile
- a second molecule of aldehyde serves as electrophile.

Model 3: Aldol Reaction with Two Different Aldehydes

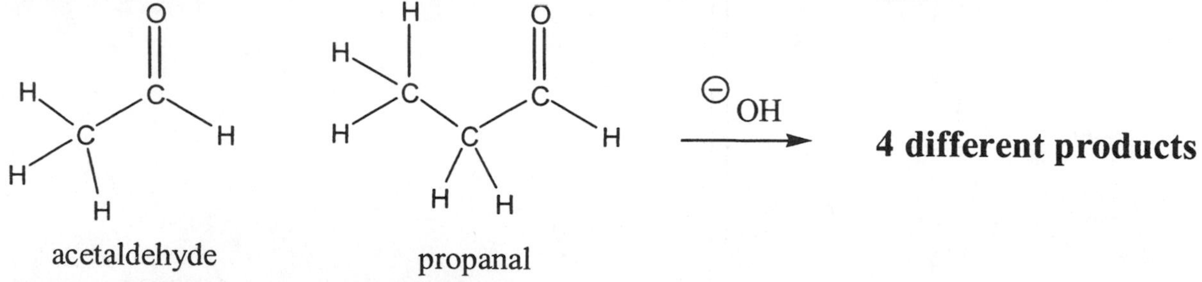

Critical Thinking Questions

9. Consider the mixture in Model 3.
   a) Circle each α-hydrogen on acetaldehyde and propanal.
   b) T or F: Both acetaldehyde and propanal can serve as nucleophile in an aldol.
   c) T or F: Both acetaldehyde and propanal can serve as electrophile in an aldol.
   d) Draw the **two aldol products** and **two mixed aldol products** that result from the mixture in Model 3. Do not draw the mechanisms.

e) For each product on the previous page state whether…
- **acetaldehyde enolate ion** or **propanal enolate ion** acted as the **nucleophile** to form that product.
- **acetaldehyde** or **propanal** acted as **electrophile** to form that product.

Model 4: "Aldol" Reactions Involving Ketones ("Ketols"?)

- Ketones are also known to participate in aldol reactions.
- The yields of such reactions are generally very low.

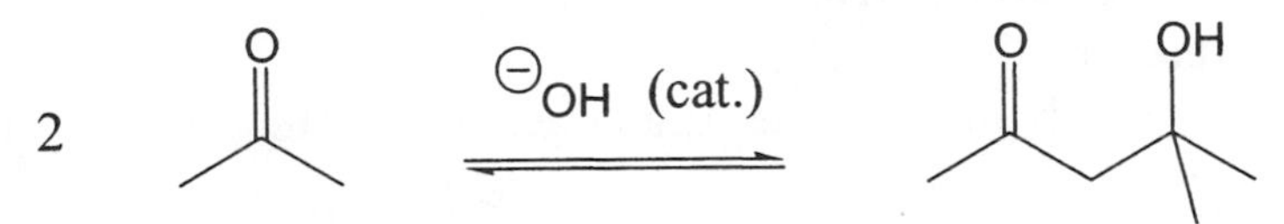

Critical Thinking Questions

10. Chemists usually call the product of the reaction in Model 4 an **aldol product**. A student rightly objects to this, saying he thinks it should be called a **"ketol."** Explain this student's reasoning. (Hint: What functional groups are in the product?)

11. The energy diagram for the reaction in Model 4 is shown below.
    a) According to this diagram is this reaction up-hill or down-hill in energy?
    b) Is your answer consistent with the information given in Model 4? Explain.

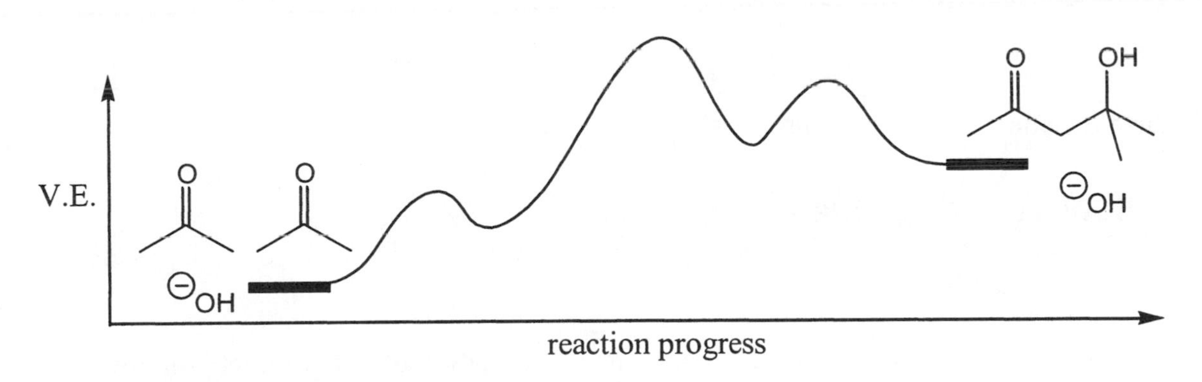

12. Aldehydes and ketones can both serve as good nucleophiles in aldol reactions (as long as they have an α-H to remove).
    a) Which type of carbonyl compound do you expect will be a **better electrophile** in an aldol or mixed aldol reaction: **aldehyde** or **ketone** [circle one]?

    b) Explain your reasoning.

    c) Is your answer consistent with the energy diagram above?

Model 5: Mixed Aldol with an Aldehyde and a Ketone

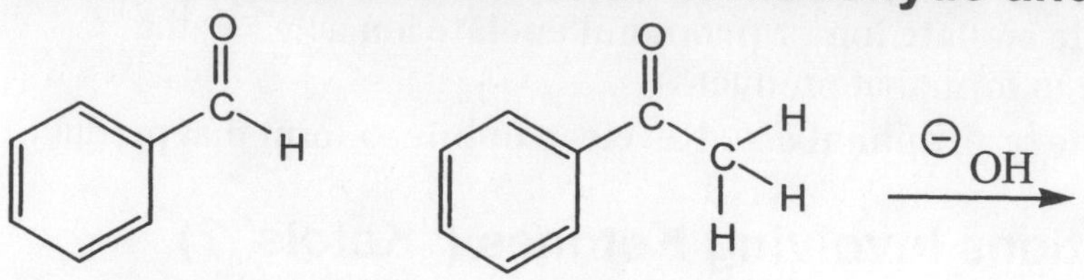

O
‖
C H
C H H
H
⊖ OH
→

benzaldehyde acetophenone

Critical Thinking Questions

13. Consider the reactants in Model 5.
    a) Explain why benzaldehyde cannot act as a nucleophile in an aldol reaction.

    b) Which species will better serve as an **electrophile** in an aldol reaction: **benzaldehyde** or **acetophenone** [circle one]?

    c) Complete the reaction in Model 5 by drawing the **single** most likely aldol or mixed aldol product.

Model 6: Controlling Product Formation in a Mixed Aldol

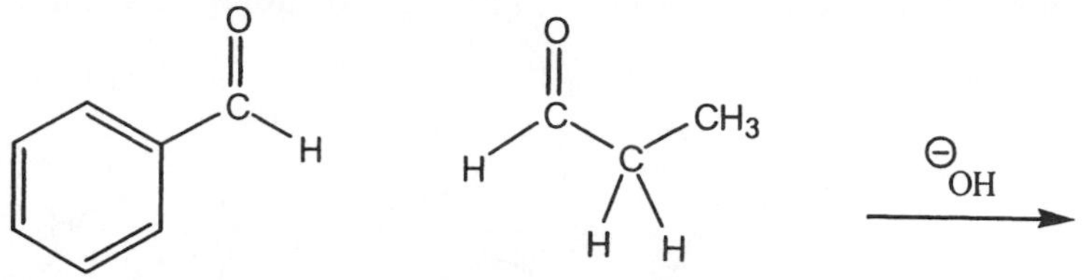

O
‖
C H
H C CH$_3$
H H
⊖ OH
→

benzaldehyde propanal

Critical Thinking Questions

14. The reaction in Model 6 normally yields two different aldol/mixed aldol products. Draw them both.

15. One of the following procedures maximizes formation of one product you drew in Model 6, and nearly eliminates formation of the other.

    I. Benzaldehyde is added very slowly to a mixture of propanal and base.
    II. Propanal is added very slowly to a mixture of benzaldehyde and base.

    a) Circle the letter of the successful procedure.
    b) Circle the product in Model 6 that will form preferentially using this procedure, and cross out the other one.
    c) Explain why the procedure works.

Part C: "Condensation Reactions"

(Why are aldol reactions run at high temperature called "condensation" reactions?)

Model 7: Dehydration, the Elimination of Water (HOH)

- When a suitable alcohol is heated in base, the result is an E2 reaction.

Cyclohexanol (ring: CH(OH), CH(H), CH_2, CH_2, CH_2, CH_2) + $^{\ominus}OH$ $\xrightarrow{E2}$ cyclohexene (ring: CH=CH, CH_2, CH_2, CH_2, CH_2) + HOH + $^{\ominus}OH$

Critical Thinking Questions

16. Review: What does the "**E**" in "**E2**" stand for?
17. Review: Is an E2 reaction, **one step** or **two steps** [circle one]?
18. Use curved arrows to show the mechanism of the reaction in Model 7.
19. Construct an explanation for why chemists say the alcohol in Model 7 has being **"de-hydrated."**
20. Is hydroxide considered a catalyst in this reaction? Explain.

Model 8: Dehydration of an Aldol Product

- At low temperature (below Room T) an aldol product is stable in base.
- At high temperature (above Room T) an aldol product undergoes dehydration.

CH_3CHO $\underset{\text{low Temp.}}{\overset{^{\ominus}OH \text{ catalyst}}{\rightleftharpoons}}$ $H_3C-CH(OH)-CH_2-CHO$ **LOW TEMP. rxn stops at the aldol.**

CH_3CHO (aldehyde) $\underset{\text{high Temp.}}{\overset{^{\ominus}OH \text{ catalyst}}{\rightleftharpoons}}$ $H_3C-CH(OH)-CH_2-CHO$ (aldol) $\xrightarrow[\text{high Temp. rxn continues}]{^{\ominus}OH \text{ catalyst}}$ $H(CH_3)C=CH-CHO$ (α,β-unsaturated aldehyde)

Critical Thinking Questions

21. At high temperature an aldol product will react further with base to form an α,β-unsaturated aldehyde shown above, right.
    a) Use curved arrows to show a mechanism for this further reaction at high T.

b) Construct an explanation for the name "α,β-unsaturated aldehyde." (Hint: Which C's will get an H if you treat this product with H_2/Pt so as to "saturate" it with the maximum number of H's?)

c) Researchers first studying aldol reactions noticed that during the reaction droplets of water ("condensation") would form on the inside of the flask. For this reason chemists call aldols, mixed aldols and similar reactions "**condensation reactions.**"

What is the source of these water droplets in an aldol reaction? (Hint: they only form at high temperatures.)

Model 9: Dehydration "Drives" a Condensation Reaction

Below is an energy diagram for the aldol reaction from Model 4, run at high temperature.

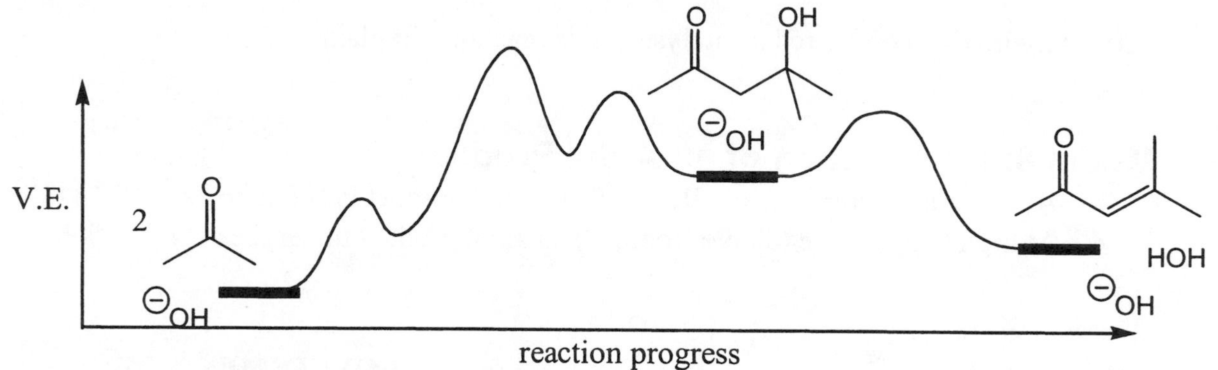

Critical Thinking Questions

22. Is the reaction in Model 9, overall, **up-hill** or **down-hill** in terms of potential energy?

 a) Do you expect a high yield of α,β-unsaturated product in this reaction?

 b) Researchers have found that the reaction in Model 9 (and other condensation reactions) can be driven to the right to give nearly 100% yield of α,β-unsaturated product by boiling off water as steam. Explain.

Review:

- The alpha (α) carbon of a carbonyl compound can be made nucleophilic by the addition of a base to form the **enolate ion**.

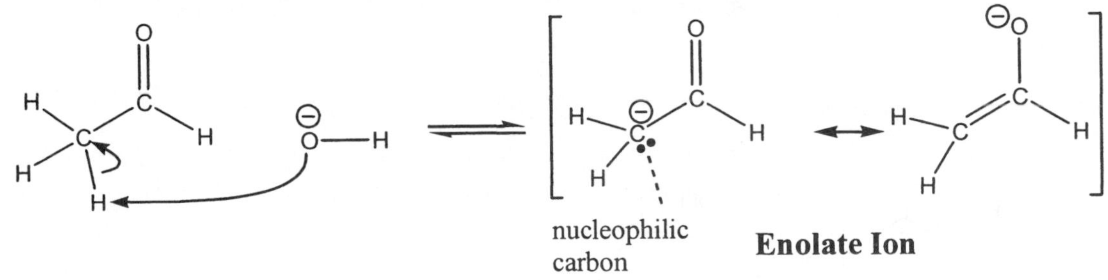

nucleophilic carbon

Enolate Ion

- In acid or base, a **keto** compound is in equilibrium with its **enol** form.
- Using 2nd order resonance structures we can demonstrate that the alpha (α) carbon of an enol is nucleophilic (as shown below).

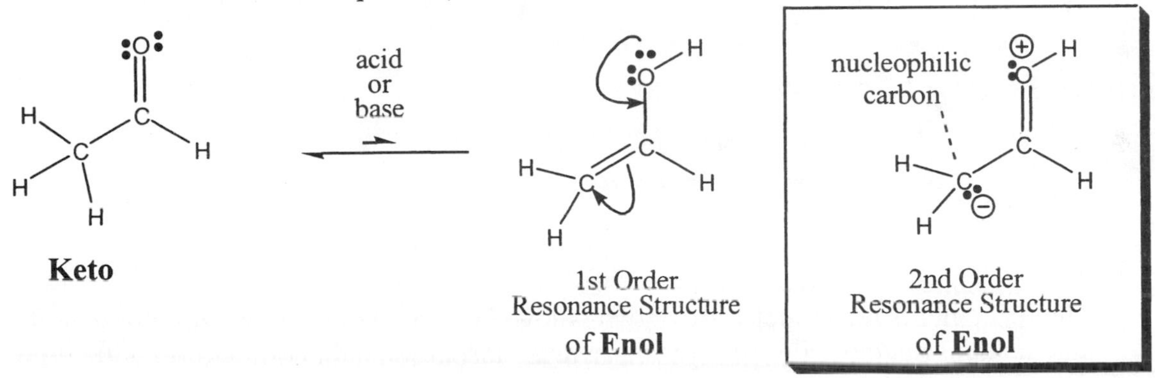

acid or base

Keto

1st Order Resonance Structure of **Enol**

nucleophilic carbon

2nd Order Resonance Structure of **Enol**

Model 10: α,β-Unsaturated Carbonyl Compounds

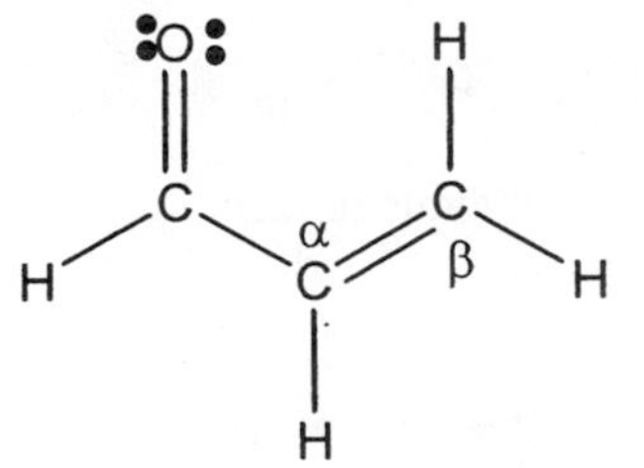

Note: <u>Oxygen must always have an octet</u>, even in 2nd order resonance structures!

Critical Thinking Questions

23. Draw any reasonable 2nd order resonance structures for the α,β-unsaturated carbonyl compound in Model 10.
    a) Use these to predict whether the alpha (α) carbon of an α,β-unsaturated carbonyl compound will be **nucleophilic** <u>or</u> **electrophilic**.

    b) Use these to predict whether the beta (β) carbon of an α,β-unsaturated carbonyl compound will be **nucleophilic** <u>or</u> **electrophilic**.

Exercises for Part A

1. For each of the following aldehydes, draw the aldol product that will result when the aldehyde is treated with a catalytic amount of base (e.g. KOH).

2. The reactants from Model 1 are drawn below. Hydroxide is both a strong base and an excellent nucleophile. As a result, hydrate formation is in competition with aldol formation.

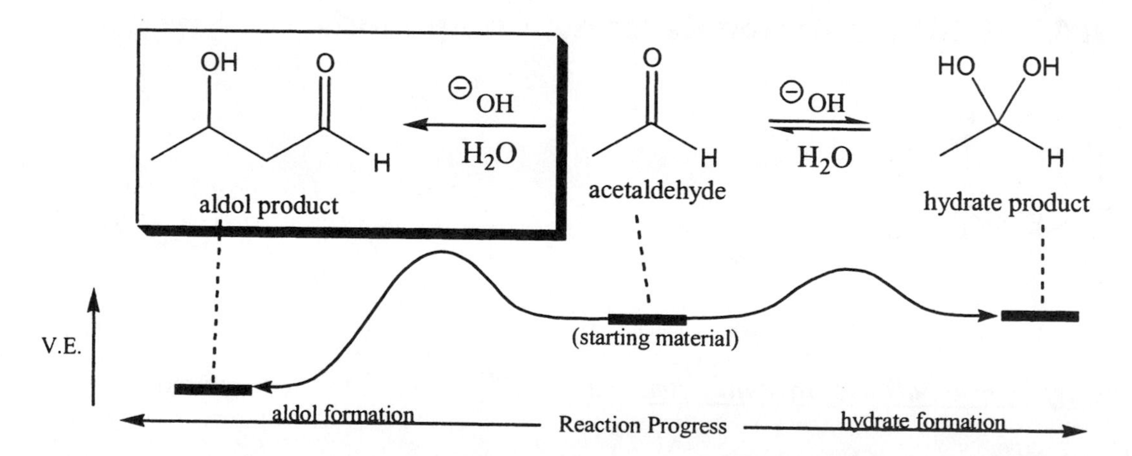

a) Use curved arrows to show the mechanism of hydrate formation.

b) Hydrate formation is faster than aldol formation (as shown by the energy diagram), yet the amount of hydrate found in the reaction at any given time is below the detection limit. Explain this observation.

3. Draw the aldol product that results from an intramolecular aldol reaction involving a base catalyst and the following dialdehyde.

    a) Explain why the intramolecular product you drew above is more likely to form than the following product (assume a dilute solution of dialdehyde).

    b) In the lab you treat a concentrated solution of this dialdehyde with hydroxide and the result is long polymer chains. Explain this finding.

4. Draw the <u>two</u> different aldol products that result from intramolecular aldol reactions involving the di-aldehyde 3-methylhexandial and base.

5. In an aldol reaction one molecule of aldehyde reacts with base and becomes a nucleophile, and one molecule of aldehyde serves as electrophile. For each product you drew in the previous question indicate which aldehyede of the dialdehyde acted as Nuc$^-$ and which acted as Elec$^+$. Call the ends the **1-aldehyde** and the **6-aldehyde**.

6. Read the assigned pages in your text and do the assigned problems.

Exercises for Part B

7. Draw the mechanism of the reaction in Model 4.

8. Draw the aldehydes or ketones that could be used to make the following aldol products. (Note: Some of these products may not be formed in high yield due to the fact that ketones are poor electrophiles.) **Use Retrosynthesis-described in Model 2.**

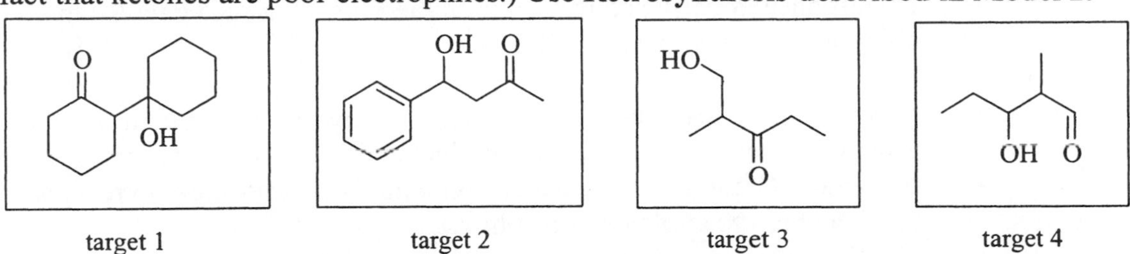

9. Draw a new example of two different reactants (use aldehydes or ketones or one of each) that would give a high yield of a single aldol product regardless of the way they are mixed. And **draw the product**.

10. Draw a new example of two different reactants (use aldehydes or ketones or one of each) that would give a high yield of a single aldol product only when one is slowly added to the other.
    a) Specify the correct procedure (e.g. add A to B or B to A)
    b) **Draw the product.**

11. Draw a new example of two different reactants (use aldehydes and/or ketones) that would give a mixture of four different aldol products and draw each product.

12. The structures shown below are called β-keto carbonyl compounds. Construct an explanation for this name.

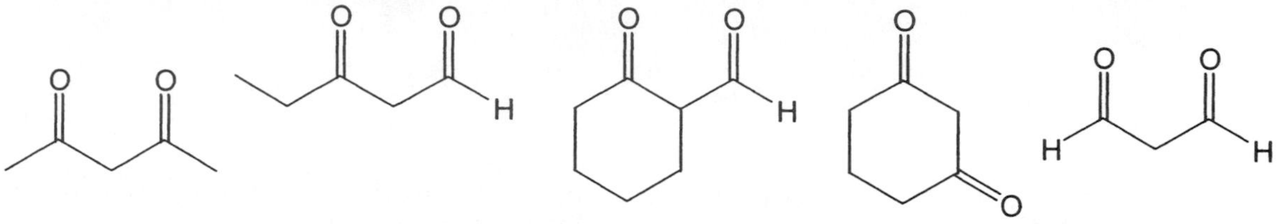

    a) Which H is the most acidic on the following β-keto carbonyl compound?

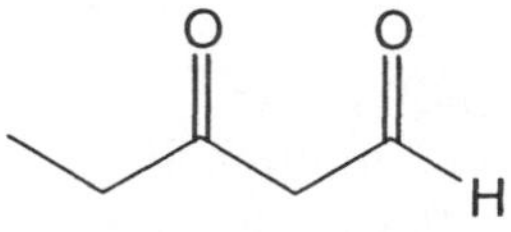

    b) Draw the conjugate base of the compound from a) including all important resonance structures.

    c) Which of the following compounds do you expect to be more acidic? Explain your reasoning.

Cmpd X =

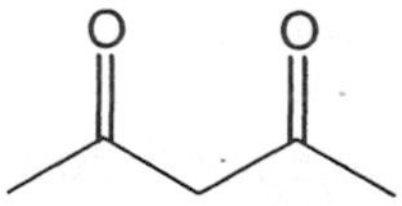

Cmpd Y =

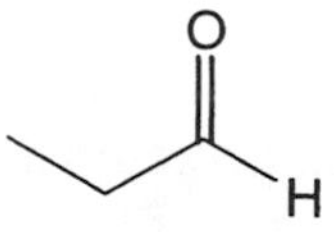

    d) Is your answer above consistent with the following pKa data for these compounds: $pKa_x = 9$ and $pKa_y = 17$?

    e) When compounds X and Y are mixed with hydroxide, one mixed aldol-type product is generated. Draw this product and show the mechanism of formation. (Hint: consider which compound is more likely to react with base and become the enolate nucleophile.)

13. Read the assigned pages in your text and do the assigned problems.

Exercises for Part C

14. Explain the following results. When cyclohexanone is treated with base at room temperature a product with molecular formula $C_{12}H_{20}O_2$ is recovered in 22% yield. However, when cyclohexanone is treated with acid at room temperature a product with molecular formula $C_{12}H_{18}O$ is recovered in 92% yield.

15. In some cases the dehydration step is very down hill so it is nearly impossible to isolate the aldol product. The reaction below is such an example:

O H O $^{\ominus}$OH

dilute, room temp.

(or acid)

benzaldehyde acetone

   a) Draw the aldol product and the α,β-unsaturated carbonyl products-E and Z (dehydration products).
   b) Construct an explanation for why the α,β-unsaturated carbonyl products are much more stable than the aldol product. (The aldol product cannot be isolated in this case, even if the reaction is run at very low temperature.)

16. Even in harsh basic conditions the following pair of carbonyl compounds gives a very low yield of a single aldol product.

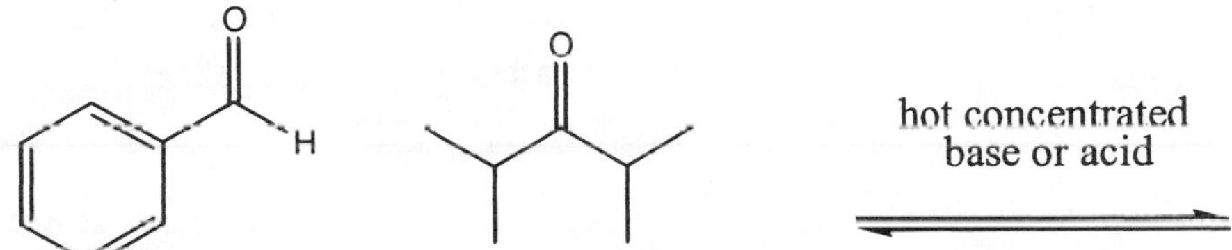

benzaldehyde 2,4-dimethyl-3-pentanone

   a) Draw the most likely aldol product that results from this mixed aldol reaction.
   b) Construct an explanation for why, in this case, the α,β-unsaturated carbonyl compound cannot form under any conditions.
   c) Explain why this reaction give as very low yield of aldol product.
   d) Explain why, in this case, attempting to drive the reaction to the right by boiling off water does not work.

17. The preparation of α,β-unsaturated carbonyl compounds is a subject of great interest among synthetic organic chemists. The aldol method is perhaps the most elegant way to form a new carbon-carbon bond, but there are other ways. Propose a synthesis of the following target from the starting material given.

O O

H H

starting material target

18. Certain compounds re-arrange in acidic or basic basic solution to form α,β-unsaturated carbonyl compounds. Use curved arrows to construct a mechanism for the following re-arrangement in base (HO^-), then in acid (H_3O^+).

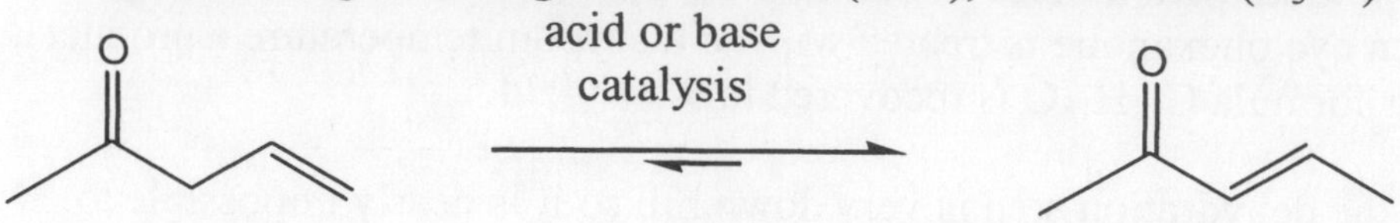

acid or base catalysis

19. According to the equilibrium arrows, the rearrangement above is down hill. What is the driving force for this reaction? (Explain why the product is lower in potential energy/more stable than the reactant.)

20. Draw products that form when 2-butanone is heated in base at 50° C (RT = 25°C).

21. Design an efficient synthesis of each of the following target molecules. All carbon atoms must come from the starting material/s given. Be sure to specify the conditions of any aldol reactions.

Target	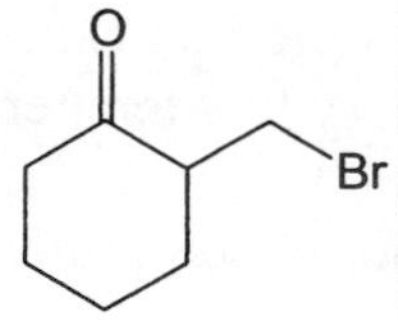 O, Br	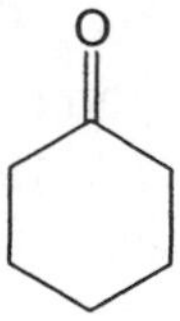 O	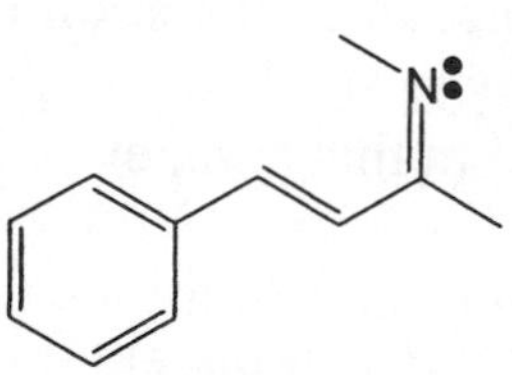 N:	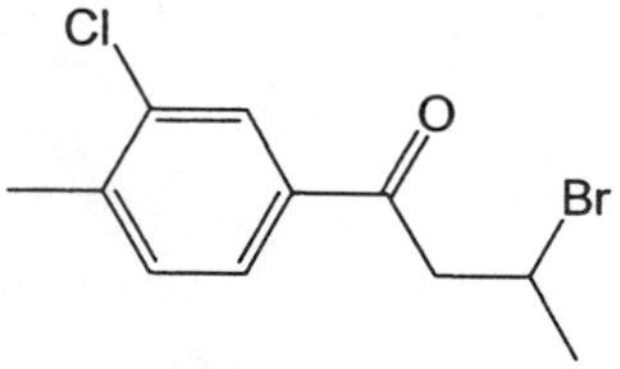 Cl, O, Br
Starting material	cyclohexanone formaldehyde	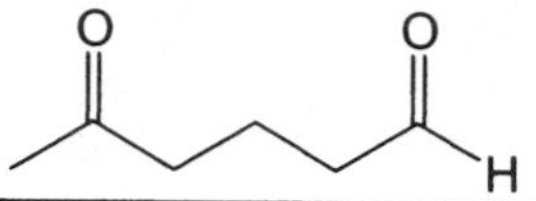 O, O, H	acetone benzaldehyde methyl amine	toluene acetaldehyde

hint: an acid chloride can be made from a carboxylic acid

22. Read the assigned pages in your text and do the assigned problems.

ChemActivity 35

Acid Catalyzed Aldol Reactions

(What species serves as the nucleophile in an acid catalyzed aldol reaction?)

Review:

- In **basic solution**, a carbonyl compound is in equilibrium with its **enolate ion**.
- In **acidic solution**, a carbonyl is in rapid equilibrium with its **enol** form.
- An enol has a nucleophilic α-carbon, as demonstrated by a 2nd order R.S..

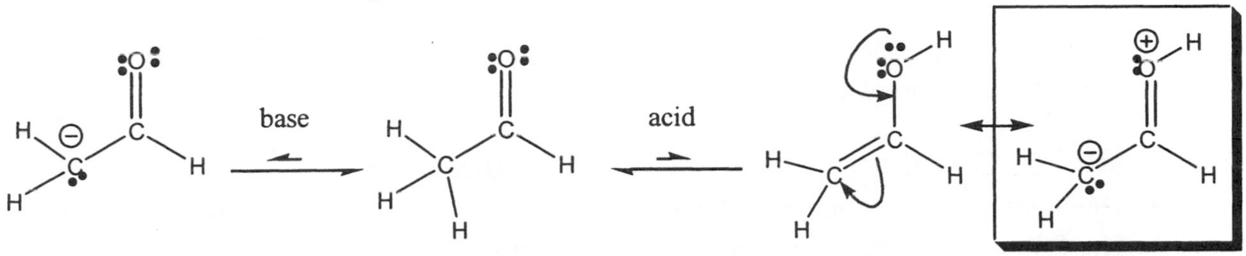

base

acid

Enolate Ion **Keto** **Enol** 2nd Order R.S. of **Enol**

Model 1: Acid Catalyzed Aldol Reaction

- An aldol reaction also works in acidic conditions.
- It is usually **impossible to isolate the aldol product** because in acid (even at low T) the reaction proceeds all the way to the α,β-unsaturated carbonyl product.

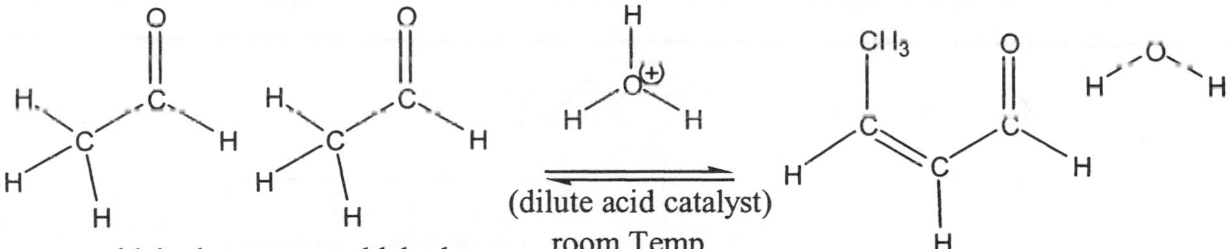

acetaldehyde acetaldehyde (dilute acid catalyst) room Temp. α,β-unsaturated aldehyde

Critical Thinking Questions

1. Use curved arrows to show a mechanism for the reaction in Model 1. (Hint: the first step is a tautomerization to turn one of the aldehydes into a nucleophilic enol, and it may help to draw the 2nd order resonance structure of this enol.)

Model 2: Analyzing Condensation Products with Retrosynthesis

Reactants —C-C bond formation, acid or base catalyst→ Aldol Product —dehydration, acid or base catalyst→ α,β–unsaturated Product

Critical Thinking Questions

2. Thinking backwards, determine the aldol product that was dehydrated to form the α,β-unsaturated carbonyl product shown above, right.

3. Use retrosynthesis to identify the starting materials that give rise to the aldol product you drew above.

Model 3: C-C Bond Formation in Biological Systems

- The aldol reaction is one of just a few ways to make carbon-carbon σ bonds.
- The following are the other ways we have learned to make carbon-carbon σ bonds.

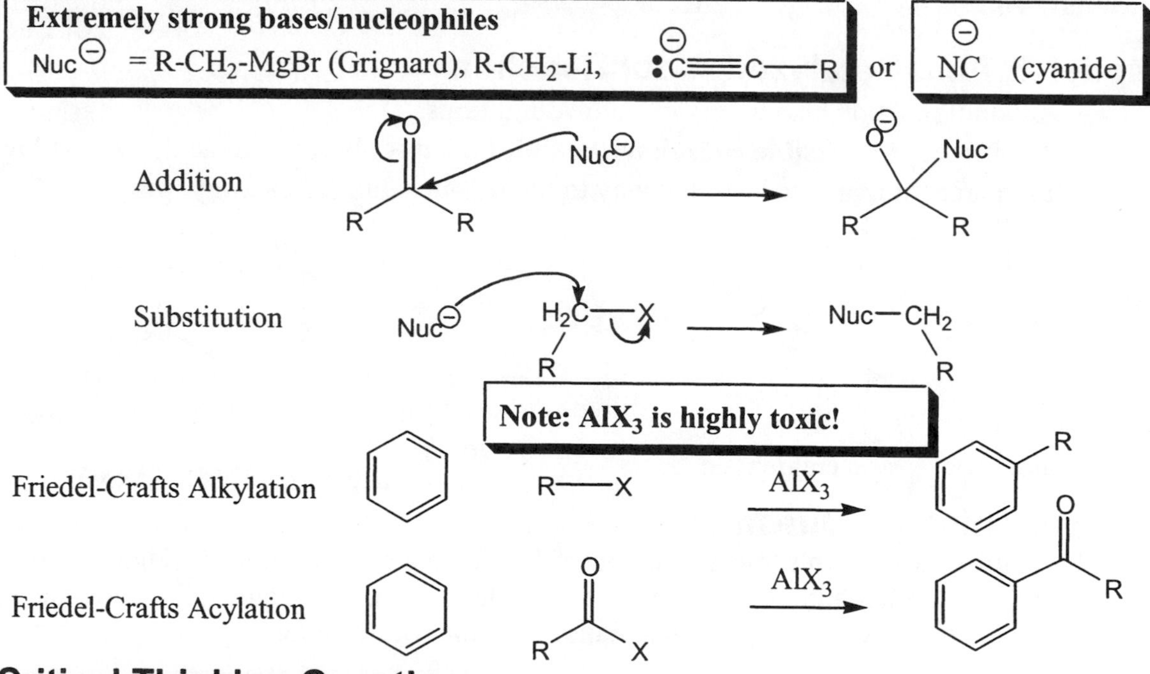

Critical Thinking Questions

4. Which reagents in the reactions above are non-toxic?

5. Does an aldol reaction require concentrated acids, bases, or toxic reagents?

6. Biological systems are constantly performing organic syntheses involving carbon-carbon bond formation. Which of the C-C bond forming reactions we have studied is most likely to take place in a living system?

Model 4: Reversibility of the Aldol Reaction

- One very important thing to remember about the aldol reaction is that it can be **un-done**. (It is a reversible reaction.)

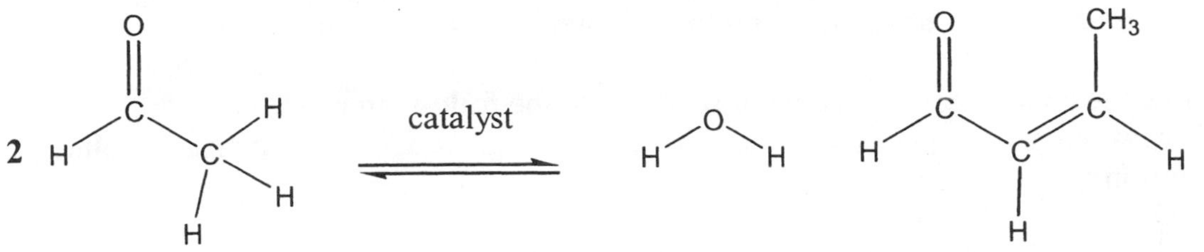

2 acetaldehyde $\underset{}{\overset{\text{catalyst}}{\rightleftharpoons}}$ H_2O + α,β-unsaturated aldehyde

acetaldehyde — α,β-unsaturated aldehyde

Critical Thinking Questions

7. What do you think happens if you add a large amount of water to the α,β-unsaturated aldehyde shown above (in the presence of acid or base catalyst)?

8. Plants use aldol-type condensation reactions to build fructose (and other 6-carbon sugars) from smaller pieces.

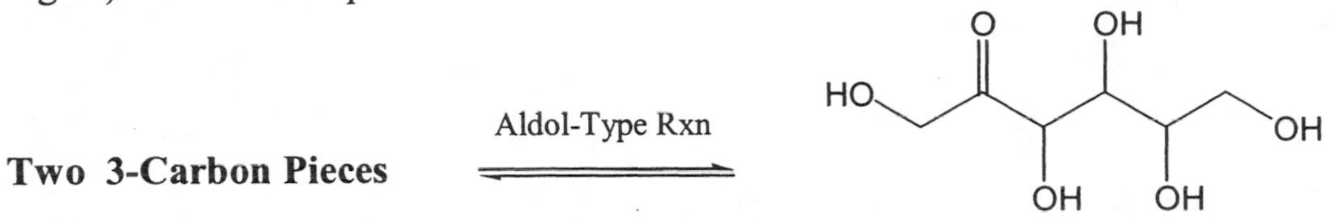

Two 3-Carbon Pieces $\overset{\text{Aldol-Type Rxn}}{\rightleftharpoons}$ **Fructose**

We humans eat the plants and **begin** the digestion of fructose by doing a **reverse aldol reaction** to break it back into 3-carbon pieces.

a) Fructose is a complicated looking molecule, but really **it is an aldol in disguise**. Circle the carbonyl and the one OH group on fructose (below) that are positioned like the C=O and OH of an aldol.
b) Cross out the other 4 OH groups (on the structure below).
c) **DIGEST fructose into two pieces!** That is, use retrosynthesis to figure out the structures of the two 3-carbon pieces into which we break it down.

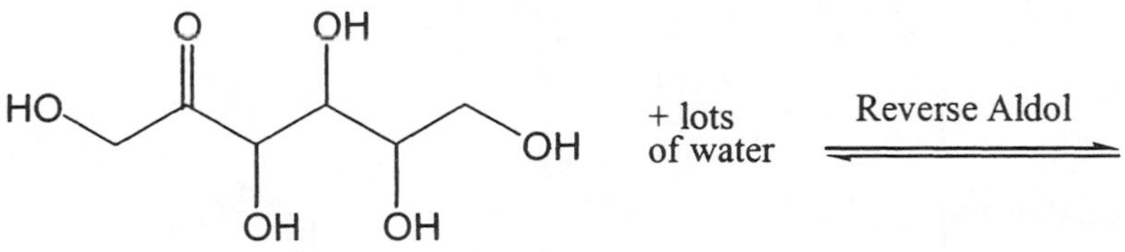

Fructose + lots of water $\overset{\text{Reverse Aldol}}{\rightleftharpoons}$ **Two 3-Carbon Pieces**

Exercises

1. Careful analysis of the product mixture in Model1 shows two isomeric products. A researcher isolates them and analyzes each by mass spectrometry finding they each have a parent ion of 70 amu. Propose a structure for the product not shown in Model 1. (Hint: Is the α,β-unsaturated product shown in Model 1 E or Z?)

2. Draw the condensation product or products of the following reaction mixtures: (Unless otherwise specified you can assume all organic reactions are run in dilute solution.)

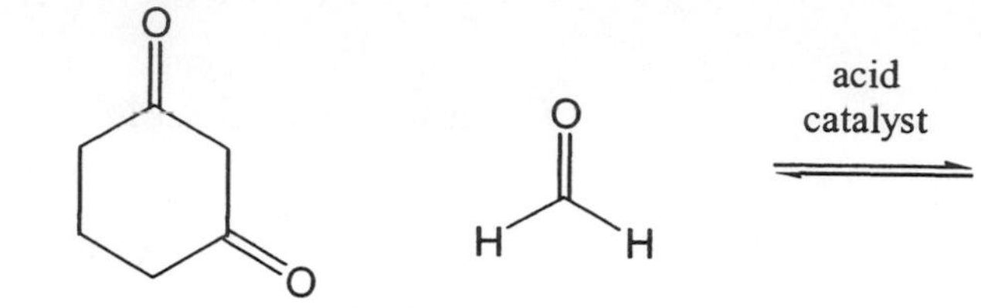

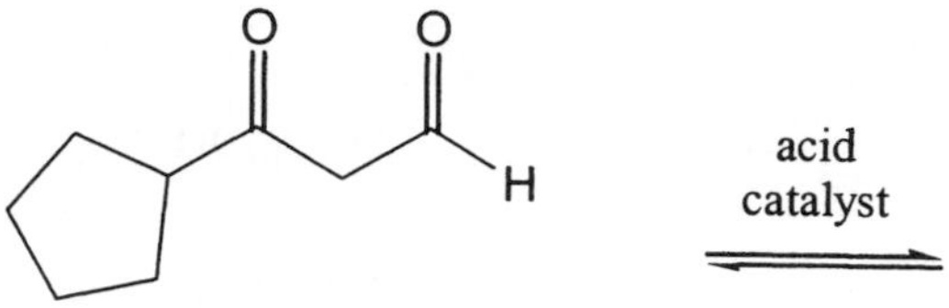

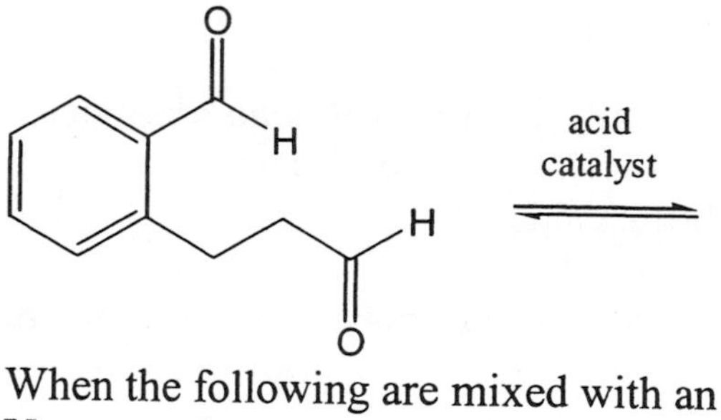

acid catalyst

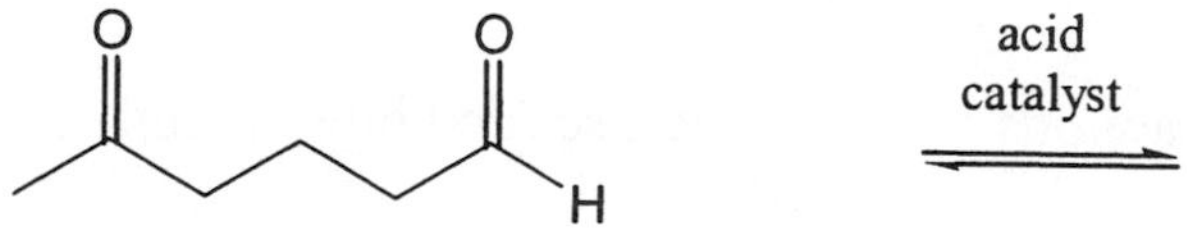

acid catalyst

acid catalyst

acid catalyst

3. When the following are mixed with an acid catalyst, one product is acetophenone. Use curved arrows to construct a reasonable mechanism for this reaction. (One possible intermediate product is shown in the box.)

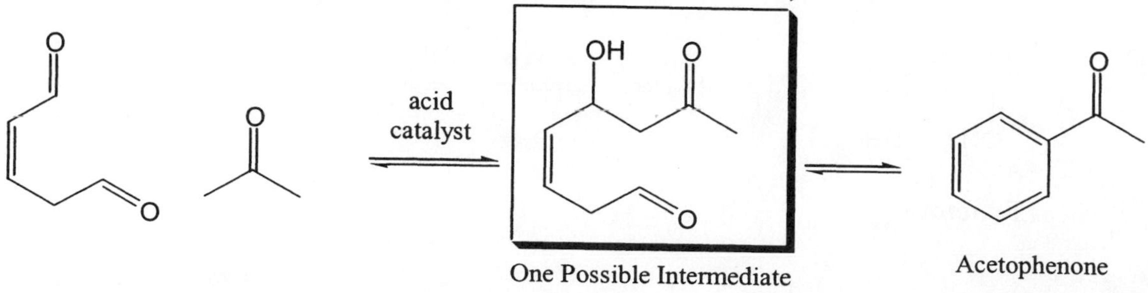

acid catalyst

One Possible Intermediate

Acetophenone

4. Use retrosynthesis to determine starting materials that will give rise to each of the following condensation products.
   - The first step in such an analysis should always be to figure out the aldol product that preceded dehydration (see Model 2).
   - If the target is made using a mixed aldol condensation specify the procedure that will give the highest yield of the desired target.

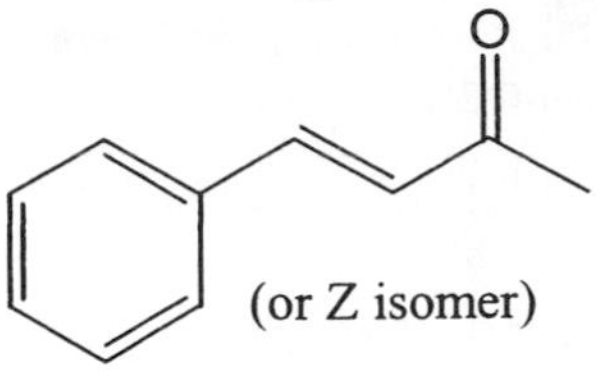

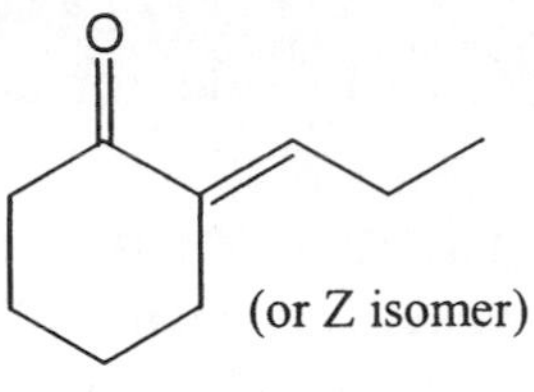

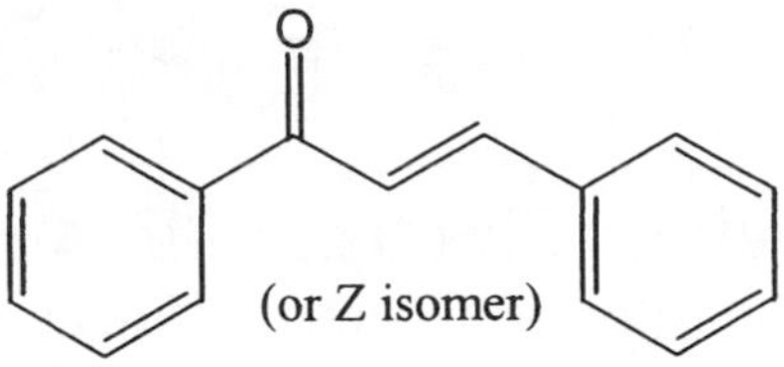

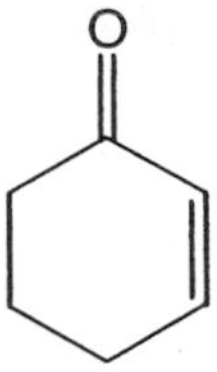

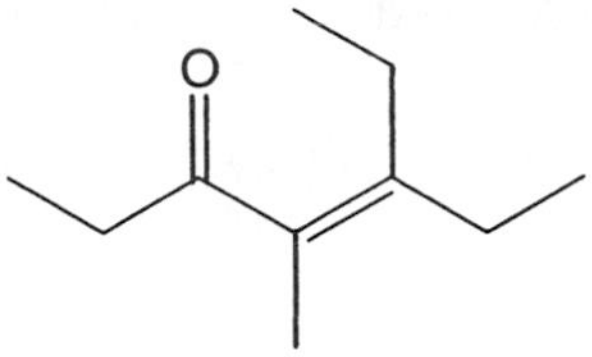

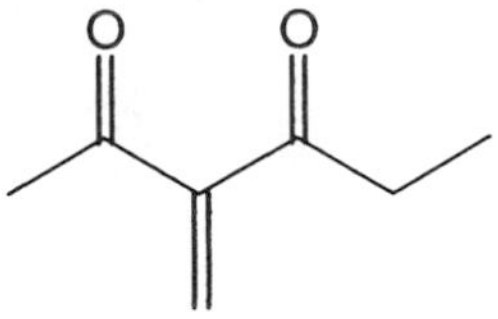

(or Z isomer) (or Z isomer) (or Z isomer)

5. Draw the enol of each of the following molecules.

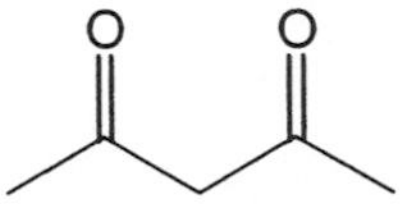

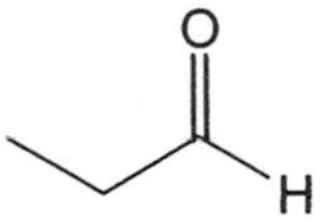

   a) Which enol do you expect to be more favorable/lower in P.E.? (This is also the enol that is more likely to form.)

   b) Explain your reasoning.

   c) Is your answer above consistent with the observation that the β-keto compound on the left enolizes faster than the aldehyde on the right (both in the presence of acid catalyst)?

   d) Draw the condensation product that is most likely to form if these two are mixed in the presence of an acid catalyst. (Which one is more likely to act as nucleophile?)

6. Read the assigned pages in your text and do the assigned problems.

ChemActivity 36

Part A: 1,4-Addition to α,β-Unsaturated Carbonyl Compounds

(What is the difference between a 1,4-addition and a 1,2-addition?)

Review: Electrophilic Additon to a Diene

In the following **electrophilic addition reactions** we used D-Cl instead of H-Cl so we can follow the placement of both the electrophile (D^+) and the nucleophile (Cl^-).

Figure A: 1,2-Addition of D–Cl to 1,3-Butadiene

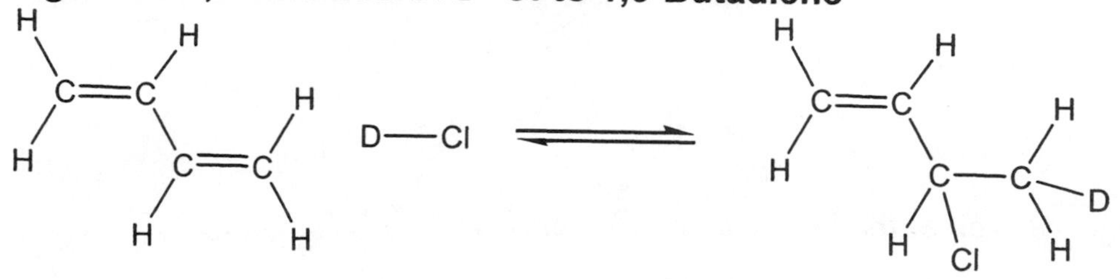

Figure B: 1,4-Addition of D–Cl to 1,3-Butadiene

Critical Thinking Questions

1. Explain why the reaction in Figure A is called a "**1,2-addition**," while the reaction in Figure B is called a "**1,4-addition**."

2. Use curved arrows to show a likely mechanism for each reaction (Fig. A & B).

Model 1: Electrophilic Additon to an α,β-Unsaturated Carbonyl

- You can think of an α,β-unsaturated carbonyl compound as a di-ene (*see below*).
- **Most** nucleophile/electrophile pairs will do **1,4-addition** to an α,β-unsaturated carbonyl.

Critical Thinking Questions

3. Which atom of an α,β-unsaturated carbonyl compound is most nucelophilic/basic?
   a) Use curved arrows to show the most likely reaction between the α,β-unsaturated carbonyl compound shown below and the electrophile, D^+.
   b) Draw the cation that results including **all three resonance structures.**

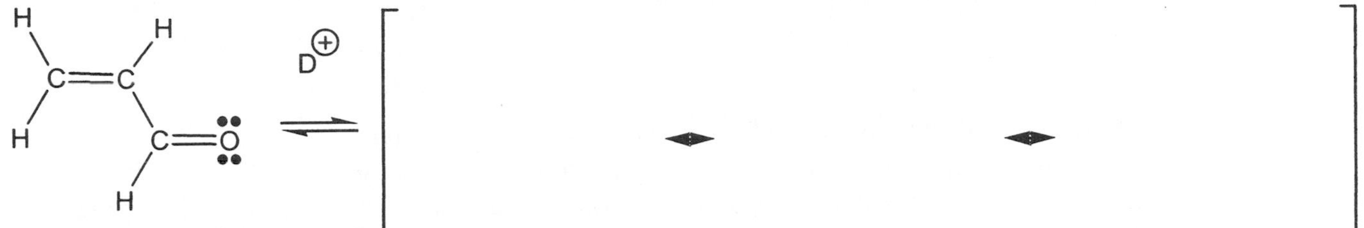

α,β-Unsaturated Carbonyl

4. Cl^-, like most nucleophiles, preferentially adds to the **4 position** (β position) of an α,β-unsaturated carbonyl (instead of the 2 position/α position).
   a) Add Cl^- to the drawing above and use curved arrows to show it reacting with the "4" position of any of the resonance structures (also called the β position).
   b) Draw the 1,4-additon product of this reaction.

D—Cl

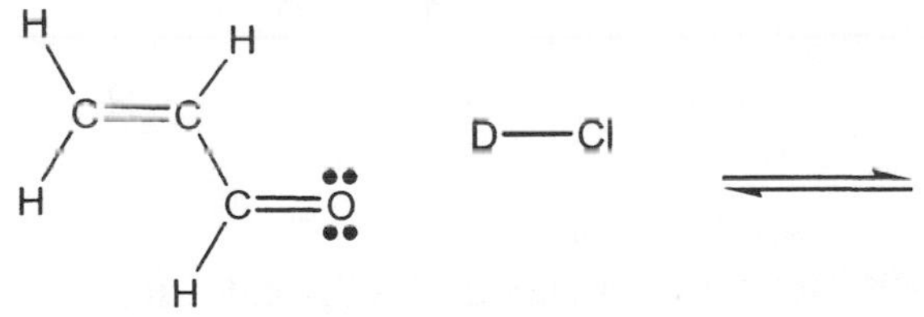

1,4-Addition Product

5. Interestingly, researchers have found that this reaction actually yields the product shown below, right!
   a) What is the relationship between the two products (above and below)?
   b) Does it fit with your current understanding that (in the presence of a catalyst) the product above will spontaneously change into the product shown below?

D—Cl

1,4-Addition Product

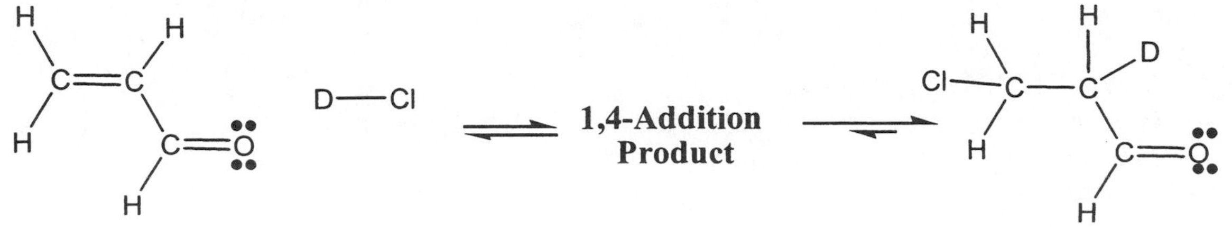

   c) You might want to call the product of this reaction a **3,4-addition product**. Explain. (Note that chemists call it a 1,4-addition product because of the original positions of the electrophile and nucleophile.)

Information: Tautomerization

From now on in the course, unless you are explicitly asked to do so, you are not responsible for showing the mechanism of a tautomerization. If you encounter an enol in a mechanism, AND there is a polar molecule capable of helping with H transfer, you can simply write "**tautomerize**" and draw the corresponding keto form (as shown below).

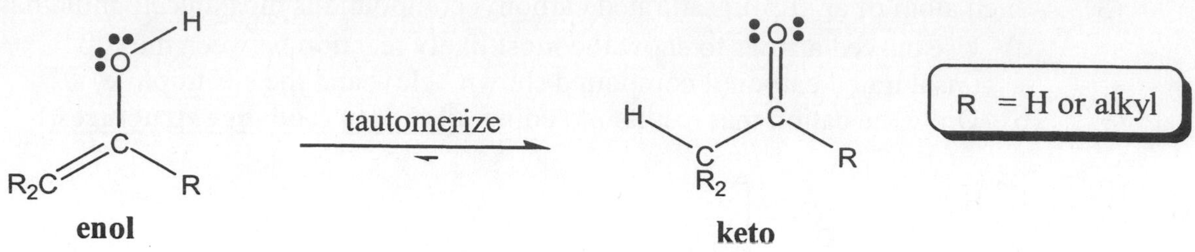

enol —tautomerize→ keto

R = H or alkyl

Model 2: Amine Nucleophile in Acid Solution

Amines are bases, but as long as a solution is not too acidic, there will be some "**free amine**" (= unprotonated amine), available to act as a base or nucleophile in an acidic solution.

Figure 2: "1,4-Addition" to an α,β-Unsaturated Carbonyl Compound

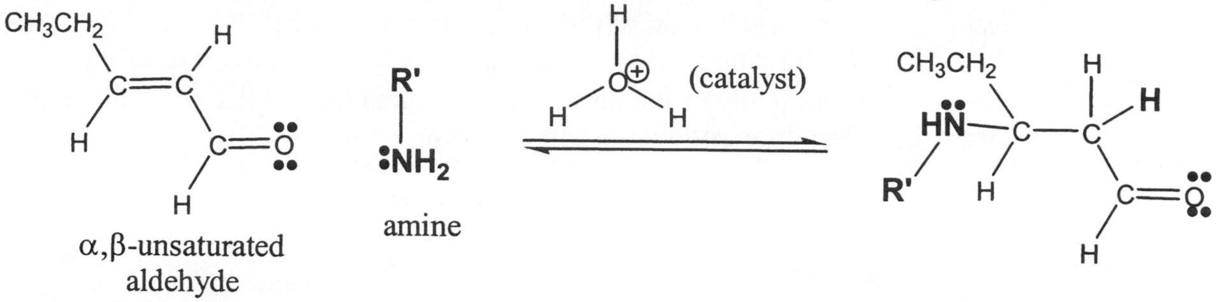

α,β-unsaturated aldehyde + amine ($R'NH_2$) —H_3O^+ (catalyst)⇌ product

Critical Thinking Questions

6. Why is the product above called a 1,4-addition product instead of a 3,4-addition product?
7. Use curved arrows to show a mechanism for the reaction in Figure 2.

Part B: Nucleophiles that Prefer 1,2-Addition to an α,β-Unsaturated Carbonyl

(Which nucleophiles tend to undergo addition to the 2-position?)

Model 3: 1,2-Addition to an αβ-Unsaturated Carbonyl

Most nucleophiles, including almost all nitrogen, oxygen, halogen and cyanide nucleophiles, react with the β-position of an αβ-unsaturated carbonyl compound. After tautomerization, this results in a stable carbonyl compound. High-energy Grignard and organolithium reagents such as R–CH_2–Li, $LiAlH_4$ and R-CH_2-MgBr buck this trend and react with the carbonyl carbon.

1,2-Addition

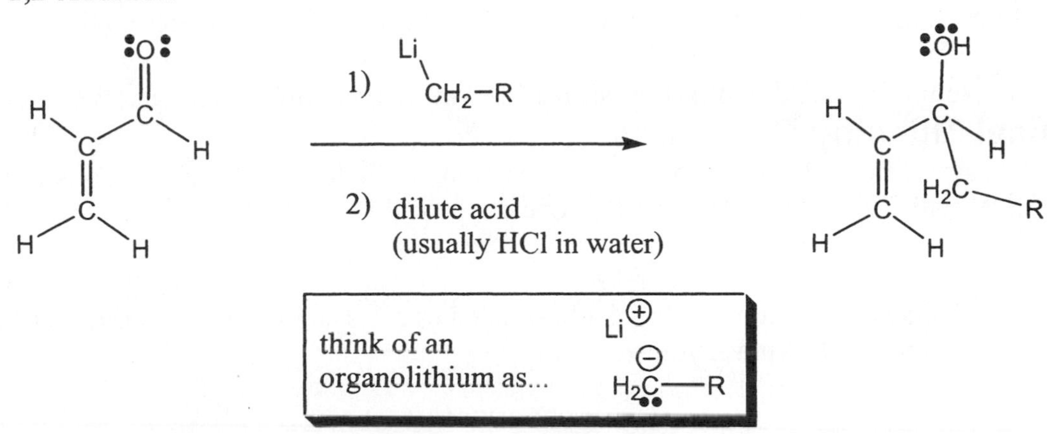

Critical Thinking Questions

8. Show the mechanism for the reaction in Model 3.

Model 4: Reversible vs. Irreversible Reactions

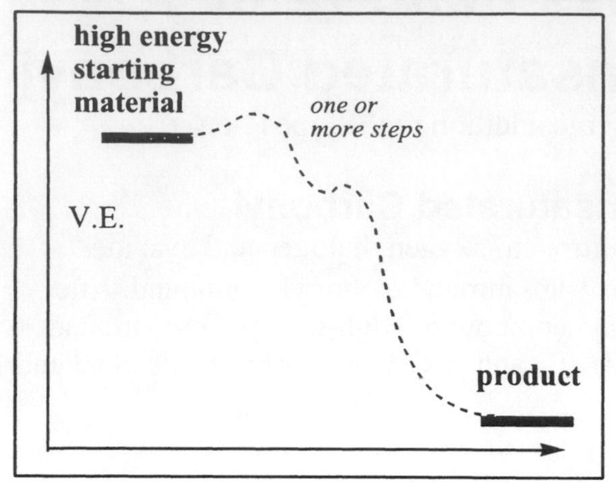

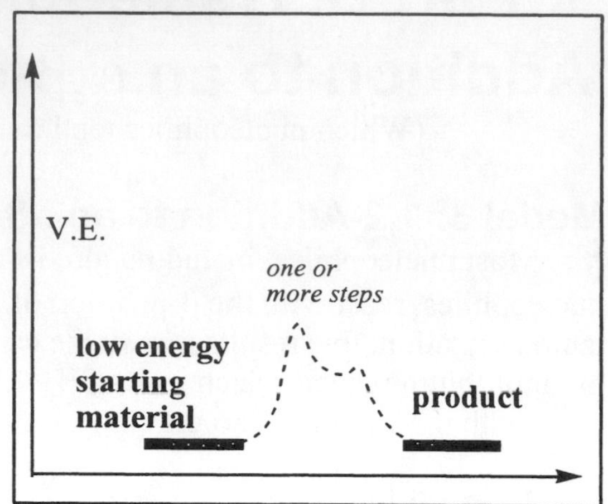

- A reaction is **reversible** if the reverse reaction is favorable compared to the forward reaction.
- A reaction is considered **irreversible** if the reverse reaction is very unfavorable.

Critical Thinking Questions

9. Construct an explanation for why the reaction on the left in Model 4 is considered essentially "irreversible," while the one on the right is "reversible?"

10. Which energy diagram in Model 4 best matches a 1,2-addition with a Grignard or lithium reagent? Explain your reasoning.

Model 5: 1,2-Addition vs. 1,4-Addition

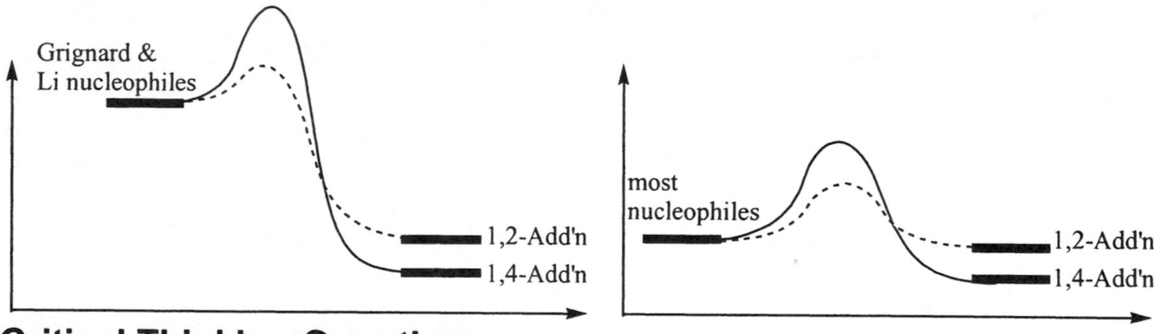

Critical Thinking Questions

11. Addition of most nucleophiles to an αβ-unsaturated carbonyl compound results in a 1,4-addition product. However, sometimes there is an initial reaction to form a 1,2-addition product. On the energy diagram above, **right**, trace the path of an αβ-unsaturated carbonyl compound that first becomes a 1,2-addition product, and then eventually becomes a 1,4-addition product.

12. Addition of a high energy nucleophile (Grignard or Li reagent) to an αβ-unsaturated carbonyl compound results in a 1,2-addition product. Explain why this product **cannot** turn into a lower energy 1,4-addition product.

Exercises for Part A

1. Either of the two different double bonds of an α,β-unsaturated carbonyl compound can react with an electrophile (D^+). **Draw** **all important resonance structures** of the **most likely** carbocation that results from the two possible reactions.

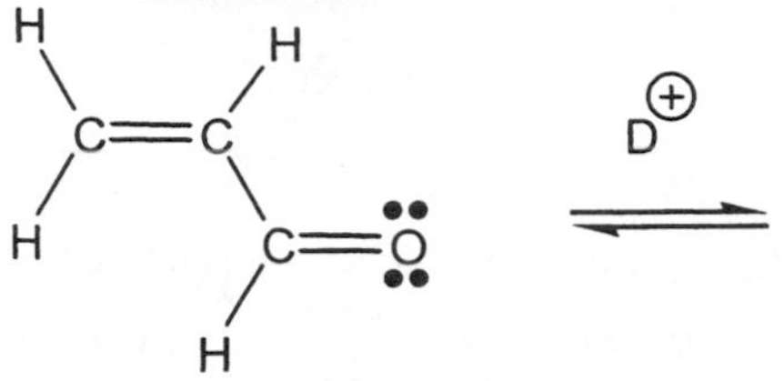

α,β-Unsaturated Carbonyl

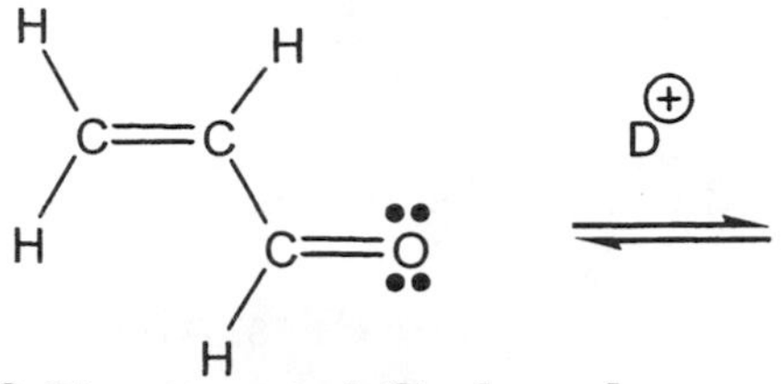

α,β-Unsaturated Carbonyl

   a) Recall that, as a rule, oxygen must always have an octet. Based on this, write "**much more likely**" next to one of the electrophilic additions in the previous question?
   b) Is your conclusion consistent with your answer to CTQ 3?

2. Show the mechanism and most likely product that results from 1,4-addition of water to an α,β-unsaturated carbonyl in acidic solution (acid catalyzed). Water, like most nucleohphiles, prefers to react with the 4 position rather than the 2 position.

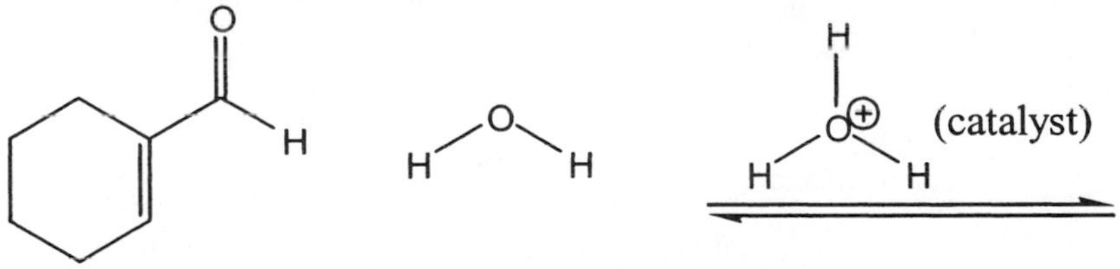

(catalyst)

3. Draw the mechanism and most likely product that results from a 1,4-addition of ethanol in ethoxide ($CH_3CH_2O^-$) to this same α,β-unsaturated carbonyl in basic solution. (With a strong base/nucleophile, there can be no acid catalyst to first protonate the carbonyl oxygen. Fortunately, with a strong nucleophile in solution, this is unnecessary.)

4. Is ethoxide considered a catalyst in this reaction? Explain.

5. Draw the products that result from addition of 1) $NaNHCH_3$ followed by 2) dilute HCl to this same α,β-unsaturated carbonyl.

6. Draw the product that results when HCN (and a catalytic amount of NaCN) are added to the α,β-unsaturated carbonyl in the previous question. (Cyanide anion also prefers to react with the 4-position.)

7. Read the assigned pages in your text and do the assigned problems.

Exercises for Part B

8. For the reaction in Model 3, the organolithium reagent is added and allowed to react BEFORE dilute acid is added. Predict what side-products would form if these two reagents were added at the same time.

9. Show the mechanism of the imine formation reaction when an aldehyde is treated with an amine with an acid catalyst.

10. Lithium aluminum hydride ($LiAlH_4$) is the functional equivalent of a hydride ion (H^-). Because it adds to the 2 position, it can be used to selectively reduce the carbonyl of an an α,β-unsaturated carbonyl without affecting the C=C. (The opposite can be achieved with palladium (Pd°) on carbon with H_2 gas.) Draw the product of the reaction between
    a) The α,β-unsaturated aldehyde in Exercise 2 and $LiAlH_4$.
    b) The α,β-unsaturated aldehyde in Exercise 2 and Pd/C with H_2 gas.

11. Design an efficient synthesis of each of the following target molecules. All carbon atoms must come from the starting material/s given.

Target	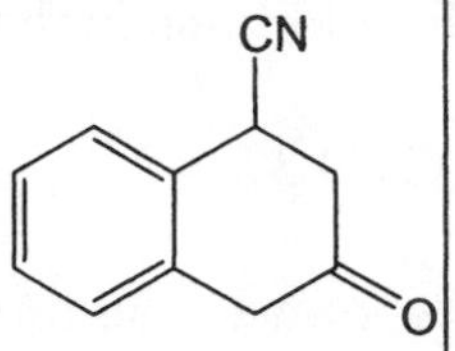	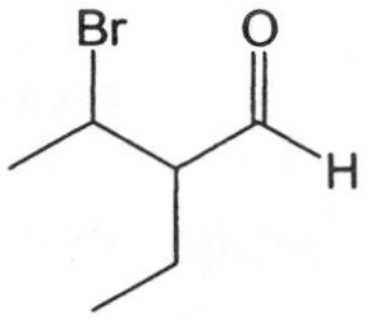	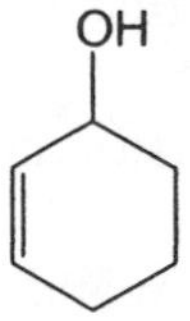	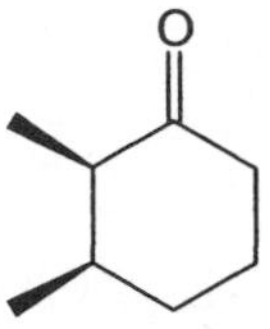 cis ONLY
Starting material	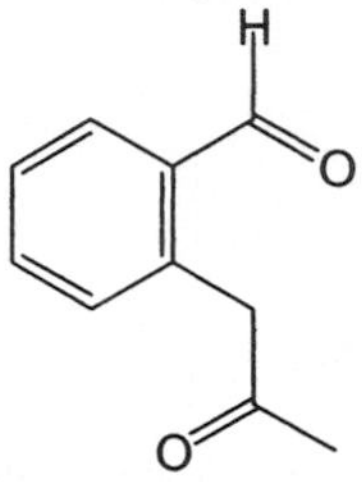 potassium cyanide	butanal bromoethane	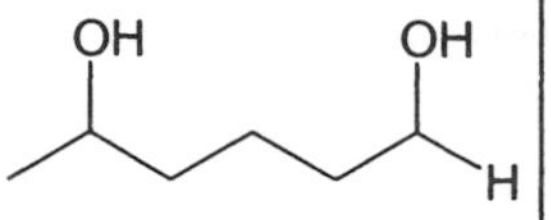	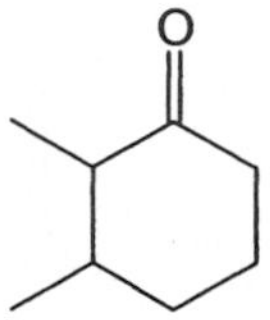 mixture of cis/trans

12. Read the assigned pages in your text and do the assigned problems.

ChemActivity 37

Michael Addition

(What is the mechanism of the Michael Addition?)

Model 1: 1,4 vs 1,2 Addition to an α,β-Unsaturated Carbonyl

The common carbon nucleophiles we have studied so far (Li and Grignard reagents) "prefer" to add to the 2-position because…

- the 1,2-addition is faster than the 1,4-addition and
- it is very downhill to the right so as to be irreversible.

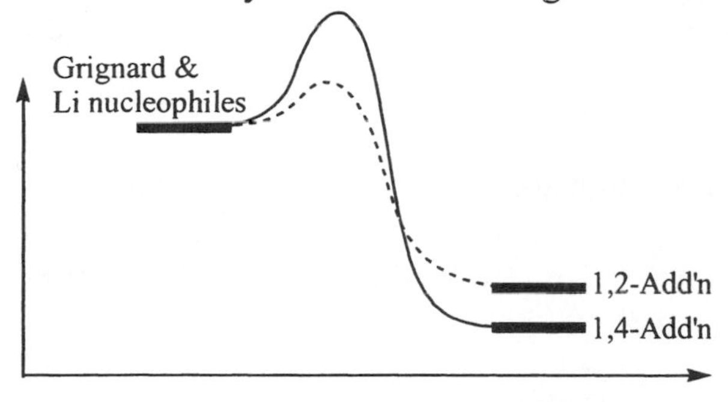

Critical Thinking Questions

1. One of the following synthetic targets is not easily accessible from the starting material using Grignard or lithium nucleophiles.

   a) Indicate the new C-C sigma bond in each Target.
   b) Cross out the more inaccessible target and explain your reasoning.

O

Starting Material

$CH(CH_3)_2$

OH

Target A

$CH(CH_3)_2$

O

Target B

   c) T or F: Using reactions we have studied, C-C bond formation at the 4-position of an α,β-unsaturated carbonyl compound is difficult or impossible, while C-C bond formation at the 2-position is easy.

Model 2: Michael Addition Reaction

Professor Arthur Michael (1853-1942) of Harvard University developed a method for making carbon-carbon sigma bonds at the 4-position of an α,β-unsaturated carbonyl compound. The reaction now bears his name.

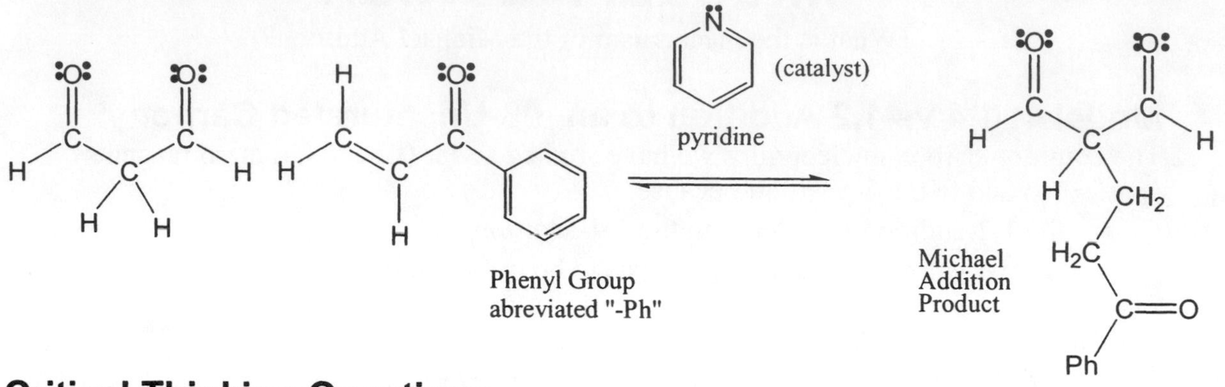

(catalyst)

pyridine

Phenyl Group abreviated "-Ph"

Michael Addition Product

Critical Thinking Questions

2. Indicate the new C-C sigma bond in the reaction above.

3. On the product, indicate the carbon that was originally the 4-position (or β-position) of the α,β-unsaturated carbonyl compound.

4. What does the Michael Reaction accomplish that cannot be done with other reactions we have studied?

5. Pyridine is a moderate base and a poor nucleophile.
   a) Use curved arrows to show the most likely acid-base reaction (proton transfer) involving pyridine and one of the reactants in Model 2.

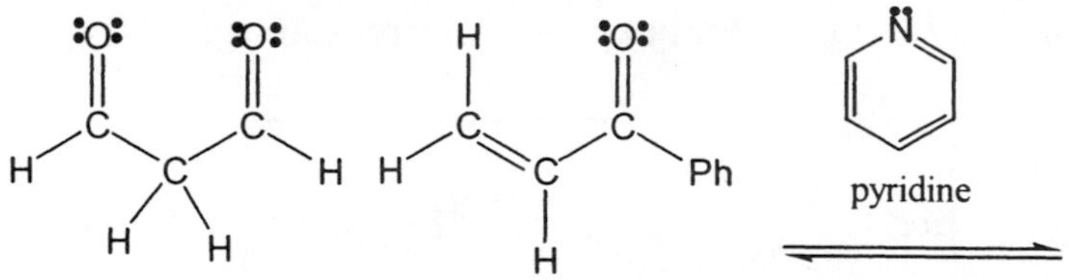

pyridine

   b) Draw the conjugate base, including all important resonance structures.
   c) Explain why the proton you pulled off with pyridine is the most acidic proton among those on the reactant molecules.

6. Use curved arrows to show a mechanism for the Michael Addition in Model 2.

Exercises

1. Can the 1,2-addition product shown in Model 1 tautomerize to a lower energy keto form? Explain.

2. The 2nd order resonance structure below explains why the carbonyl carbon is electrophilic and the carbonyl oxygen is nucleophilic.

very bad resonanc structure (O must have an octet)

composite structure

Draw similar resonance structures for the α,β-unsaturated aldehyde below.

   a) Complete the following composite structure for this aldehyde by adding δ+ and δ– where necessary. [A composite structure uses partial charges (δ+) and partial bonds (---) to represent multiple resonance structures at the same time.]

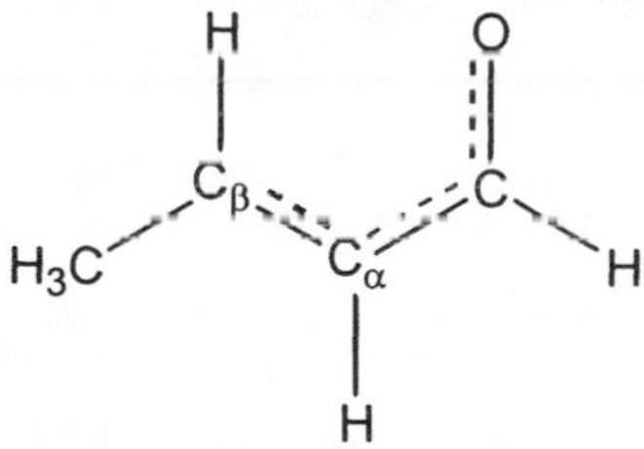

   b) Is the β-carbon of an α,β-unsaturated carbonyl compound **electrophilic** or **nucleophilic** [circle one]?
   c) Based on the pattern of charge distribution on the α,β-unsaturated aldehyde above, do you expect the α-carbon to have a δ+ or δ– charge [circle one]?
   d) Is the α-carbon of an α,β-unsaturated carbonyl compound **electrophilic** or **nucleophilic** [circle one]?

3. For each of the following reactions…
   i) Identify each α-carbon that will act as a nucleophile and whether it will be an enol or an enolate nucleophile.
   ii) Identify each β-carbon that will act as an electrophile
   iii) Identify each carbonyl carbon that will act as an electrophile.
   iv) Draw the keto form of any enol product shown.
   v) Identify each 1,2-addition and each 1,4-addtion (some are neither).

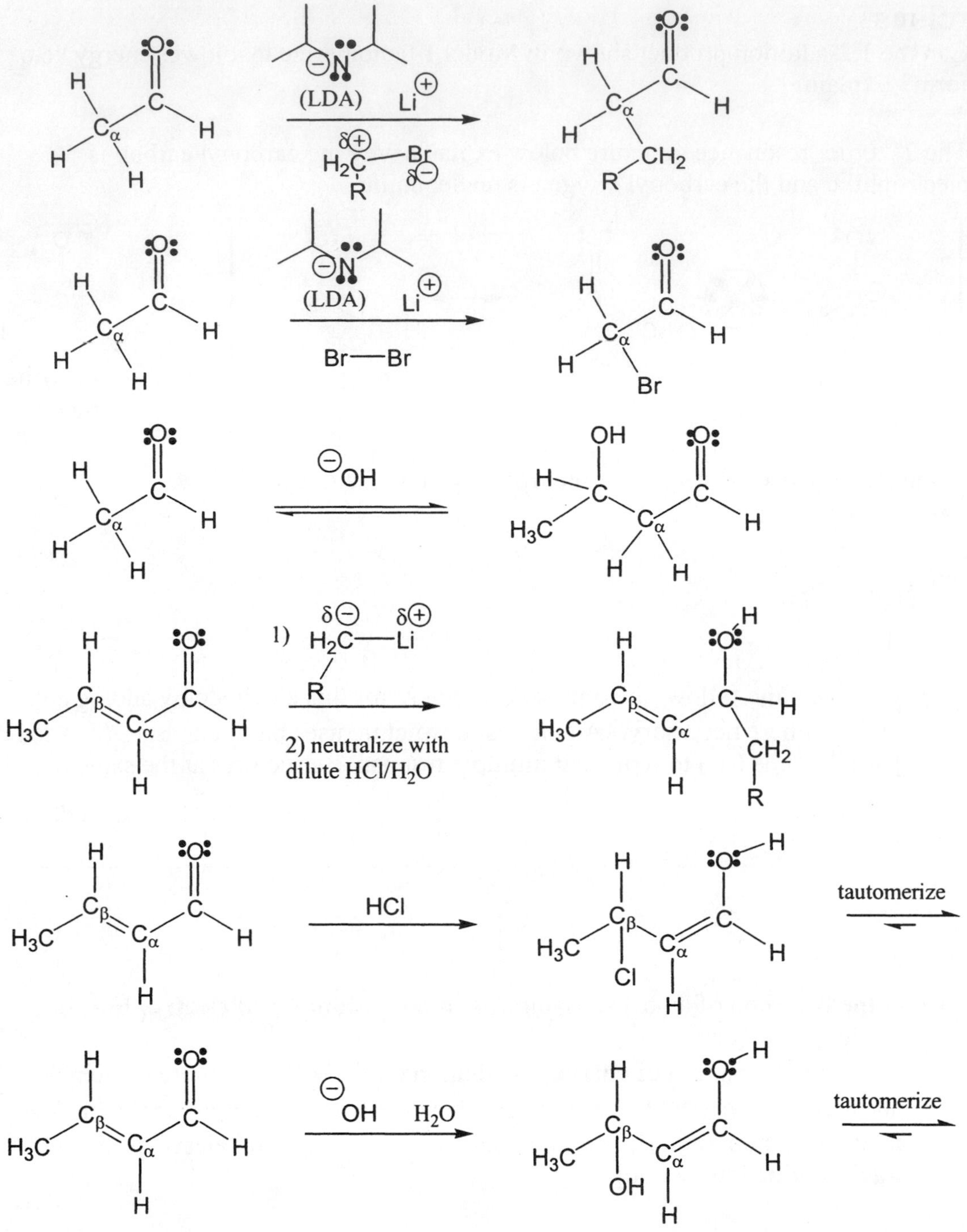

(LDA) Li^+

H_2C—Br, R

$Br—Br$

^-OH

1) H_2C—Li, R

2) neutralize with dilute HCl/H_2O

HCl

tautomerize

^-OH H_2O

tautomerize

4. The di-ketone below can undergo an intramolecular Michael Addition. Draw the mechanism of this reaction.

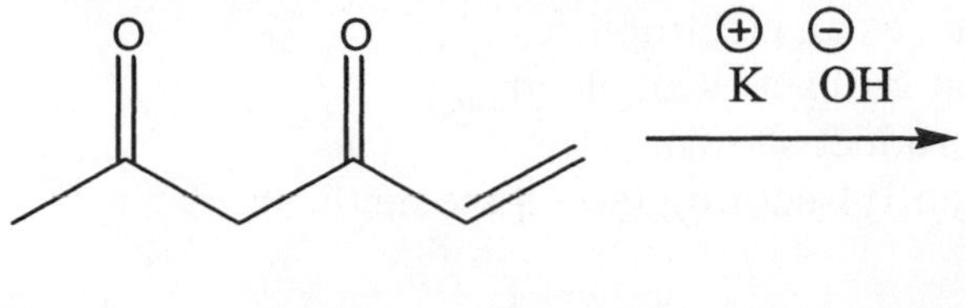

K^+ ^-OH

5. The reactants from the Michael Reaction in Model 2 are drawn below. Describe reaction conditions that would minimize the formation of the aldol product shown below.

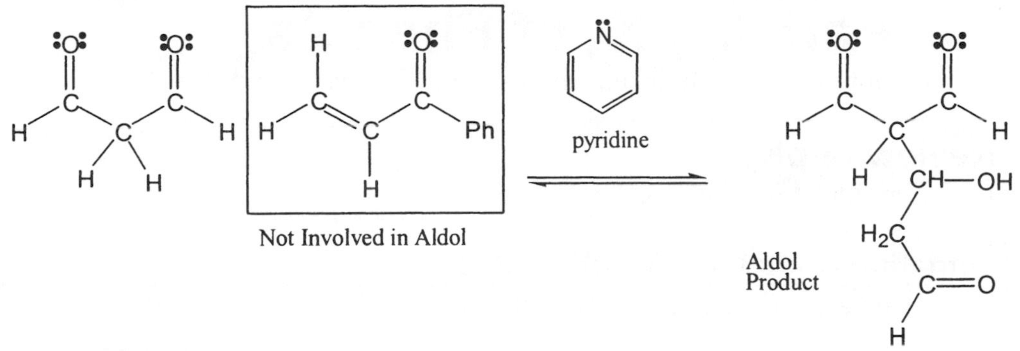

Not Involved in Aldol

pyridine

Aldol Product

6. H_2NNH_2 in base selectively reduces an aldehyde C=O to a CH_2. (At higher temperatures it can reduce a ketone C=O to a CH_2.)

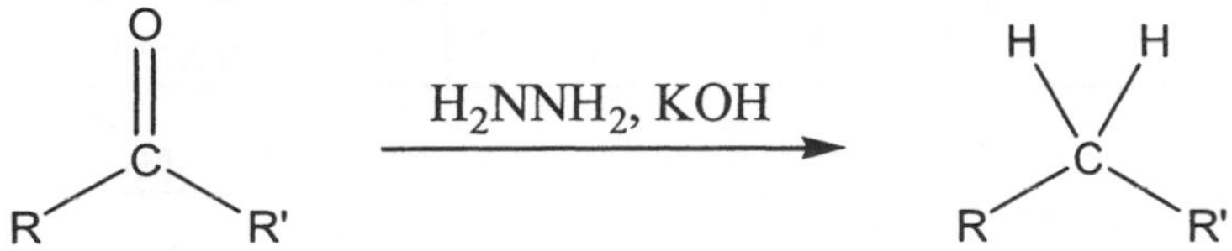

H_2NNH_2, KOH

Design a synthesis of Target B in CTQ 1 from the starting material shown there.

7. Show the mechanism of each reaction in Exercise 3.

8. Design an efficient synthesis of each of the following target molecules. All carbon atoms must come from the starting material/s given.

Target	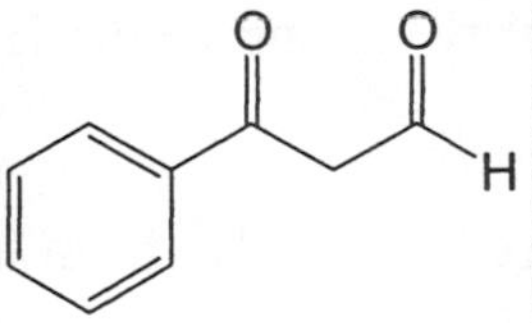	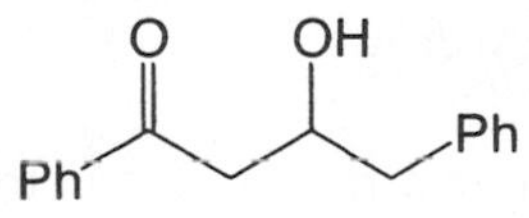	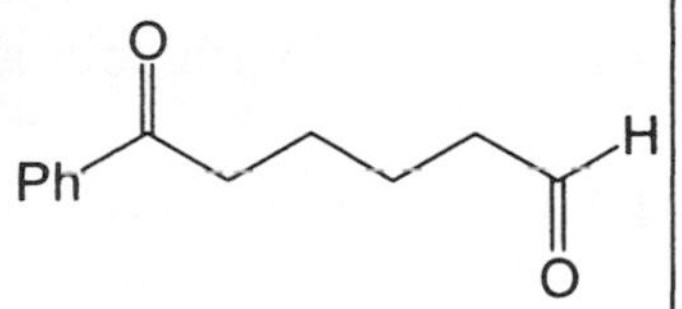	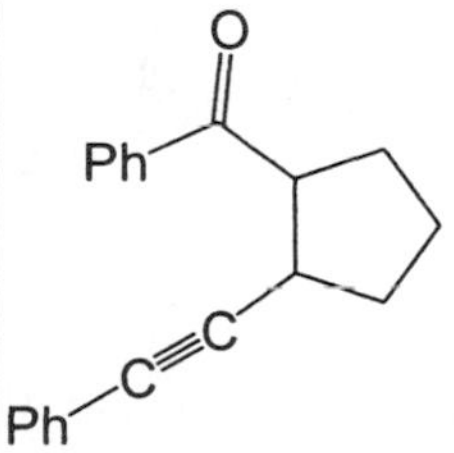
Starting material	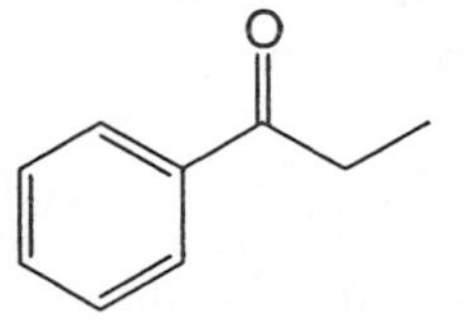	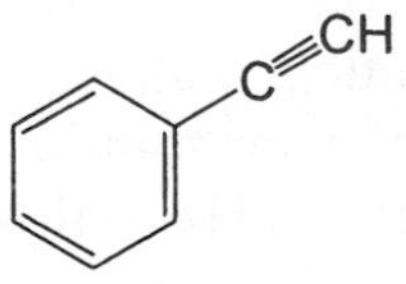 Hint: consider Markov. & Anti-Markov. hydration	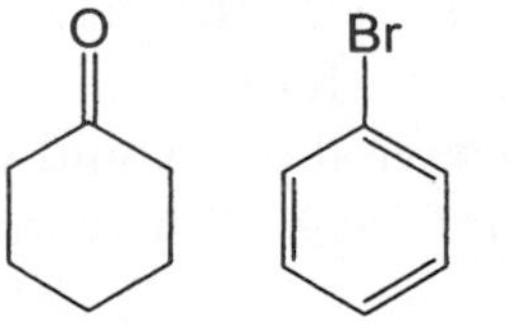 Hint: this synthesis involves a Grignard reagent in one step and ozone in another.	Use any of the starting materials OR PRODUCTS listed so far on this table.

9. Read the assigned pages in the text and do the assigned problems.

ChemActivity 38

Part A: pK_a of Phenols

(What can you tell about a molecule and its conjugate base from its pKa?)

Model 1: Review of pK_a

Consider the hypothetical acids H–A and H–B…

Table 1: Information Contained in a pKa Value

	Acid	pKa	Conjugate Base	pH when [HX] = [X⁻]	Energy Required to Break H–X Bond
Acid #1	H–A	6	A^-	6	6 pKa units of energy
Acid #2	H–B	9	B^-	9	9 pKa units of energy

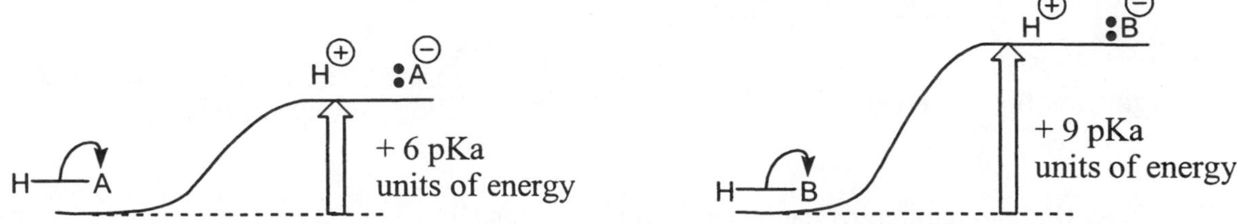

Acid base reactions are very fast, and usually come to equilibrium within seconds.

Critical Thinking Questions

1. According to the pKa data, which is a stronger acid: **HA** or **HB** [circle one]?

2. According to Model 1, which acid requires **more** energy to remove the H^+? (Is this consistent with your answer to CTQ 1?)

3. The pKa of an acid can tell you how difficult it is to remove an H^+ from that acid. Explain.

4. A **buffer** "sets" the pH of a solution and keeps it at that pH. In the lab, you can choose a buffer to "set" your solution to almost any pH between pH 2 and 12.
   a) To what pH would you "set" your solution of HA in water if you want $[HA]=[A^-]$ (the concentration of HA = the concentration of A^-)?

Information:

- At **higher pH** there is **a higher $[HO^-]$** and lower $[H^+]$ in solution
- At **lower pH** there is **a higher $[H^+]$** and a lower $[HO^-]$ in solution. (recall that for all solutions $[H^+] \times [HO^-]$ = a constant = 10^{-14})

   b) You put a sample of HA into water buffered to pH 7, which is greater when the solution comes to equilibrium (in a few seconds): **[HA]** or **$[A^-]$**? Explain.

Model 2: Acidity of Cyclohexanol vs. Phenol

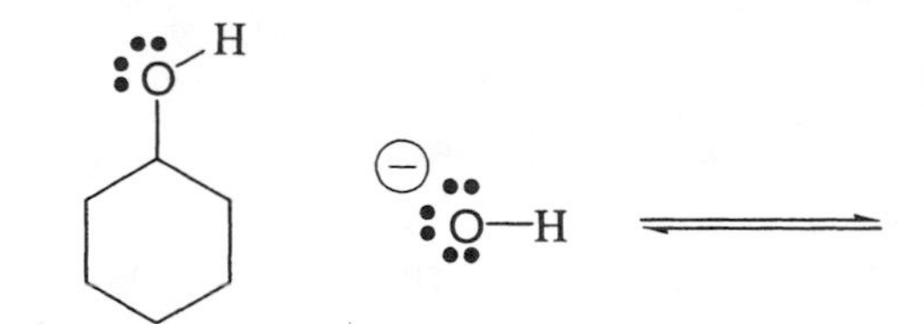

cyclohexanol

conjugate base of cyclohexanol (w/ resonance structures–if any)

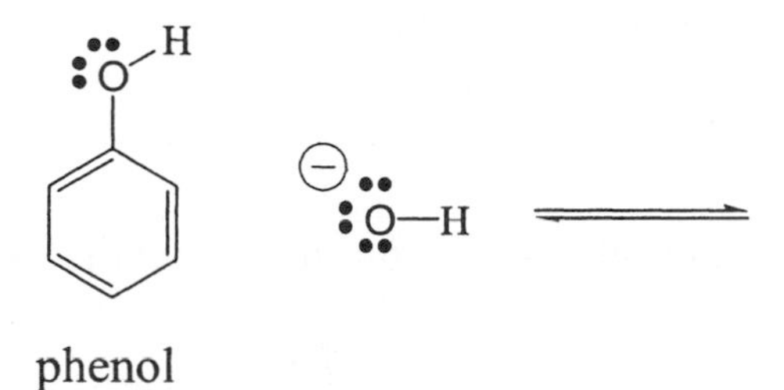

phenol

conjugate base of phenol (w/ resonance structures–if any)
"phenoxide ion"

Critical Thinking Questions

5. Show the most likely **acid-base reaction** between each pair of reactants in Model 2, and draw the conjugate base that results from the reaction, **including all important resonance structures** (if any).

   a) Which conjugate base do you expect to be **lower in potential energy**? Explain your reasoning.

   b) Sketch an energy diagram showing these two reactions. Assume phenol and cyclohexanol have about the same energy.

   c) Predict which alcohol (cyclohexanol or phenol) is more acidic. (More acidic = easier to remove an H^+.)

6. The pKa of phenol is 10. The pKa of cyclohexanol is 17. Is this consistent with your conclusions on the previous page?

   a) You put phenol in water buffered to pH 11. A minute later, which do you expect to be greater: **[phenol]** or **[phenoxide ion]** [circle one]?

   b) Shifting the pH by 1 changes the ratio of [H–A]/[A^-] by a factor of 10. What is the following ratio at pH 14?
   (Hint: first think, which dominates at pH 14.)

$$\frac{[\text{phenoxide}]}{[\text{phenol}]} =$$

   c) At what pH do you expect a buffered solution to be ≈99% phenol (and ≈1% phenoxide)?

7. A nitro group (abbreviated —NO_2) is an electron withdrawing group. Circle the ion you expect to be lower in potential energy?

NO_2

meta-nitrophenoxide ion
(one of several resonance structures)

phenoxide ion
(one of several resonance structures)

Explain your reasoning.

8. Below is an energy diagram for removal of an H^+ from two different acids. Assume the two acids have about the same starting energy. Write the name ***meta*-nitrophenoxide ion** in one box, and **phenoxide ion** in the other box.

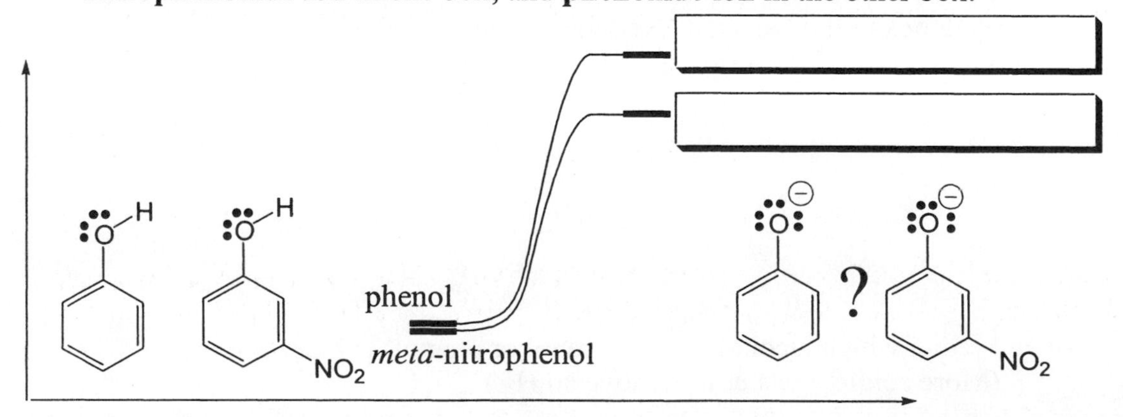

9. Predict which acid has a lower pH (is more acidic): ***meta*-nitrophenol** or **phenol**.

Part B: pK_a's of Substituted Phenols

(How does the position of an electron withdrawing or donating group affect pKa?)

Model 3: Resonance Description of Phenoxide Ion

Critical Thinking Questions

10. Use curved arrows to show generation of the set of resonance structures in Model 3.
11. Based on the description of phenoxide ion above, complete the composite structure below. Place δ– where appropriate. Lone pairs not shown. Ion has overall –1 charge.

Composite Structure of Phenoxide Ion *(please complete)*

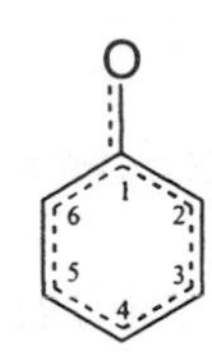

12. At which positions on the phenyl ring do you expect an **electron withdrawing group** to have the greatest impact on the potential energy of the phenoxide ion?

2(ortho) ***3(meta)*** ***4(para)*** ***5(meta)*** ***6(ortho)*** [circle all that apply]

13. Do you expect an **electron donating group** (such as an alkyl group) to stabilize (lower the P.E. of) or destabilize (raise the P.E. of) a phenoxide ion? Explain.

CH_2CH_3

Electron Donating Group

14. At what position/s on the ring would you expect such an electron donating group to have the greatest impact?

15. Number the following phenols **from most acidic to least acidic** and explain your reasoning. (1 = most acidic…4 = least acidic).

NO_2 NO_2 CH_3

Model 4: pK_a Values for Various Substituted Phenols

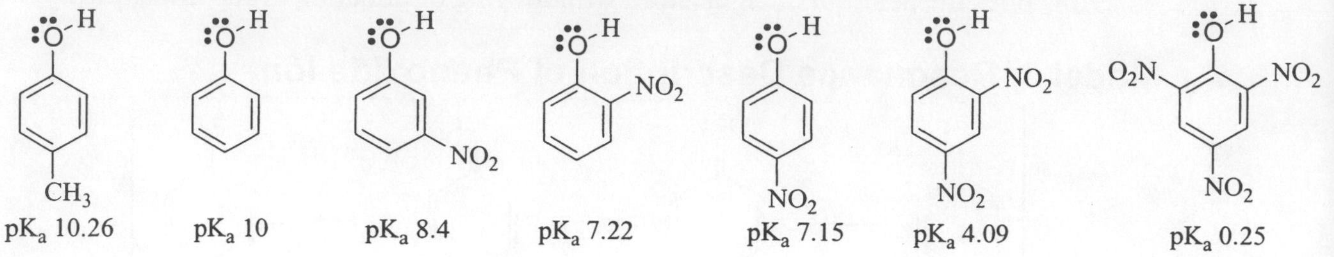

CH_3 — pK_a 10.26

pK_a 10

NO_2 — pK_a 8.4

NO_2 — pK_a 7.22

NO_2 — pK_a 7.15

NO_2, NO_2 — pK_a 4.09

O_2N, NO_2, NO_2 — pK_a 0.25

Critical Thinking Questions

16. Are your predictions in the previous question consistent with the pKa data above?

17. Below the tri-nitro phenol, draw its conjugate base.
    a) Is this conjugate base the highest P.E. or the lowest P.E., among the conjugate bases of structures in Model 4?

    b) Explain your reasoning.

18. Below are four resonance structures of the conjugate base of *para*-nitrophenol, with the nitro group drawn as a full Lewis structure. Draw a fifth resonance structure that shows that nitro **withdraws electrons from the ring via resonance effects.**

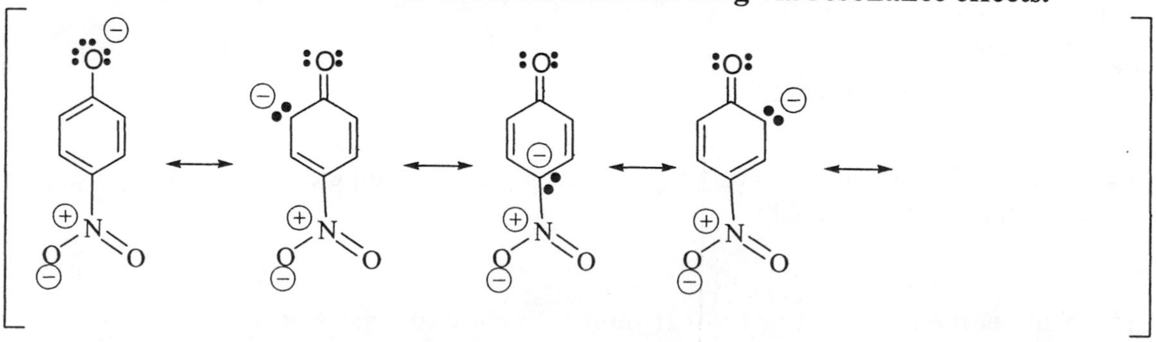

conjugate base of *para*-nitrophenoxide ion

Model 5: Carbanions (Opposite of Carbocations)

- A carbanion is a structure with a negative charge on carbon.
- In many ways, carbanions are the opposite of carbocations.

Carbocations

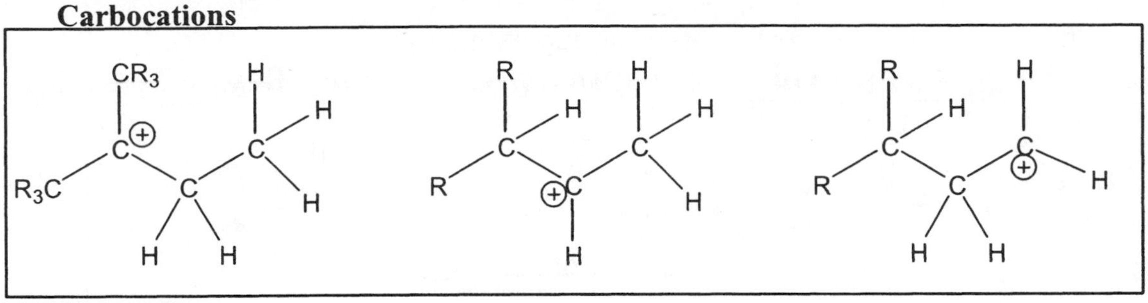

Carbanions

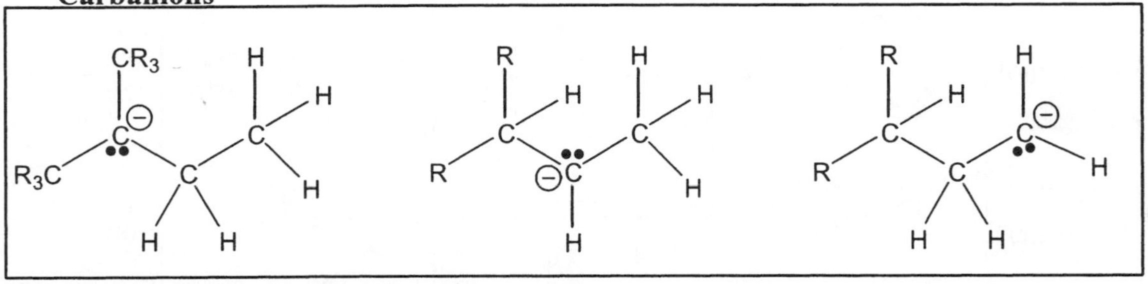

Critical Thinking Questions

19. Label each **carbocation** and each **carbanion** in Model 5 as being primary (1°), secondary (2°), or tertiary (3°).

20. Label one of the carbocations "**most stable**" and another "**least stable**," and explain your reasoning.

21. Label one of the carbanions "**most stable**" and another "**least stable**," and explain your reasoning.

22. Which of the following is higher in potential energy and therefore a stronger base? (Hint: consider which carbons on each structure hold a partial negative charge.)

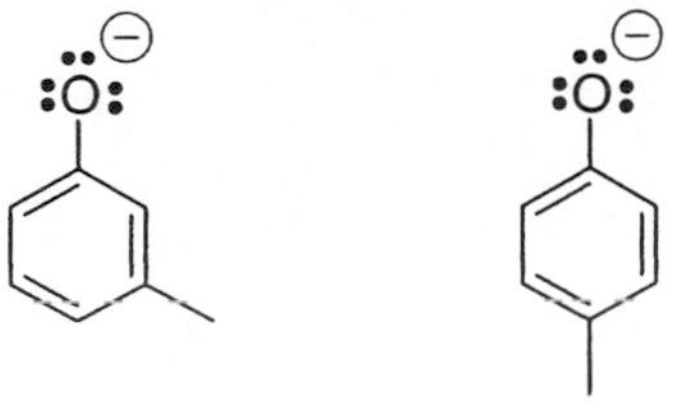

Exercises for Part A

1. You put 1 mole of acid #2 (from Model 1) into an aqueous solution and exactly ½ of it (0.5 mole) dissociate into H_3O^+ and B^-, and the other half of it stays as H–B. What is the pH of this solution?

2. It is useful to memorize the following approximate pKa's:

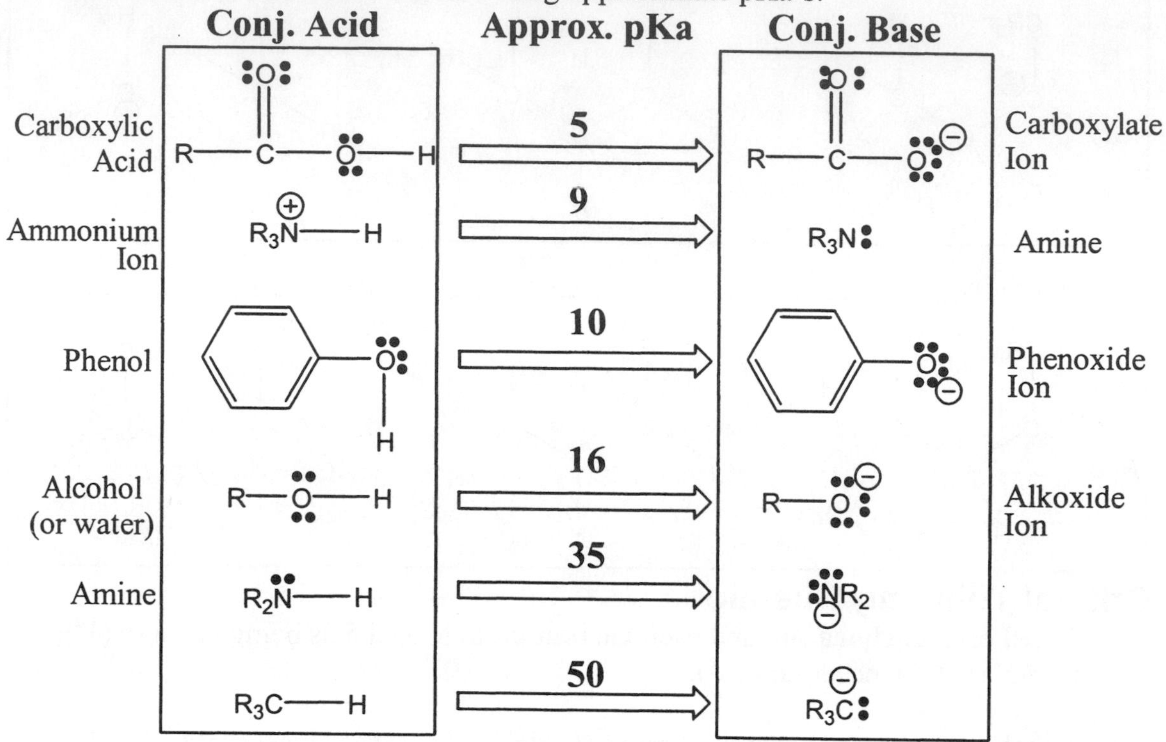

3. For the species in the table above, label the strongest base, strongest acid, weakest base and weakest acid.

4. What is the ratio of [carboxylic acid]/[carboxylate] in a solution buffered to pH 8?

5. A typical aldehyde such as propanal has a pKa of 17, while a typical ketone such as 3-hexanone has a pKa of 19.
   a) Draw the conjugate bases of propanal and 3-hexanone.
   b) According to the pKa's which compound is more acidic?
   c) Construct an explanation for why this compound is more acidic? Hint: it is mostly due to electronic (not steric) factors.

6. It is recommended that the following nucleophilic substitution reaction be run with at least a 0.001 M (molar) concentration of ammonia (NH_3).

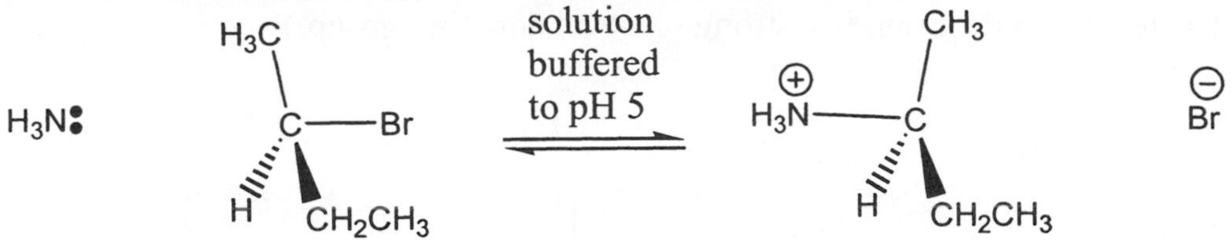

H_3N: + H_3C–C(H)(CH_2CH_3)–Br $\xrightarrow{\text{solution buffered to pH 5}}$ H_3N^{+}–C(CH_3)(H)(CH_2CH_3) + Br^{-}

a) Use curved arrows to show the mechanism of this reaction (assume it is a one-step reaction).
b) What is the name of this type of reaction mechanism?
c) Label the reactants and products with their absolute configuration (R or S).
d) You place 1 mole of NH_4Cl into a 1 L solution buffered to pH 5. Is the reaction within the recommended NH_3 concentration?
e) What is the minimum pH at which this reaction can be run?
f) What is the problem with running the reaction below this pH?

7. Cyanide ion has a pKa of 9. For each of the following reactions, sodium cyanide (NaCN) is added to a solution containing an α,β-unsaturated aldehyde. The solution is buffered to the pH given below the arrow. Explain the results of each reaction below. (Hint: both H_3O^+ and NC^- are needed for the reaction.)

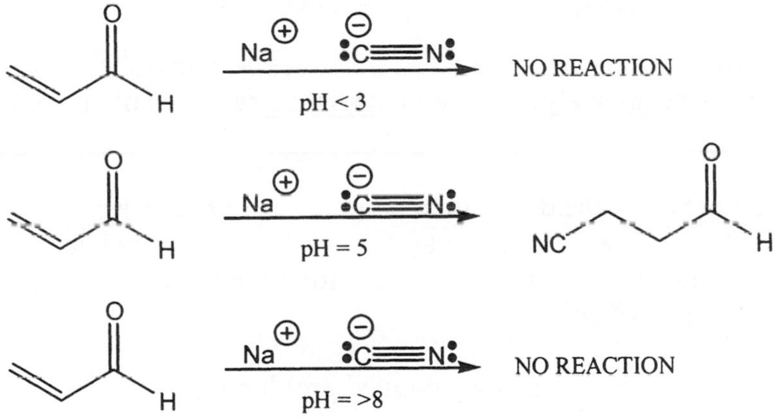

$CH_2{=}CHCHO$ $\xrightarrow[\text{pH} < 3]{Na^{+}\ ^{-}:C{\equiv}N:}$ NO REACTION

$CH_2{=}CHCHO$ $\xrightarrow[\text{pH} = 5]{Na^{+}\ ^{-}:C{\equiv}N:}$ $NCCH_2CH_2CHO$

$CH_2{=}CHCHO$ $\xrightarrow[\text{pH} = >8]{Na^{+}\ ^{-}:C{\equiv}N:}$ NO REACTION

8. Review the reactions on Reaction Sheets R1 and R4. You may want to add the following reactions to R4.

Aldehyde or Ketone: $R_2C{=}O$ $\xrightarrow{H_2NNH_2}$ R_2CH_2 Alkane

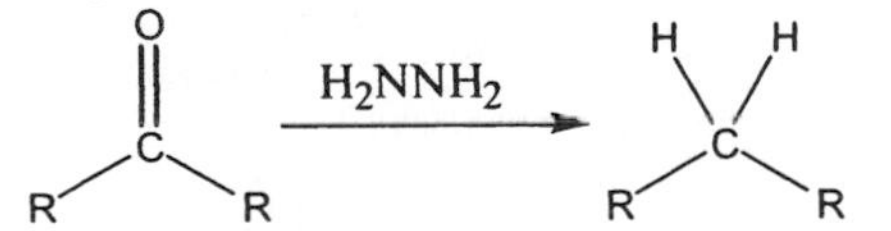

Aldehyde: $RCHO$ $\xrightarrow{KMnO_4\ (\text{mild})\ \text{or}\ CrO_3}$ $RCOOH$ Carboxylic Acid

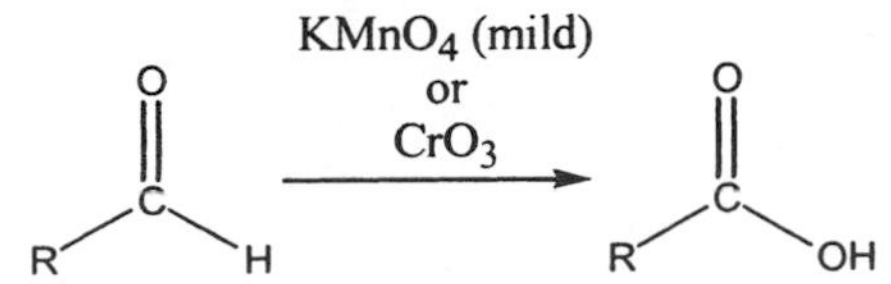

9. Read the assigned pages in your text and do the assigned problems.

Exercises for Part B

10. Number the following molecules from **least acidic** (1) to most acidic (8). (Note: an —OR group is a strong electron donating group.)

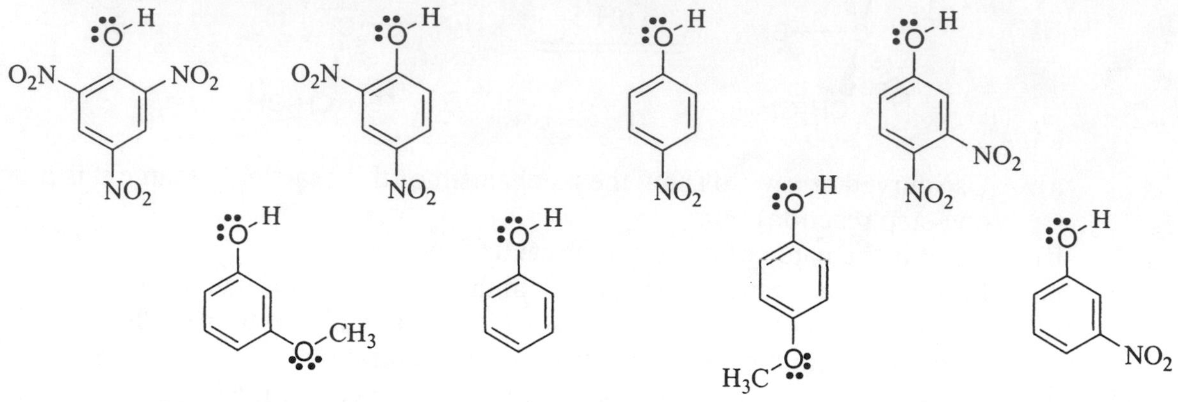

11. 2-nitrophenol is slightly <u>less</u> acidic than 4-nitrophenol. This is because the ortho-nitro group can form a hydrogen bond to the alcohol H.
    a) Draw the full Lewis structure of 2-nitrophenol showing this hydrogen bond as a ---- (dotted line).
    b) Explain how this hydrogen bond can increase the pKa of 2-nitrophenol relative to 4-nitrophenol.

12. Draw all four resonance structures of the conjugate base of *para*-methylphenol, circle the **least important (least favorable) resonance structure** and explain your reasoning.

13. Carbocations and carbanions follow the opposite trend in terms of the energy of primary, secondary and tertiary. However, an allyl (allyl) carbanion is much lower energy than a primary carbanion, and a benzyl (benzyl) carbanion is even lower in energy. Explain using resonance structures.

14. Read the assigned pages in your text and do the assigned problems.

ChemActivity 39

Nucleophilic Aromatic Substitution (NAS)

(Why do some aryl halides undergo NAS reactions while others do not?)

Review: Nucleophilic Addition-Elimination at the Carbonyl C

Addition-Elimination at the Carbonyl Carbon

Rxn A

Nucleophilic Addition **Elimination** ester

Model 1: Nucleophilic Addition-Elimination at the β-Carbon

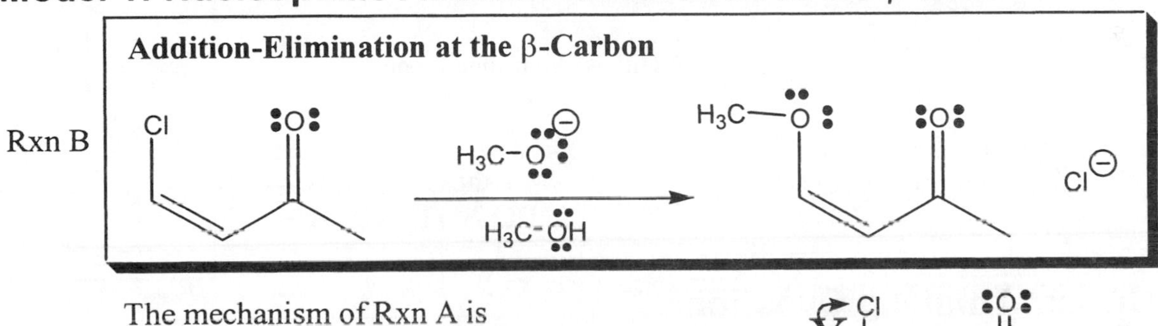

The mechanism of Rxn A is
NOT an S_N2 or S_N1 reaction.
(S_N2 and S_N1 reactions can only take place at sp^3 hybridized carbons)

Critical Thinking Questions

1. According to Model 1, why can't Rxn A be an S_N1 or S_N2 reaction?
2. Rxn A begins with a Nucleophilic Addition to the carbonyl carbon. Rxn B begins with a Nucleophilic Addition to the **β-carbon**. Devise a mechanism for Rxn B.

Model 2: Addition-Elimination on an Aromatic Ring

Rxn B is in competition with the following 1,4-addition reaction. A mixture of the product below and the product in Model 1 will form.

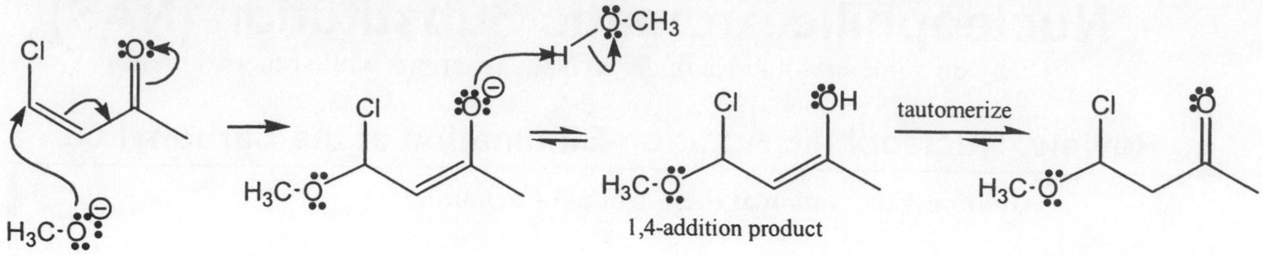

Cl :O: H_3C-O: 1,4-addition product tautomerize

However, if the β-carbon is part of an aromatic ring, the **addition-elimination product is strongly preferred**! (The 1,4-addition product in the box below does not form.)

Rxn C

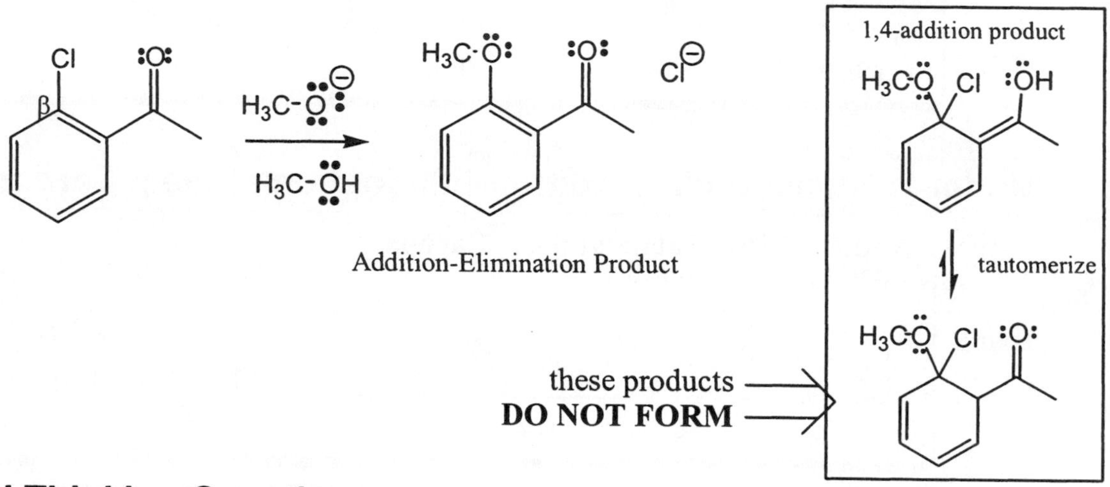

H_3C-O: H_3C-OH

Addition-Elimination Product

1,4-addition product

tautomerize

these products
DO NOT FORM

Critical Thinking Questions

3. Construct an explanation for why, in Rxn C, the addition-elimination product is formed preferentially over the 1,4-addition products in the box above.

4. Construct a mechanism for Rxn C.

5. Three resonance structures of the intermediate in Rxn C are drawn below. Draw the fourth (and most important) resonance structure.

6. Explain why the resonance structure you drew is more important than the other three.

7. Three resonance structures for the following intermediate are shown below. Can you draw a fourth resonance structure with the negative charge delocalized onto oxygen? (Try to move the negative charge onto oxygen using curved arrows.)

8. Is your conclusion above consistent with the following energy diagrams? Explain.

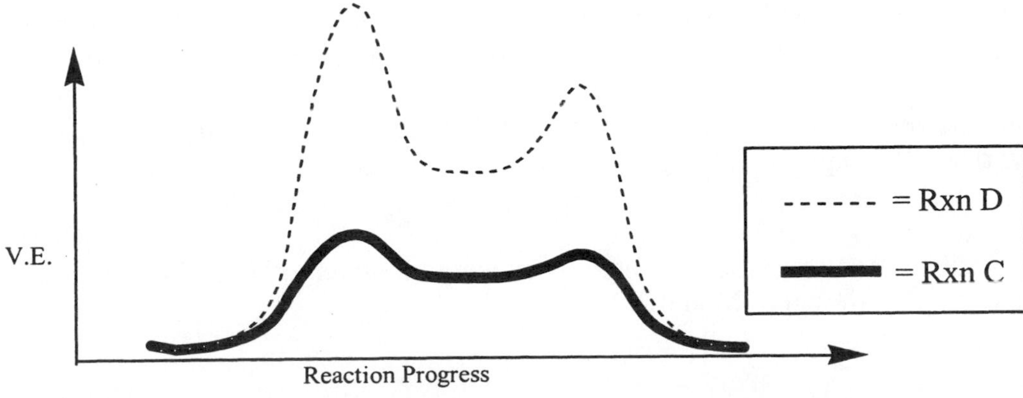

Model 3: Nucleophilic Aromatic Substitution

- Addition-elimination on an aromatic ring is called Nucleophilic Aromatic Substitution (**NAS,** for short).
- There must be a leaving group on the ring (usually a halogen).
- There must be a resonance electron withdrawing group *ortho* or *para* to this leaving group. (Usually a nitro group or a carbonyl group.)

X = F, Cl, Br or I

CH_3 ⊖ Nuc → NO REACTION

⊖ Nuc, Rate = very very slow → basically no product observed

NO_2 ⊖ Nuc, Rate = very slow → Nuc, NO_2, :X:⊖ Extremely Low yield

NO_2 ⊖ Nuc, Rate = slow → Nuc, NO_2, :X:⊖ Low Yield

NO_2, NO_2 ⊖ Nuc, Rate = fast → Nuc, NO_2, NO_2, :X:⊖ High Yield

Critical Thinking Questions

9. In an NAS reaction is the aromatic ring acting as a **nucleophile** (-) or as an **electrophile** (+)?

10. Based on your answer above, do you expect the reaction to be faster when the ring is substituted with **electron withdrawing groups** or **electron donating groups**?

11. Is your answer above consistent with the fact that the first reaction in Model 3 does not work at all? Explain.

12. Draw a substituted aromatic ring that you expect will undergo NAS faster than any of the examples shown in Model 3.

Review of EAS (Electrophilic Aromatic Substitution)

- In EAS the ring acts as a nucleophile and reacts with an electrophile (E^+).
- An electron donating group (such as an alkyl group or any atom with a lone pair) on the ring activates the positions *ortho* and *para* to it.
- An strong electron withdrawing group (such as nitro or a carbonyl) deactivates the positions *ortho* and *para* to it, causing substitution at the *meta* position.
- No leaving group is required in EAS (the E^+ substitutes for an H on the ring.)

Exercises

1. Write "**Ok NAS**" or "**great NAS**" under compounds that are ok or great candidates for nucleophilic aromatic substitution. Write "**good EAS**" or "**great EAS**" under compounds that are good or great candidates for electrophilic aromatic substitution. Write "**neither**" under compounds that are poor candidates for either NAS or EAS and explain your reasoning.

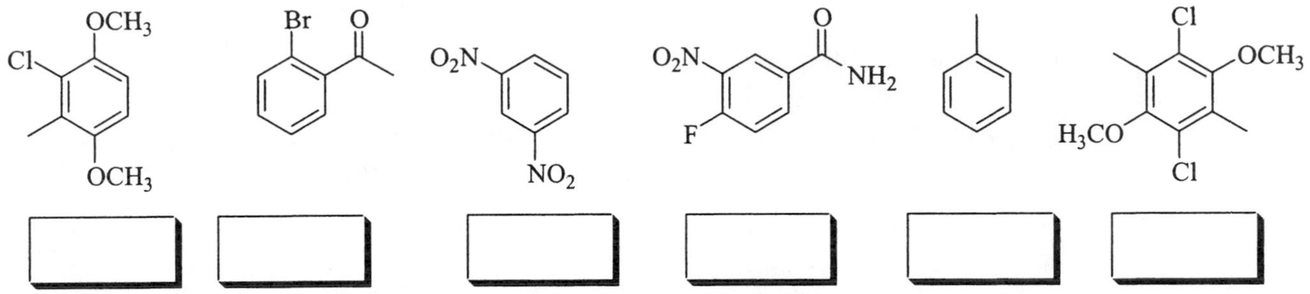

OCH₃ Cl OCH₃ | Br O | O_2N NO_2 | O_2N F O NH_2 | | Cl OCH_3 H_3CO Cl

2. The product shown does not form. Draw the correct product and explain why it is formed instead of the one shown.

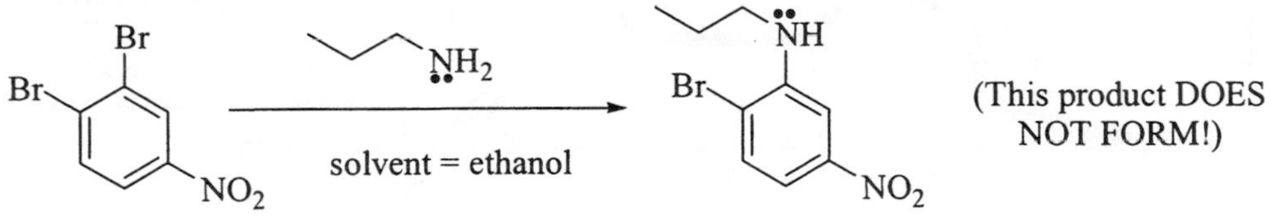

Br Br NO_2 → (NH_2; solvent = ethanol) → NH Br NO_2 (This product DOES NOT FORM!)

3. Show a reasonable mechanism for the following reaction.

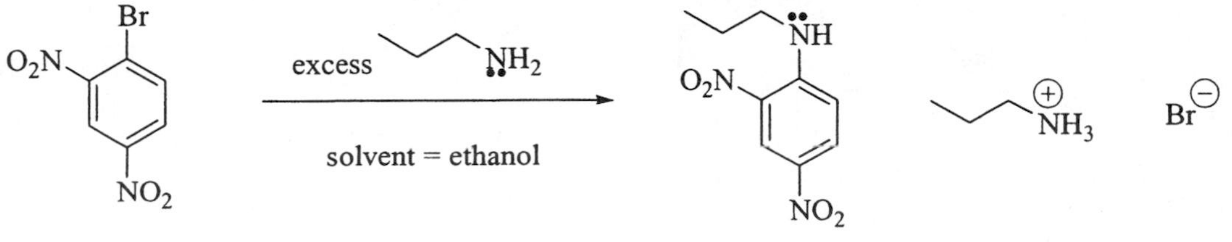

Br O_2N NO_2 → (excess NH_2; solvent = ethanol) → NH O_2N NO_2 NH_3^+ Br^-

4. Consider the following NAS reactions (only the first one yields significant product):

Rxn C — $H_3C\text{-}O^{-}$ / $H_3C\text{-}OH$

Rxn E — $H_3C\text{-}O^{-}$ / $H_3C\text{-}OH$

Rxn F — $H_3C\text{-}O^{-}$ / $H_3C\text{-}OH$

a) Draw the intermediate for each reaction.

b) Match each reaction to an energy profile line on the energy diagram below. Hint: consider the potential energy of each intermediate.

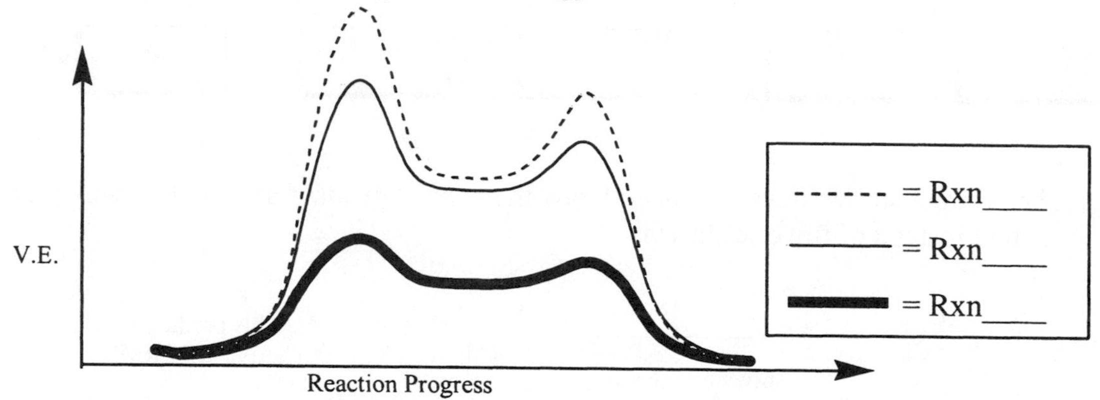

c) Construct an explanation for why Rxn E is slightly better than Rxn F (even though both are very very slow and yield almost no detectable products).

d) Draw all five resonance structures of the intermediate of the following reaction.

Rxn G — $H_3C\text{-}O^{-}$ / $H_3C\text{-}OH$

e) Add a line to the energy diagram above for Rxn G.

5. Consider the following benzene derivatives.

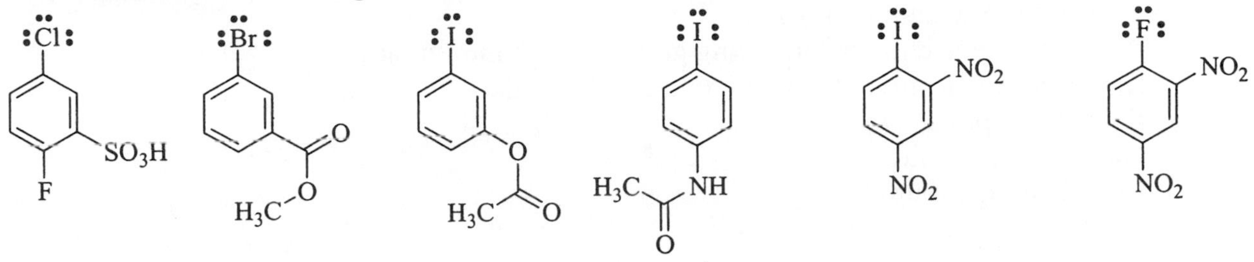

a) Circle each substituent that is an electron withdrawing group.
b) F is the strongest electron withdrawing group among the halogens. Order the molecules above from fastest (1) to slowest (6) in terms of rate of NAS with a given nucleophile.

6. The following is a summary of how the position and type of substituents on a phenyl ring affect the rate of nucleophilic aromatic substitution (NAS) vs. electrophilic aromatic substitution (EAS).

R_w= a strong electron withdrawing group such as nitro or acyl.
R_D= a strong electron donating group such as amino or hydroxy.
E^+ = electrophile
Nuc^- = nucleophile
LG = leaving group
o/p = *ortho* or *para*
m = *meta*

Reaction Type	Group/s on Ring	Product Description	Reaction Rate	Explanation
EAS Aromatic ring acts as **nucleophile** intermediate is a **carbocation**	R_w	E^+ adds *m*	Slow	R_w deactivates *o/p* positions via resonance (**destabilizes <u>carbocation</u>** intermediate)
	R_D	E^+ adds *o/p*	Fast	R_D activates *o/p* positions via resonance (stabilizes **<u>carbocation</u>** intermediate)
NAS Aromatic ring acts as **electrophile** intermediate is an **anion**	R_w *o* or *p* to a LG	Nuc^- subs for LG	Slow	R_w stabilizes **<u>anion</u>** intermediate through resonance
	Two R_w's *o*<u>&</u> *p* to a LG	Nuc^- subs for LG	Fast	Both R_w's stabilize **<u>anion</u>** intermediate through resonance
	R_w *m* to a LG	Nuc^- sub for LG (low yield)	Very Slow	R_w inductively stabilizes **<u>anion</u>** <u>but no resonance stabilization</u>
	R_D *o/p* or *m* to a LG	No Reaction	–	R_D **destabilizes <u>anion</u>** intermediate

7. Read the assigned pages in your text and do the assigned problems.

ChemActivity 40

Part A: Reactions of Carboxylic Acids

(How does the acidic O–H group affect nucleophilic addition to the carbonyl carbon?)

Model 1: Acid-Base Reactivity of Carboxylic Acids

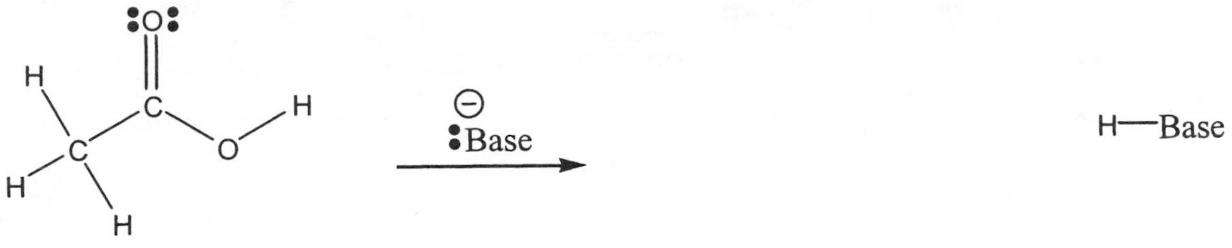

CH_3COOH + $:Base^{-}$ → H—Base

acetic acid
(ethanoic acid)

Critical Thinking Questions

1. How many different kinds of hydrogens are there on acetic acid?
   a) Draw both possible conjugate bases **INCLUDING ANY IMPORTANT RESONANCE STRUCTURES**.
   b) Circle the conjugate base that is lower in potential energy and explain your reasoning.

2. Methoxide ion ($^{-}OCH_3$) is an excellent nucleophile, yet the following nucleophilic addition does not occur.

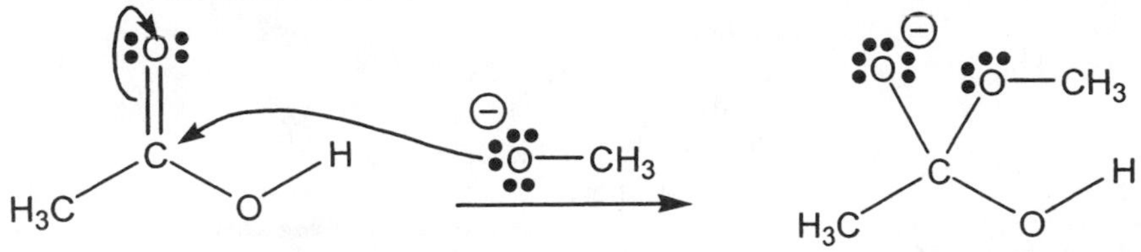

H_3C–COOH + $^{-}:O–CH_3$ → Nucleophilic Addition Product

   a) Which is higher in energy: the **nucleophilic addition product above** or **the conjugate base of acetic acid?** [Cross out the one that is higher in energy.]

   b) Is your answer in part a) consistent with the fact that with methoxide the acid base reaction is favored over the nucleophilic addition reaction? Explain.

Model 2: Carboxylic Acid Derivatives

- Several classes of compounds can be made from carboxylic acids. These are known as **carboxylic acid derivatives**. The most important of these are esters, amides and alkanoyl chlorides (a.k.a. acid chlorides).
- Conversions between these classes of compounds can be accomplished with acid catalysis (acidic conditions), and in most cases with base catalysis (basic conditions).

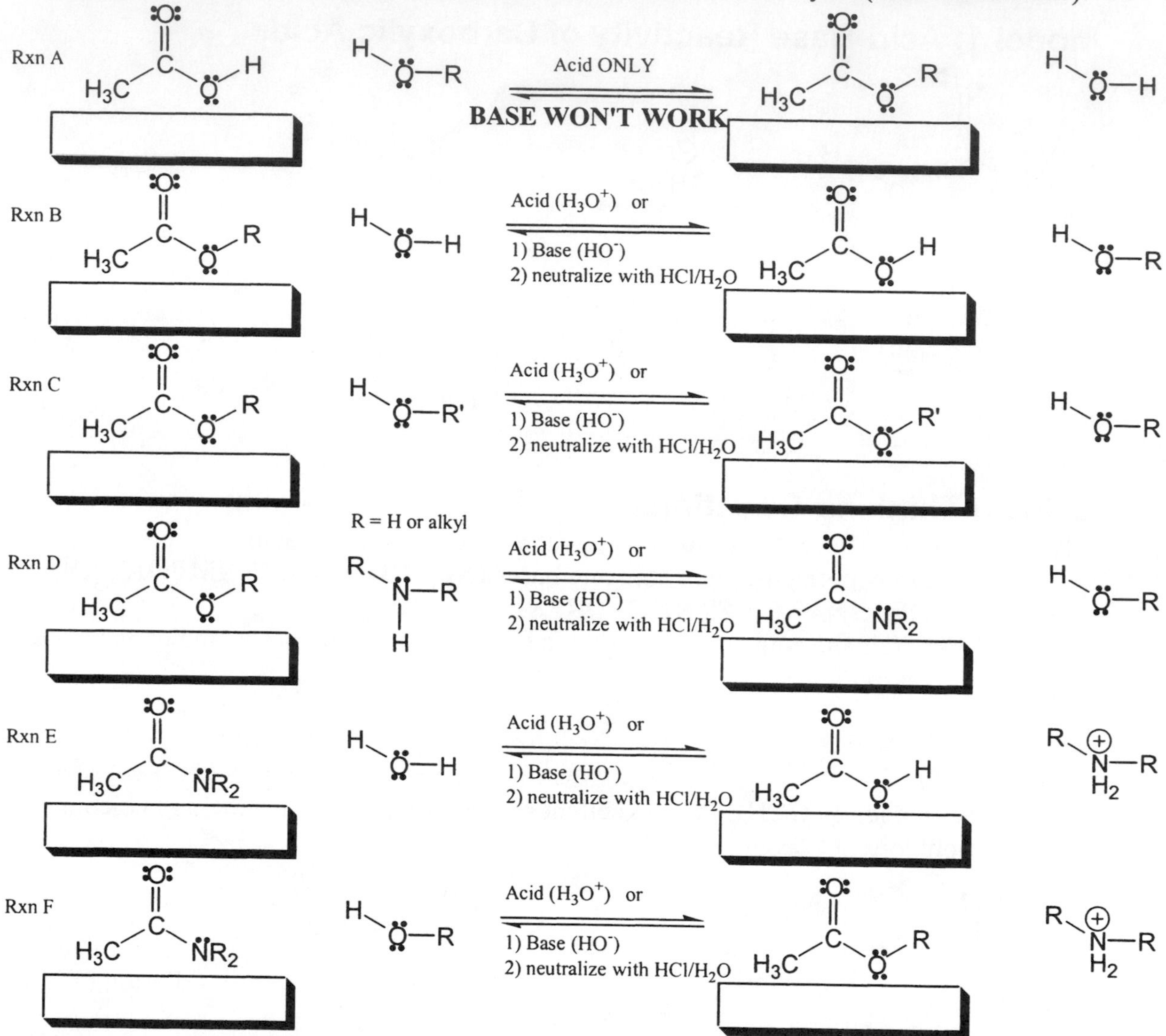

Note: The conversion of an amide to a carboxylic acid can be accomplished in acid or base, but only by carefully controlling the pH of the solution. For now, we will not discuss this more complicated case.

Critical Thinking Questions

3. Consider the transformations above.
   a) In each box write the name **carboxylic acid, ester** or **amide**, as appropriate.

   b) Identify each example of water, alcohol, amine or ammonium ion that appears in Model 2.

4. Under basic conditions Rxn B in Model 2 has two parts. The first part involves mixing the ester with hydroxide. Use curved arrows to show the mechanism of this **three-step** reaction.

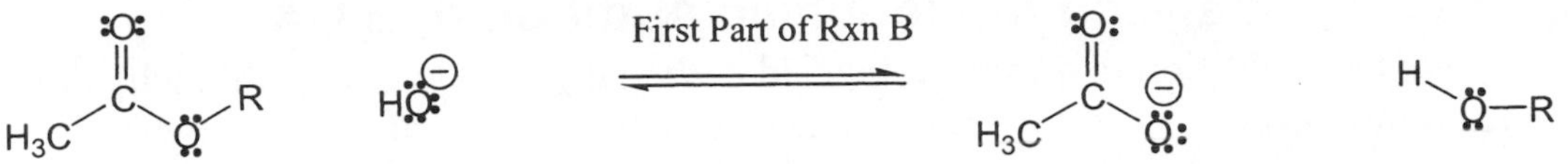

First Part of Rxn B

5. Show the mechanism that occurs when the reaction above is neutralized with dilute HCl in water. (This is the second part of base catalyzed Rxn B.)

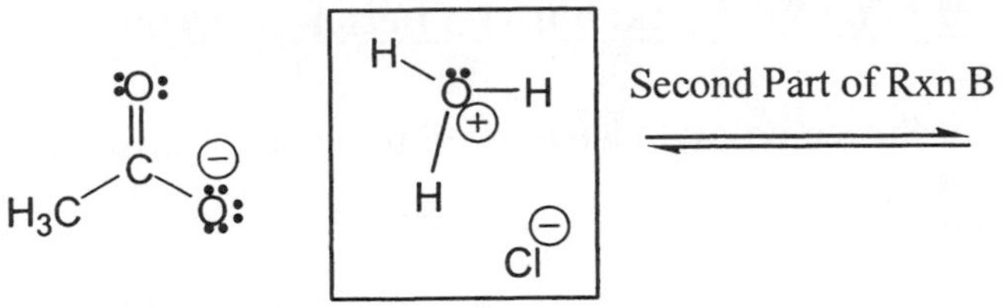

Second Part of Rxn B

dilute HCl in water

6. Rxn A in Model 2 does not work in basic conditions and the product in the box below does not form. Use curved arrows to show the mechanism and product of the side reaction that would result if Rxn A were attempted in basic conditions.

Rxn A attempted in base

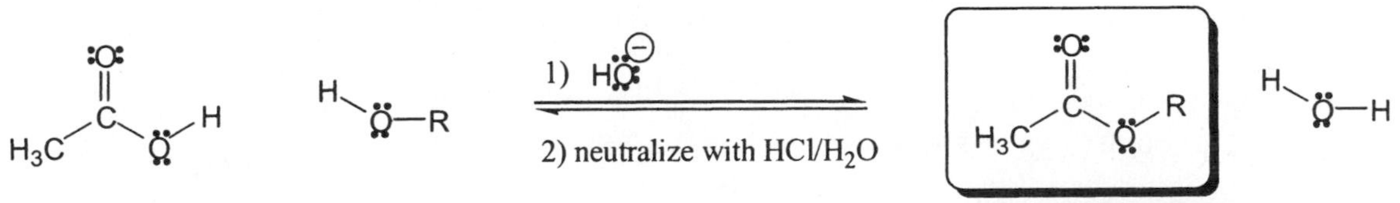

1) HO^-

2) neutralize with HCl/H_2O

THIS PRODUCT DOES NOT FORM!!

7. Explain why Rxn A does not work in basic conditions, but Rxns B-F do work in basic conditions.

Part B: Reactions of Carboxylic Acids

(How can we convert a carboxylic acid to an ester?)

Model 3: Acid Catalyzed Nucleophilic Addition to a CA

- Nucleophilic addition to a carboxylic acid carbonyl group is impossible when the nucleophile is also a good base. (Nuc^- acts as a base and removes the acidic H^+ instead.)

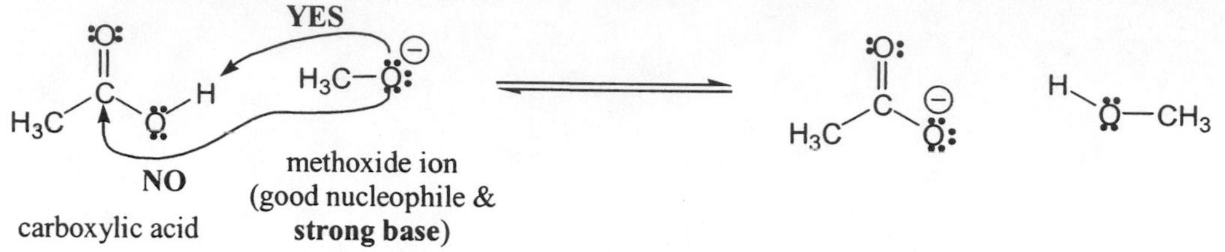

YES

NO

carboxylic acid

methoxide ion (good nucleophile & **strong base**)

- To solve this problem we must use a nucleophile that is a very weak base. For example: use methanol ($HOCH_3$) instead of methoxide ion ($^-OCH_3$).
- Recall that methanol is a weak nucleophile. To get any weak nucleophile to react with a carbonyl **we must add acid to make the carbonyl a better electrophile.**

Critical Thinking Questions

8. The first part of Rxn A in Model 2 is shown below. Use curved arrows to construct a mechanism for this part of the reaction.

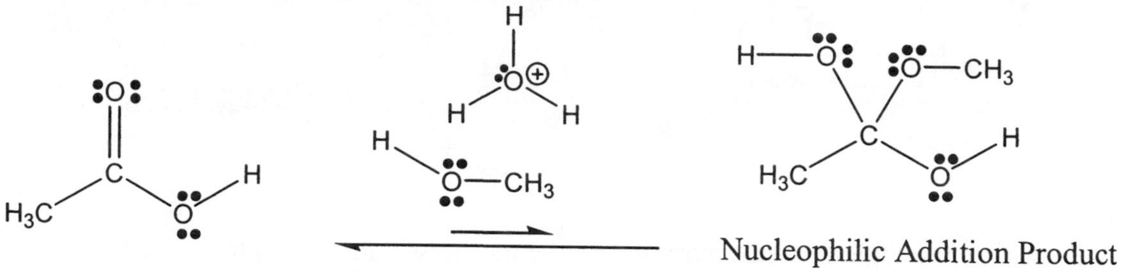

Nucleophilic Addition Product

9. Use curved arrows to show the mechanism of the second part of Rxn A in Model 2.

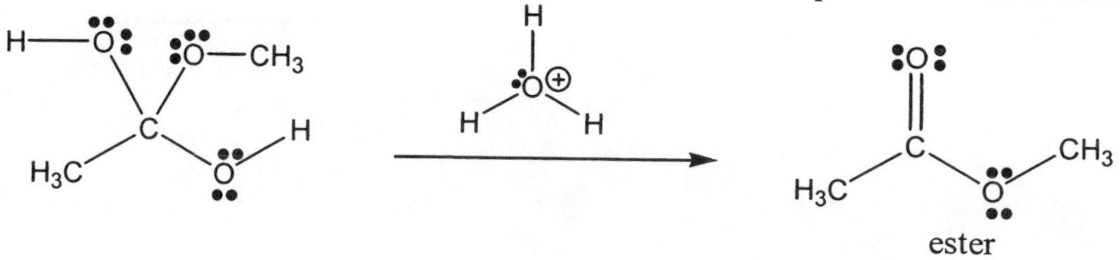

ester

10. A student says: "So what if an acid base reaction happens between methoxide and the carboxylic acid. Why can't you add two equivalents of methoxide, and have the second molecule of methoxide react with the carbonyl after the H has been removed?"

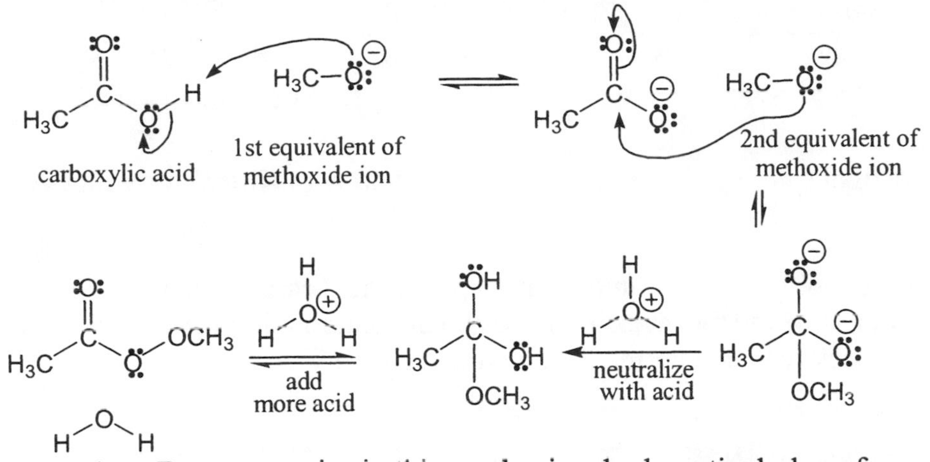

carboxylic acid

1st equivalent of methoxide ion

2nd equivalent of methoxide ion

add more acid

neutralize with acid

a) Do any species in this mechanism look particularly unfavorable and high in energy?

b) Explain why a carboxylate ion is a terrible electrophile.

c) Are your answers above consistent with the fact that the mechanism above does not occur?

Exercises for Part A

1. Write appropriate reagents in the boxes over the reaction arrows in Rxns G - I. You are not responsible for the mechanisms of any of these oxidation reactions.

Rxn G

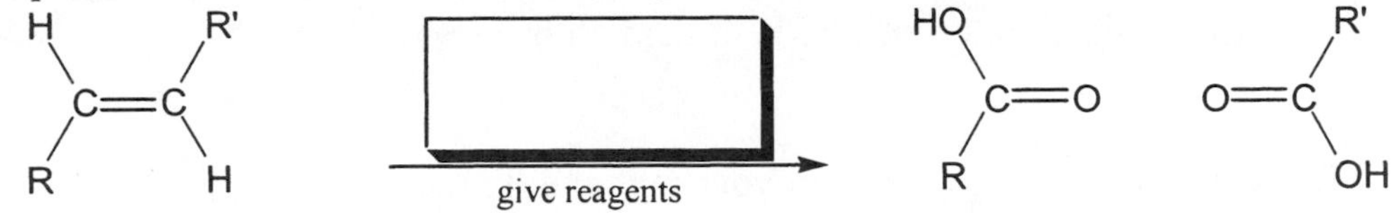

give reagents

Rxn H

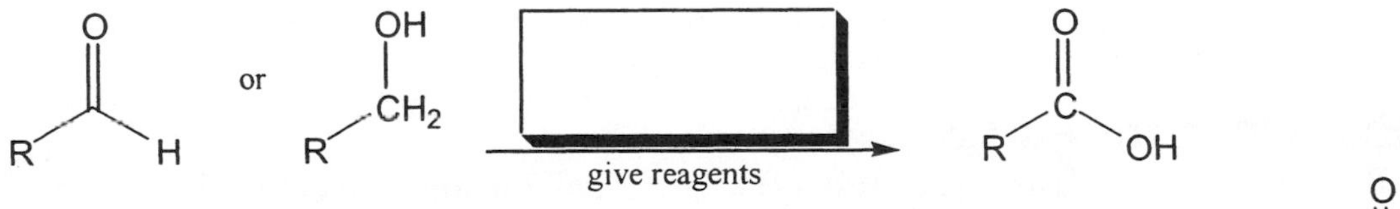

or

give reagents

Rxn I

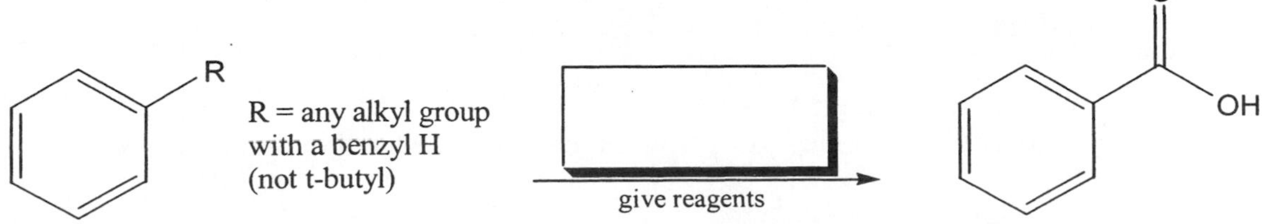

R = any alkyl group with a benzyl H (not t-butyl)

give reagents

2. Show the mechanisms of Rxns B-F in Model 2 using base catalysis (basic conditions).
3. New Reaction: Reduction of a carboxylic acid to an alcohol.

carboxylic acid — 1) $LiAlH_4$ reduction → — 2) H_3O^+ neutralize → primary alcohol

(Note: a primary alcohol can be oxidized to a carboxylic acid with $KMnO_4$ or with CrO_3.)

4. The following is another very common way of preparing a carboxylic acid. Use curved arrows to show the mechanism of Rxn J. Hint: carbon dioxide is similar to, but even more reactive than a normal carbonyl compound.

Rxn J: $R-CH_2-MgBr$ or $R-CH_2-Li$ — 1) O=C=O; 2) neutralize with dilute HCl/H_2O → $R-CH_2-C(=O)OH$

5. Carboxylic acids can be made via acid catalyzed hydrolysis of nitriles (cyanide compounds). Use curved arrows to show the mechanism of Rxn K (6 steps). Hint: the C=N bond is similar to a C=O bond in terms of polarization and reactivity.

Rxn K: R—C≡N: + H_2O — H_3O^+ (catalyst) ⇌ $R-C(=O)OH$ + NH_4^+

6. Review Nomenclature Worksheet II and be sure you can recognize and name a carboxylic acid, ester, amide or acid chloride.

7. Read the assigned pages in your text and do the assigned problems.

Exercises for Part B

8. Show the mechanisms of Rxns B-F in Model 2 using acid catalysis (acidic conditions).

9. Rank the following compounds from…
"1 = best electrophile" to "4 = worst electrophile" and explain your reasoning.

$H_3C-C(=O)-OCH_3$ $H_3C-C(=O)-O^-$ $H_3C-C(=O^+H)-H$ $H_3C-C(=O^+H)-OH$

10. Design an efficient synthesis of each of the following target molecules. All carbon atoms must come from the starting material/s given.

Target	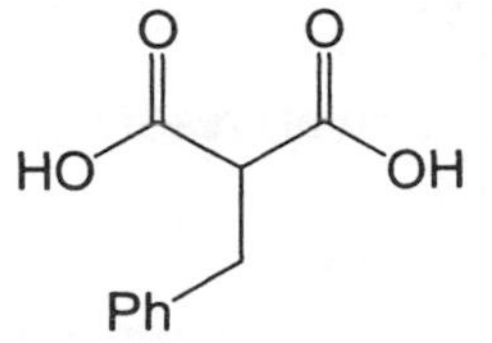 O O HO OH Ph	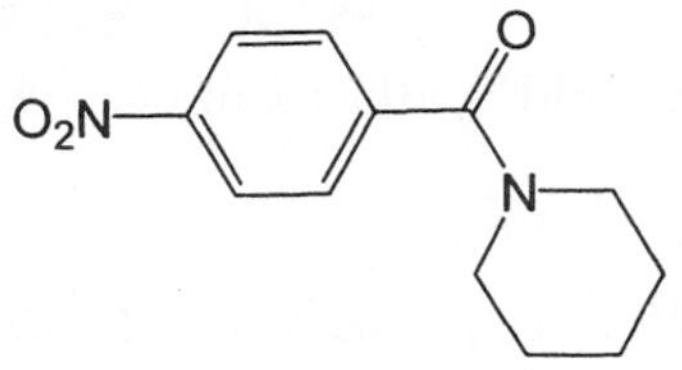 O_2N O N	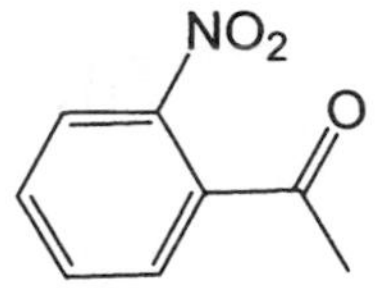 NO_2 O
Starting material	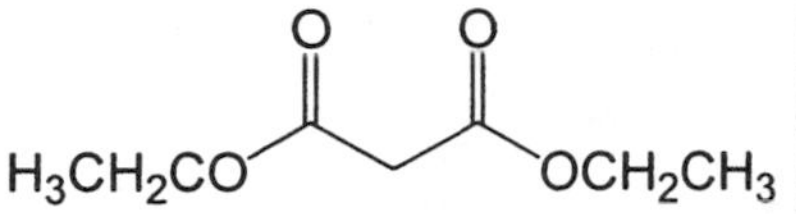 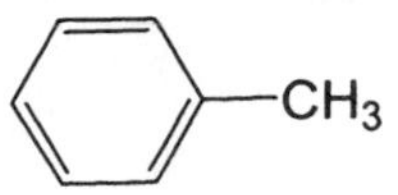 O O H_3CH_2CO OCH_2CH_3 CH_3	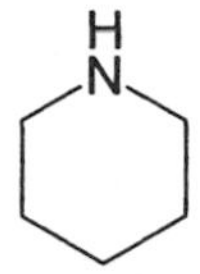 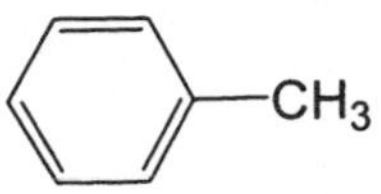 H N CH_3	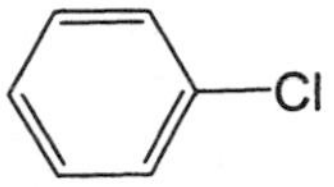 Cl HC≡CH

11. Read the assigned pages in your text and do the assigned problems.

ChemActivity 41

Part A: Reactions of Alkanoyl Chlorides (Acid Chlorides)

(How can high yield synthesis of carboxylic acid derivatives be accomplished?)

Model 1: Equilibrium between Carboxylic Acid and Ester

$$RCOOH + H{-}O{-}CH_2CH_3 \xrightleftharpoons[]{H_3O^+ \text{ catalyst}} RCOOCH_2CH_3 + H{-}O{-}H$$

- The reaction above involves five steps.
- Each intermediate has approximately the same potential energy, as shown in the reaction diagram below.

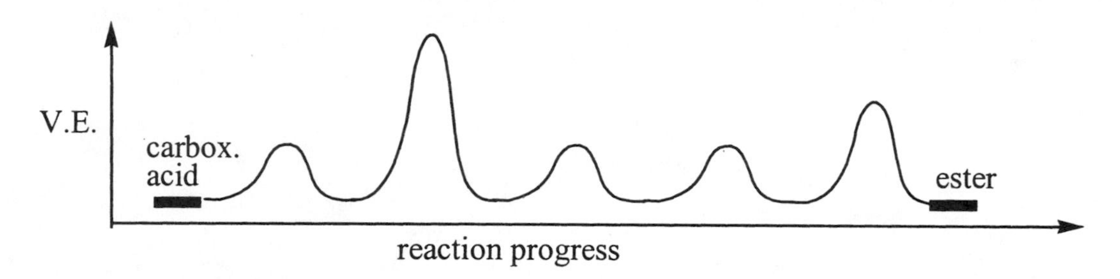

Critical Thinking Questions

1. For the reaction above…
   a) **Assume [water] = [ethanol]** and estimate the ratio of products/reactants at equilibrium (K_{eq})

$$K_{eq} = \frac{[\text{ester}]}{[\text{carbox. acid}]} =$$

   b) This K_{eq} would seem to doom this reaction to a maximum yield of 50%. How, in the lab, could we push the reaction toward ester or toward carboxylic acid?

Model 2: Synthesis of Alkanoyl Chlorides (a.k.a. Acid Chlorides)

- Chemists have devised a way of increasing the yield of reactions such as the one in Model 1 to near 100% without having to drive the reaction using LeChatelier's Principle (adding large amounts of ethanol or removing water).
- This method involves replacing the troublesome OH of a carboxylic acid with a Cl.
- The resulting compound is called an alkanoyl chloride or an **acid chloride**.

carboxylic acid + thionyl chloride ⇌ alkanoyl chloride (acid chloride) + SO_2 (a gas) + H—Cl

The mechanism of this reaction is a complicated version of others you have seen. You will have an opportunity to go through it as part of the Exercises for this activity.

Critical Thinking Questions

2. Use curved arrows to show the mechanism of Step B of the following synthesis. **Recall that S_N1 and S_N2 reactions cannot take place at sp^2 hybridized carbons.**

RCOOH —($SOCl_2$, Step A)→ RCOCl —(Na^+ $^-O-CH_2CH_3$, Step B)→ $RCOOCH_2CH_3$ + Na^+ Cl^-

3. When excess of a lithium or Grignard reagent is added to an acid chloride **two equivalents** of the nucleophile are used up. Devise a mechanism for this reaction. (Hint: it goes through a ketone intermediate.)

RCOCl —(H_3C—Li)→ $RC(CH_3)_2O^-$ 2 Li^+ Cl^- —(neutralize with dilute acid)→ $RC(CH_3)_2OH$

4. The transformation of a carboxylic acid into an ester (as shown in CTQ 2) can be accomplished without the strong base, ethoxide ion. Show a mechanism for Step B of the synthesis below.

$SOCl_2$, Step A; $HO{-}CH_2CH_3$, Step B

$RCOOH \xrightarrow[\text{Step A}]{SOCl_2} RCOCl \xrightarrow[\text{Step B}]{HO-CH_2CH_3} RCOOCH_2CH_3 + H{-}Cl$

Part B: Reactions of Acid Anhydrides

(In what ways is an acid anhydride similar to an acid chloride?)

Model 3: Synthesis of an Acid Anhydride

- An acid anhydride can be made from an acid chloride and a carboxylic acid.
- Acid anhydrides are **less reactive than acid chlorides**.
- Acid anhydrides are used in place of acid chlorides when a slower, more controlled reaction is necessary.

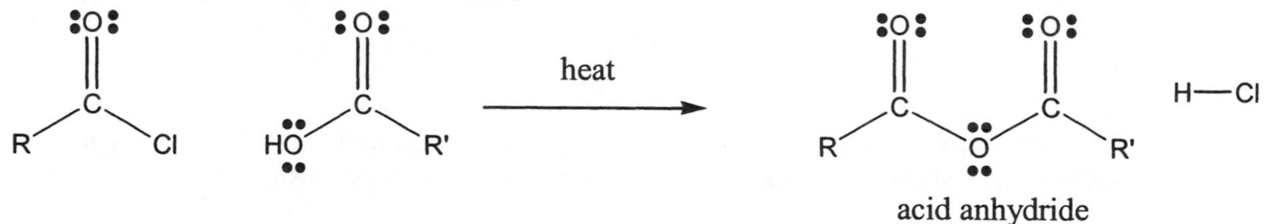

acid anhydride

Critical Thinking Questions

5. As shown in the acid chloride formation mechanism in the homework exercises for Part A, a carboxylic acid can act as a nucleophile when reacting with an excellent electrophile like an acid chloride. Devise a mechanism for the reaction in Model 3.

6. Acid anhydrides react exactly like acid chlorides (but in a slower and more controlled fashion). For example, the ester synthesis in CTQs 2 & 4 can bc accomplished via an acid anhydride, as shown below. Show the mechanism of **Step B** in this synthesis. (What is the other product formed in this reaction?)

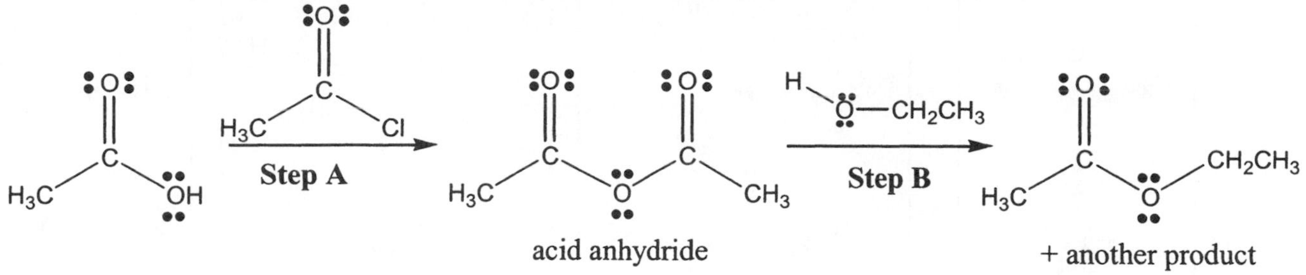

$H_3C-C(=O)-OH \xrightarrow[\text{Step A}]{H_3C-C(=O)-Cl} H_3C-C(=O)-O-C(=O)-CH_3 \xrightarrow[\text{Step B}]{H-O-CH_2CH_3} H_3C-C(=O)-O-CH_2CH_3$

acid anhydride

+ another product

Model 4: Reactions of Acid Chlorides and Anhydrides

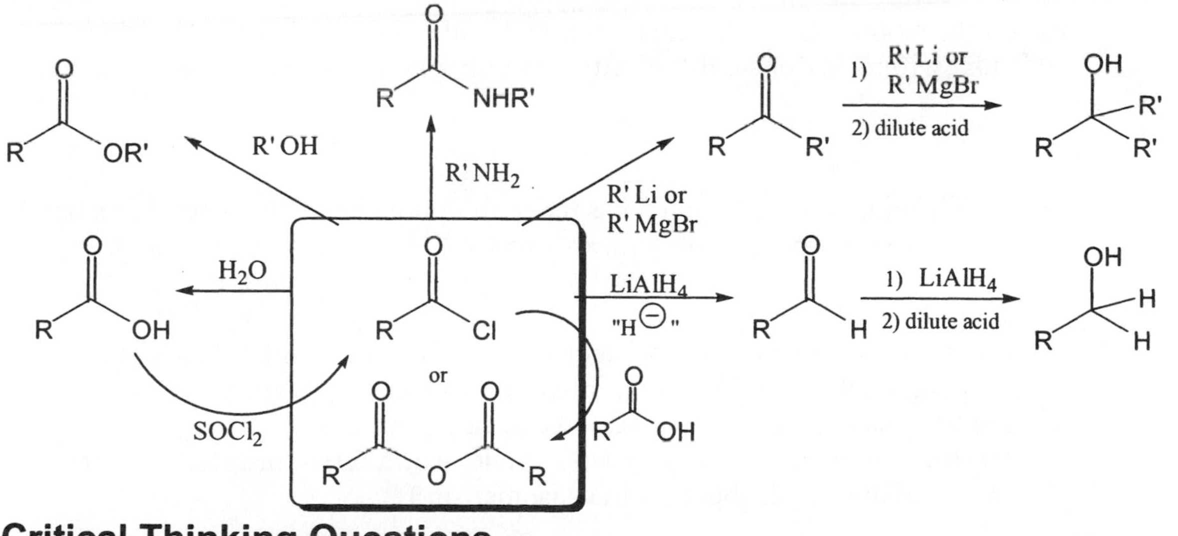

R'OH → RC(=O)OR'

R'NH₂ → RC(=O)NHR'

R'Li or R'MgBr → RC(=O)R' → 1) R'Li or R'MgBr, 2) dilute acid → $RC(OH)R'_2$

H_2O → RC(=O)OH

RC(=O)OH → $SOCl_2$ → RC(=O)Cl

RC(=O)Cl or RC(=O)OC(=O)R

$LiAlH_4$ "$H^{\ominus}$" → RC(=O)H → 1) $LiAlH_4$, 2) dilute acid → RCH_2OH

RC(=O)Cl → RC(=O)OH → RC(=O)OC(=O)R

Critical Thinking Questions

7. Label the compounds in Model 4 according to their functional group category. Use **primary alcohol, tertiary alcohol, aldehyde, ketone, carboxylic acid, amide, ester, carboxylic acid,** and **acid anhydride**.

Model 5: Reactivity of Carboxylic Acid Derivatives

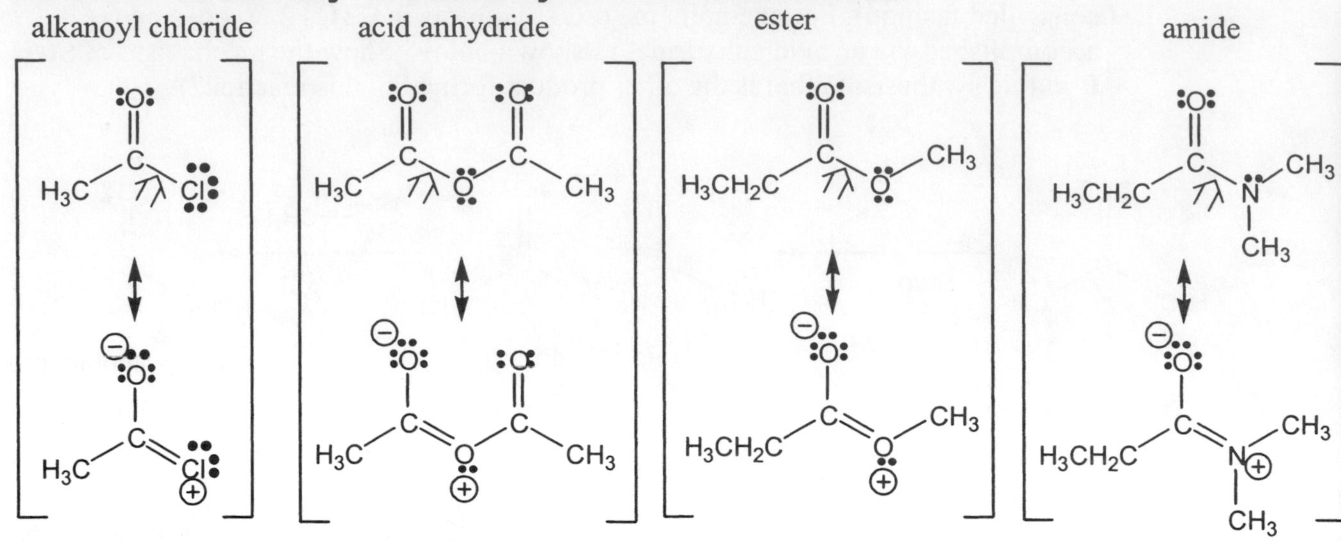

For each carboxylic acid derivative, we can draw a second order resonance structure showing donation of electron density to the carbonyl carbon.

Critical Thinking Questions

8. For each set in Model 5, draw curved arrows showing conversion of the top resonance structure into the bottom one.
   a) Of the atoms shown donating electron density to the carbonyl, which one is **least** willing to donate this electron pair? Explain your reasoning.

   b) Of the atoms shown donating electron density to the carbonyl, which one is **most** willing to donate this electron pair? Explain your reasoning.

   c) Which compound in Model 5 has the most double bond character in the bond marked in the top structure with an arrow =>? Explain your reasoning.

   d) Is your answer in part c) consistent with the following NMR data: at room temperature the ^{1}H NMR of the amide above shows 2 slightly different methyl singlets (each integrates to 3 H's). At elevated temperatures, the NMR spectrum of this compound shows 1 singlet worth 6 H's for these two methyl groups. (Hint: think about cis/trans isomerism.)

Exercises for Part A

1. Show the mechanism of a reaction between an acid chloride and an amine to form an amide.

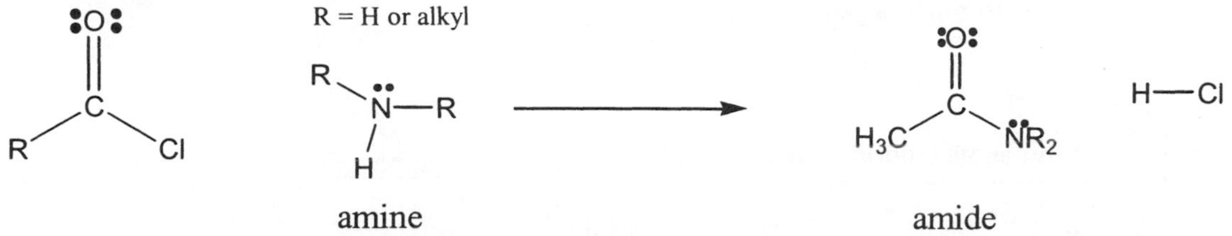

R = H or alkyl

amine amide

2. The mechanism of the reaction in Model 2 is similar to many you have seen. You will not be held responsible for this mechanism but it is useful to go through it at least once. Each intermediate of the mechanism is drawn below. **Add curved arrows to show electron movement in each step**. (It may be helpful to draw in lone pairs.)

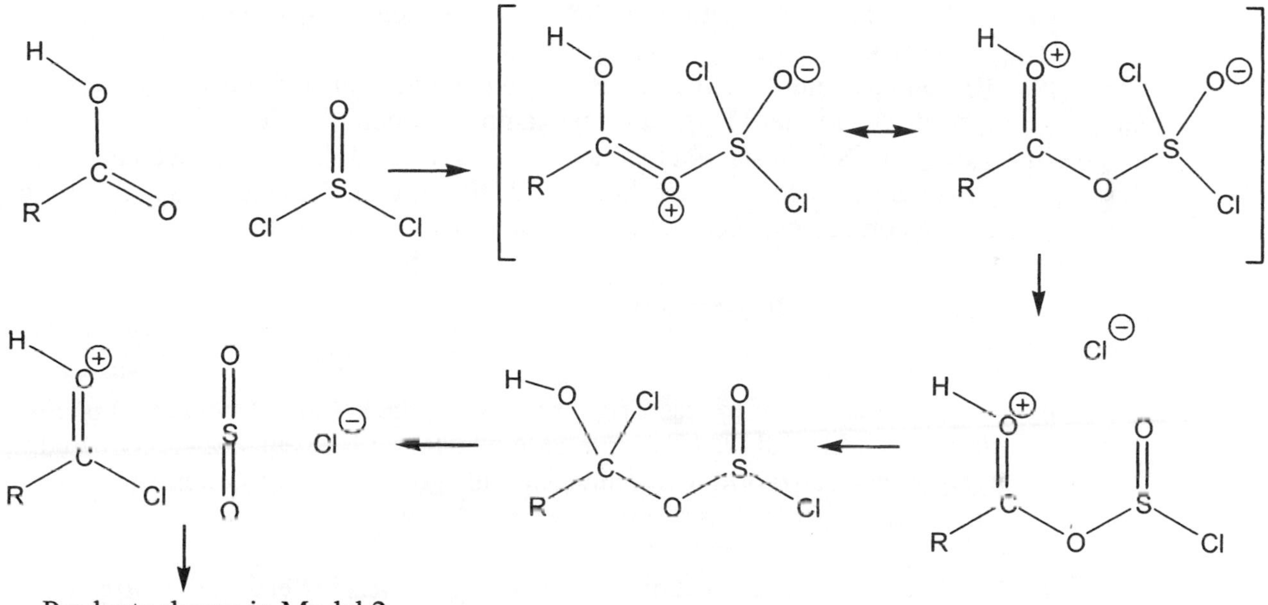

Products shown in Model 2

3. $LiAlH_4$ (and $NaBH_4$) often react as though they are a hydride anion (H^-). Hydride anion is a very strong base and an even better nucleophile. Show the mechanism of a reaction between an acid chloride and one equivalent of $LiAlH_4$ to form an aldehyde. Then show the mechanism of the reaction that occurs if a second equivalent of $LiAlH_4$ is added followed by neutralization with dilute acid.

4. Read the assigned pages in your text and do the assigned problems.

Exercises for Part B

5. Rank the compounds in Model 5 according to how readily each will react with water to form a carboxylic acid. Explain your reasoning.

6. Not all acid anhydrides are symmetrical.
   a) The product of Step A, below is an asymmetrical acid anhydride. Draw it.
   b) In Step B four different products will form. Draw each of these (two esters and two carboxylic acids.)

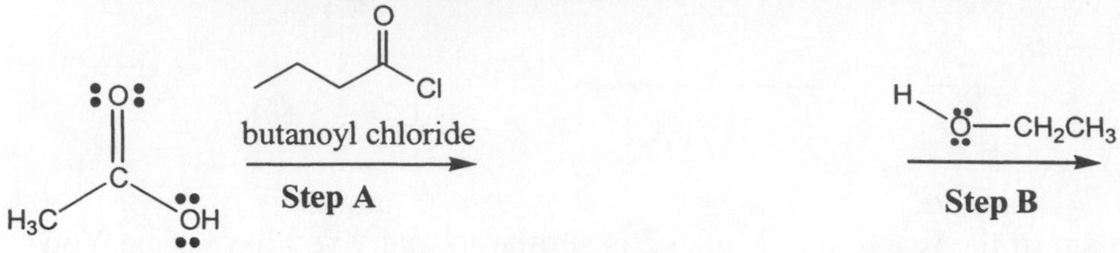

H_3C–C(=O)–OH —(butanoyl chloride, Step A)→ —(H–O–CH_2CH_3, Step B)→

7. Alkanoyl chlorides and acid anhydrides allow convenient synthesis of carboxylic acid derivatives because they each have a good leaving group attached to the carbonyl carbon.
   a) What is this leaving group in the case of an alkanoyl chloride?
   b) What is this leaving group in the case of an acid anhydride? (Hint: the leaving group is not just a single atom.)
   c) If you use a nucleophile that is strong enough, you don't need a good leaving group. A Grignard reagent is an example of such a nucleophile. The following synthesis of a ketone is very high yield. Show the mechanism of this reaction. (Remember that OH or OR are good leaving groups in strongly basic conditions, such as in a Grignard or Lithium reaction.)

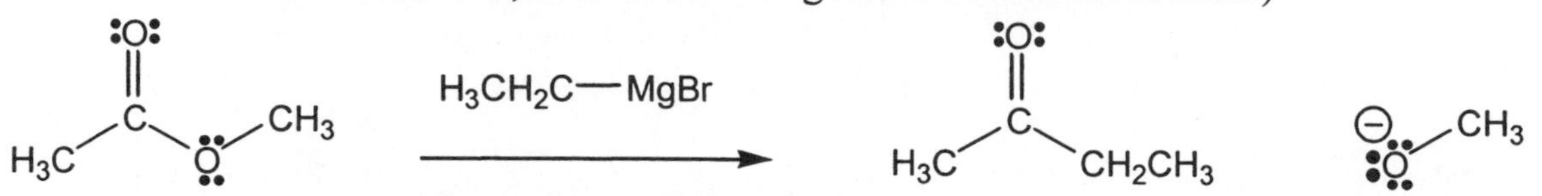

H_3C–C(=O)–O–CH_3 —(H_3CH_2C—MgBr)→ H_3C–C(=O)–CH_2CH_3 + $^{\ominus}$O–CH_3

   d) In the presence of <u>excess</u> Grignard reagent, the ketone (2-butanone) is not observed. Instead the conjugate base of 3-methyl-3-pentanol is recovered. (Upon neutralization with dilute acid this gives 3-methyl-3-pentanol.) Show the mechanism of this reaction.

8. Repeat parts c and d from question 7 above using $LiAlH_4$ instead of the Grignard reagent. Recall that lithium aluminum hydride reacts as though it is an H^- (hydride anion). Hint: 1 equivalent of $LiAlH_4$ gives an aldehyde, and 2 gives the conjugate base of ethanol.
9. The yield of a reaction with an acid chloride can be pushed from good to near 100% with the addition of a non-nucleophilic base such as 2,6-dimethylpyridine.

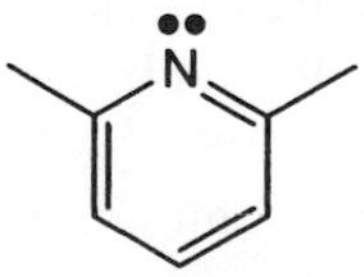

   a) Construct an explanation for why 2,6-dimethylpyridine is a poor nucleophile.
   b) Construct an explanation for why 2,6-dimethylpyridine drives a reaction like the one in CTQ 4, Step B to the right (toward products).
   c) What side product would result if pyridine (instead of 2,6-dimethylpyridine) were added to the reaction in CTQ 4, Step B?
10. Review Nomenclature Worksheet II and be sure you can name an alkanoyl chloride. Read the assigned pages in your text and do the assigned problems.

ChemActivity 42

Part A: Claisen Reaction

(What products result when esters undergo an aldol-like reaction?)

Model 1: Most Acidic H On an Ester

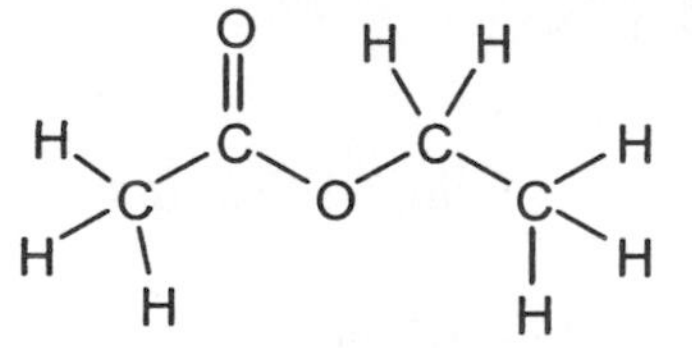

ethyl acetate

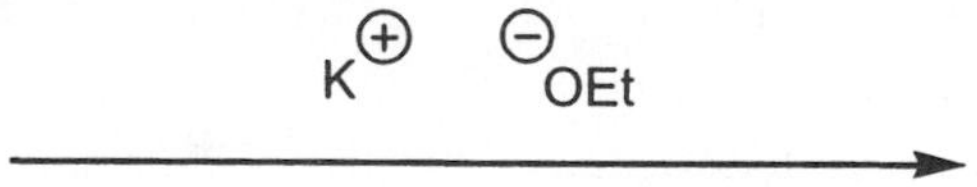

$K^{\oplus}$ $^{\ominus}OEt$

you can represent an ethyl group as "Et" so $^{\ominus}OCH_2CH_3$ is $^{\ominus}OEt$

Critical Thinking Questions

1. Circle the most acidic H on the drawing of ethyl acetate in Model 1 and draw mechanism and product of an acid-base reaction with ethoxide ion (^{-}OEt).

   Hint: If you're not sure which H is most acidic, draw the different possible conjugate bases and decide which one is lowest in potential energy (easiest to form).

2. In the presence of ethoxide ion, two esters can undergo an Aldol-like reaction. Show the first two steps of this reaction mechanism using curved arrows.
   Step I is an acid-base reaction.
   Step II is a carbon-carbon sigma bond formation reaction.

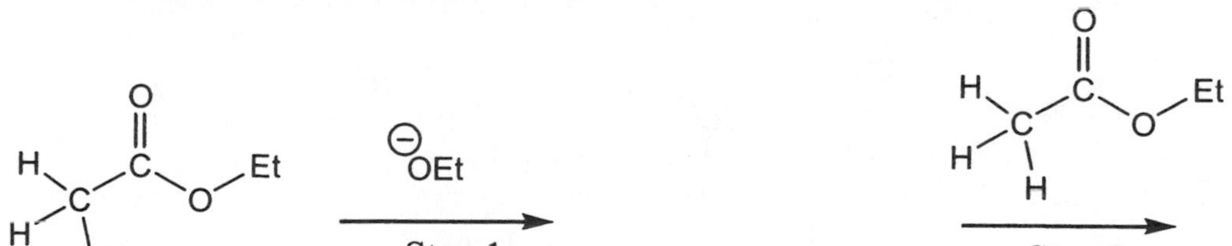

ethyl acetate

$^{\ominus}OEt$

Step 1 (acid-base)

Step 2 (C-C bond formation)

Model 2: Claisen Reaction

- An Aldol-like reaction with esters is called a Claisen Reaction or a Claisen Condensation (named for German chemist Ludwig Claisen, 1851-1930).
- The final product of a Claisen reaction is the resonance-stabilized anion below, right.

2 H–CH₂–C(=O)–O–R $\xrightarrow[\text{Part A}]{^{\ominus}OR}$ H₃C–C(=O)–CH₂–C(=O)–O–R + HOR $\xrightarrow[\text{Part B}]{^{\ominus}OR}$ H₃C–C(=O)–$\ddot{C}^{\ominus}$H–C(=O)–O–R + HOR

Claisen Product

Critical Thinking Questions

3. Use curved arrows to show the mechanism of **Part A** of the Claisen reaction in Model 2.

4. Show the Mechanism of **Part B** of the Claisen Reaction in Model 2.

H₃C–C(=O)–CH₂–C(=O)–O–R $\xrightarrow[\text{Part B}]{^{\ominus}OR}$ [H₃C–C(=O)–$\ddot{C}^{\ominus}$H–C(=O)–O–R ↔] + HOR

Claisen Product

↕

a) Draw the other two resonance structures of the Claisen Product above.
b) Do you expect **Part B** to be **uphill** or **downhill** [circle one] in energy? Hint: Is the negative charge more stable in the reactants or products?
c) Is your answer in b) consistent with the fact that the reaction goes all the way to the Claisen Product (the anion at the end of Part B)?
d) Is the base (⁻OR) acting as a catalyst in a Claisen reaction or is it a reactant that is consumed over time in the reaction?

Model 3: Choosing the Right Base for a Claisen Reaction

$H_3C-C(=O)-O-CH_2CH_3$ + $^{\ominus}OCH_2CH_3$ $K^{\oplus}$ ⇌ $H_3C-C(O^{\ominus})(OCH_2CH_3)-O-CH_2CH_3$ ⇌ $H_3C-C(=O)-O-CH_2CH_3$ + $^{\ominus}OCH_2CH_3$

Critical Thinking Questions

5. Add curved arrows to the reaction in Model 3.

6. Is ethoxide ion ($CH_3CH_2O^-$) acting as a **nucleophile** or as a **base** [circle one] in this reaction?

7. A student uses potassium ethoxide in the following Claisen Reaction. The result is the unwanted side-product shown in the box.

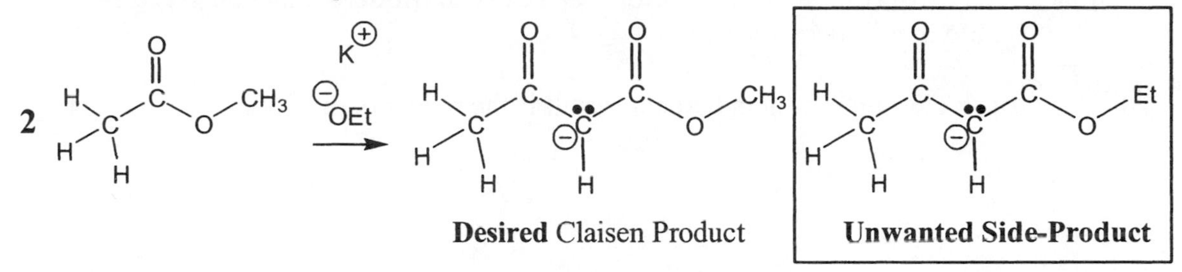

a) Explain how this side product came about.

b) What base should he have used instead of ethoxide ($^-$OEt) to ensure that only the desired Claisen product is produced?

c) Explain why is it important that the R group on the base match the R group on the ester?

Part B: Mixed Claisen Reactions

(What products result when two different esters undergo a Claisen Reaction?)

Model 4: Mixed Claisen Reaction

- As with aldol reactions, mixed Claisen reactions are possible.

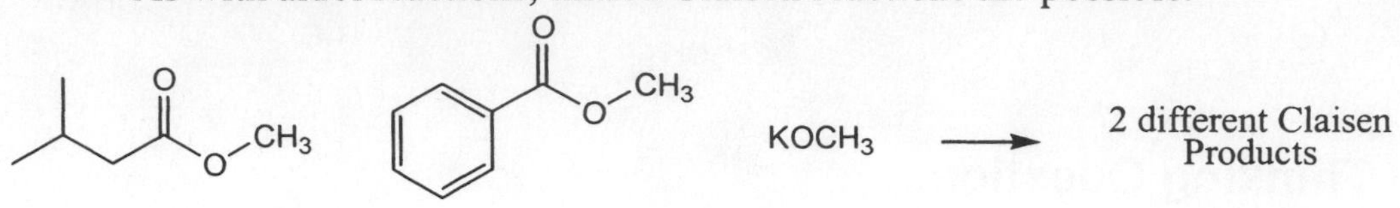

CH_3 CH_3 $KOCH_3$ ⟶ 2 different Claisen Products

Critical Thinking Questions

8. Place the label **"can be Nuc$^-$"** below each ester in Model 4 that can react with base to become a nucleophile for a Claisen reaction.
9. Place the label **"can be Elec$^{\delta+}$"** below each ester in Model 4 that can serve as the electrophile in a Claisen reaction.

10. Draw **both** Claisen products that can result if the reagents in Model 4 are mixed.

11. Suggest a laboratory procedure that ensures a high yield of the mixed Claisen product in Model 4 and minimizes the formation of the other possible Claisen product (made from two molecules of methyl-3-methylbutanoate).

Model 5: Protonation or α-Alkylation of a Claisen Product

2 $H_3C-C(=O)-O-R$ + $^{\ominus}OR$ ⟶ Claisen Product —(H—Cl dilute; **Synthetic Pathway #1**)⟶ β-keto ester + $^{\ominus}Cl$

Claisen Product —(1-bromopropane, Br; **Synthetic Pathway #2**)⟶

Critical Thinking Questions

12. Show the mechanism of the reaction in Synthetic Pathway #1.

13. Explain why the product on the right of Model 5 is called a **β-keto ester**.

14. Draw the mechanism and product that would result if the Claisen product were to act as a carbon nucleophile and react with an alkyl halide (such as 1-bromopropane) (**Synthetic Pathway #2**).

Part C: Aldol vs. Claisen

(Which is a better electrophile: a ketone or an ester?)

Model 6: OCH_3 is a Stronger Electron Donating Group than CH_3

Strong Electron Donating Group

Explains why OCH_3 is a strong *ortho/para* director in EAS.

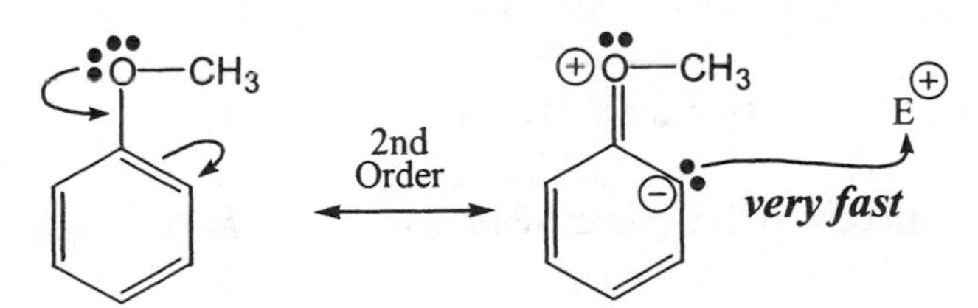

CH3 E slower

By comparison CH_3 is a weak (inductive) donating group.

Critical Thinking Questions

15. According to Model 6, which is a stronger **electron donating group**: an OR group or an R group? (Where R is an alkyl group such as methyl, ethyl etc.)

16. Consider the following acid base reactions:

Ester (methyl acetate)

Ketone (acetone)

a) Label the OCH_3 and the CH_3 group on the right of each molecule as being a strong or weak electron donating group.

b) Which reaction is more favorable? Explain your reasoning.

c) Which is more acidic: an ester or a ketone?

d) Is your answer above consistent with the fact that acetone has a pKa of 19 and methyl acetate has a pKa of 25?

17. Consider the following nucleophilic addition reactions:

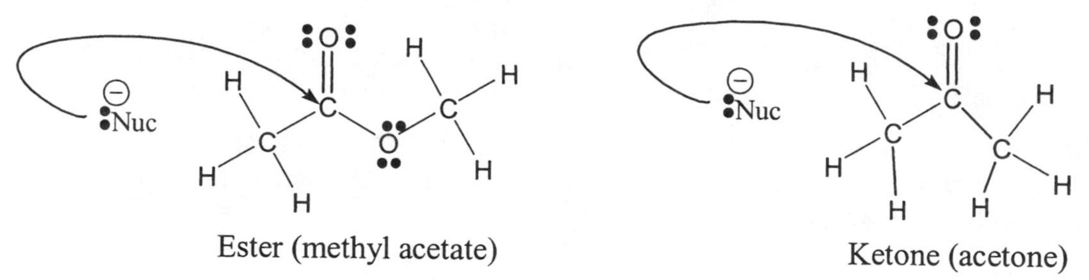

Ester (methyl acetate)

Ketone (acetone)

a) In terms of sterics, both carbonyl carbons above are very similar. Explain.

b) Based on electronic factors (electron donation and withdrawal), which carbonyl carbon above will have a larger δ+ charge: **ester** <u>or</u> **ketone**?

c) Which is a better electrophile: and **ester** <u>or</u> a **ketone** [circle one]?

18. If acetone and methyl acetate were mixed with a base in order to do a Claisen-type reaction…
    a) which compound would end up acting as a nucleophile: **ester** <u>or</u> **ketone**?
    b) which compound would end up acting as a electrophile: **ester** <u>or</u> **ketone**?

Model 7: Competition Between Claisen and Aldol

- A ketone is more acidic than an ester. This means that in the reaction mixture below, acetone is most likely to be deprotonated thus becoming an enolate ion nucleophile.
- This enolate nucleophile can react with another ketone forming an aldol product, or it can react with an ester, leading to a β-dicarbonyl product (via a Claisen-type mechanism).

Acetone

$^{\ominus}OCH_3$

methyl acctatc (no acid-base rxn)

Enolate Ion (nucelophile)

Rxn A

Rxn B

β-dicarbonyl Product

Aldol Product

Critical Thinking Questions

19. Acetone is more likely to act as nucleophile AND it's a better electrophile. Based on this which product in Model 7 do you expect to be favored?

20. This is an energy diagram for Rxn B in Model 7. Consider the negative charge in the reactants and the products and explain why this reaction is up-hill in terms of potential energy (endothermic).
Hint: is the negative charge more stable in the reactants or in the products?

V.E.

resonance stabilized

reaction progress

NOT resonance stabilized

Aldol Product

Model 8: Energy Diagram for Claisen vs. Aldol

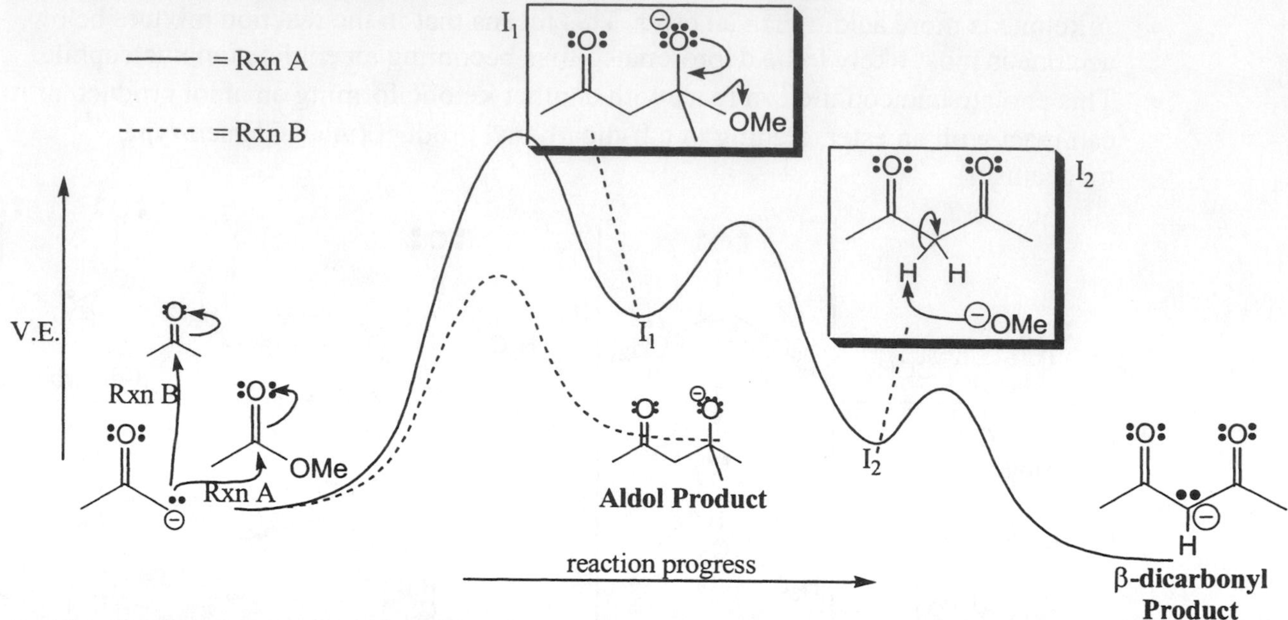

Critical Thinking Questions

21. Based on the energy diagram in Model 8…
    a) Which reaction is faster? Explain how you can tell from the energy diagram.

    b) Your answer in part a) is another way of saying that a ketone is a better electrophile than an ester. Explain.

    c) Assuming there is plenty of energy to surmount either energy barrier, which product do you expect to dominate the product mixture? Explain how you can tell from the energy diagram.

    d) Explain the following statement: "Even though a ketone is the better electrophile, the ester ends up acting as electrophile because the final product is more favorable."

22. During the reaction in Model 8, we can measure small amounts of aldol product.
    a) Discuss the following statement:

       "The aldol product serves as a stored source of acetone, as acetone is used up making the β-dicarbonyl product, a reverse aldol reaction replenishes the supply of acetone starting material."

    b) On the reaction diagram in Model 8, trace the path of a molecule of acetone that is first incorporated into an aldol product, and subsequently ends up as part of a β-dicarbonyl product.

Exercises for Part A

1. What base would you add to each of the following esters to cause a Claisen Reaction?

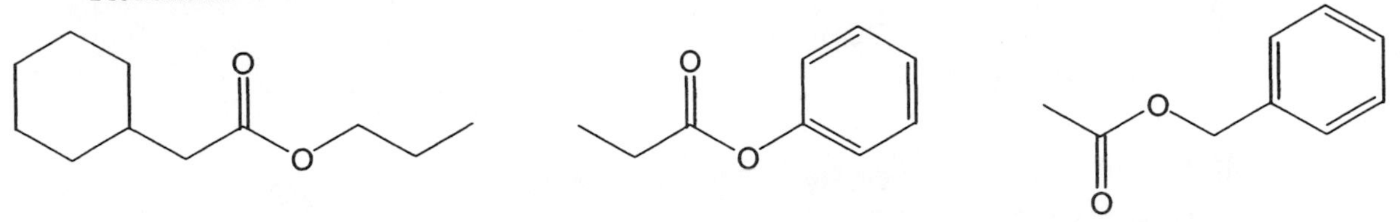

2. Show the mechanism of a Claisen reaction involving the ester above, right.

3. Draw the Claisen product that would result with each ester in Exercise 1.

4. The following reactants yield a major product that includes a five or six member ring. For each, draw the best mechanism leading to ring formation.

a) $\xrightarrow{^{\ominus}OCH_2CH_3}$

b) H_3CO OCH_3 H_3CO OCH_3 $\xrightarrow{^{\ominus}OCH_3}$

5. Read the assigned pages in your text and do the assigned problems.

Exercises for Part B

6. An aldehyde or ketone can act as a nucleophile in a Claisen-type reaction.
   a) Draw the mechanism and product of a Claisen-type reaction between acetaldehyde and ethyl formate.

   $\xrightarrow{^{\ominus}OEt}$

   b) Draw the aldol product that will also form from this reaction mixture.

   c) Suggest a laboratory procedure that will nearly eliminate formation of the aldol product from the reaction mixture.

7. For each of the following reaction mixtures, list the number of different products that could occur via carbon-carbon bond formation (aldol, Claisen or some mixture).

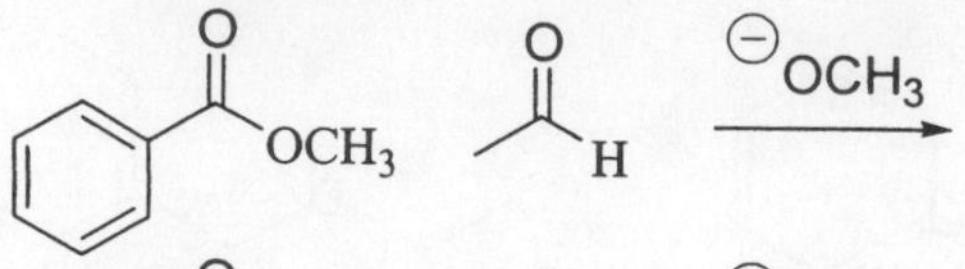

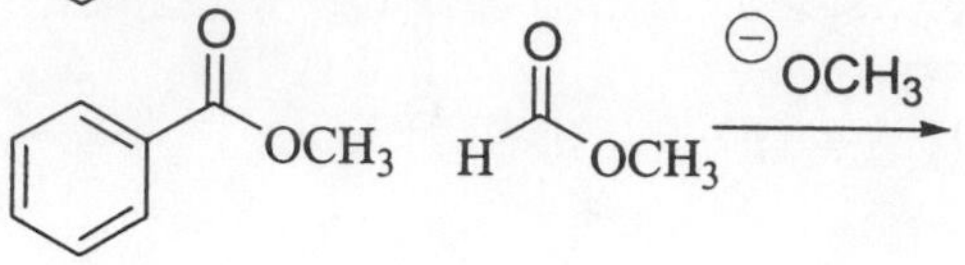

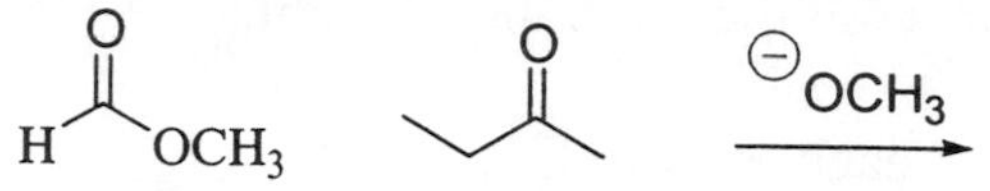

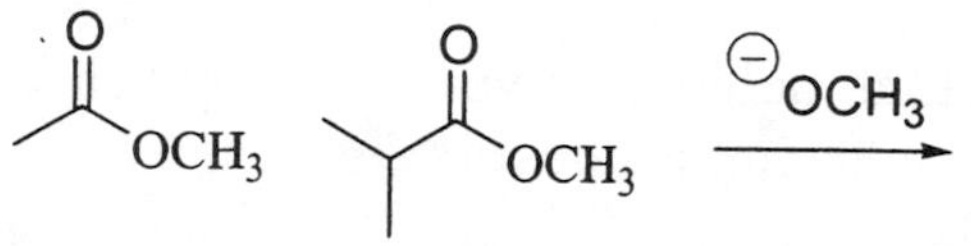

8. Draw the products in the previous question in which an ester acts as the electrophile.

9. Apply retrosynthetic analysis to each of the following products. Each is the product of an aldol or Claisen-type reaction that has been neutralized with dilute acid. (As in **Synthetic Pathway #1** in Model 5.)
   i) Start by marking the new bond that was formed in the aldol/Claisen reaction.
   ii) Draw the starting aldehydes, ketones, esters or some combination of these that would give rise to the product shown. There are several possible correct answers for some targets.
   iii) Choose a base that would be suitable for such an aldol/Claisen reaction.

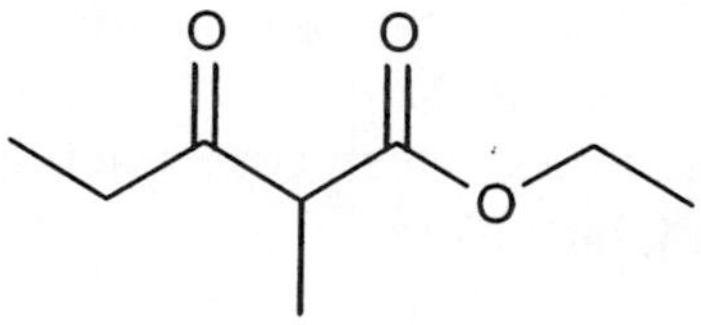

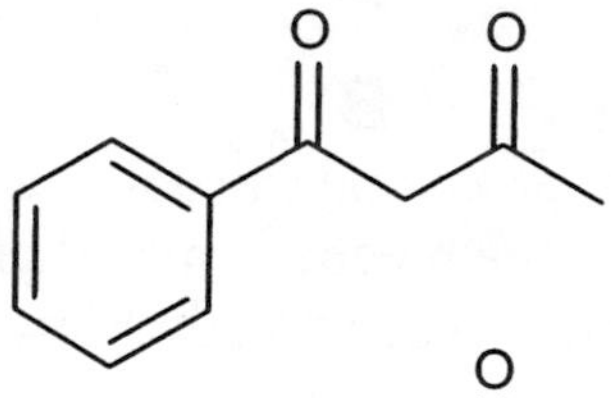

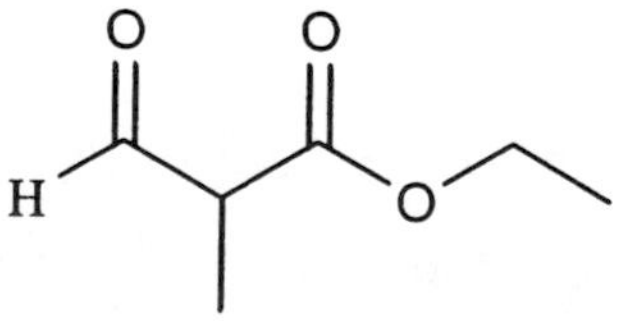

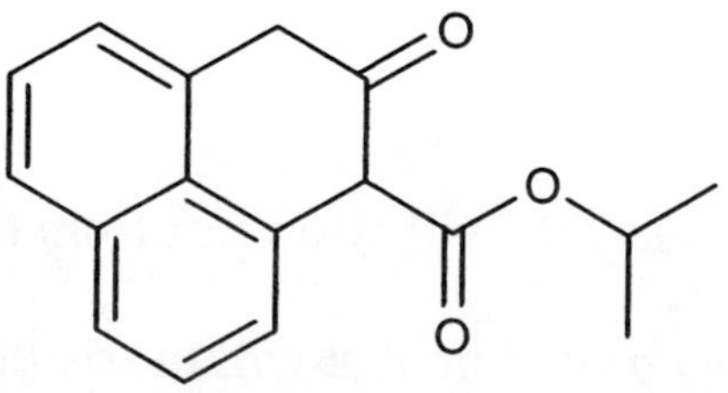

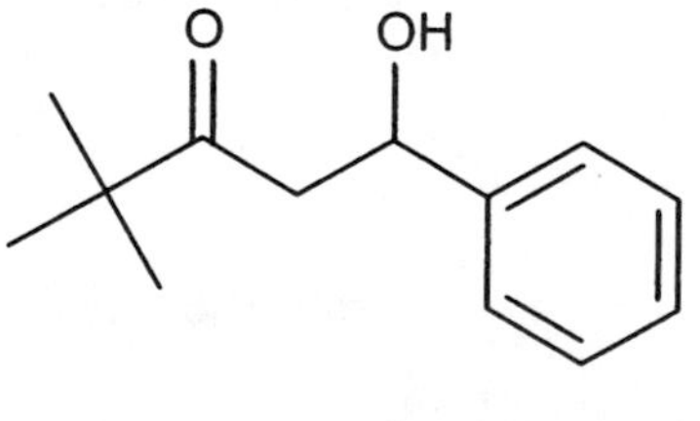

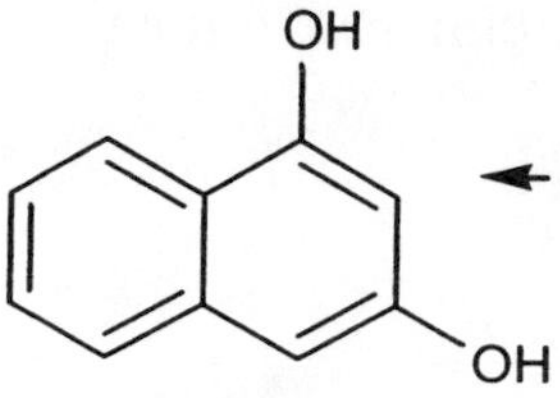

(Hint: This molecule adopts a "double enol" form so as to be aromatic. Draw the "double keto" form and consider how this could have been made.)

10. Each of the following targets is the product of a Claisen reaction between two <u>identical</u> esters, followed by addition of an alkyl halide. (As in **Synthetic Pathway #2** in Model 5.) Apply retrosynthetic analysis…
    i) Start by marking the new bond in the target molecule so as to identify the part of the target that was an alkyl halide and the part that was a Claisen product.
    ii) Draw the Claisen product and alkyl halide that could be mixed to give the target shown.
    iii) Draw the one starting ester (none of these are mixed Claisen reactions) that could give rise to the Claisen product you drew in ii).

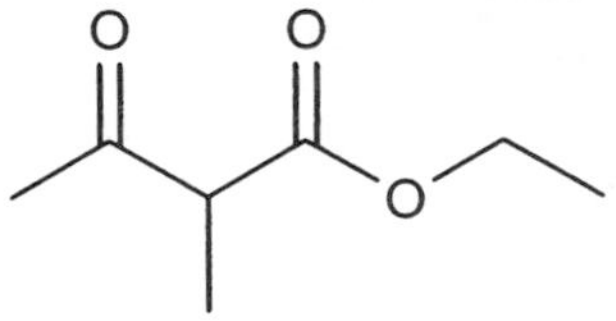

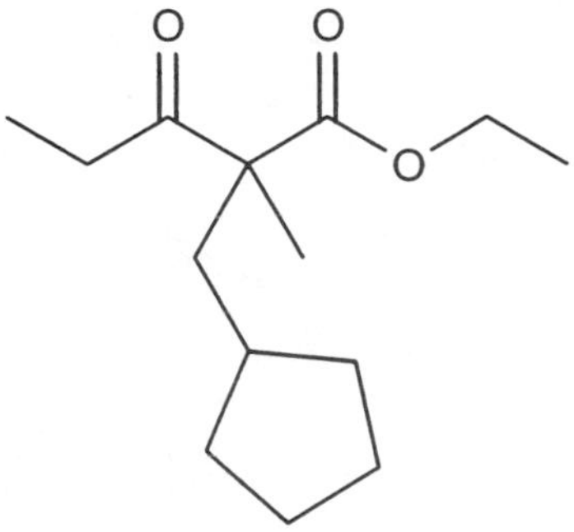

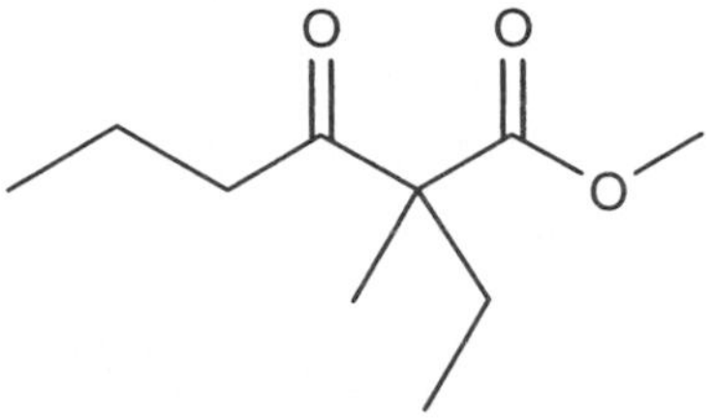

11. Read the assigned pages in your text and do the assigned problems.

Exercises for Part C

12. According to the reaction diagram in Model 8, the last step of Rxn A is down-hill in terms of potential energy. Is this consistent with the placement of the negative charge on Intermediate 2 (I_2) vs. the β-dicarbonyl product? Explain.

13. Explain the following statement: "If a molecule has no alpha H's between the two carbonyl groups, it cannot undergo the last step of a Claisen-type reaction (Rxn A in Model 8). This means the reaction will be endothermic, and therefore, unfavorable."

14. Give an example of a Claisen product with no alpha H's between the two carbonyl groups.

15. Draw the mechanism and most likely product of a Claisen reaction between the following two esters.

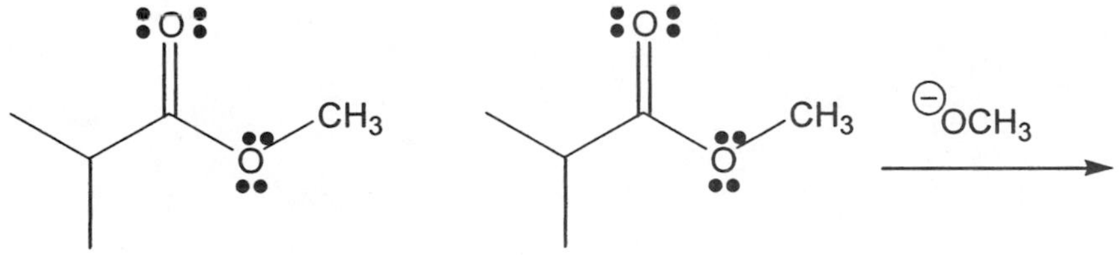

CH_3 CH_3 $^{\ominus}OCH_3$

16. Explain why the Claisen reaction above will be endothermic, and therefore up-hill in terms of energy.

17. Rank the following carbonyl compounds from the best electrophile (1) to the worst electrophile (6) and explain your reasoning.

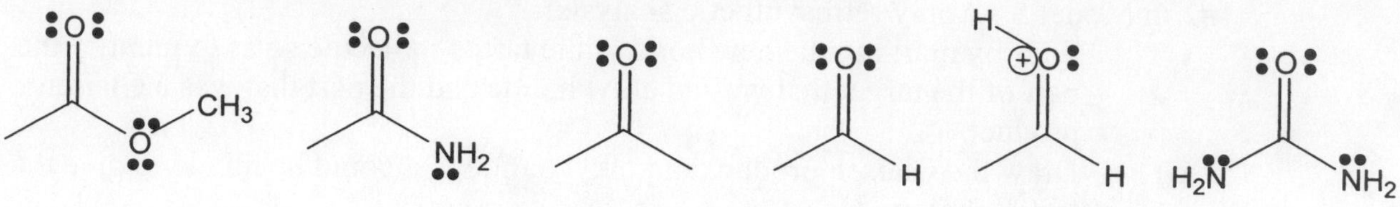

CH_3 NH_2 H H H_2N NH_2

18. Devise a mechanism for the following reaction known as a Robinson Annulation. Hint: it is a Michael reaction followed by an intramolecular aldol reaction.

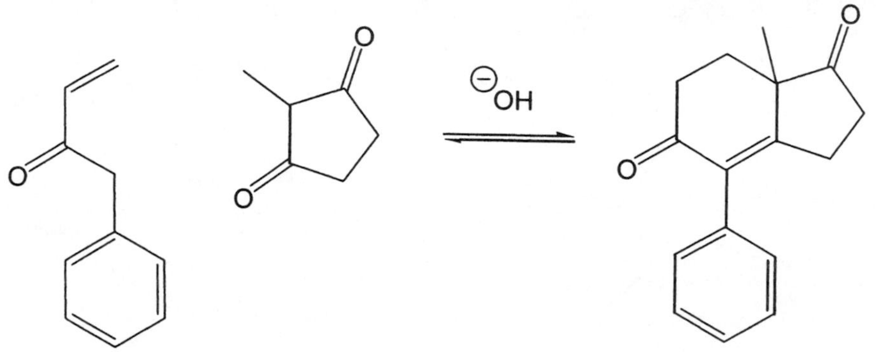

O O $^{\ominus}OH$ O O O

19. Read the assigned pages in your text and do the assigned problems.

ChemActivity 43

Decarboxylation

(By what mechanism do certain compounds give off carbon dioxide upon heating?)

Model 1: Decarboxylation (Loss of CO_2)

- A **carboxylic acid with a carbonyl two carbons away** (β-keto-carboxylic acid) will **lose CO_2** upon heating.

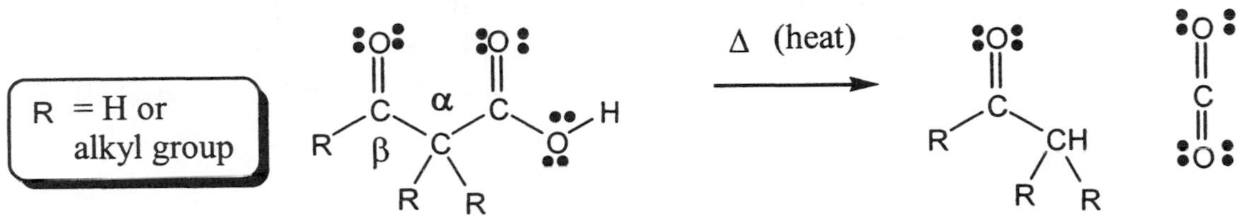

R = H or alkyl group

Δ (heat)

β-keto carboxylic acid

1. Construct a mechanism for loss of carbon dioxide (CO_2) in Model 1.
- The reaction occurs without involvement from any other molecules including solvent or catalysts.
- It helps to draw the starting material in the conformation shown below.
- The final stage of the mechanism is an enol-to-keto tautomerization (shown).

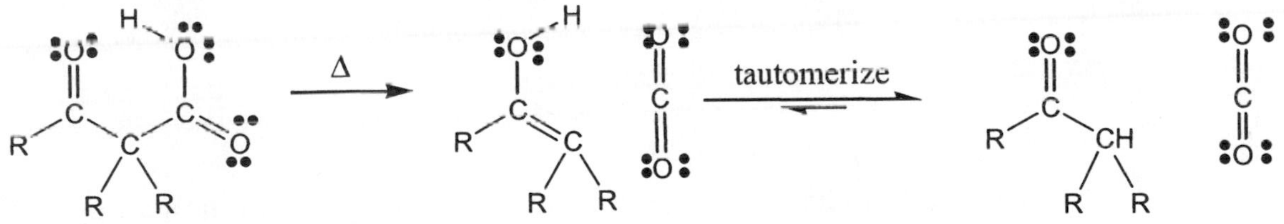

Δ

tautomerize

2. Above each reaction arrow, write the reagent/s or conditions necessary for that step in the synthesis. (Note: For some steps, not all products are shown.)

OCH₂CH₃ —Step 1→ [OEt ↔ OEt ↔ OEt] —Step 2→ OEt (H R) —Step 3→ [OEt (R) ↔ OEt (R) ↔ OEt (R)] —Step 4→ OEt (R' R) —Step 5→ OH (R' R) —Step 6→ H (R' R)

3. The following ketone can be made using the synthetic strategy in the previous question.

a) Circle each carbon that in the target that was originally part of the starting esters.

b) Ethyl acetate has four carbons. If two molecules of ethyl acetate are used in this synthesis, and the product only ends up with three of these carbons, this means five carbons were lost. Detail how each of the missing five carbons were lost. (Choose from: left as a RO^- leaving group, or left as CO_2 in a decarboxylation).

c) Put a square around each of the R groups that were added.

d) Draw the alkyl halide that would have been used to add each of these R groups (as in Steps 2 and 4 above).

4. Give an example, not appearing in this ChemActivity, of a ketone that could be synthesized starting from ethyl acetate and ending with a decarboxylation. Then show a synthesis of this ketone like the one in CTQ 2.

5. The strategy in CTQ 2 can also be used starting with longer esters such as propionic esters.

   a) Review: Draw the Claisen product that results from a propionic ester such as methyl propionate (shown below).

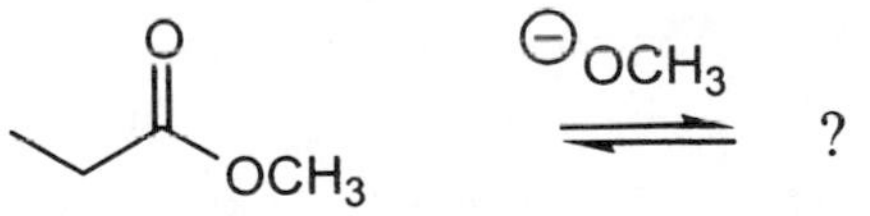

O
OCH_3 $\ominus OCH_3$ $\rightleftharpoons$?

   b) The following target can be made using a strategy similar to the one in CTQ 2, except that the starting ester is a propionic ester (three carbons in main chain instead of two). Circle each carbon in this target that was originally part of the starting esters.

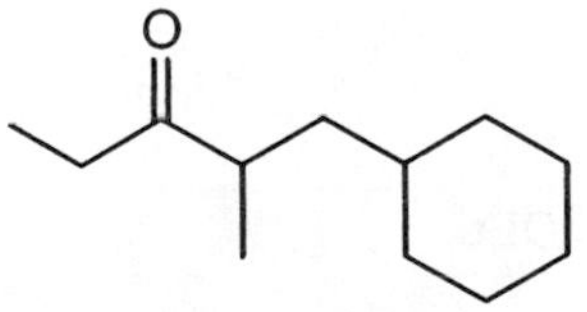

O

   c) Put a square around each of the R groups that were added.

   d) Draw the alkyl halide that would have been used to add each of these R groups (as in Steps 2 and 4 in CTQ 2).

Exercises

1. The molecule below is a methyl ketone that can be made starting with a Claisen reaction involving an acetic ester. Show a synthesis of this target.

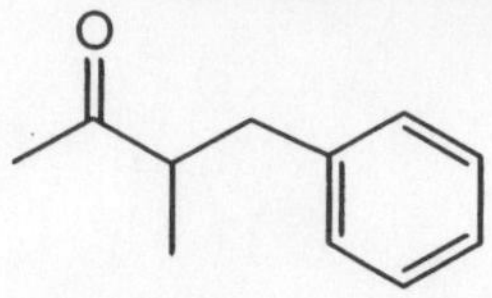

Target 1

2. None of the following can be made using the strategy above. **Note: only primary alkyl halides can effectively be used to alkylate the α-position.** Briefly explain why each uncircled product cannot be made from an ester using the strategy in CTQ 2 called "acetoacetic ester synthesis of ketones."

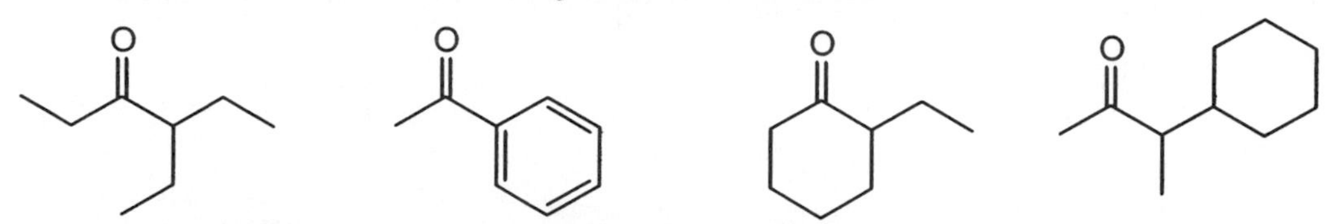

3. Would **methyl acetate** or **ethyl acetate** work equally well as starting materials in the synthesis in CTQ 2 and Exercise 1? Explain.
4. Predict which of the following will evolve carbon dioxide upon heating. 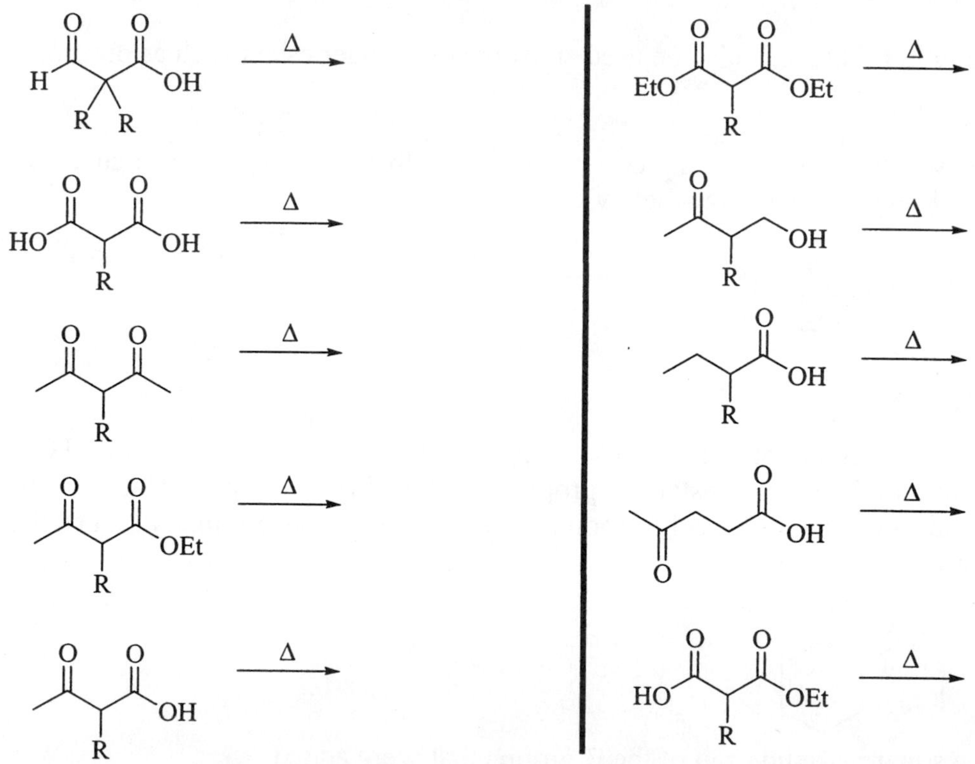
5. Draw an example (not appearing in this activity) of a dicarbonyl that will decarboxylate upon heating, and one that will not.
6. Show the mechanism of each reaction in CTQ 2.
7. Read the assigned pages in your text and do the assigned problems.

ChemActivity 44

Synthesis Workshop II

(How do I use retrosynthesis to determine the starting ester in a ketone synthesis?)

Review: Retrosynthesis

- Start from the starting material and work forward **AND think backwards from the target.**
- Organic chemists call this "thinking backwards" **retrosynthesis**.
- A key step is always the identification of a carbon-carbon bond in the target that you must make.
- Decide which of the two C's in this bond is the electrophilic C and which is the nucleophilic C.

Model 1: *Example Target = a methyl ketone*

Note: most methyl ketones can be made using the acetoacetic ester synthesis described in the previous activity. Recall that the last step in this synthesis involves decarboxylation!

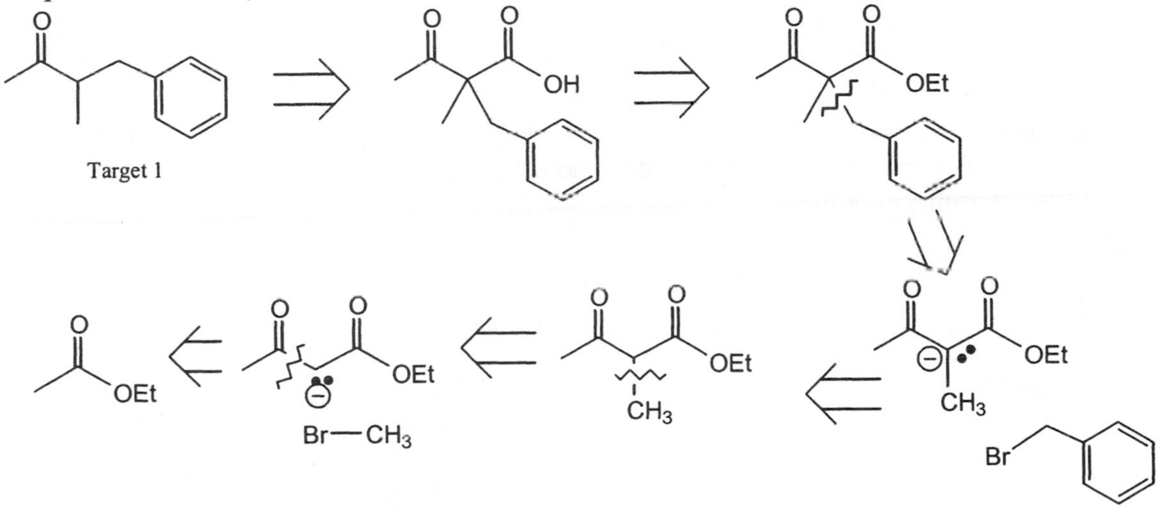

Target 1

OH

OEt

OEt

OEt

OEt

OEt

Br—CH_3

CH_3

CH_3

Br

Critical Thinking Questions

1. Going forward toward the target in a synthesis usually involves making new C-C bonds and going backwards from the target usually involves breaking C-C bonds. This example contains an exception.
   a) In which forward step is a C-C bond breaking?
   b) Indicate the C-C bond that is being broken (going forward).

2. Over each retrosynthesis arrow in Model 1 write the reagents necessary to accomplish this forward step.

3. Design a synthesis of the following target. Note: this target is not a methyl ketone. You may find it helpful to employ retrosynthesis. All carbons in the product must come from the reagents in the box.

Legal Carbon Containing Starting Materials

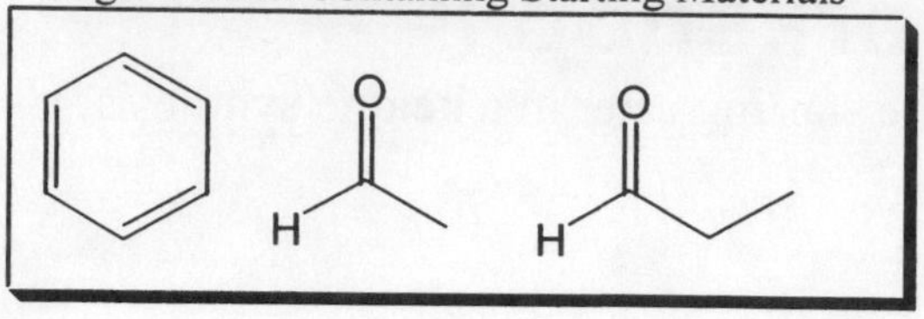

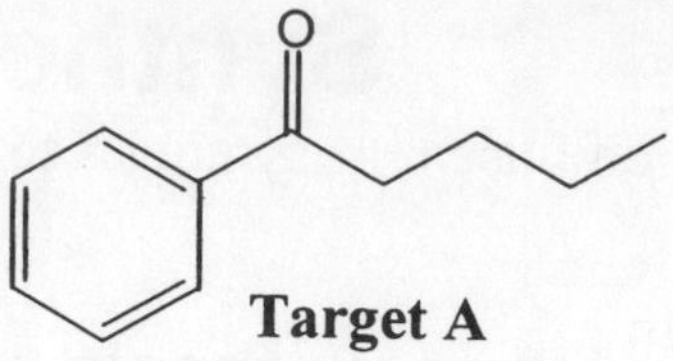

H H

Target A

4. Design a synthesis of the following target using any alkyl halide (with no other functional groups) and any ester (with no other functional groups). Hint: is the target a methyl ketone?

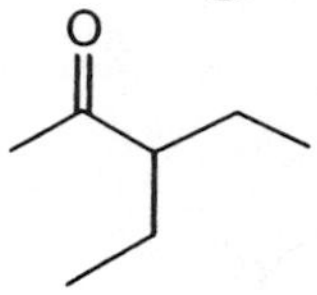

O

Target B

Exercises

1. Use retrosynthesis to design a way of making each of the following target molecules. The starting materials must include an acetic ester such as ethyl acetate.

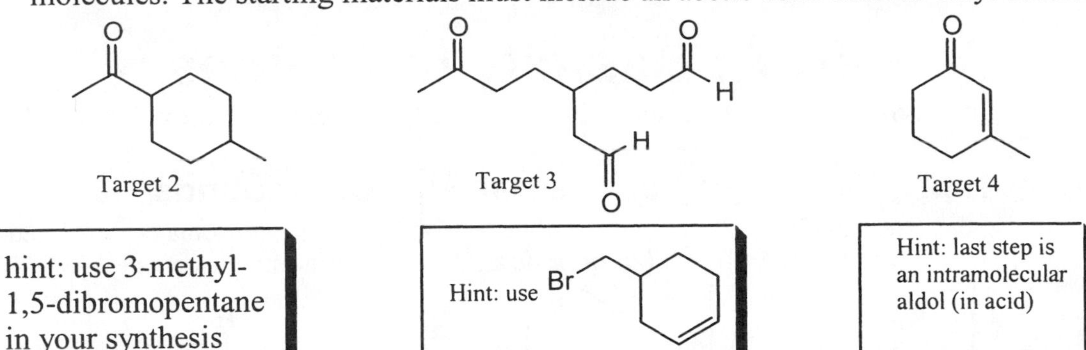

2. Use retrosynthesis to design a way of making each of the following target molecules. The starting materials must include ethyl malonate (shown below).

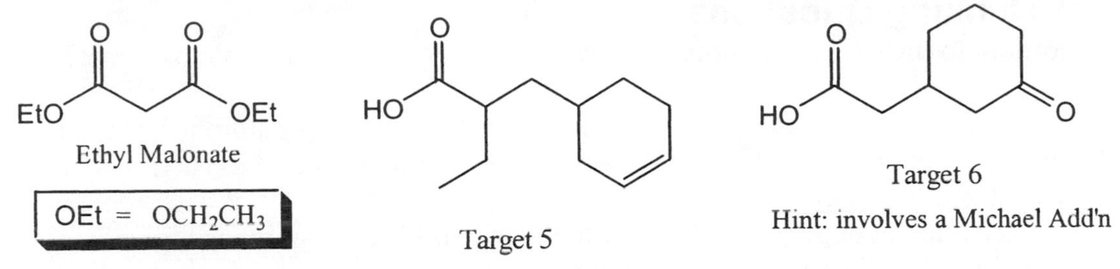

3. If you use very high concentrations of 3-methyl-1,5-dibromopentane in the synthesis of Target #2 at the top of the page, the following side-product will also occur. Explain why Target #2 is favored over this side-product at normal concentrations.

4. Read the assigned pages in your text and do the assigned problems.

ChemActivity 45

Part A: Basicity of Amines

(Why is dimethyl amine a better nucleophile than methyl amine even though it's bulkier?)

Model 1: pK_a Values for Various Nitrogen Compounds

	acetamide (an amide)	aniline (an aromatic amine)	ammonia (the simplist amine)	ethyl amine (a 1° amine)
	$CH_3C(=O)\ddot{N}H_2$	$C_6H_5\ddot{N}H_2$	$H-\ddot{N}(H)-H$	$CH_3CH_2\ddot{N}H_2$
pKa of conjugate acid	0	4.6	9.3	10.8

Critical Thinking Questions

1. According to the pKa data in Model 1, which nitrogen compound is most basic?

2. A good way to think about base strength is: The lone pair on an atom is basic if it "wants" to make bond to H^+. A lone pair "wants" to make a bond to H^+ if it is **localized** and **high energy**.
   a) Construct an explanation for why the lone pair on ethylamine is higher in energy than the lone pair on ammonia.

   b) Draw diethyl amine, a secondary (2°) amine, and predict whether the pKa of its conjugate acid will be higher or lower than 10.8.

   c) Diethyl amine is a **stronger base** or **weaker base** [circle one] than ethyl amine.

3. Construct an explanation for why an aromatic amine is a much weaker base than a non-aromatic amine. (Hint: draw a set of "**second order**" resonance structures.)

a) Predict if the pKa of the conjugate acid of the aromatic compound below is **close to 10** or **close to 5** [circle one]? Explain your reasoning.

"benzyl amine" ($C_6H_5{-}CH_2{-}\ddot{N}H_2$)

b) Explain why the aromatic molecule in part a) is not considered an "aromatic amine."

4. Explain why an amide is an even worse base than an aromatic amine. (Recall that amides are unusual in that they have a 2nd order resonance structure that is nearly equivalent to the 1st order resonance structure.)

Model 2: Alkylation of Amines

$$2\ H_2\ddot{N}{-}H + Br{-}CH_3 \xrightarrow[\text{Rxn 1}]{\textbf{fast}} H_2\ddot{N}{-}CH_3 + H{-}N^{\oplus}H_2{-}H\ \ Br^{\ominus}$$

$$2\ H_2\ddot{N}{-}CH_3 + Br{-}CH_3 \xrightarrow[\text{Rxn 2}]{\textbf{faster}} H(CH_3)\ddot{N}{-}CH_3 + H{-}N^{\oplus}H_2{-}CH_3\ \ Br^{\ominus}$$

$$2\ H(CH_3)\ddot{N}{-}CH_3 + Br{-}CH_3 \xrightarrow[\text{Rxn 3}]{\textbf{fastest}} H_3C(CH_3)\ddot{N}{-}CH_3 + H{-}N^{\oplus}H(CH_3){-}CH_3\ \ Br^{\ominus}$$

Critical Thinking Questions

5. Use curved arrows to show the mechanism of Rxn 1 in Model 2. What is the general name for this reaction type?

6. Construct an explanation for why the rate of Rxn 3 > rate of Rxn 2 > rate of Rxn 1.

Part B: Synthesis of Amines/pK_a of AA's

(How can we add just one alkyl group to an amine?)

Model 3: Messy Alkylation of Amines

- Recall from Part A that a 2° amine is a better nucleophile than a 1° amine.
- This means the reaction below is a messy way to add an alkyl group to an amine.
- Once a little bit of 2° amine product is formed, it will **out-compete the starting material**, and quickly react with methyl bromide to form the 3° amine product.

Br—CH_3 Br—CH_3

RNH_2 + Br—CH_3 → $RNHCH_3$ / $RN(CH_3)_2$ / $RN^+(CH_3)_3$ Br^-

1° amine "primary"

2° amine "secondary"

3° amine "tertiary"

"quaternary (4°) ammonium ion"

mixture of products

Critical Thinking Questions

7. Explain why the reaction in Model 3 gives a messy mixture of all the products shown.

Model 4: Selective Alkylation of Amines

Gabriel Synthesis of Primary Amines:

Named for Prof. Sigmund Gabriel (1851-1924) Univ. of Berlin

phthalic acid —($:NH_3$, heat (acid catalyst))→ phthalimide (N—H) + 2 H_2O —(even a weak base)→ phthalimide anion ($:N:^-$)

—(any 1° alkyl halide, Br—CH_3)→ N-methylphthalimide (N—CH_3) Br^- —(H_2SO_4, H_2O)→ $H_3N^+CH_3$ + phthalic acid —(neutralize w/ base)→ H_2NCH_3

methylamine (ONLY!)

Model 4 continued on next page...

Reduction of Amides, Imines and Nitriles to form 1° and 2° Amines:

amide —($LiAlH_4$; R' = alkyl or H)→ 1° or 2° amine

aldehyde or ketone —(H_2N–R'; R' = alkyl or H)→ imine —(H_2, metal catalyst)→ 1° or 2° amine

nitrile (R—C≡N:) —($LiAlH_4$ or H_2 /metal catalyst)→ $R-CH_2-NH_2$ 1° amine

Critical Thinking Questions

8. What problem plagues direct alkylation of amines (Model 3), but does not plague each of the reactions in Model 4?

9. Amines are very weak acids. It takes a huge amount of energy (30+ pKa units) to pull an H off a neutral N (e.g. R_2NH) to make a N with a negative charge (R_2N^-). Explain why only a weak base (a small amount of energy) is necessary to prepare the N with a negative charge in Step 2 of the Gabriel Synthesis.

Model 5: pK_a Values for Carboxylic Acids and Amines

carboxylic acid (pK_a 5) + H_2O ⇌ conjugate base (carboxylate ion) + H_3O^+

ammonium ion (pK_a 9) + H_2O ⇌ conjugate base (amine) + H_3O^+

Critical Thinking Questions

10. What form of the nitrogen functional group in Model 5 will be most abundant at biological pH (near pH 7)? **ammonium ion** or **amine** [circle one].

11. At what pH do you expect the ratio of [carboxylate ion]/[carboxylic acid] to be 100/1?

Model 6: Amino Carboxylic Acids (amino acids)

- An **amino acid** is a multi-functional group molecule with an amine end and a carboxylic acid end.
- There are twenty common amino acids found in biological systems.
- Proteins in all living organisms are polymers (long chains) of these amino acids.
- By convention, amino acids are drawn with the amine/ammonium end on the left (as shown below).

$H_3N^{\oplus}$ O H R

Each amino acid differs in the identity of the R group.

Critical Thinking Questions

12. Water solutions are limited to a pH range of about 0-15. Based on the information in Model 5…
    a) What is the likely pH (give a range) of the aqueous solution containing the amino acid shown in Model 6?

    b) Draw the dominant form of an amino acid at pH 7.

    c) Over what pH range would the form you drew in part b) be dominant?

Exercises for Part A

1. Use wedge and dash bonds to draw a 3-D representation of ammonia.

   a) Label your drawing with the word used to describe the shape of this molecule. Note that the shape of a molecule is the geometric shape described by the positions of the nucleii (not the electron pairs).

   b) What is the **hybridization state** of the N in ammonia?

2. Amines cannot be chiral. This is because the lone pair does not hold a fixed position around the N for very long. An amine is constantly "flipping" back and forth like an umbrella inverting and un-inverting in a strong wind.

   a) However, ammonium salts can be chiral. Draw a wedge and dash representation of the **R** enantiomer of the conjugate acid of *N*-methyl-*N*-ethyl propylamine (N with 1 methyl, 1 ethyl and 1 propyl group).

   b) Consider the following reaction: If a sample of the pure **R** ammonium salt you drew above is treated with base, and then acid, will the resulting sample be **pure R, pure S,** or **a racemic mixture** [circle one]?

3. Draw all three 2nd order resonance structures of aniline and explain why aromatic amines are weak bases and poor nucleophiles (as compared to non-aromatic amines).

4. Place the following labels on the molecules below. Some with have more than one label, some will have no labels. **Aromatic Molecule Aromatic Amine Amide**

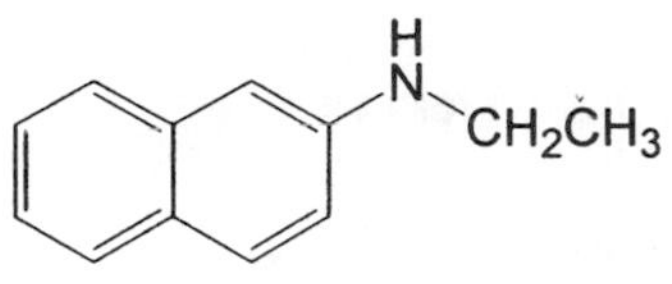

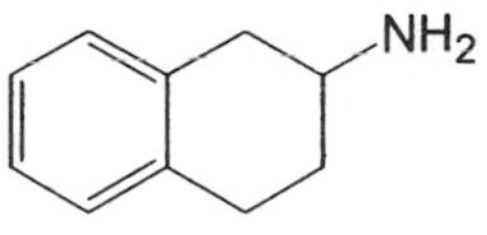

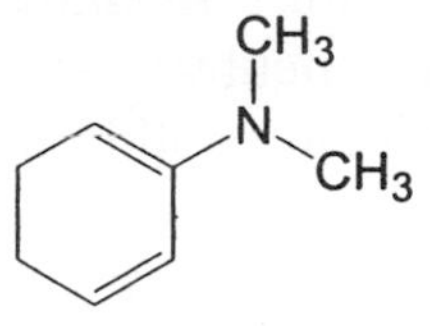

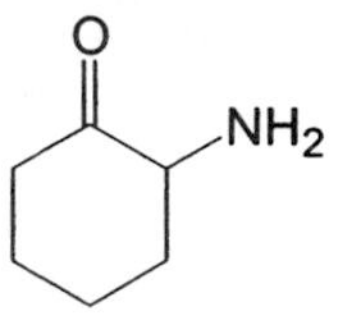

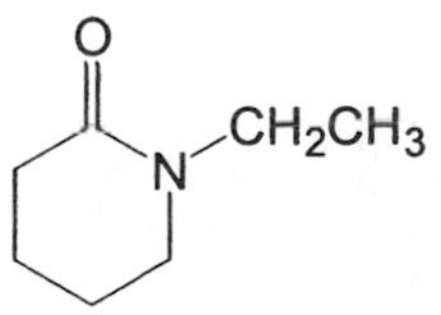

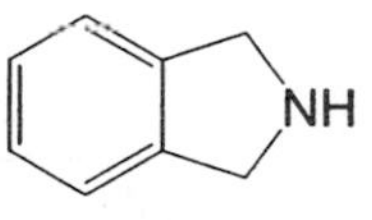

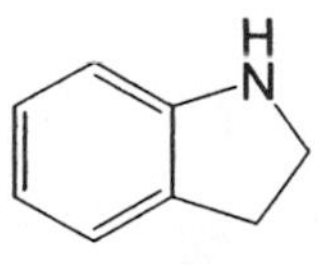

5. Indicate the approximate pKa of the conjugate acid of each compound above by writing one of... **pKa about 0** **pKa about 5** **pKa about 10.**

6. Which of the following aromatic compounds is more basic? Explain your reasoning.

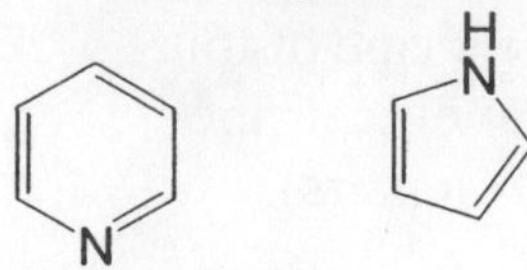

7. Read the assigned pages in your text and do the assigned problems.

Exercises for Part B

8. Show three different ways of making benzyl amine (below) using ammonia (NH_3) as your original nitrogen source.

NH_2

9. On the pH scale below, draw all the possible forms of an amino acid and indicate the pH range through which each form is most likely to be dominant.

10. In what pH range (if any) is an amino acid likely to...
    a) have no + or – formal charges on any of its atoms?
    b) be overall neutral?

(Note: A neutral molecule with one + and one – charge is called a **zwitterion**.)

11. Show the mechanism of Steps 1 and 4 in the Gabriel Synthesis in Model 4.

12. Consider the reaction below.

NH_2 O → NH_2

   a) Write appropriate reagents over the reaction arrow.
   b) As a result of this reaction, is the N made more nucleophilic or less nucleophilic?
   c) Draw "second order" resonance structures to support your answer in part b).

13. Read the assigned pages in your text and do the assigned problems.

ChemActivity 46

Part A: Amino Acids and Peptides

(Is the peptide IAG the same as the peptide GAI?)

Model 1: The 20 Amino Acids at Biological pH

See diagram of AA structures at the end of this ChemActivity.

Side chains are classified into three general types: those with...

i) **nonpolar** sidechains
ii) **polar** sidechains
iii) **charged** sidechains (at biological pH)

A working definition of **nonpolar** = the sidechain has more than 2 C's per neutral heteroatom. ("**hetero**" = "other" and refers to atoms "other" than C or H.)

Critical Thinking Questions

1. Each amino acid has a distinct side chain attached to a carbon referred to as the "**alpha carbon**."
   a) Why is this carbon called the "alpha carbon."

   b) Identify a example of an amino acid with a **nonpolar** sidechain...a **polar** sidechain...a **charged** sidechain.

2. All but one amino acid is chiral.
   a) Which amino acid is not chiral.

   b) What is the absolute configuration (R or S) of alanine?

   c) Chiral, naturally occurring amino acids are called "**L-amino acids**." Confirm that, at the alpha carbon, all L-amino acids have the same spatial arrangement of the ammonium, carboxylate, H and sidechain groups at the alpha C.

   d) Use wedge and dash bonds to draw a "D-amino acid" (one not found in nature).

Model 2: Peptide Bonds/Dipeptides/Polypeptide Chains

- Specially designed enzyme catalysts can hold two amino acids in the right conformation and covalently link them together, forming a **dipeptide**.
- Enzyme catalysts can add more amino acids to the carboxy terminal end, one at a time, building a long unbranched chain of amino acids called a **polypeptide**.

O CH_2OH

H_3N^{+} N H O^{-}

"Amino Terminus" CH_2SH O **"Carboxy Terminus"** (chain extended by adding amino acids to carboxy end)

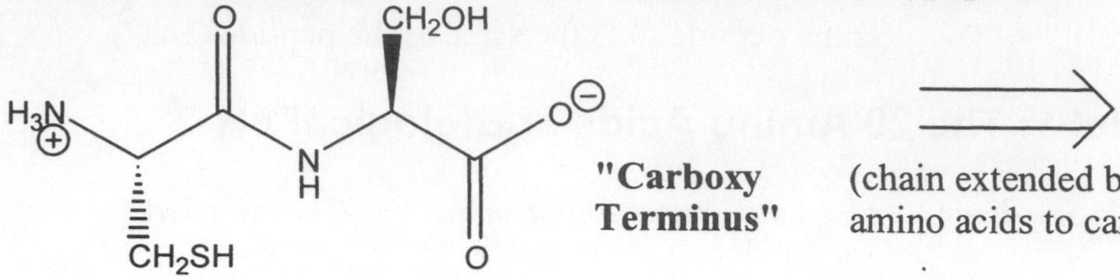

Critical Thinking Questions

3. A **peptide bond** is the bond formed with the help of an enzyme that connects two amino acids.
   a) Put a slash through the peptide bond in Model 2, as if you were retrosynthesizing this molecule from its constituent amino acids.

   b) When a peptide bond is made, a carboxylic acid functional group and an amine functional group are combined to form a functional group called an "**amide**." Put a box around the atoms comprising the amide functional group in the dipeptide above.

   c) On the structure in Model 2, circle together all the atoms that came from an individual amino acid, and identify each of these amino acids by name.

4. The **dipeptide** shown in Model 2 is named "Cys-Ser" or "CS".
   a) Explain each of these names.

   b) Draw the **tri-peptide** IAG.

5. A **polypeptide chain** is a long chain of amino acids.
   a) Write the **name** (using the three-letter codes) of a 5 amino-acid polypeptide chain that you expect to be very **insoluble** in water. Such polypeptides are called **hydrophobic** because they appear to "run away from water."

   b) Write the **name** (using the three-letter codes) of a 5 amino-acid polypeptide chain that you expect to be very **soluble** in water. Such polypeptides are called **hydrophilic** because they appear to "love water."

Part B: Proteins

(What is the difference between the 1°, 2°, 3° and 4° structures of a protein?)

Model 3: α-Helix and β-Pleated Sheet

- part of a polypeptide chain can coil up into a hollow rod-like structure called an α-**helix.** This coil can be 10 amino acids long or hundreds of amino acids long.
- A part of a polypeptide chain can line up with other parts of the chain to form a fabric-like array called a **β-pleated sheet.**
- Both α-helices and β-pleated sheets are held together by hydrogen bonding between different amino acids in the polypeptide **backbone** (backbone shown in **bold**, below).

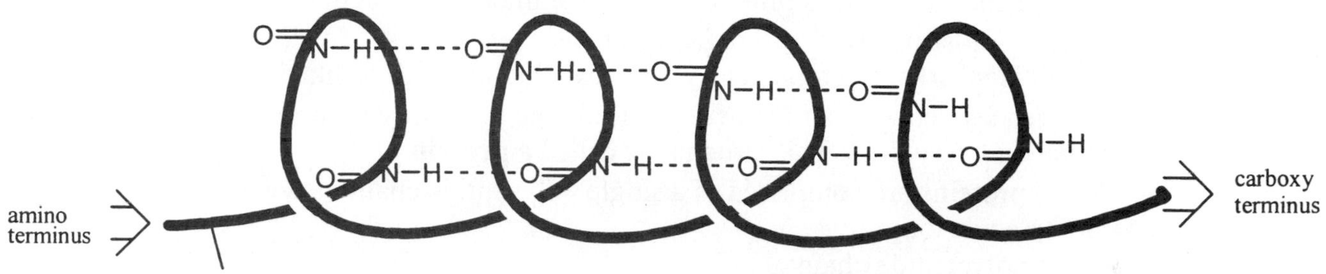

amino terminus

carboxy terminus

Schematic of polypeptide backbone (only atoms involved in hydrogen bonding are shown)

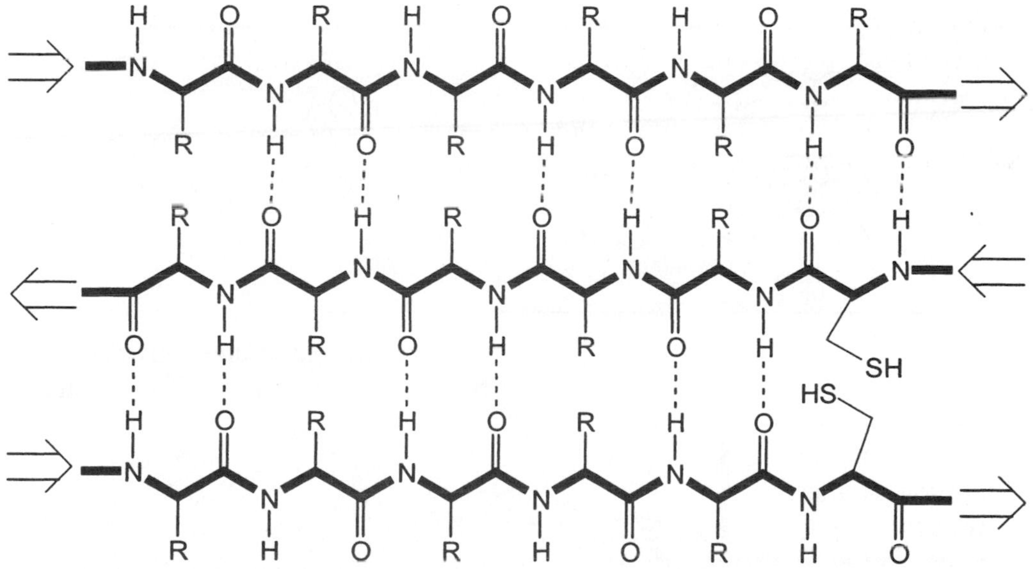

Critical Thinking Questions

6. Label the two structures in Model 3 as being either an α-helix or a β-pleated sheet, based on the descriptions at the top of the page.

7. What do the dotted lines on these structures represent?
8. A thiol is like an alcohol, but with S replacing O. Two thiols can be oxidized (loss of H_2) to form a **disulfide bond** (shown below).

$$R{-}S{-}H \quad H{-}S{-}R \xrightarrow{\text{oxidize}} R{-}S{-}S{-}R \quad H_2$$

thiol thiol disulfide

a) What amino acid has a thiol sidechain?

b) Modify the drawing of the β-pleated sheet in Model 3 to show how the sidechains shown at the bottom right of the drawing can be oxidized to covalently link two strands of the β-pleated sheet.

Model 4: Protein Folding

- Parts of a large polypeptide chain will spontaneously organize into α-helices, β-pleated sheets and sometimes other less common sub-structures.
- These "sub-structures" will organize themselves into a specific 3D super-structure, which is mostly held together by non-covalent interactions like hydrogen bonds.
- A given sequence of amino acids will fold the same way to form the same 3D structure every time. This structure is called a **protein**.
- Many proteins are comprised of a single polypeptide chain between 50 and 500 amino acids long.

Linear polypeptide chain:

ALIVFYWMGPSTNGCFPCHEGQVIMNDRTWLAYSKRDEHCGNTSPQMWYFVILA...etc.

spontaneously forms helices and sheets

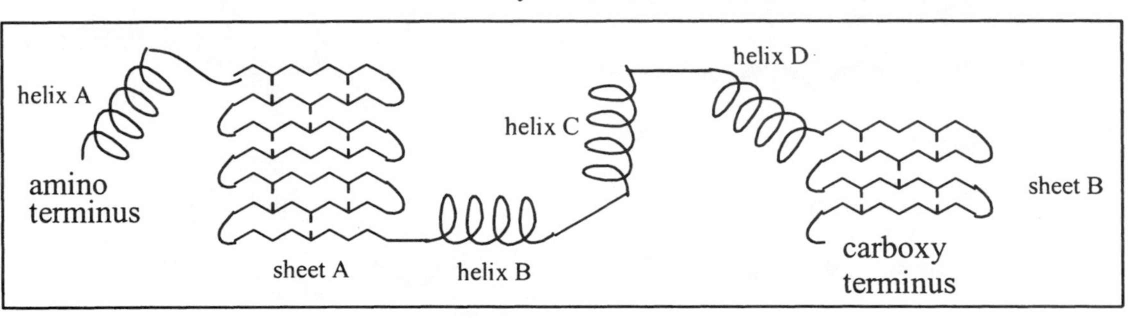

sub-structures spontaneously and repeatably organize into a very specific 3D structure

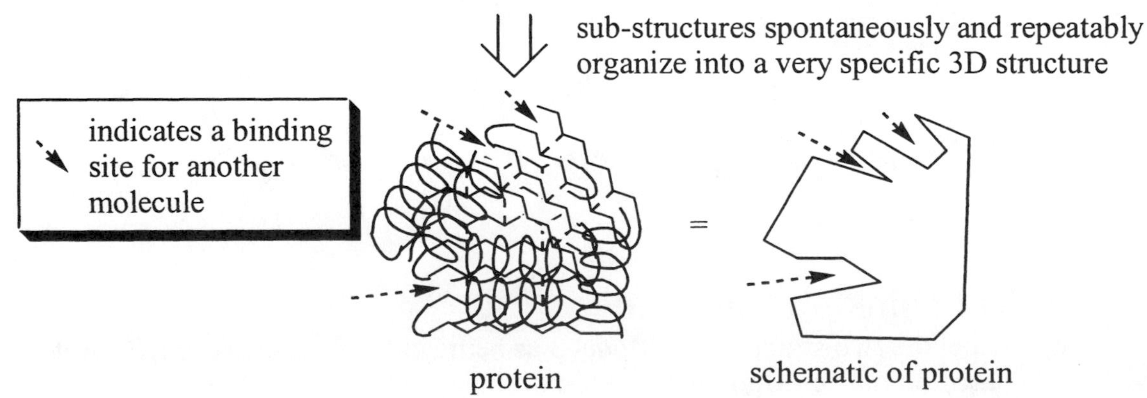

Critical Thinking Questions

9. Model 4 describes three levels of protein structure, usually called primary (1°), secondary (2°), and tertiary (3°) structure.
   a) If you know the primary structure of a protein, what do you know?

   b) If you know the secondary structure of a protein, what do you know?

   c) If you know the tertiary structure of a protein, what do you know?

Model 5: Protein Function

Proteins (on average) make up 50% of the dry weight of a cell.
They serve a myriad of chemical and structural functions including...

- structure (hair, skin, connective tissue, muscle etc.)
- enzymes (very specific and efficient catalysts such as those in Model 2)
- antibodies
- hormones
- transport for smaller molecules (e.g. O_2/CO_2 to and from the lungs and mitochondria = the tiny combustion engine that provides power to each of our cells).

Critical Thinking Questions

10. The chicken and the egg paradox speaks to the problem of how the first chicken got here. Explain how this parable relates to proteins. (Hint: see Model 2).

Exercises for Part A

1. There is a difference between the tripeptides IAG and GAI. Explain.

2. Simply mixing the three amino acids shown below results in an extremely low yield of the tripeptide IAG.

CH_3 H_3N^+ O^- O H_3C CH CH_2CH_3 H_3N^+ O^- O H_3N^+ O^- O

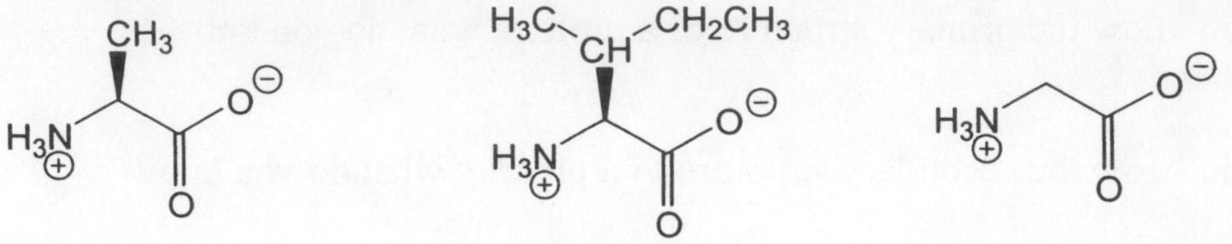

This is partly because a carboxylate ion is a terrible electrophile and an ammonium ion is a terrible nucleophile. The enzymes involved in peptide synthesis overcomes these problems by activating the amino and carboxy ends of each amino acid.

R H_3N^+ O----EnzymeA withdraws e- donates e- O

withdraws e- EnzymeB------N H donates e- R' O^- O

a) EnzymeA and EnzymeB, above, either donate or withdraw electron density. For each enzyme, circle the correct function and cross out the wrong one.

b) What other function must the enzyme perform to ensure a 100% yield of the tripeptide IAG?

c) Name one possible peptide sideproduct that would form if the protein simply activated both ends of the three amino acids shown above.

Exercises for Part B

3. In the 1960's Linus Pauling and E.J. Corey discovered that certain sequences of amino acids tend to organize themselves into rod-like structures (α-helix) while other sequences organize into sheet-like structures (β-pleated sheet). Still other sequences form neither. For example, a sequence that has many proline amino acids will double back on itself too often to be able to form either an α-helix or a β-pleated sheet. Among all amino acids, what is unique about proline?

4. Some very large proteins are actually made up of smaller protein sub-units. For example, the transport protein, hemoglobin is comprised of four nearly identical polypeptide chains. Each chain undergoes folding to produce a specific secondary (2°) then tertiary (3°) structure. Finally, the four 3° structures come together to form the quaternary (4°) structure of the protein hemoglobin. What is the difference between tertiary structure and quaternary structure of a protein?

5. Explain the following statement: "At this point in time, we are very bad at predicting the tertiary structure of a protein based on its linear amino acid sequence. Nevertheless, the amino acid sequence of a protein determines the tertiary structure of the protein."

6. The following pathway is an example of the trouble a chemist has to go through to create a single peptide bond in the laboratory. Describe at least two things this cumbersome synthesis accomplishes that simply adding two amino acids together cannot accomplish.

STEP 1

neutralize

STEP 2

STEP 3 $SOCl_2$

electrophile

nucleophile

STEP 4

This is an amino acid with the carboxy end protected to make it a poor nucleophile.

unprotected amino acid

STEP 5 H_3O^{+}

dipeptide

The 20 Amino Acids at Biological pH (with three-letter and one-letter abbreviations)

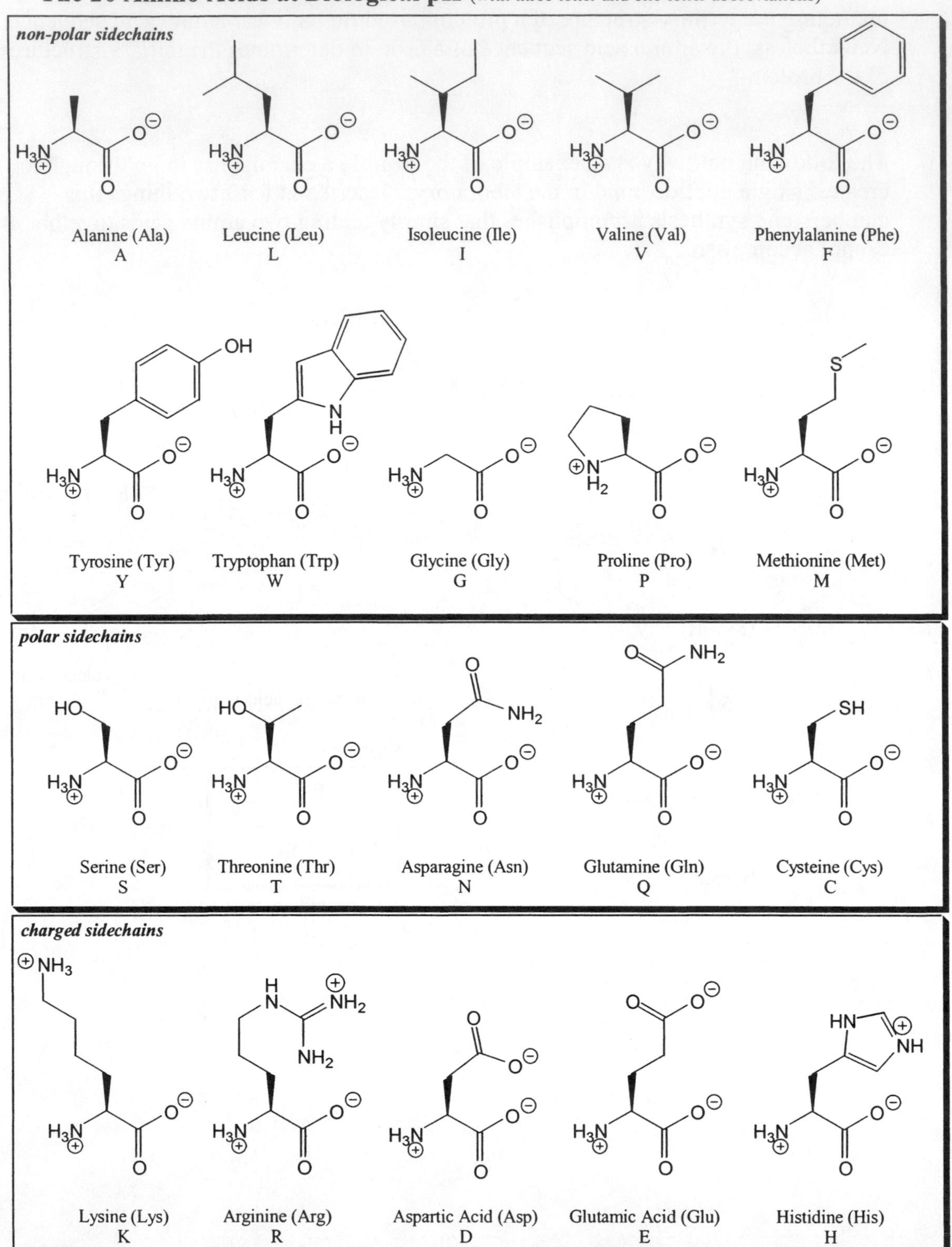

Nomenclature Worksheet 1

Alkanes, Alcohols, Ethers, Amines, Alkyl Halides, Alkenes, Alkynes

This system of naming molecules was designed so that organic chemists could quickly and easily learn to talk and write the language of chemistry. Like any "foreign language" the more you use it the easier it will be to learn and remember.

I. Alkanes (C and H with single bonds)

A. Straight Chain Alkanes

Names consist of a prefix telling the number of carbon atoms in the chain followed by the suffix "ane." Starting with five carbons ("pent"), the prefixes are derived from the Greek names for numbers.

Table 1: Straight Chain Alkanes

#C atoms	Condensed Structure	Name
1	CH_4	**meth**ane
2	$CH_3\ CH_3$	**eth**ane
3	$CH_3\ CH_2\ CH_3$	**prop**ane
4	$CH_3\ CH_2\ CH_2\ CH_3$	**but**ane
5	$CH_3\ CH_2\ CII_2\ CH_2\ CH_3$	**pent**ane
6	$CH_3\ CH_2\ CH_2\ CH_2\ CH_2CH_3$	**hex**ane
7	$CH_3\ CH_2\ CH_2\ CH_2\ CH_2\ CH_2CH_3$	**hept**ane
8	$CH_3\ CH_2\ CH_2\ CH_2\ CH_2\ CH_2\ CH_2CH_3$	**oct**ane
9	$CH_3\ CH_2\ CH_2\ CH_2\ CH_2\ CH_2\ CH_2\ CH_2CH_3$	**non**ane
10	$CH_3\ CH_2\ CH_2\ CH_2\ CH_2\ CH_2\ CH_2\ CH_2\ CH_2CH_3$	**dec**ane

B. Alkyl Groups

An alkyl group is a fragment of an alkane which can from a branch off the main trunk in complex alkanes (*see* part C). Alkyl groups are named by replacing "ane" in the parent name with "yl." Know all the following normal (*n*) and branched alkyl groups.

Table 2: Alkyl Groups

# C atoms	Condensed Structure	Alkyl Group Name
1	$-CH_3$	**meth**yl
2	$-CH_2CH_3$	**eth**yl
3	$-CH_2CH_2CH_3$	**prop**yl or *n*-**prop**yl
3	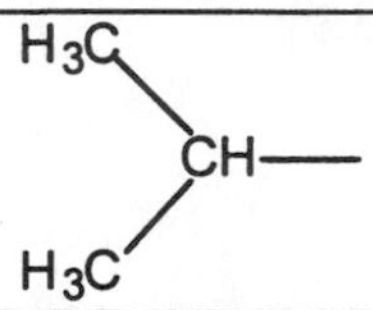 H_3C CH— H_3C	iso**prop**yl or *i*-**prop**yl
4	$-CH_2CH_2CH_2CH_3$	**but**yl or *n*-**but**yl
4	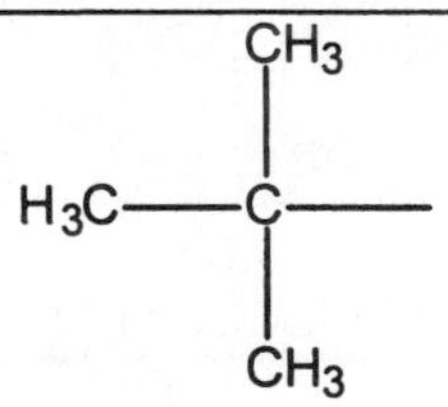 CH_3 H_3C—C— CH_3	tert**but**yl or *t*- **but**yl (looks like a T!)

C. Branched Alkanes

Alkyl groups form the branches of a branched alkane.

For Example

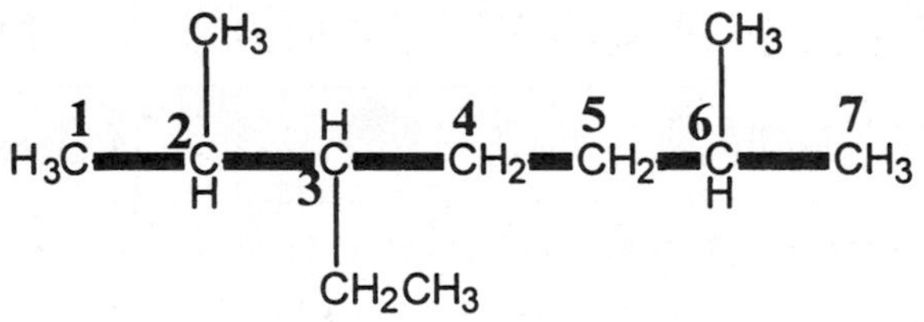

$$\begin{array}{ccccccc} & CH_3 & & & & CH_3 & \\ 1 & 2 & 3 & 4 & 5 & 6 & 7 \\ H_3C- & CH- & CH- & CH_2- & CH_2- & CH- & CH_3 \\ & & CH_2CH_3 & & & & \end{array}$$

3-**ethyl**-2,6-di**methyl**heptane

1. Find the longest parent chain (in this case it is 7 carbons long). If there is a ring, it is considered the parent chain.
2. Number the atoms of the parent chain starting from the end with more branches (in this case the left end)
3. Write the name as a word, listing the alkyl groups alphabetically (ethyl goes before methyl). Use numbers, separated by commas, to designate the position of each group and separate numbers from words using hyphens (3-ethyl). If there are two or more identical groups use the prefixes di, tri, tetra etc. (as in **di**methyl).

D. Cyclic Alkanes

Cyclic alkanes are named by appending the prefix "cyclo" to the beginning of the parent name.

Examples

CH_3 / CH / H_2C CH_2 / H_2C CH_2 / C H_2

CH_3 / CH / H_2C CH—CH_3 / H_2C—CH / CH_3

H_3C CH_2CH_3

methyl**cyclohexane** 1,2,3-trimethyl**cyclopentane** 1-ethyl-2-methyl**cyclobutane**

Practice Problems

1. Draw structures that correspond to the following names:

   a) 3-methylhexane
   b) 1,1-diethylcyclobutane
   c) 1-*t*-butyl-4-methylcyclohexane

2. Write a correct name for each of the following structures:

a)

H_3C—CH_2 CH_3

H_3C—C(H)—C—CH_3

CH_3

b)

CH_3

H_2C

CH—C—CH_3

CH

CH_3

H_3C

c)

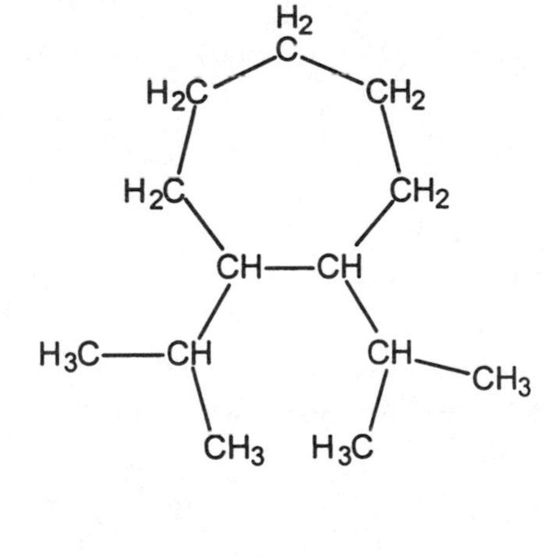

3. Why is methylcyclohexane not called 1-methylcyclohexane?

II. Heteroatom Functional Groups

(“hetero” is Greek for “other”—in this case it means “atoms other than C and H”)

A. Alcohols

An alcohol contains an –OH group. There are two naming strategies. You can use the suffix “ol” at the end of the parent name, or you can use the word “hydroxy” in the name.

Examples

CH_3OH
methanol

CH_3CH_2OH
ethanol

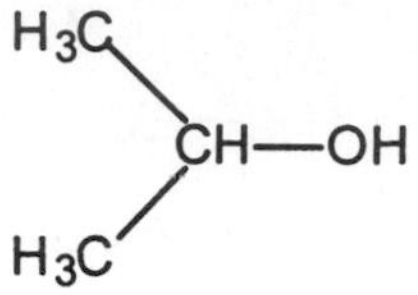

H_3C, H_3C — CH—OH
iso**propan**ol

H_3C—CH(OH)—CH(CH_2CH_3)—CH_3

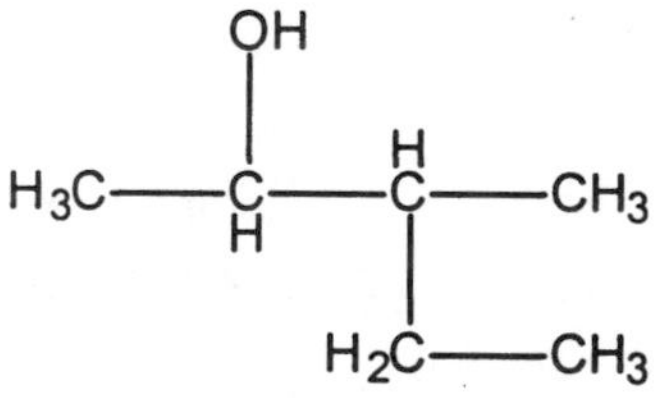

3-methyl-2-pentanol
or
2-hydroxy-3-methylpentane

B. Ethers

An ether is an oxygen bound to two alkyl groups. Simple ethers are named using the alkyl group names followed by the word “ether,” all separated by spaces.

Examples

$CH_3OCH_2CH_3$
ethyl methyl ether

$CH_3CH_2OCH_2CH_3$
diethyl ether

More complex ethers that contain a methyl or ethyl group can be named by incorporating the word “methoxy” or “ethoxy” in the same way that alkyl group names are used.

Examples

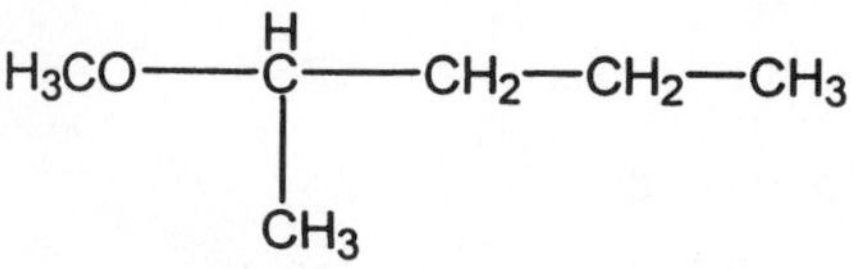

H_3CO—CH(CH_3)—CH_2—CH_2—CH_3
2-**methoxy**pentane

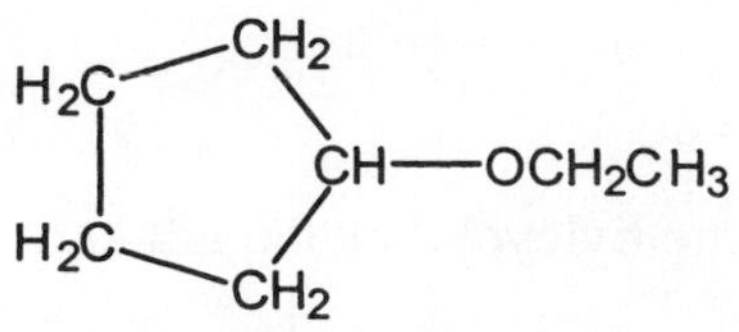

cyclo-(H_2C—CH_2—CH—CH_2—H_2C), CH—OCH_2CH_3
ethoxycyclopentane

C. Amines

An amine consists of a nitrogen bound to carbon or hydrogen. There are several naming strategies: For primary amines you can add the suffix "amine" to the end of the parent name, or you can use the word "amino" in the name. For secondary or tertiary amines use amine at the end of the name, listing each alkyl group at the beginning, or choose the longest parent amine and list the other groups attached to nitrogen after a capital "N."

Examples

CH_3NH_2
methanamine

$CH_3CH_2NH_2$
ethanamine

$H_3C-CH(NH_2)-C(CH_3)_2-CH_2-CH_3$

3,3-dimethyl-2-pentanamine
or
2-amino-3,3-dimethylpentane

Amines are classified according to the number of carbon atoms the nitrogen is bound to. Primary = one C, secondary = 2 C's, tertiary = 3 C's (quarternary = 4 C's).

Table 3: Amine Nomenclature

# C atoms bound to N	Class	Example	Name
1	primary (1°)	$H_2N-CH_2CH_2CH_3$	1-**propan**amine or 1-amino**propane**
2	secondary (2°)	$H_3C-NH-CH_2CH_2CH_3$	**methyl propyl** amine or N-**methyl**-1-aminopropane or N-**methyl**-1-propanamine
3	tertiary (3°)	$H_3C-N(CH_3)-CH(CH_3)-CH_3$	**dimethyl isopropyl** amine or N,N-di**methyl**-2-propanamine or N,N-di**methyl**-2-aminopropane

D. Alkyl Halides

Alkyl halides consist of a halide (fluorine, chlorine, bromine or iodine) attached to a carbon. Naming is done by using the words fluoro, chloro, bromo or iodo in the name in the same way you would use an alkyl group name.

Examples

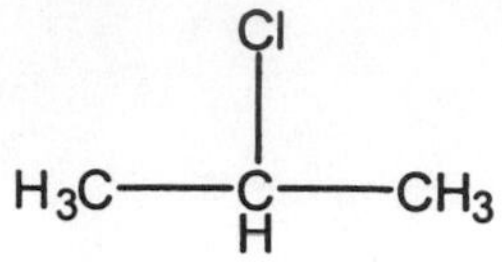

CH_2Cl_2

2-chloropropane bromocyclopropane dichloromethane

Practice Problems

4. Draw structures that correspond to the following names:

   a) 2-bromo-1-methoxypropane
   b) 3-amino-2,2-dimethylpentane
   c) cyclohexanol

5. Write an IUPAC name for each of the following structures:

a) 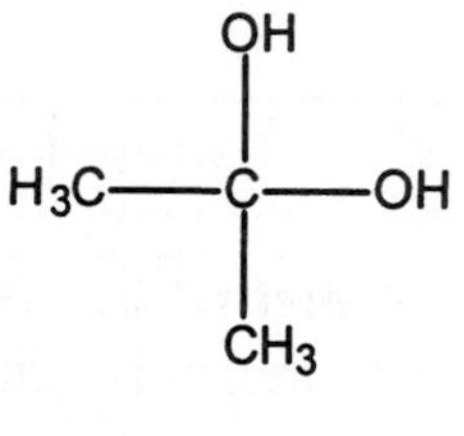 b) 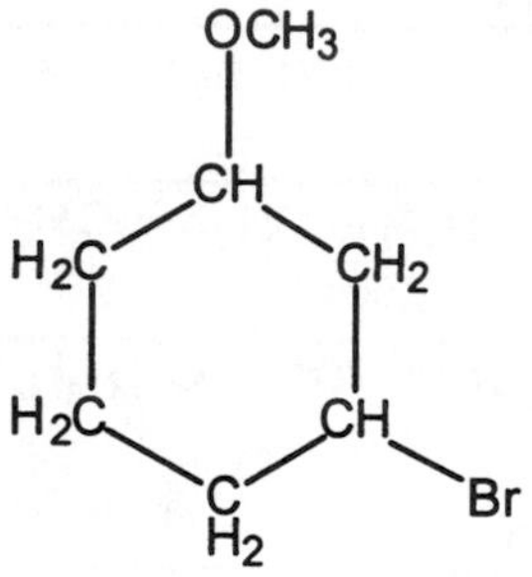 c)

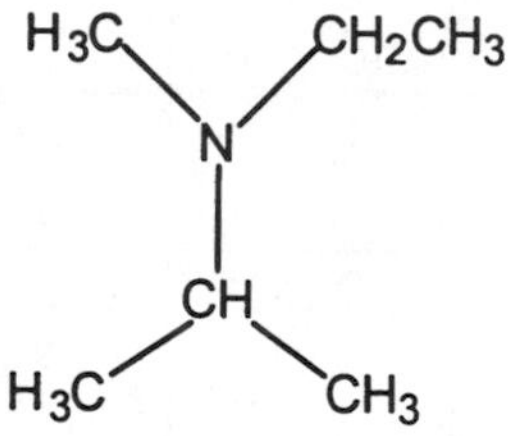

III. Alkenes

An alkene is a molecule with a carbon-carbon double bond. Naming an alkene is the same as naming an alkane except the suffix "ane" is replaced with the suffix "ene." The parent chain must contain the double bond (which is given the lowest possible number).

Examples

$H_2C{=}CH_2$

ethene

$H_2C{=}C(CH_3)_2$

methylpropene
(no 2 is needed sincethere is not other placeto put the methyl group)

4-methylcyclohexene

$H_3C{-}CH{=}C(CH_3){-}CH(Br){-}CH_3$

4-bromo-3-methyl-2-pentene

$H_3C{-}CH_2{-}CH_2{-}CH(CH{=}CH_2){-}CH_2{-}CH_2{-}CH_3$

3-propyl-1-hexene

IV. Alkynes

An alkyne is a molecule with a carbon-carbon triple bond. Naming an alkyne is the same as naming an alkene except the suffix "yne" is used.

Examples

$HC{\equiv}C{-}CH_3$

propyne

$H_3C{-}C{\equiv}C{-}CH_3$

2-butyne

$H_3C{-}C(CH_3)_2{-}C{\equiv}C{-}CH_3$

4,4-dimethyl-2-pentyne

$HC{\equiv}CH$

ethyne
common name: **acet**ylene

(Acetylene is an important non-IUPAC name to remember. The prefix "acet" tells you that this is a 2 carbon group. It is a very old word derived from **acet**ic acid (vinegar), which is a 2 carbon acid. Later in this course we will make frequent use of the prefix "acet" to mean a 2 carbon group.)

Practice Problems

6. Draw structures that correspond to the following names:

   a) cyclopentene
   b) 2,3-dimethyl-2-butene
   c) 3,3-dimethy-1-butyne

7. Write an IUPAC name for each of the following structures:

   a) $H_2C{=}C(CH_3)-CH_2-Br$

   b) $H_3C-C{\equiv}C-CH_3$

   c) $CH{=}CH$, H_2C, $C(OCH_3)(OCH_3)$, H_2C, CH_2, H_2C-CH_2 (eight-membered ring)

8. Why is 2-methylpropene not called 2-methyl-1-propene?

Nomenclature Worksheet 2

Part A: Benzene Derivatives

Simple benzene derivatives are named using a prefix added to the word benzene, as in the first three examples below. Alternatively, the substituent can be considered the parent chain and the benzene ring can be considered a group (like a methyl group). In such cases the ring is called a **phenyl group**, as in the fourth example below.

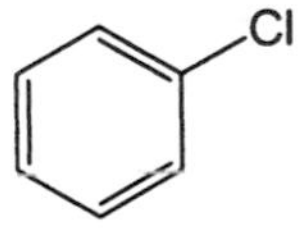

Cl NO$_2$ OH

Chlorobenzene

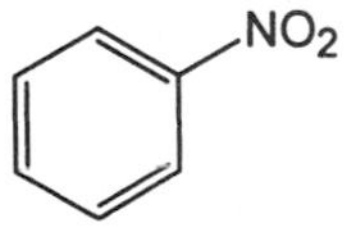

Nitrobenzene

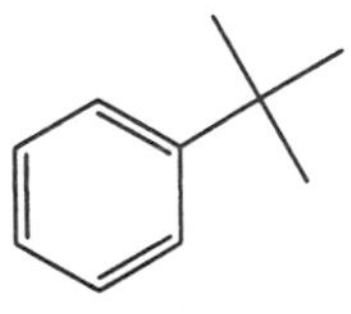

tert-butylbenzene (1,1-dimethylethylbenzene)

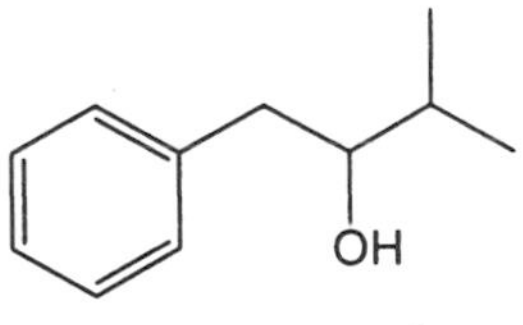

2-hydroxy-3-methylbutylbenzene (1-phenyl-2-hydroxy-3-methylbutane)

The following benzene derivatives are almost always called by their common names. Please memorize these common names. (The IUPAC names shown in parentheses are rarely used.)

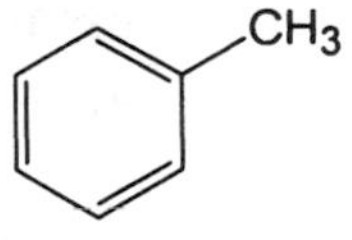

CH$_3$ OH NH$_2$ O H O OH

toluene (methylbenzene)

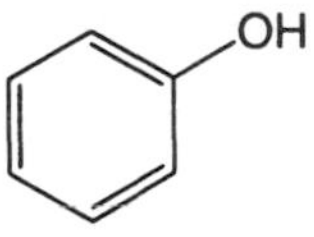

phenol (hydroxybenzene)

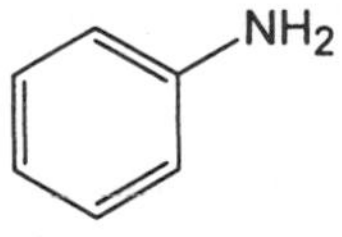

aniline (aminobenzene)

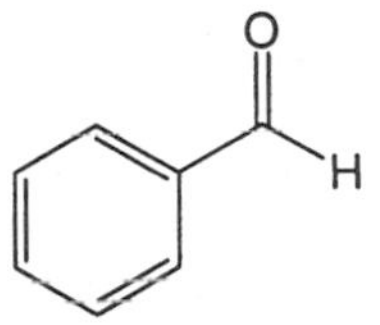

benzaldehyde

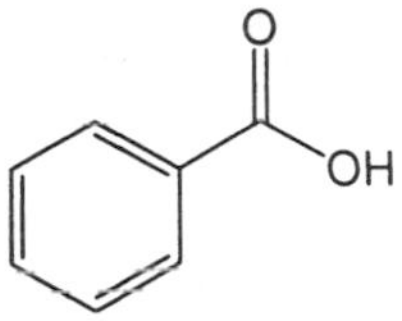

benzoic acid

When there are two substituents on thc ring, either the ortho, meta, para system or the numbering system may be used. When there are three or more substituents the numbering systems must be used. For example:

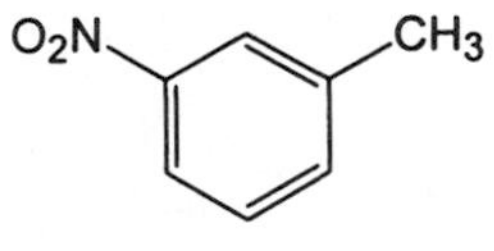

O$_2$N CH$_3$ OH NH$_2$ NH$_2$ H$_3$C O H H$_3$C CH$_3$ OH O OH H$_3$C CH$_3$

meta-nitrotoluene or 3-nitrotoluene

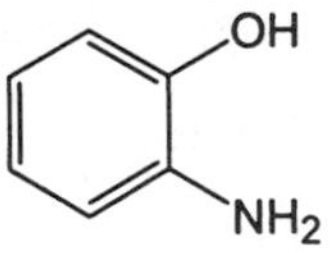

ortho-aminophenol or *ortho*-hydroxyaniline or 2-aminophenol etc.

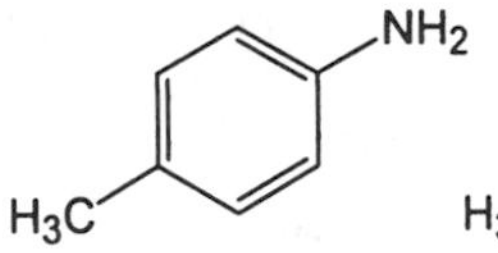

para-methylaniline etc.

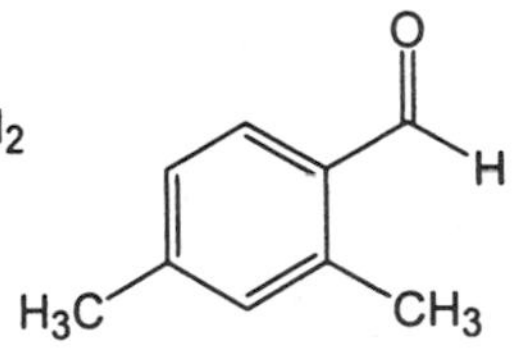

2,4-dimethyl-benzaldehyde

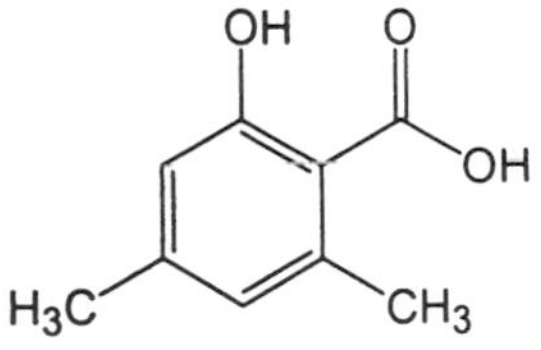

2-hydroxy-4,6-dimethyl-benzoic acid

Practice Problems

1. Write a correct name (not appearing in the example above) for 2-aminophenol and *para*-methylaniline.

2. Draw the structure that corresponds to the name: 2-phenyl-3-hydroxytoluene.

3. Explain why the following is not a correct name: 4,5-dimethylphenol.

Part B: Carbonyl Compounds

There are three basic categories of carbonyl compounds: aldehydes, ketones, and carboxylic acid derivatives. The last category contains several types of compounds. We will learn to name the main ones: carboxylic acids, esters, amides and acid chlorides.

Certain small molecules are referred to so often by their common names that you should memorize the short list of common names below.

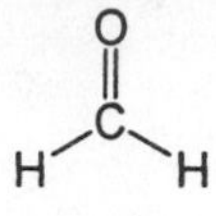

formaldehyde (methanal)

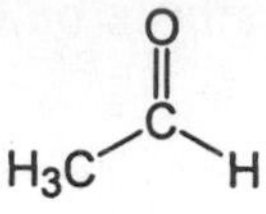

acetaldehyde (ethanal)

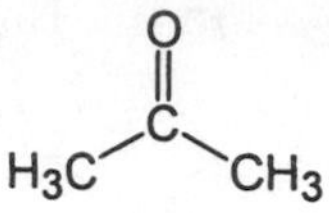

acetone (2-propanone)

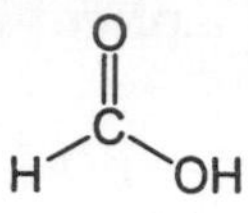

formic acid (methanoic acid)

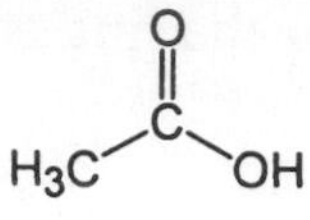

acetic acid (ethanoic acid) **aka: vinegar**

In general the prefix "acet–" is used in place of the prefix "ethan–" as in acetic acid; and the prefix "form–" is used in place of "methan–" as in formic acid.

Aldehydes and Ketones

- To name an aldehyde or ketone, use the parent chain name without the final "e" (e.g. "butane" becomes "butan")
- Aldehydes are named by appending the suffix "–al"
- Ketones are named by appending the suffix "-one"
- Always assign the aldehyde or ketone carbon the smallest possible number
- With multiple functional groups that require suffixes (e.g. –ene, –one, –al) one suffix can be separated from the parent name as in the 5th example below.

For example:

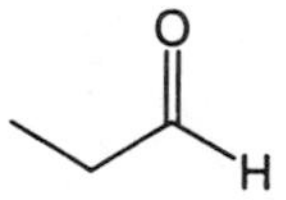

propanal

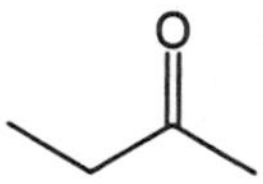

2-butanone

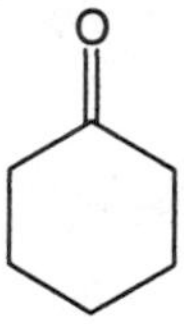

cyclohexanone

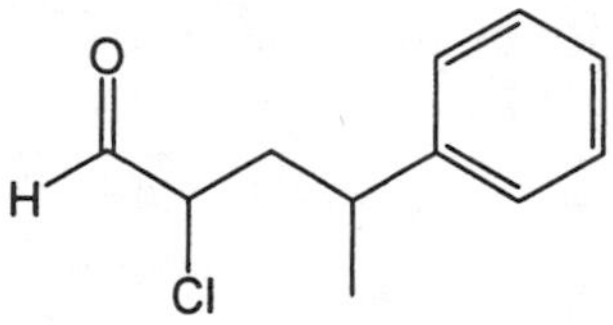

2-chloro-4-phenylpentanal

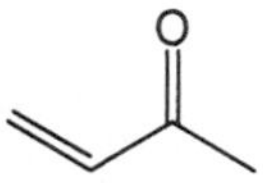

3-buten-2-one or 3-ene-2-butanone

Carboxylic Acids

- Carboxylic acids are named using the suffix "–oic" appended to the truncated parent chain name (e.g. butane is truncated to "butan") followed by "–oic" and the word "acid."

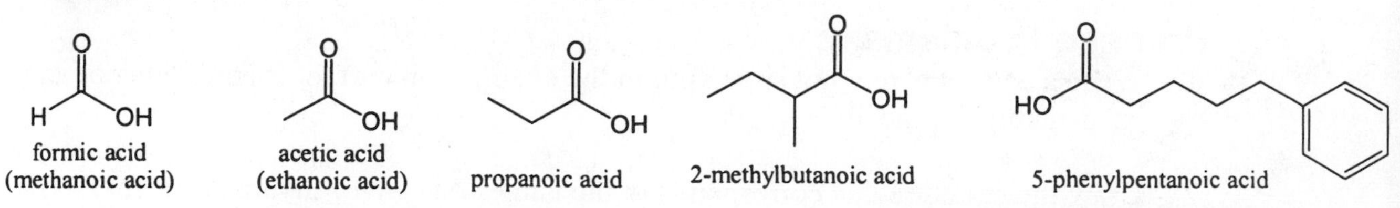

formic acid (methanoic acid) acetic acid (ethanoic acid) propanoic acid 2-methylbutanoic acid 5-phenylpentanoic acid

Esters

- Esters are a cross between an alcohol and a carboxylic acid.
- They are named by writing the branch name associated with the alcohol (e.g. methyl, ethyl, etc.)
- Followed by the name of the acid parent compound with "-oic" replaced with "-oate"

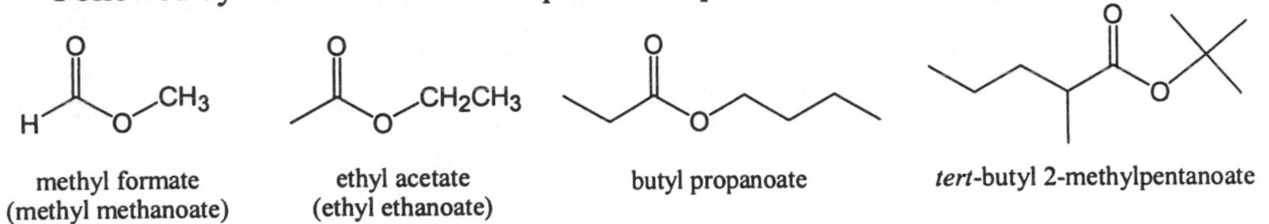

methyl formate (methyl methanoate) | ethyl acetate (ethyl ethanoate) | butyl propanoate | *tert*-butyl 2-methylpentanoate

Amides

- Amides are a cross between an amine and a carboxylic acid.
- They are named by writing the branch name of any groups attached to the N. Note: the branch names are written after an italicized capital *N* (e.g. "*N*-methyl")
- Followed by the name of the parent acid with "-oic" replaced with "-amide"

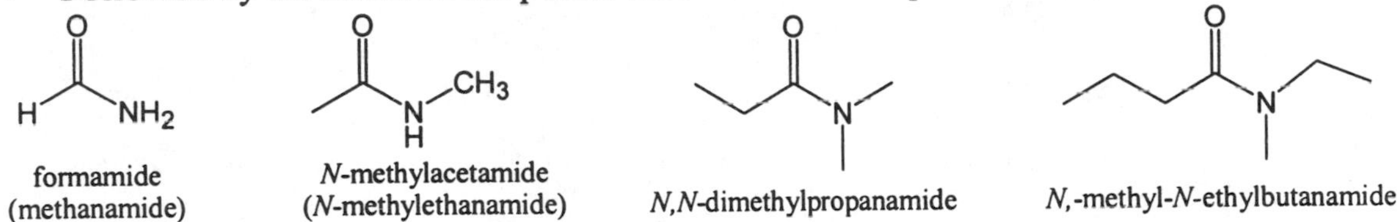

formamide (methanamide) | *N*-methylacetamide (*N*-methylethanamide) | *N,N*-dimethylpropanamide | *N*,-methyl-*N*-ethylbutanamide

Carboxylic Acid Chlorides (a.k.a. Alkanoyl Chlorides)

- Acid chlorides are named by writing the name of the parent acid with "-oic" replaced with "-oyl" followed by the word chloride.

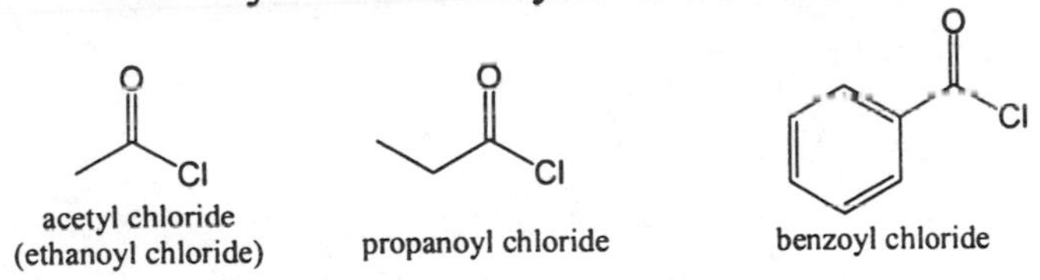

acetyl chloride (ethanoyl chloride) | propanoyl chloride | benzoyl chloride

Practice Problems

4. Draw the structure which corresponds to each of the following IUPAC names.
   - a) butanal
   - b) 3-chloropentanoyl chloride
   - c) cyclobutanone
   - d) methyl acetate (or methyl ethanoate)
   - e) 4-ethyl-2,3-dimethylhexanamide
   - f) acetamide (or ethanamide)
   - g) *N*-propyl-*N*-*iso*-propylhexanamide

5. Assign a correct name to each of the following structures:

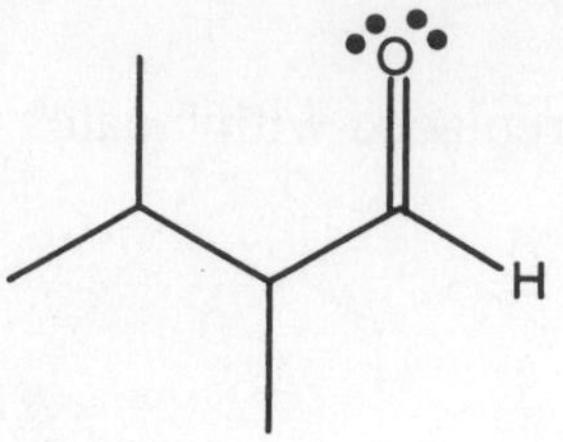

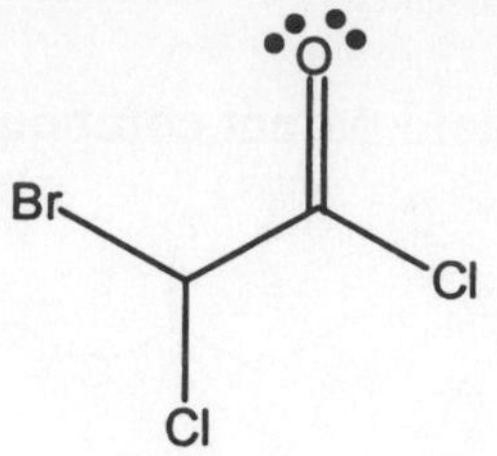

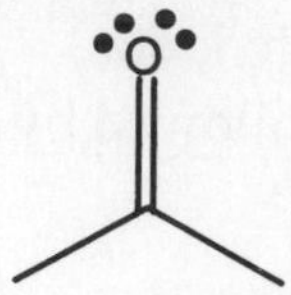

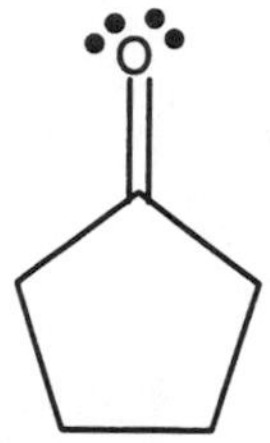

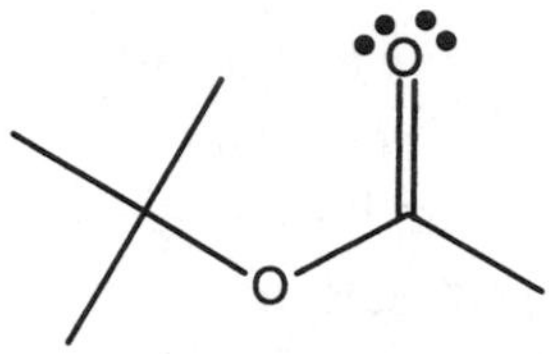

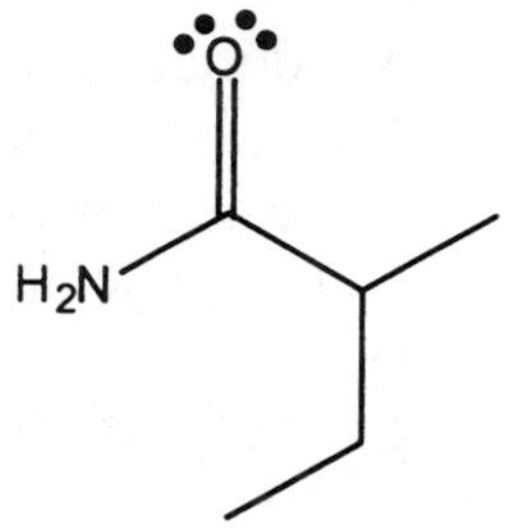

H

Br

Cl

Cl

O

H_2N

Preparation of Alkyl Halides

Acid catalyzed nucleophilic substitution:

$$\text{R–OH} \xrightarrow{\text{H–X}} \text{R–X}$$

Reactions of Alkyl Halides

Nucleophilic substitution:

$$\text{R–X} \xrightarrow{\text{Nuc:}^{\ominus}} \text{R–Nuc}$$

Elimination:

$$\text{H–C–C–X} \xrightarrow{\text{Base:}} \text{C=C}$$

Preparation of carbon nucleophile (Grignard Reagent = $C^{\ominus}$)

$$\text{R–X} \xrightarrow{\text{Mg metal}} \text{R–MgX}$$

Preparation of Alcohols

Nucleophilic substitution:

$$\text{R–L.G.} \xrightarrow{\text{HO}^{\ominus}} \text{R–OH}$$

L.G. = good leaving group

Nucleophilic addition to a Carbonyl ($C^{\ominus}$nucleophile): a.k.a Reduction of a Carbonyl

$$\text{C=O} \xrightarrow[\text{2) } H_3O^{\oplus}]{\text{1) R–MgX}} \text{R–C–OH}$$

1) and 2) are incompatible reagents so they must be added sequentially (not at the same time).

Nucleophilic addition to a Carbonyl ($H^{\ominus}$ nucleophile): a.k.a Reduction of a Carbonyl

$$\text{C=O} \xrightarrow[\text{2) } H_3O^{\oplus}]{\text{1) } LiAlH_4} \text{H–C–OH}$$

$LiAlH_4 = Li^{\oplus}\ H^{\ominus}$

Reactions of Alcohols

Acid Catlyzed Nucleophilic Substitution or Elimination

$$\text{H–C–C–OH} \xrightarrow[\text{Nuc:}^{\ominus}]{H^{\oplus}\text{ (cat.)}} \text{H–C–C–Nuc}$$

$$\text{H–C–C–OH} \xrightarrow[\text{Base:}]{H^{\oplus}\text{ (cat.)}} \text{C=C}$$

Preparation of Ethers

Nucleophilic Substitution:

$$Na^{\oplus}\ {}^{\ominus}O{-}R \quad X{-}R' \longrightarrow R{-}O{-}R'$$

Acid Catalyzed Electrophilic Addition of Water:

$$\text{C=C} \xrightarrow[\text{HOR}]{H^{\oplus}\text{ (cat.)}} \text{H–C–C–OR}$$

Reactions of Ethers

Acid Catalyzed Nucleophilic Substitution:

$$R{-}O{-}R' \xrightarrow{X{-}H} R{-}\overset{\oplus}{O}(H){-}R' \quad X^{\ominus} \longrightarrow HO{-}R \quad X{-}R'$$

Ring Opening of Epoxide (cyclic ether) by Strong Nucleophile

$$\text{epoxide} \xrightarrow[\text{2) } H^{\oplus}]{\text{1) Nuc}^{\ominus}} \text{HO–C–C–Nuc}$$

(Nuc anti to OH)

Acid Catalyzed Ring Opening of Epoxide (cyclic ether) by Weak Nucleophile

$$\text{epoxide} \xrightarrow[\text{Nuc}^{\ominus}]{H^{\oplus}\text{ (cat.)}} \text{HO–C–C–Nuc}$$

(Nuc anti to OH)

Preparation of Amines

Reduction of Nitro Group:

$$R{-}NO_2 \xrightarrow[\text{(Reducing Agent)}]{\text{Zn(Hg)/HCl}} R{-}NH_2$$

Nucleophilic Substitution:

$$R{-}L.G. \xrightarrow{NH_3} R{-}NH_2 \quad H{-}L.G.$$

Reactions of Amines

As a Nucleophile:

$$R{-}NH_2 \xrightarrow{X{-}R'} R{-}NH{-}R' \xrightarrow{X{-}R'} R{-}N(R')_2$$

Preparation of Alkenes

Dehydrohalogenation (Elimination of HX):

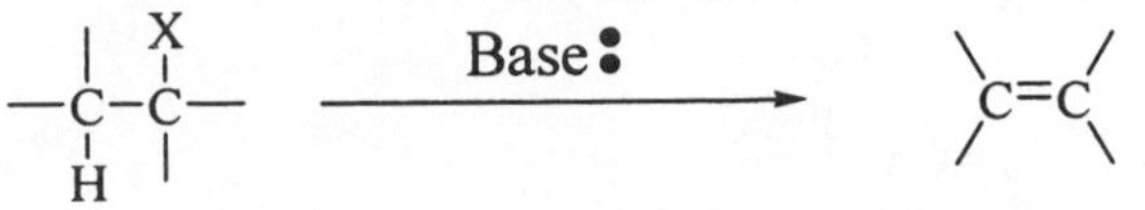

—C(H)—C(X)— —(Base:)→ C=C

Dehydration (Elimination of Water):

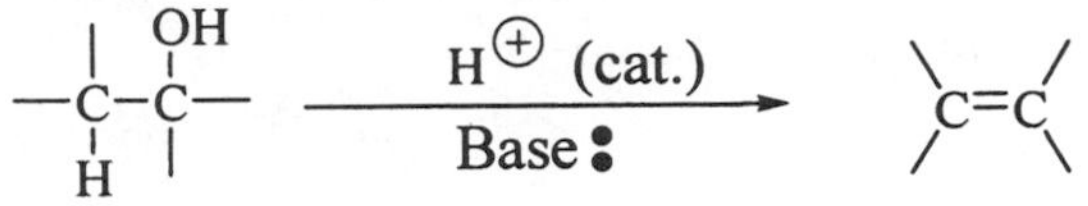

—C(H)—C(OH)— —($H^{\oplus}$ (cat.), Base:)→ C=C

Reactions of Alkenes

Electrophilic Addition:

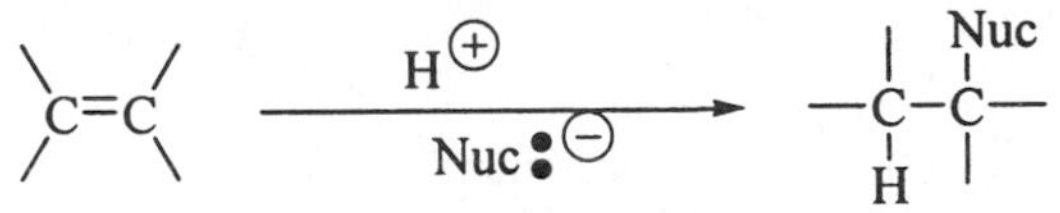

C=C —($H^{\oplus}$, Nuc:$^{\ominus}$)→ —C(H)—C(Nuc)—

Electrophilic Addition of Halogen (X_2):

syn = same side; **anti** = opposite side

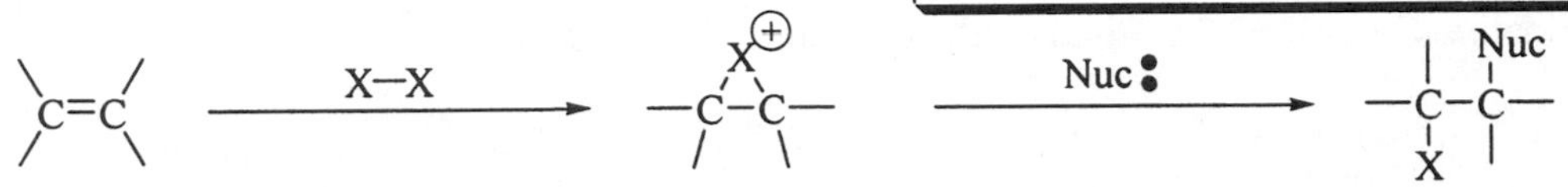

C=C —(X—X)→ —C—C— (bridging $X^{\oplus}$) —(Nuc:)→ —C(X)—C(Nuc)—

X = Cl or Br   Nuc = $X^{\ominus}$, H_2O, ROH   *(anti addition)*

Metal Catalyzed Hydrogenation:

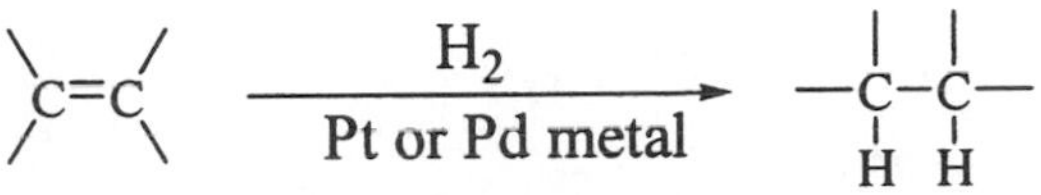

C=C —(H_2, Pt or Pd metal)→ —C(H)—C(H)—   *(syn addition of H's)*

Addition of HBr to Alkenes

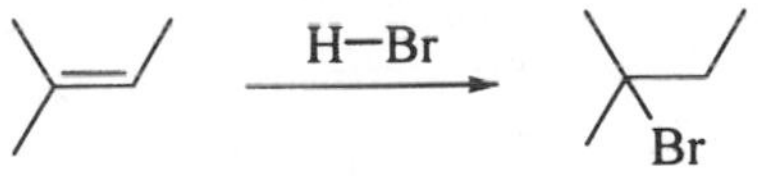

H—Br → Br product — Markovnikov

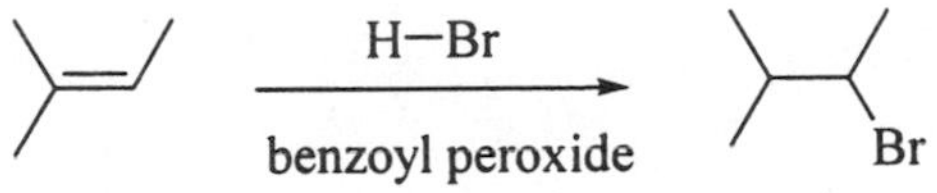

H—Br, benzoyl peroxide → Br product — Anti - Markovnikov

Addition of H_2O to Alkenes

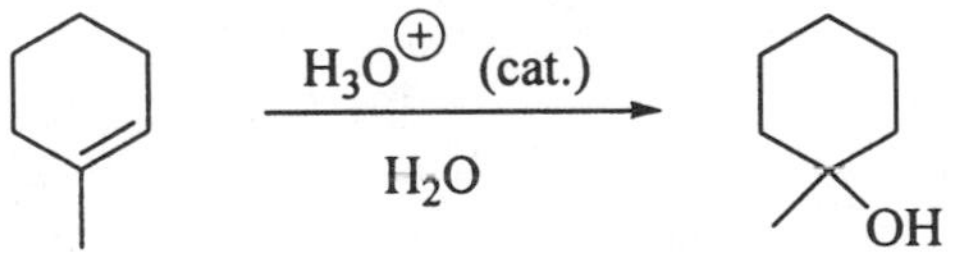

$H_3O^{\oplus}$ (cat.), H_2O → OH — Markovnikov

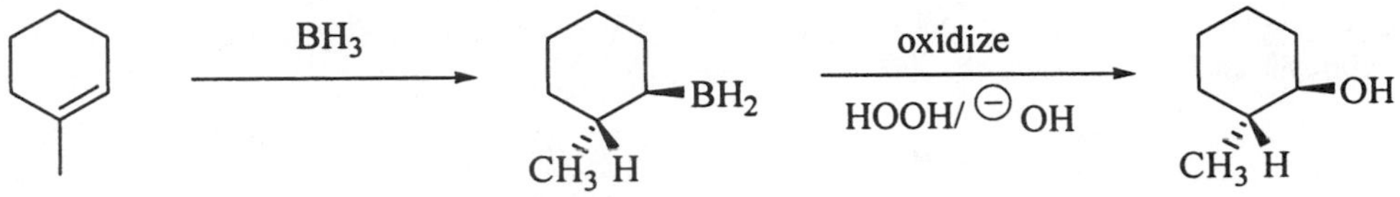

BH_3 → BH_2, CH_3 H — syn (cis) addition — oxidize, HOOH/$^{\ominus}$OH → OH, CH_3 H — syn (cis) — Anti-Markovnikov + enantiomer

Oxidative Cleavage of Alkenes

Complete Oxidative Cleavage of Double Bond (milder conditions)

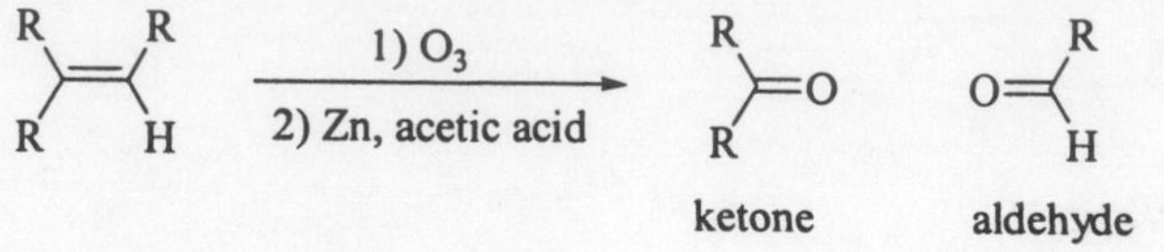

1) O_3
2) Zn, acetic acid

ketone aldehyde

Complete Oxidative Cleavage of Double Bond (harsh conditions)

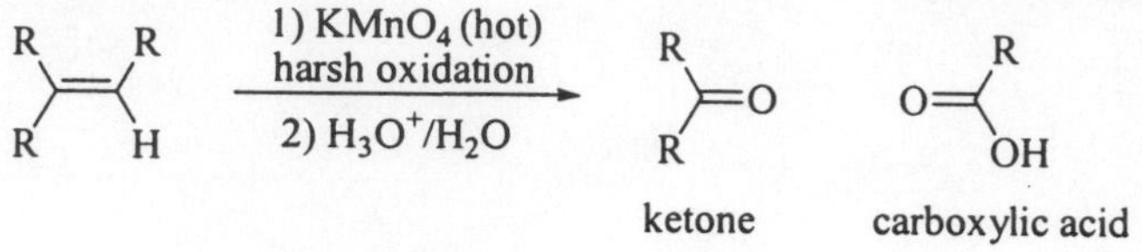

1) $KMnO_4$ (hot) harsh oxidation
2) H_3O^+/H_2O

ketone carboxylic acid

Oxidative Cleavage of π Bond - syn Addition of Two OH's

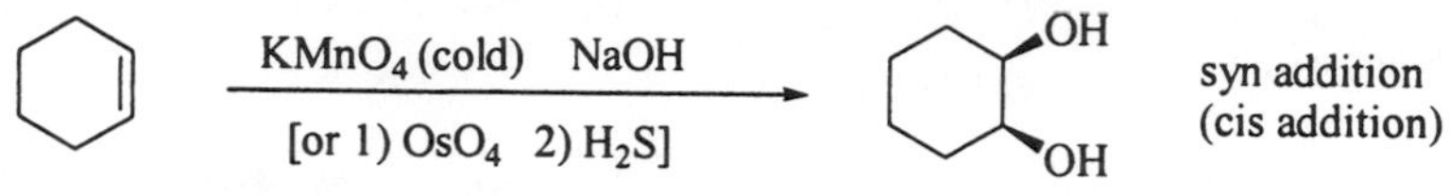

$KMnO_4$ (cold) NaOH
[or 1) OsO_4 2) H_2S]

OH
OH

syn addition
(cis addition)

Oxidative Cleavage of π Bond - anti Addition of Two OH's

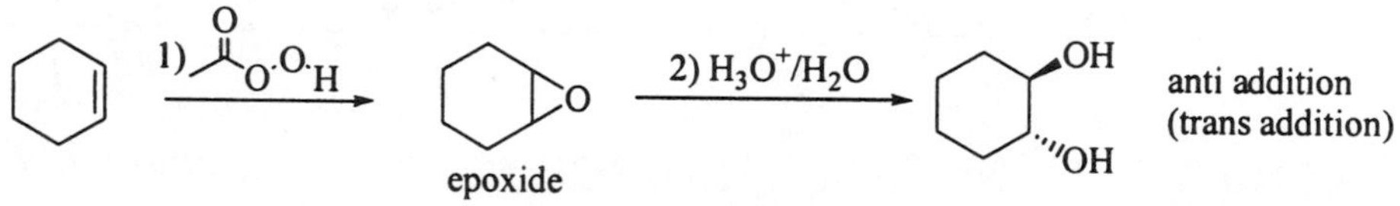

1)
2) H_3O^+/H_2O

O
epoxide

OH
OH

anti addition
(trans addition)

Oxidation of Alcohols

Mild and Harsh Oxidation of 1° Alcohol

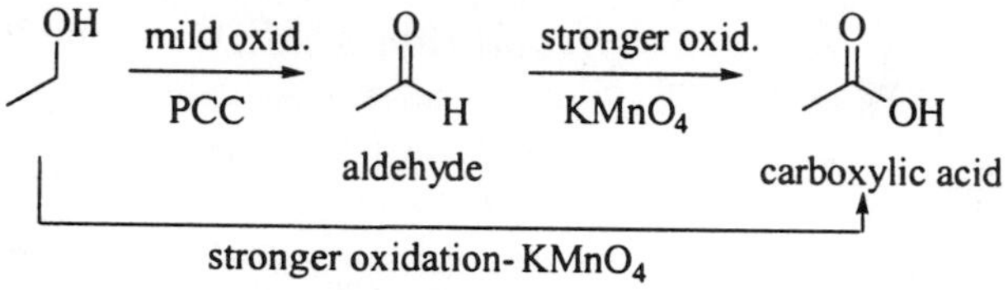

OH mild oxid. PCC
H aldehyde
stronger oxid. $KMnO_4$
OH carboxylic acid

stronger oxidation- $KMnO_4$

Oxidation of 2° Alcohol

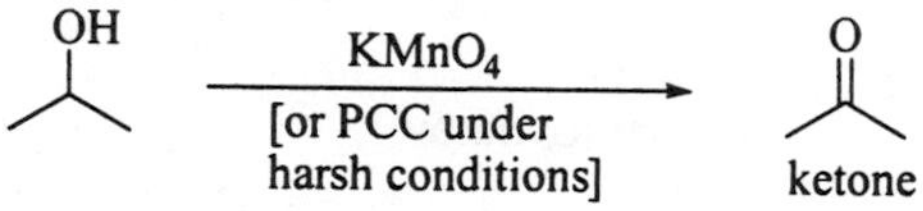

OH
$KMnO_4$
[or PCC under harsh conditions]
ketone

Special Reactions

Radical Bromination (other halogenations less selective)

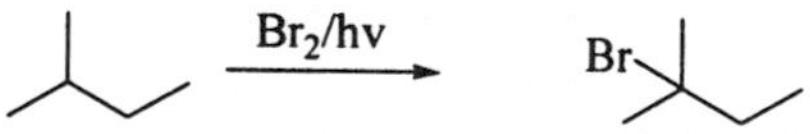

Br_2/hv
Br

Preparation of an **Acid Chloride**

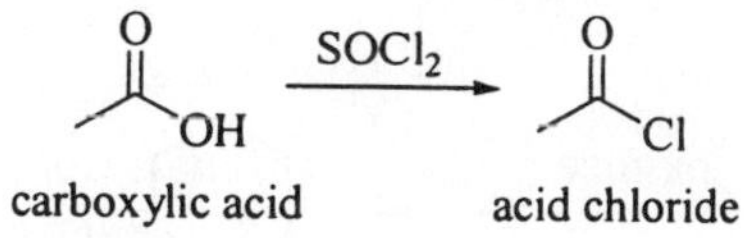

$SOCl_2$
OH Cl
carboxylic acid acid chloride

Preparation of Alkynes

Dehydrohalogenation (Elimination of HX):

$$R-C(H)(H)-C(X)(X)-R \text{ (geminal) or } R-C(X)(H)-C(X)(H)-R \xrightarrow{KOH\ (HO^{\ominus})} RC(H)=C(X)R \text{ and } RC(H)=C(R)X \xrightarrow{NaNH_2\ (H_2N^{\ominus})} R-C\equiv C-R$$

(geminal)

Reactions of Alkynes

Electrophilic Addition of HX (Markovnikov):

$$H-C\equiv C-R \xrightarrow[\text{dark, no peroxides}]{H-X} H_2C=C(X)R \text{ (Markov.)} \xrightarrow[\text{dark, no peroxides}]{H-X} H-C(H)(H)-C(X)(X)-R \quad \text{Markov.}$$

Radical Addition of HBr (Anti-Markovnikov):

$$H-C\equiv C-R \xrightarrow[\text{ROOR/heat}]{H-Br} BrC(H)=C(H)R \text{ and } HC(Br)=C(H)R \text{ (Anti-Markov.)} \xrightarrow[\text{ROOR/heat}]{H-Br} H-C(Br)(Br)-C(H)(H)-R \quad \text{Anti-Markov.}$$

Hydration (Markovnikov)

$$H-C\equiv C-R \xrightarrow[H_2O]{H^{\oplus}\text{ (cat.)}} H_2C=C(OH)R \text{ (Markov. enol)} \xrightleftharpoons{\text{tautomerize}} H-CH_2-C(=O)-R \quad \text{Markov.}$$

keto form (ketone)

Hydration (Anti-Markovnikov)

$$H-C\equiv C-R \xrightarrow{1)\ BH_3;\ 2)\ H_2O_2/NaOH/H_2O} HOC(H)=C(H)R \text{ (Anti-Markov. enol)} \xrightleftharpoons{\text{tautomerize}} H-C(=O)-CH_2-R \quad \text{Anti-Markov.}$$

keto form (aldehyde)

Acid/Base Reaction:

$$R-C\equiv C-H \xrightarrow{NaNH_2} R-C\equiv C:^{\ominus}$$

Nucleophilic Substitution:

$$R-C\equiv C-H \xrightarrow{NaNH_2} R-C\equiv C:^{\ominus} \xrightarrow{R'-X} R-C\equiv C-R'$$

Reduction:

$$R-C\equiv C-R \xrightarrow{H_2\ /\ Pt\ metal} R-C(H)(H)-C(H)(H)-R \quad \textbf{complete reduction}$$

$$R-C\equiv C-R \xrightarrow{H_2\ /\ Lindlar\ Catalyst} (H)(R)C=C(H)(R) \text{ (cis)} \quad \textbf{syn} \text{ addition of 2 H's}$$

$$R-C\equiv C-R \xrightarrow[2)\ H_2O]{1)\ Na\ metal/\ liqud\ NH_3} (H)(R)C=C(R)(H) \text{ (trans)} \quad \textbf{anti} \text{ addition of 2 H's}$$

Electrophilic Aromatic Substitution (EAS) Reactions

Nitration: benzene + H_2SO_4 HNO_3 → nitrobenzene (NO_2) + $HSO_4^{\ominus}$ $H_3O^{\oplus}$

Sulfonation: benzene + SO_3 (conc. sulfuric acid) ⇌ (SO_3H) ; reverse: dilute acid — ***Reversible***

Chlorination: benzene + Cl_2 $FeCl_3$ (cat.) → chlorobenzene (Cl) + HCl

Bromination: benzene + Br_2 $FeBr_3$ (cat.) → bromobenzene (Br) + HBr

Friedel-Crafts Alkylation: benzene + R—X AlX_3 (cat.) → R-benzene + HX

Friedel-Crafts Acylation: benzene + RC(=O)X AlX_3 → PhC(=O)R + HX

> X = Cl or Br
> R = alkyl group like methyl, ethyl, cyclohexyl, benzyl etc.
> ***Do not work with deactivated rings.***

Oxidation/Reduction Reactions with Aromatic Molecules

m dir. NO_2 (nitrobenzene) → Zn(Hg), HCl (reduction) or H_2, Pd (reduction) → NH_2 (aniline) *o/p dir.* ; reverse: CF_3COOOH (oxidation)

m dir. O=C–R (aryl ketone) → Zn(Hg), HCl (reduction) or H_2, Pd (reduction) → H_2C–R *o/p dir.* ; reverse: CrO_3, H_2SO_4, H_2O (oxidation)

o/p dir. R → $KMnO_4$ (oxidation), R = alkyl group → C(=O)OH *m dir.*

Y-benzene → H_2/Pt at high pressure, or H_2/Rh (reduction) → Y-cyclohexane — **Y also reduced unless impossible**

Preparation of Aldehydes and Ketones

Complete Oxidative Cleavage of Double Bond (milder conditions)

$R_2C{=}CHR \xrightarrow[\text{2) Zn, acetic acid}]{\text{1) } O_3} R_2C{=}O + O{=}CHR$

ketone aldehyde

Mild Oxidation of 1° Alcohol

$RCH_2OH \xrightarrow[\text{(mild oxid.)}]{\text{PCC}} RCHO$

1° alcohol → aldehyde

FYI

$ClCrO_3^{\ominus}$

PCC = pyridinium chlorochromate

Oxidation of 2° Alcohol

$R_2CHOH \xrightarrow{\text{PCC or } KMnO_4 \text{ or } CrO_3/H_3O^+} R_2C{=}O$

2° alcohol → ketone

Hydration of an alkyne (oxymurcuration):

$-C{\equiv}C- \xrightarrow{H_3O^+/H_2O \quad \text{or 1) } Hg(OAc)_2 \text{ 2) } NaBH_4}$ enol ⇌ keto

enol *keto*

Nucleophilic Addition-Elimination on an Acid Chloride

$R'COCl \xrightarrow{Nuc^{\ominus}}$

acid chloride

	aldehyde	ketone	carboxylic acid	ester	amide
Product	R'CHO	R'COR	R'COOH	R'COOR	R'CONHR
Nuc =	($LiAlH_4$ or $NaBH_4$)	(R–MgX or R-Li)	(KOH)	(KOR)	(H_2NR)

Reactions of Aldehydes and Ketones

Nucleophilic addition to a Carbonyl ($H^{\ominus}$ or $C^{\ominus}$ nucleophile):

aldehyde or ketone $\xrightarrow[\text{2) } H_3O^{\oplus}]{\text{1) } H^{\ominus} (LiAlH_4 \text{ or } NaBH_4) \text{ or } R^{\ominus} (R\text{–}MgX)}$ $-C(OH)(H)-$ or $-C(OH)(R)-$

Acetal/Ketal formation

$R_2C{=}O$ (ketone) or $RCHO$ (aldehyde) $\xrightarrow[\text{two equivalents R'–OH}]{H^{\oplus} \text{ (cat.)}}$ $R-C(OR')_2-R$ (ketal) or $R-C(OR')_2-H$ (acetal)

Imine Formation

aldehyde or ketone $\xrightarrow{RNH_2}$ $C{=}N-R$ (imine)

Special Reactions of Aldehydes and Ketones

Cyanohydrin Formation

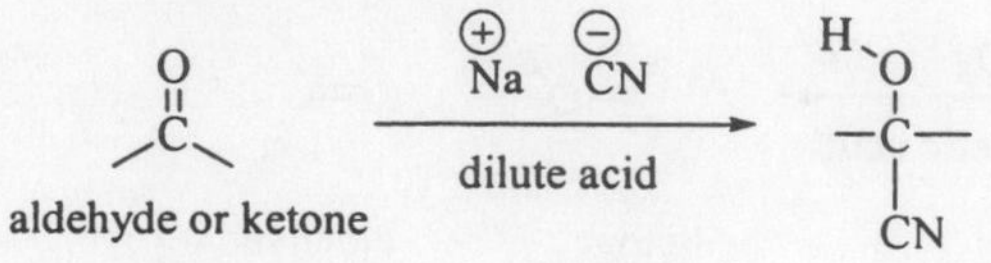

aldehyde or ketone —($Na^{\oplus}$ $CN^{\ominus}$, dilute acid)→ H–O–C(–CN)

Cyclic Ketal as a Protecting Group

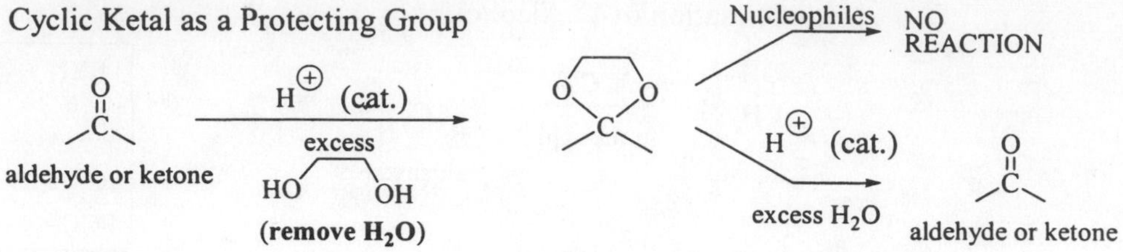

aldehyde or ketone —($H^{\oplus}$ (cat.), excess HO–CH$_2$CH$_2$–OH, **(remove H_2O)**)→ cyclic ketal

cyclic ketal —Nucleophiles→ NO REACTION

cyclic ketal —($H^{\oplus}$ (cat.), excess H_2O)→ aldehyde or ketone

Cyclic Hemi-acetal formation by 6-Carbon Sugars

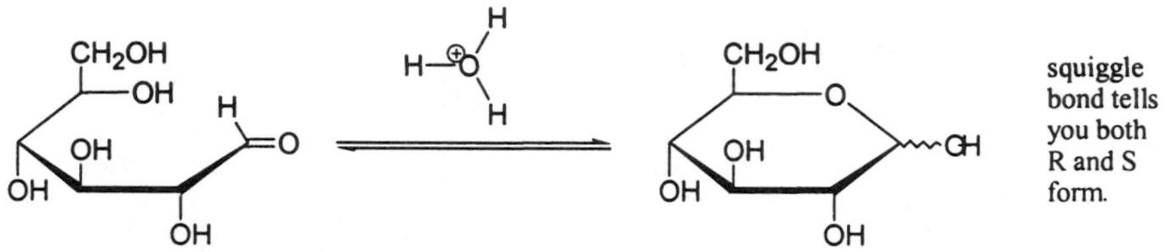

CH$_2$OH, OH, H, OH, OH, OH ⇌ ($H_3O^{\oplus}$) CH$_2$OH, O, OH, OH, OH, OH

squiggle bond tells you both R and S form.

The following are optional reactions that may simplify some syntheses:

Wolff-Kishner

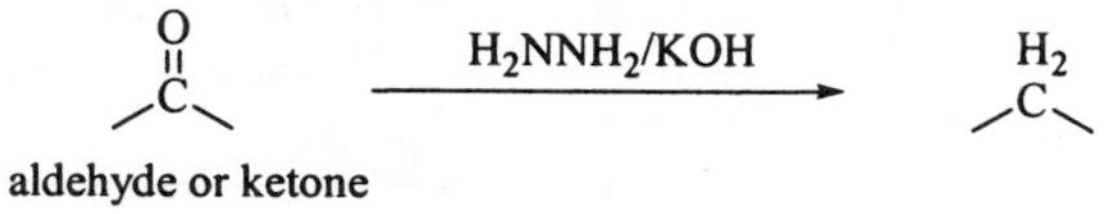

aldehyde or ketone —(H_2NNH_2/KOH)→ CH_2

Wittig Reaction

R(R')C(H)X —(1) PPh_3; 2) $CH_3CH_2CH_2CH_2^{\ominus}$ $Li^{\oplus}$)→ R(R')C=PPh_3 (ylide) + aldehyde or ketone → R'(R)C=C (alkene)

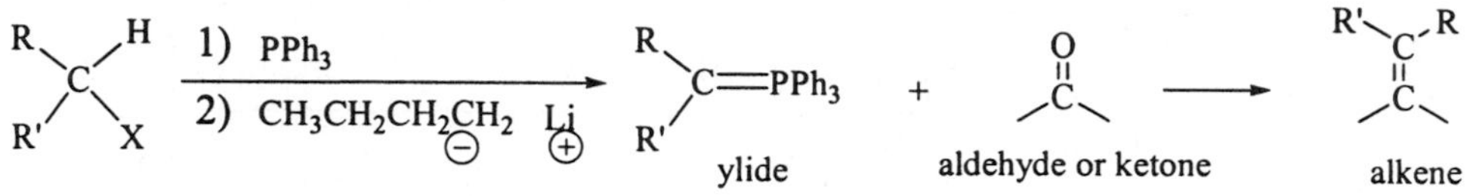

Baeyer-Villiger Reaction

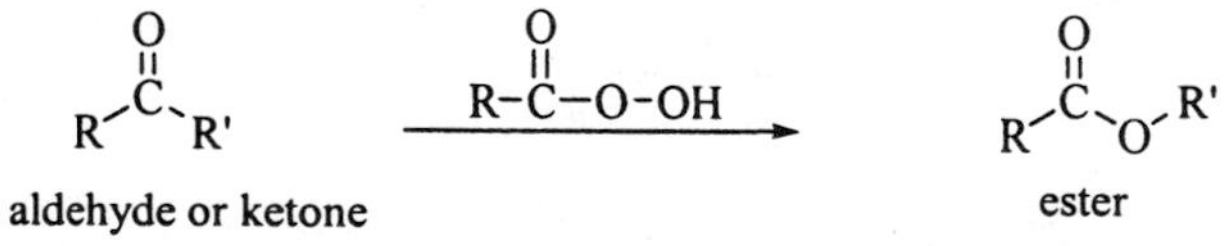

R–C(=O)–R' (aldehyde or ketone) —(R–C(=O)–O–OH)→ R–C(=O)–O–R' (ester)

Reactions of α Carbon Nucleophiles

Acid-Base

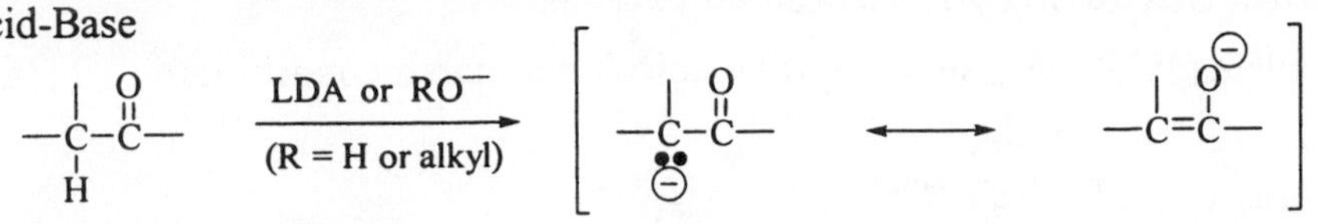

LDA or RO^- (R = H or alkyl)

α–Halogenation with acid

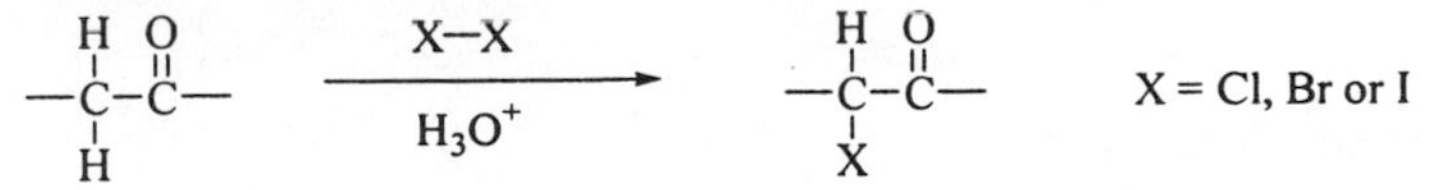

X–X, H_3O^+

X = Cl, Br or I

α–Halogenation with base

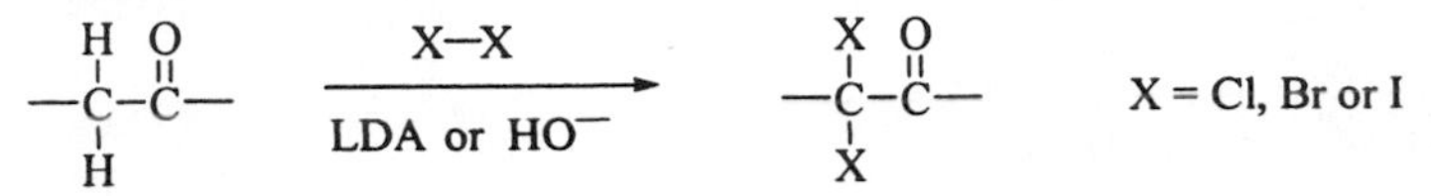

X–X, LDA or HO^-

X = Cl, Br or I

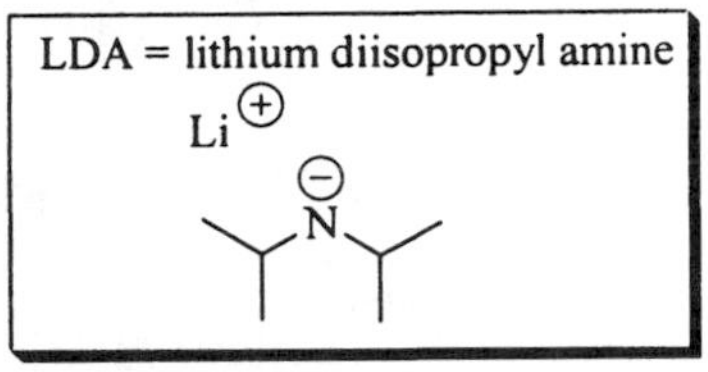

LDA = lithium diisopropyl amine

$Li^{\oplus}$ $N^{\ominus}$

α–Alkylation

R–X, LDA

R–X must be 1°

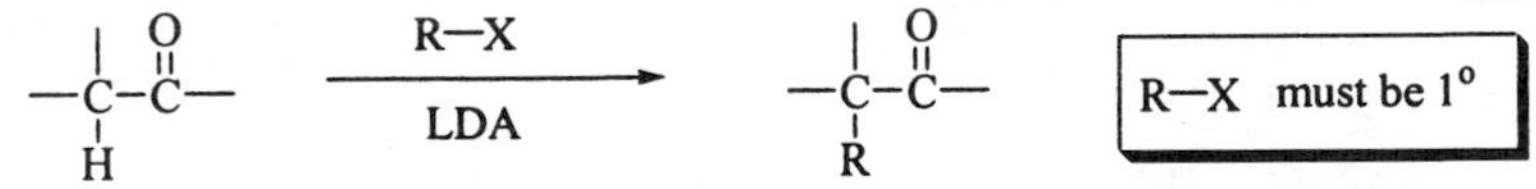

Base Catalyzed Aldol Reaction

NEW C-C BOND

H_α–C_α(R)(H_α)–C_2(=O)–R' + R–C_1(=O)–R" (electrophile, best when R" = H) $\xrightarrow{HO^-}$ R–C_1(OH)(R")–C_α(R)(H_α)–C_2(=O)–R' $\xrightarrow[\text{heat}]{HO^-}$ (R–C_1(R")=C_α(R)–C_2(=O)–R')

β-hydroxy aldehyde ("Aldol"): R' = H

β-hydroxy ketone: R' = alkyl

α,β-unsaturated carbonyl likely when one R = phenyl

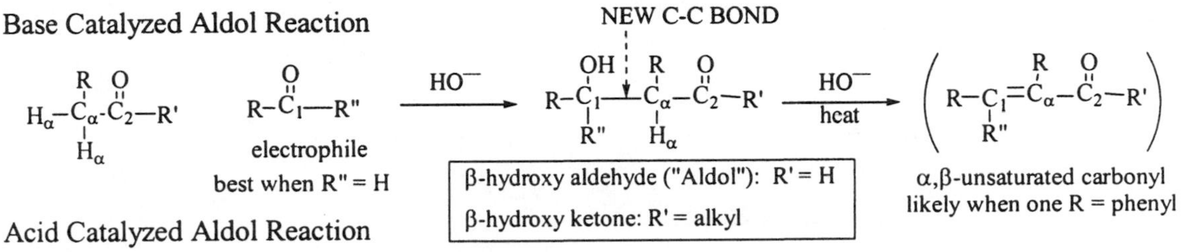

Acid Catalyzed Aldol Reaction

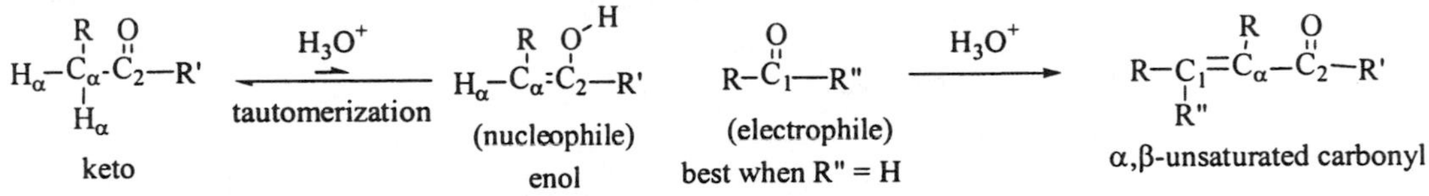

H_α–C_α(R)(H_α)–C_2(=O)–R' (keto) $\underset{\text{tautomerization}}{\overset{H_3O^+}{\rightleftharpoons}}$ H_α–C_α(R)=C_2(OH)–R' (nucleophile) enol + R–C_1(=O)–R" (electrophile) best when R" = H $\xrightarrow{H_3O^+}$ R–C_1(R")=C_α(R)–C_2(=O)–R' α,β-unsaturated carbonyl

Michael Addition

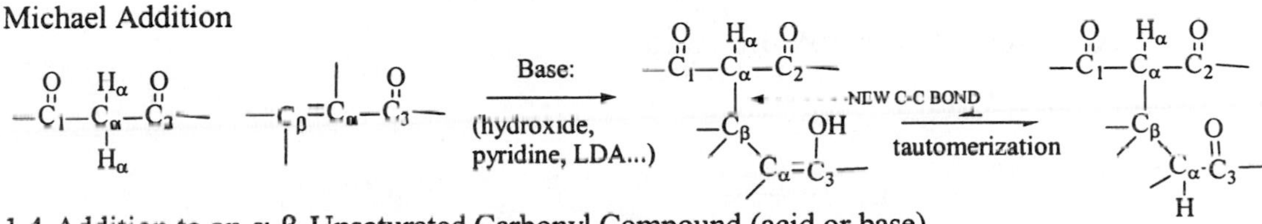

–C_1(=O)–C_α(H_α)(H_α)–C_2(=O)– + –C_β=C_α–C_3(=O)– $\xrightarrow{\text{Base:}}$ (hydroxide, pyridine, LDA...)

–C_1(=O)–C_α(H_α)–C_2(=O)– bonded to C_β–C_α=C_3(OH)– (NEW C-C BOND) $\underset{\text{tautomerization}}{\rightleftharpoons}$ –C_1(=O)–C_α(H_α)–C_2(=O)– bonded to C_β–C_α(H)–C_3(=O)–

1,4-Addition to an α,β-Unsaturated Carbonyl Compound (acid or base)

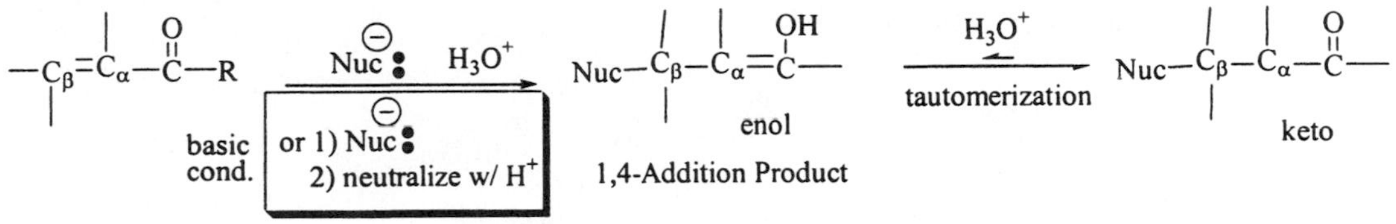

–C_β=C_α–C(=O)–R $\xrightarrow{Nuc^{\ominus},\ H_3O^+}$ Nuc–C_β–C_α=C(OH)– (enol, 1,4-Addition Product) $\underset{\text{tautomerization}}{\overset{H_3O^+}{\rightleftharpoons}}$ Nuc–C_β–C_α–C(=O)– (keto)

basic cond.: or 1) $Nuc^{\ominus}$ 2) neutralize w/ H^+

1,2-Addition to an α,β-Unsaturated Carbonyl Compound

–C_β=C_α–C(=O)– $\xrightarrow{Li^{\oplus}\ ^{\ominus}R"}$ –C_β=C_α–C($O^{\ominus}$)(R")– $\xrightarrow{\text{neutralize w/ } H^+}$ –C_β=C_α–C(OH)(R")–

1,2-Addition Product

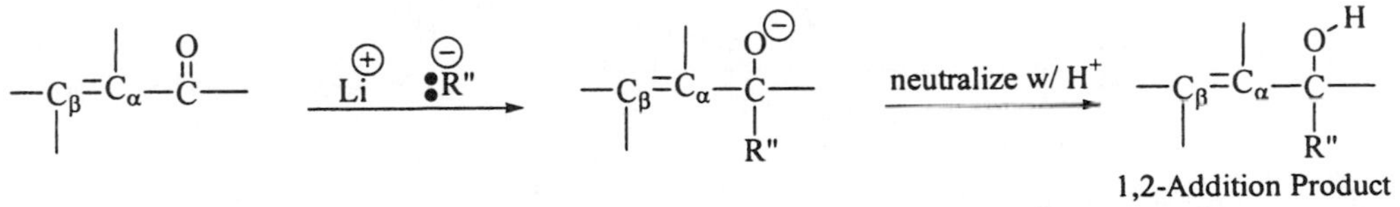

Preparation of Carboxylic Acids

Complete Oxidative Cleavage of Double Bond (harsh conditions)

$R_2C{=}CHR \xrightarrow[\text{2) } H_3O^+/H_2O]{\text{1) } KMnO_4 \text{ (hot) harsh oxidation}} R_2C{=}O + RCOOH$

ketone carboxylic acid

Harsh Oxidation of 1° Alcohol

$R{-}CH_2{-}OH \xrightarrow[KMnO_4]{\text{harsh oxid.}} R{-}C(=O){-}OH$

carboxylic acid

Grignard Addition to Carbon Dioxide

$R{-}MgX \xrightarrow[\text{2) } H_3O^+]{\text{1) } O{=}C{=}O} R{-}C(=O){-}O{-}H$

Hydrolysis of a Nitrile (R–CN)

$R{-}C{\equiv}N \xrightarrow{H_3O^+/H_2O} R{-}C(=O){-}NH_2 \xrightarrow{H_3O^+/H_2O} R{-}C(=O){-}O{-}H \quad NH_3$

nitrile amide carboxylic acid

Oxidation at the Benzyl Position

$C_6H_5{-}R \xrightarrow{KMnO_4} C_6H_5{-}C(=O){-}OH$

benzoic acid

Reactions of Carboxylic Acids

Acid-Base

$R{-}C(=O){-}O{-}H \xrightarrow{^{\ominus}OH} R{-}C(=O){-}O^{\ominus} \quad H_2O$

Acid Chloride Formation

$R{-}C(=O){-}O{-}H \xrightarrow{SOCl_2} R{-}C(=O){-}Cl$

acid chloride

Reduction

$R{-}C(=O){-}O{-}H \xrightarrow[\text{2) } H_3O^+]{\text{1) } H^{\ominus} \; (LiAlH_4)} R{-}CH_2{-}O{-}H \quad H_2O$

1° alcohol

Acid Catalyzed Ester Formation

$R{-}C(=O){-}O{-}H \quad R'{-}OH \xrightarrow[\text{Base doesn't work!}]{H_3O^{\oplus} \text{ (catalyst)}} R{-}C(=O){-}O{-}R' \quad H_2O$

ester

Preparation of Esters

From Acyl Chlorides

$$R{-}\overset{O}{\overset{\|}{C}}{-}Cl \xrightarrow{R'OH} R{-}\overset{O}{\overset{\|}{C}}{-}OR' \quad H{-}Cl$$

From Carboxylic Acids

$$R{-}\overset{O}{\overset{\|}{C}}{-}OH + R'OH \underset{}{\overset{H_3O^{\oplus} \text{ or } HO^{\ominus} \text{ (catalyst)}}{\rightleftharpoons}} R{-}\overset{O}{\overset{\|}{C}}{-}OR' + H_2O$$

Reaction of Esters

To form carboxylic acid

$$R{-}\overset{O}{\overset{\|}{C}}{-}OR' \quad H_2O \overset{H_3O^{\oplus} \text{ (catalyst)}}{\rightleftharpoons} R{-}\overset{O}{\overset{\|}{C}}{-}OH + R'OH$$

$H_3O^{\oplus}$ or $HO^{\ominus}$

To form conjugate base of carboxylic acid

$$R{-}\overset{O}{\overset{\|}{C}}{-}OR' \quad HO^{\ominus} \rightleftharpoons R{-}\overset{O}{\overset{\|}{C}}{-}O^{-} + R'OH$$

To form amide

$$R{-}\overset{O}{\overset{\|}{C}}{-}OR' \quad NH_3 \overset{H_3O^{\oplus} \text{ or } HO^{\ominus} \text{ (catalyst)}}{\rightleftharpoons} R{-}\overset{O}{\overset{\|}{C}}{-}NH_2 + R'OH$$

To form a different ester ("trans-esterification")

$$R{-}\overset{O}{\overset{\|}{C}}{-}OR'' + R'OH \overset{H_3O^{\oplus} \text{ or } HO^{\ominus} \text{ (catalyst)}}{\rightleftharpoons} R{-}\overset{O}{\overset{\|}{C}}{-}OR' + R''OH$$

Reduction to form 1° alcohol

$$R{-}\overset{O}{\overset{\|}{C}}{-}OR' \xrightarrow[2)\ H_3O^+]{1)\ LiAlH_4} R{-}\overset{O}{\overset{\|}{C}}{-}H + R'OH \xrightarrow[2)\ H_3O^+]{1)\ LiAlH_4} R{-}\overset{H_2}{C}{-}OH$$

Grignard Addition to form 3° alcohol

$$R{-}\overset{O}{\overset{\|}{C}}{-}OR'' \xrightarrow[2)\ H_3O^+]{1)\ R'MgX} R{-}\overset{O}{\overset{\|}{C}}{-}R' + R''OH \xrightarrow[2)\ H_3O^+]{1)\ R'MgX} {-}\underset{R'}{\overset{OH}{C}}{-}R'$$

Preparations of Amides

From Acyl Chlorides

$$R{-}\overset{O}{\overset{\|}{C}}{-}Cl \xrightarrow{HNR'_2} R{-}\overset{O}{\overset{\|}{C}}{-}NR'_2 \quad H{-}Cl$$

R' = H, alkyl, aryl

From Esters

$$R{-}\overset{O}{\overset{\|}{C}}{-}OR \quad HNR'_2 \overset{H_3O^{\oplus} \text{ or } HO^{\ominus} \text{ (catalyst)}}{\rightleftharpoons} R{-}\overset{O}{\overset{\|}{C}}{-}NR'_2 + ROH$$

From Nitriles

$$R{-}C{\equiv}N \xrightarrow[H_2SO_4]{H_2O} R{-}\overset{O}{\overset{\|}{C}}{-}NH_2$$

(Incomplete hydrolysis)

Nucleophilic Aromatic Substitution (NAS)

X = F, Cl, Br, I

Z = electron widrawing group in ortho or para position

$Nu^{\ominus}$ ⟶ $X^{\ominus}$

Claisen Reaction

R'–C₁–OR (electrophile) + H_α–C_α(R'')(H_α)–C_2–OR —$RO^{\ominus}$⟶ R'–C_1(=O)–C_α(R'')(H_α)–C_2(=O)–OR —$RO^{\ominus}$⟶ R'–C_1($O^{\ominus}$)=C_α(R'')–C_2(=O)–OR

R = alkyl group; R' and R'' = H or alkyl group

Special Ketone Synthesis: alkylation --> saponification --> decarboxylation

$RO^{\ominus}$, X–R' (alkylation) ⟶ 1) $HO^{\ominus}$ 2) H_3O^+ (saponification (ester to carboxylic acid)) ⟶ heat (decarboxylation) ⟶ + CO_2

Preparation of Amines

Direct Alkylation

1 or 2 equiv. Br—R' ⟶ gives mixture of...

excess Br—R' and a non-nucleophilic base

$Br^{\ominus}$

Gabriel Synthesis of 1° Amines

NH_3, heat (acid catalyst) ⟶ N–H ⟶ 1) K_2CO_3 (base) 2) X–R any 1° alkyl halide ⟶ N–R ⟶ H_2SO_4, H_2O ⟶ + H_2N–R

Reduction of Amides, Imines and Nitriles to form 1° and 2° Amines

amide —$LiAlH_4$⟶ 1° or 2° amine

R' = alkyl or H

aldehyde or ketone —H_2N–R' (acid cat. optional)⟶ imine —H_2 metal catalyst⟶ 1° or 2° amine

R–C≡N: nitrile —$LiAlH_4$ or H_2/metal catalyst⟶ R–CH_2–NH_2 1° amine